AMERICAN
ECONOMIC HISTORY

HARPER'S HISTORICAL SERIES
Under the Editorship of
DEAN GUY STANTON FORD

AN INTRODUCTION TO ECONOMIC HISTORY
By N. S. B. Gras

AMERICAN ECONOMIC HISTORY
By Harold U. Faulkner

THE FAR EAST, A POLITICAL AND DIPLOMATIC HISTORY
By Payson J. Treat

HISTORY OF ENGLAND
By W. E. Lunt

SHORT HISTORY OF CHINA (*In Press*)
By E. T. Williams

MEDIAEVAL FOUNDATIONS OF WESTERN CIVILIZATION
By G. C. Sellery, and A. C. Krey
(*Ready in the fall of 1928*)

(*Other Volumes in Preparation*)

RELIEF MAP OF NORTH AMERICA

AMERICAN ECONOMIC HISTORY

BY

HAROLD UNDERWOOD FAULKNER, Ph.D.

Associate Professor of History in Smith College

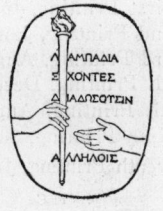

HARPER & BROTHERS PUBLISHERS
NEW YORK AND LONDON

AMERICAN
ECONOMIC HISTORY

Copyright, 1924
By Harper & Brothers
Printed in the U. S. A.

First Printing, July, 1924
Second Printing, January, 1925
Third Printing, August, 1925
Fourth Printing, December, 1925
Fifth Printing, August, 1926
Sixth Printing, July, 1927
Seventh Printing, June, 1928

F—C

TO
MY MOTHER
HELEN UNDERWOOD FAULKNER

CONTENTS

COLONIAL BEGINNINGS

CHAPTER		PAGE
	Editor's Introduction	xi
	Preface	xiii
I.	Physiographic Factors and Natural Resources	1
II.	Social and Economic Aspects of Colonization	27
III.	Colonial Agriculture and Labor	57
IV.	Colonial Industry	78
V.	Colonial Commerce	102
VI.	The Westward Movement Before the Revolution	117
VII.	The Economic Causes of the Revolution	140

ECONOMIC INDEPENDENCE AND THE ADVANCE TO THE PACIFIC

VIII.	The Revolution and the Constitution	160
IX.	The Westward Movement from the Revolution to the Civil War	188
X.	Agriculture from the Close of the Revolution to the Civil War	216
XI.	The American Merchant Marine and the Development of Foreign Commerce to 1860	242
XII.	The Industrial Revolution in America	266
XIII.	Finance and Tariff	296
XIV.	Transportation and Communication to 1860	308

CONTENTS

CHAPTER		PAGE
XV.	SOCIAL BACKGROUND OF THE FORMATIVE PERIOD	338
XVI.	ECONOMIC CAUSES OF THE CIVIL WAR	359

INDUSTRIAL EXPANSION AND ECONOMIC DEVELOPMENT

XVII.	THE CIVIL WAR	381
XVIII.	THE LAST FRONTIER	403
XIX.	THE AGRARIAN REVOLUTION	423
XX.	INTERNAL TRANSPORTATION AND COMMUNICATION SINCE 1860	452
XXI.	FINANCIAL HISTORY SINCE 1860	491
XXII.	BUSINESS CONSOLIDATION	515
XXIII.	MANUFACTURING SINCE 1860	546
XXIV.	THE LABOR MOVEMENT	584
XXV.	WORLD TRADE AND THE NEW IMPERIALISM	618
XXVI.	RECENT ECONOMIC TENDENCIES	652
XXVII.	THE WORLD WAR AND RECONSTRUCTION	678
	INDEX	695

MAPS AND GRAPHS

	PAGE
RELIEF MAP OF NORTH AMERICA . . . . . *Frontispiece*	
PRINCIPAL PETROLEUM AND NATURAL-GAS FIELDS OF THE UNITED STATES	14
REGIONAL DISTRIBUTION OF PRODUCTS IN THE UNITED STATES	21
MEDIÆVAL TRADE ROUTES ACROSS ASIA	30
TRADE BETWEEN GREAT BRITAIN AND THE AMERICAN MAINLAND COLONIES, 1745-1776	150
THE WEST, 1775-1782	167
DISTRIBUTION OF VOTES IN RATIFICATION OF THE CONSTITUTION—NEW ENGLAND, 1787-1790	182
DISTRIBUTION OF VOTES IN RATIFICATION OF THE CONSTITUTION—MIDDLE AND SOUTHERN STATES, 1787-1788 . .	183
DISTRIBUTION OF POPULATION, 1790	195
DISTRIBUTION OF POPULATION, 1800	195
DISTRIBUTION OF POPULATION, 1820	203
DISTRIBUTION OF POPULATION, 1840	203
DISTRIBUTION OF POPULATION, 1860	211
PERCENTAGE OF IMPORTS AND EXPORTS CARRIED IN AMERICAN VESSELS, 1789-1921	255
RAILROAD LINES IN ACTUAL OPERATION, OCTOBER, 1860 .	330
IMMIGRATION INTO THE UNITED STATES, 1820-1860 . . .	343
AVERAGE ANNUAL NEW YORK COTTON PRICES FOR MIDDLING UPLANDS, 1790-1921	393
PRODUCTION OF COTTON IN THE UNITED STATES BY 500-POUND BALES, 1865-1921	448
MILES OF RAILROAD IN OPERATION, 1850-1919	453

MAPS AND GRAPHS

	PAGE
GOLD VALUE OF THE GREENBACK, 1862-1879	499
BULLION VALUE OF 371¼ GRAINS OF SILVER EACH YEAR, 1866-1921	506
WAGES AND PRICES, 1840-1891	606
IMMIGRATION INTO THE UNITED STATES, 1860-1921	612
UNITED STATES IN THE CARIBBEAN	633
TONNAGE OF THE UNITED STATES MERCHANT MARINE ENGAGED IN FOREIGN COMMERCE, 1789-1921	643
MERCHANDISE IMPORTED AND EXPORTED, 1850-1921	646
TREND OF PRICES, 1890-1922	674

EDITOR'S INTRODUCTION

THE trailing clouds of glory that lay around our nation's youth have begun to fade away. With the war we reached middle age at a bound. We are now counting our resources, human and material, as we never did before. The results of the inventory are not wholly reassuring, and we are checking up our waning natural resources, our political institutions, our education, our social philosophy, to find where there is waste and lost motion. Mankind may not be at the crossroads, but it is dimly conscious that the road ahead is not the broad and happy highway of the past. It is more necessary than ever before that we should study our national history from every standpoint, and especially the economic. I think this book will be counted among the most useful of the aids yet provided for such study.

It may be well to point out that he who writes the economic history of any age or land undertakes a difficult task. If it be an economic history of the United States, as this volume is, the task is not less but rather more difficult. His subject is a country where nature has been bountiful and the exploitation of natural wealth has been less trammeled by old institutions and social customs than in Europe. Here the political individualism of a pioneer people has given freer play than ever before in human history to all the acquisitive impulses of men and groups of men. The faith of a youthful people that it lived in a land of inexhaustible resources and that nothing could happen to it that had happened to older lands where soil and forests became exhausted and the mineral wealth dissipated has opened the door for a material development unparalleled in the history of nations. How easy then for him who starts with the point of view implied in the title "economic history" to forget that the history of significant men and nations is, in its end results, but the charted field of a battle between their inherited ideas and unrealized ideals, on the one hand, and the material circumstances of their physical environment, on the other. I believe the author has written an account of our economic history free from the errors of one-sided materialism.

INTRODUCTION

It is the achievement of this book that it is American history seen whole and sturdily, though from the given standpoint of its title. The reader has presented to him the picture of success and error in the discharge of an implied trusteeship. It is done dispassionately, without disproportion or vain glorification. Those who study it can draw their own conclusions and, if they master it, will find themselves in possession of the essential equipment necessary to a citizenship that will face in the next generation more complicated social and economic problems than it has in the past.

The student of economic history should be warned that he must face and master many facts. It is not a simple matter to grasp such data and wring their meaning from them. I believe that here, too, the author has given skillfully all the aid that any self-respecting student should require. He has gathered and integrated into an account that does not halt a remarkable and significant mass of tables, figures, and graphs. He has labored patiently that the student may read intelligently and be armed to test old conclusions or draw new ones. This is a difficult task for any author, but an essential one in an economic history. If an editor may not voice an opinion he can at least express his confidence that classroom use will prove that the author has achieved a large measure of success in writing a narrative that makes statistics an asset to teaching.

Neither teacher nor text writer is content to have a student think one book is the subject. A bibliography of material that will amplify, supplement, and enforce the text is essential. Such an aid to scholarship has been, as I can testify, one of the writer's chief concerns.

GUY STANTON FORD.

PREFACE

the book in innumerable places. The manuscript was also read in its entirety by Professor Henry G. Pearson, head of the Department of English and History at the Massachusetts Institute of Technology, and by Mr. Paul Tyler Kepner of the Department of History of the High School in Brookline, Massachusetts, from whose suggestions it has greatly profited. Certain chapters in their earlier form were read and criticized by Professor J. Montgomery Gambrill of Teachers College, Columbia University, and by my father, Professor J. Alfred Faulkner of Drew Theological Seminary. Professor Davis R. Dewey of the Massachusetts Institute of Technology also examined the manuscript and made suggestions, especially relating to the bibliography. The completed manuscript was read by Professor J. D. Magee of New York University, who contributed constructively. The graphs are the work of one of my students at the Institute, Mr. L. E. Jenkins of the class of 1925. To all of these gentlemen I am greatly indebted. I should also like to express my gratitude for the courteous assistance given me by the librarians of the public and private institutions in which I worked in Cambridge, Boston, and New York. It is impossible adequately to express the debt I owe my wife, Ethel Webb Faulkner, for the constant and untiring aid which she rendered at all stages of the work.

HAROLD UNDERWOOD FAULKNER.

CAMBRIDGE, MASSACHUSETTS,
March 15, 1924.

NOTE TO THE SEVENTH PRINTING.

In making certain alterations and corrections for the seventh printing the author takes this opportunity to express his appreciation to those teachers who have taken time from a busy life to write him regarding matters of controversy or error in the book. He desires especially to thank Professor Fred A. Shannon of the Kansas State Agricultural College for a number of improvements which have been incorporated in this printing.

H. U. F.

MADISON, N. J.
March 30, 1928.

PREFACE

ECONOMIC history is not only an absorbingly interesting subject in itself, but it is the foundation and framework upon which all history must be built. In the last analysis man lives by extracting from the earth materials for his food, clothing, and shelter. As the great part of the energy of the majority of mankind is concerned with some branch of this occupation, it is reasonable to believe that man's economic status influences greatly his political conceptions and activities. Political and military history tell chiefly how things happened; economic history, why things happened. To teach the former before the latter is like putting the cart before the horse. To teach the former without the latter is a pedagogical fallacy which leaves the student with but half of the story, and that the least important. Nevertheless, it was not until Buckle and Marx in comparatively recent years hammered home the importance of economic environment that historians have conceived the possibility of writing the story of the human race in terms of wheat, cotton, and iron, as well as in those of monarchs, politicians, and soldiers. This book is written for the double purpose of presenting history in economic terms and at the same time of adding a text-book to a field as yet but meagerly supplied.

A special effort has been made to make the bibliographies comprehensive and usable. The "Selected Readings" which follow the "Notes for Further Reference" at the end of each chapter were chosen not alone for their excellence in throwing further light on the subject-matter of the chapter, but also for their availability in the average library. In general they have been listed in the order of their value as supplementary reading, except that source books are always placed at the end.

I have been exceedingly fortunate in having this book appear in a series edited by Dean Guy Stanton Ford. His patient and scholarly criticism has saved me from many errors and enriched

AMERICAN
ECONOMIC HISTORY

AMERICAN ECONOMIC HISTORY

CHAPTER I
PHYSIOGRAPHIC FACTORS AND NATURAL RESOURCES

Geographic Influences Potent in American History.—The study of history, particularly economic history, must begin with a knowledge of the physical outline and resources of the unit under observation. Nature determines to a large extent where man shall live, what kind of work he shall do, what he may produce, and the routes over which he must travel and transport his products. Because of its influence upon his economic life, natural environment goes far to determine man's social and political point of view, his habits and desires, and even his physical frame.[1]

The history of the United States, written so largely in terms of the conquest of the continent, shows physiographic influence at every step. The contour of the coast fixed the place of the first settlements, the river valleys and mountain gaps pointed the route westward, while the formation of the soil and the nature of its products determined the occupation of the settler after he had reached the new country.

Geographic Divisions of the United States.—The North American continent forms a rough triangle perhaps three thousand miles across at the north and tapering to the width of but a few miles at the Isthmus of Panama. Facing three oceans, it is influenced by each. The Pacific sends a stream of warm water against the western coast, which makes it habitable as far as Alaska, although, because of the Cordilleras, the effect is limited

---

[1] See Boas, Franz, *Changes in Bodily Form of Descendants of Immigrants* (1912), compiled from *U. S. Immigration Commission Reports.*

to the fringe of seacoast. The Gulf Stream of the Atlantic provides rainfall for the lower Mississippi and Gulf states, and its influence can be seen as far north as New England. The Arctic, where it touches America, cut off from the currents of both Atlantic and Pacific, is icebound and so renders unfit for habitation a large part of the northern half of the continent. A vast mountain range, the Cordilleras, traverses the western portion of the continent from Alaska to Panama. At its widest point, around the fortieth parallel, this system has a breadth of about one thousand miles, with many of its peaks attaining a height of fourteen thousand feet. On the east the Appalachian system, bordering a fringe of seacoast and interspersed with fertile valleys, extends from Newfoundland to Alabama. It is nowhere as high as seven thousand feet. Between these two mountain ranges lies an immense plain which, with the exception of a few patches of low mountains, stretches from the Gulf of Mexico to the Arctic. The drainage of this great plain is carried off by three main outlets: (1) the Mississippi and its tributaries, the Missouri, Ohio, Arkansas, and Red rivers, emptying into the Gulf of Mexico; (2) the Great Lakes, draining into the St. Lawrence and the Atlantic; and (3) the MacKenzie, and the numerous streams running into Hudson Bay.

On this continent, roughly between the twenty-ninth and forty-ninth parallels, lies the United States of America. In area it contains 3,026,789 square miles—over two-thirds the size of Europe. It has been divided geographically into six more or less distinct parts:[1]

(1) The eastern lowlands, or coastal plain, lying between the shore and the Appalachians. This region includes the eastern fringe of the states facing the Atlantic. Although the soil is not so fertile as that farther west, it is suitable for ordinary garden vegetables and for wheat, corn, and tobacco. It fortunately provided the first settlers with two indigenous plants—their chief cereal, corn; and their chief export (the economic backbone of the colonial south), tobacco. As the agricultural center shifted

---

[1] Shaler, N. S., in Winsor's *Narrative and Critical History of America*, vol. iv, intro., part i, p. iii.

westward, the future of the coastal plain became more and more wrapped up in two activities, manufacturing and commerce, the former made possible by the unsurpassed water power of the fall line, and the latter by the excellent harbors of the frequently indented coast. Two strings of cities mark the boundaries of the coastal plain—on the west the cities of the fall line—Montgomery, Macon, Columbia, Raleigh, Richmond, Baltimore, Trenton, Hartford; on the east the seacoast cities—Savannah, Charleston, Wilmington, Norfolk, New York.

(2) The Appalachian region, directly to the west, composed of two parallel mountain ranges, with a broad valley between. The older mountains, geologically, are the Appalachians—steep, rugged slopes forming a narrow chain extending from Newfoundland to Alabama, and rising to heights of over six thousand feet in the White Mountains of New Hampshire and the Black Mountains of North Carolina. The Alleghany Mountains are much gentler in their slopes and valleys, and are nowhere higher than five thousand feet. They extend from the Catskills in New York to northern Alabama. Between these two ranges lies a fertile valley about forty miles in width, on the average, extending from New Jersey to Georgia—a distance of over six hundred miles. On both sides of these mountain systems, and especially to the west, are vast table-lands, merging gradually into the plains. In all, this section comprises some three hundred thousand square miles, only twelve thousand of which are untillable, and contains in its fertile piedmonts and valleys, notably the Shenandoah, Cumberland, and Tennessee, some of the finest farming lands in America. The position of natural resources has given rise to a pronounced geographic localization of industries. Thus the nearness of the mountains to the coast in New England causes a rapid fall in the streams, and produces the water power that has made of New England a manufacturing center; while the coal and iron deposits of Pennsylvania and the southern Appalachian states have given rise to the great iron and steel cities of Pittsburgh and Birmingham.

(3) Lowlands of the Gulf states. This region includes Florida, southern Georgia, Alabama, Mississippi, Louisiana, and

eastern Texas, where the rich black alluvial soil and the hot climate form an excellent combination for the staple crop, cotton.

(4) The great plain of the Mississippi Valley. The Mississippi Valley consists of a relatively small delta section of alluvial soil, some twenty to thirty thousand square miles in area, and the great table-lands of the Appalachians and the Rockies. The wide fertile prairies and river valleys of this region make it the agricultural heart of the New World. It is here that immense crops of wheat and corn are raised, while in the Mississippi delta, cotton, as in the Gulf states, is king. The "Father of Waters" and its tributaries, the Ohio and the Missouri, furnish excellent natural transportation facilities, which are augmented to the north by the Great Lakes and their connecting canals.

(5) The Cordillera region. Although fertile valleys are to be found here and bits have been made artificially arable by irrigation, the greater part (Shaler estimates nineteen-twentieths) is barren. The great value of this region in the past has come chiefly from its mineral deposits of copper, iron, silver, and gold. As contrasted with the great Mississippi Valley, which has the potential capacity to support an enormous population, the Cordillera region will probably always be sparsely populated, especially when its mineral resources are exhausted. Nevertheless, the increasing attention devoted to irrigation projects and dry farming is slowly laying the foundation for a permanent prosperity.

(6) A narrow region of low mountains on the extreme western coast. Of great fertility and extremely even and temperate climate, this section has recently developed enormously the production of fruit. Although the Pacific coast is unfortunate in that it possesses but few natural harbors, the opening of the Panama Canal and the rapidly growing commercial importance of the Far East point to the increasing use of such facilities as are offered at the Golden Gate, Puget Sound, and the Columbia River. Gold brought the first large influx of English-speaking settlers to California, and that state still ranks first in the Union in the mining of gold; but its great present and future wealth depends upon other products.

PHYSIOGRAPHIC FACTORS

Geographic Influence Upon Colonization and Settlement.
—Although the American continent is accessible on its western side and the ancestors of the aborigines undoubtedly entered it from Asia, it was most fortunate that when the white man came to these shores he approached them from the east. Had the continent been turned around, its history would have been different, for the rugged Cordilleras would have presented to the pioneer a difficult, if not impassable, barrier. After the forbidding Atlantic had once been crossed, the European found a land the ingress to which was simple. The St. Lawrence Valley connecting with the Great Lakes, the Hudson River opening through the Appalachians, and the Mississippi with its innumerable tributaries penetrating the very heart of the continent, pointed the way inland and made possible a more rapid settlement. Many smaller rivers, such as the Connecticut, the Delaware, and the James, cut into the coastal plain as far as the Appalachians and formed the natural highways for the early settlers. Along these rivers the settlements were planted, and down them were floated furs and tobacco, the two products that first linked the colonies with the markets of Europe. It is estimated that there are over 26,000 miles of navigable rivers in the United States, not counting the 2,760 (meandered length, 4,329) miles of shore line on the Great Lakes. Including indentations, the coast line on both oceans amounts to over 64,000 miles, with at least two-thirds of this directly accessible to Europe on the Atlantic and the Gulf of Mexico. With numerous rivers and an indented coast, good harbors were to be expected. The Atlantic and Gulf states show excellent examples of each harbor type: New York and Baltimore of the drowned valley; Galveston, Provincetown, and many little harbors on the Carolina, Florida, and New Jersey coasts, of harbors formed from barrier reefs; New Orleans and Philadelphia of river harbors. These and numerous other natural ports and river towns provided the points at which the raw materials for export could be gathered and the manufactured products of the mother country received and marketed.

The United States as a Habitat for Man.—The territory now embraced in the United States was eminently fitted for those

European races destined to settle and conquer the American continent. It lies between the lines of forty and seventy degrees average annual temperature, representing a climate similar to that in the portions of Europe producing the most energetic and civilized races. The average rainfall is 29.6 inches, varying from 5 inches in southern Utah to 60 inches in the western valley of California and on the Gulf coast. The Pacific coast has a damp, insular climate which becomes drier toward the mountains, until in the great arid plateau moisture is almost lacking. The rainfall gradually increases, however, as we approach the Gulf of Mexico and the Atlantic. East of the Appalachians the rainfall averages from 30 to 50 inches a year. Since 20 inches is essential for agriculture and from 30 to 50 inches for ideal soil moisture, conditions for agricultural production are here most favorable. While the variations of both temperature and rainfall are greater than in Europe, the climate as a whole is essentially the same.

The glaciated soil of the territory first settled was difficult to make arable. To clear an acre of forest and underbrush and its thick covering of stones, sixty days' labor was required. Once cleared, however, this soil was found to be of a reasonable, though not high, order of fertility, and of such a nature as to maintain its character over a long period. The soils of the Mississippi basin contain much limestone, a formation fertile and especially adapted to the raising of grains and grasses. The great Appalachian Valley also partly owes its rich fertility to underlying limestone.

Adaptability of Europeans to American climate and conditions seems well established. Observations which have been made on groups of both Teutons and Celts who have been here for perhaps two hundred years and have kept their strain practically pure, point to the fact that the Europeans have not suffered physically from transplantation to America. They are no smaller in size, are as energetic, can withstand fatigue as well, and are as long lived as their kinsmen in Europe. Moreover, the energy and creative faculties of the immigrants were apparently challenged by the mighty task of subduing a continent, and the result was a sturdy and resourceful race.

A factor contributing largely to the health and longevity of the settlers in North America was that they found here no new diseases. It was the misfortune of the aborigines, on the other hand, that the settlers brought with them Old World maladies probably unknown here, which, as no immunity had been developed among them, made rapid inroads on the Indians when once started. Especially was this true of tuberculosis, smallpox, and measles. Except for the coastal lowlands of the south, which fevers make unwholesome for white settlers, all of the Atlantic seaboard is eminently fitted for European peoples.

Influence of the Character and Distribution of American Native Products on the Early Settlers.—The greatest influence upon our history, next to topography, natural routes of travel, and climate, is the part played by the character and distribution of the vegetable products. Presence or absence of forests, fertility and adaptability of the soil, and similar factors have determined both where the settler would erect his cabin and by what method he would support himself. The great variation in climate and soil has made it possible to raise in some part of our country practically every food product of importance, whether native or imported. In fact, the majority of plants of great economic value to-day have been of foreign origin.

The most pressing immediate need of the colonists was food. Even where the motive for colonization was the discovery of gold and silver, the practical question of keeping alive until the gold could be found at once intruded. Apparently it cost the Jamestown settlers years of suffering and the loss of many lives before they realized that the food supply should be their first concern. Lack of sufficient and proper nourishment was the greatest cause of the heavy mortality of the early years of Virginia, a fact that seems almost inexplicable when we consider the richness of the native food resources. The forests held an abundance of deer, turkey, rabbits, squirrels, opossum, and other edible animals. The woods, bays, and marshes were plenteously supplied with wild fowl of every variety known to Englishmen—pheasant, partridge, woodcook, quail in the woods; snipe and curlew in the marshes; waterfowl of many kinds on the bays and inlets;

and pigeons and wild geese flying overhead on their southward journey. Fish, both salt water and fresh, was everywhere abundant. In addition, many plants capable of supplying nourishment grew wild, and others were under cultivation by the Indians. In the midst of such plenty the "starving time" in Virginia must be accounted for not by any scarcity of native food, but by the ignorance of the settlers as to how to gather and utilize the native products, and to their neglect of the cultivation of food crops—a neglect due to the hope of quick riches by other means.

Wood is vital to man's existence, especially in a primitive civilization where it provides shelter, fuel, means of conveyance on land and water, and even a considerable element of food. The early comers to the Atlantic seaboard found it thickly wooded; not an unmixed blessing, since land must be cleared of trees and underbrush before it could be made suitable for most agricultural uses. Far from unimportant was the forest as a source of food. The maple furnished sugar—in many cases the only sweetening except honey that was available. Beech, hazel, and hickory nuts, chestnuts, walnuts, and butternuts were found here. Most of the varieties of fruit trees have been imported, but the wild plum, cherry, persimmon, and mulberry were native to some part of the Atlantic seaboard.

Among the edible plants, either growing in a wild state or cultivated by the Indians when white settlers arrived, were maize, or Indian corn, pumpkins (or pompions, as they were called at first), squash, beans, rice, tomatoes, peanuts, Jerusalem artichoke, peppers, American aloe, sweet potatoes, watermelons, huckleberries, blackberries, strawberries, black raspberries, cranberries, gooseberries, and grapes. Vegetables and food-bearing plants of all kinds were imported, some of them entirely foreign and others European varieties of products native to America.[1]

To the Indian and early settler the animal life of the continent was of vast importance. Besides being a constant and oftentimes a chief source of food supply, it furnished materials for clothing, shelter, and other necessities. Of the native animals, probably the most valuable to the aborigines were the deer east of the Missis-

---

[1] Carrier, Lyman, *The Beginning of Agriculture in America*, p. 41.

sippi and the buffalo which swarmed the great western plains, neither of which have any present economic significance. After the white settler appeared and the fur trade commenced, the smaller animals, such as the weasel, sable, badger, skunk, wolverine, mink, otter, and sea-otter, became important. There was a demand also for such fur-bearing rodents as the squirrel, hare, muskrat, and beaver. With the exception of the llama and the alpaca, which were used locally in South America as beasts of burden, and the dog, the American Indian never succeeded in domesticating any of the native fauna. Almost from the beginning, however, such common farm animals as horses, cattle, sheep, and swine were imported; the climate was found suitable for European livestock and the vast grazing areas and easily grown food supported a rapid increase in numbers. Poultry of all kinds was introduced from Europe, and one variety of the innumerable wild fowl frequenting the American woods, the wild turkey, was taken to Europe, domesticated, and later brought back. The turkey was the one American contribution to domestic poultry, but the original native wild turkeys were in reality much larger than the barnyard product to-day, weighing as they did thirty or forty pounds. Such splendid birds sold for a shilling apiece; and so ruthless was their destruction that within a century after the settlement of America they had practically disappeared from the settled areas of the country.

The New World could furnish fish not only for the settlers, but for all Europe. In his *New England Rarities* (1672) Josselyn enumerated over two hundred kinds of fish that were caught in New England's waters; Gosnold records that his ships in 1602 were "pestered with cod." Not only the ocean and bays, but the rivers, lakes, and brooks teemed with fish; brook trout were struck and killed with a stick, and scooped up in frying pans. Besides their value as food and export, fish had worth to the early settler as fertilizer. As a consequence the fisheries were destined to assume a place of economic importance in the commercial life of New England.

Geographical Influence Upon Occupations.—In addition to the favorable elements for easy and rapid colonization produced by rivers, harbors, and a long indented coast line, there should be noted especially the part played by the two greatest plants which America gave to the world—maize and tobacco. Maize yielded twice as much food per acre as the smaller grains, was less dependent upon seasons, could be cultivated without plowing and with the crudest implements, and grown with a minimum of labor. It provided a new and cheaper source of food, while the stalks furnished a more valuable forage than those of other grains. This plant largely helped to fix the early settlements in North America. Although the tobacco plant did not aid settlement in the same sense as did Indian corn, the fact that it soon furnished the basis of wealth to a large part of the country must be considered among the factors which contributed to the rapid transplanting of the European race.

In later years our history was influenced by the climatic adaptability of certain regions to certain products. Thus the southern states were found suitable for cotton culture, and that plant became, after the invention of the cotton gin, the great southern staple. It fastened slavery on the south; and in the train of slavery came the many developments leading to the Civil War. The fertility and climate of the upper Mississippi Valley made it ideal for the cultivation of corn, and hastened settlement. The great western prairies were adaptable both to cattle raising and to wheat farming, thus providing an economic basis for those regions. Where mineral resources were at hand, economic life turned to them, so that in the Wyoming Valley of Pennsylvania, in the Rockies, and in the oil fields, whole communities have been built up around the extraction of minerals.

The harbors and rivers of the eastern coast not only fostered ready colonization, but gave a turn to the occupation of the people. The barren soil of New England turned the settler's interest toward an easier means of livelihood than farming, and the near-by fishing made of the colonial and nineteenth century New Englander a follower of the sea. New England became the center of colonial shipping and retained that position in later years during the heyday of the American merchant marine. With the passing of shipping as their leading industry, New Englanders turned to the abundant water power and found an outlet for their energies and a source of wealth in manufacturing. In the central Atlantic states, where fertile farm lands were combined with good harbors, rivers, and water power, the activities of the people were more diversified and spread over the major occupations.

Geographical influences have been potent not only in our economic development, but in the political and military history of our nation; subsequent chapters, it is hoped, will help to point this out.

Variety of Present Resources.—Geographic influences are as powerful to-day as they have been in our past history. Furthermore, the present natural products of the country and the undeveloped resources will determine to a large extent our future in the political as well as the economic field. Let us examine briefly the natural products and resources of the United States.

In the variety of natural resources this country is rich beyond any European nation. Where most countries have two or three such assets, the United States leads the world in many. Her size and the variations of climate and physiographic factors endow her with a rich variety of natural products, animal, vegetable, and mineral. England, for example, has sufficient coal and iron for her needs, but must import grains, meats, leather, cotton, wool, silver, gold, and many other commodities essential to her industries and to the maintenance of life in her population. Italy and Norway have water power and foodstuffs, but must import iron and coal. France, although her resources and climate are varied, cannot supply her own needs in many essential products, such as petroleum, copper, and raw cotton.

The United States, on the other hand, with the exception of one or two minor minerals, rubber, and coffee, produces everything necessary for her own consumption and manufacture, and much to export. Food materials of all kinds she has in abundance, enough to support a much larger population than her own, as was shown during the World War and as could be increasingly demonstrated with more intensive methods of cultivation. The United States exports food to many parts of the world, and although she imports various foods, such as coffee, tea, sugar, spices, and tropical fruits, nevertheless her imports are rather luxuries than absolute necessities. This country could be blockaded for centuries without fatal suffering. We produce 60 per cent of the world's copper, 66 per cent of the world's oil, 75 per cent of the corn, 60 per cent of the cotton, 52 per cent of the coal, and 40 per cent of the iron and steel; yet we have only 6 per cent of the world's population and 7 per cent of the world's land.

Mineral Resources.—The annual value of the metallic products of the United States is in the neighborhood of two billions, and of nonmetallic over three and one-third billions. Of these, coal ranks first in value. The anthracite deposits of the United States, located chiefly in Pennsylvania, are by far the most important in the world in both quality and quantity. Geologists have estimated that in Pennsylvania there were originally about 19,500,000,000 tons, of which about 1,900,000,000 have been taken out, less than one-tenth. The value of the coal mined in 1919 in this country was about $1,535,000,000. The most important bituminous fields are in the Appalachian Mountains, extending from Alabama to Pennsylvania; the region second in importance is in the interior in the Mississippi Valley states, including the coal fields of Michigan, Illinois, Indiana, Kentucky, Iowa, Kansas, Missouri, Oklahoma, Arkansas, and Texas; and North Dakota is underlined with lignite as yet unused. The aggregate coal areas of the United States approximate 500,000 square miles, or about 13 per cent of the area of the country. The fact that these beds are well distributed is significant, for the expense of transportation of this essential commodity to industrial and commercial development is a big item in its ultimate cost. It is also important that the richest deposits are within a few hours' haul of the great ports of New York and Philadelphia, and but a little farther from the manufacturing states of southern New England. If such had not been the case, the history of the northeastern United States after the Industrial Revolution might have been far different.[1]

---

[1] To these fuel resources the future historian will add the present peat swamps as yet almost wholly neglected; of these there are seven million acres

Next to coal the most valuable nonmetallic mineral is petroleum. Though the first well was not opened until 1859 (producing 2,000 barrels the first year) the oil industry has grown until the output in 1920 amounted to 443,402,000 barrels, with a valuation of $1,360,000,000. Although oil was produced originally to satisfy lighting needs, its field of usefulness has been widened until by distillation and other processes such commercial products as kerosene, benzine, gasoline, naphtha, heavy and lubricating oils, paraffin, and asphalt are manufactured. With the continued extension of the manufacture of gasoline motors and oil-burning engines, the value of petroleum in industry is constantly increasing. The chief oil fields, covering some 8,450 square miles, are as follows: (1) Appalachian field, extending from New York through Pennsylvania, southeastern Ohio, West Virginia and Kentucky into Tennessee; (2) the Lima-Indiana-Illinois field, including northeastern Ohio, a strip through middle Indiana, and southeastern Illinois; (3) the midcontinent region, including western Missouri, Kansas, and Oklahoma; (4) the Gulf fields, comprising the coastal plains of Texas and Louisiana; and (5) the California field. In addition, Wyoming, Montana, and Colorado have producing fields. About 60 per cent of the world's output comes from the United States. The original supply in this country was probably in the neighborhood of fourteen billion barrels, of which some five billion have now been used up.

The most valuable, most widely distributed, and cheapest of the metals in the United States is iron. It is found in practically every state, but the chief production districts can be grouped as follows: (1) Lake Superior district, including Minnesota, Michigan, and Wisconsin, producing, in 1919, 88.5 per cent of the whole; (2) the southern district, including Alabama, Georgia, North Carolina, Virginia, Maryland, Arkansas, Missouri, and Texas, producing 6.9 per cent; (3) the northern district, including New England, New York, New Jersey, and Pennsylvania, producing 4 per cent; and (4) the western district, including California,

in Minnesota alone. See E. K. Soper, *The Peat Deposits of Minnesota,* Bulletin 16, Minnesota Geological Survey (1919).

Washington, Idaho, Montana, Wyoming, Utah, and New Mexico, supplying .6 per cent of the whole. The Lake Superior region not only contributes over four-fifths of the iron ore, but contains at least three-fourths of the available deposits. It is, moreover, of distinctly superior grade to that of the Appalachians and Rockies. The advantageous situation of Pittsburgh and Birmingham as regards both coal and iron gave them a start in the iron and steel industry, but the recent predominance of the Lake

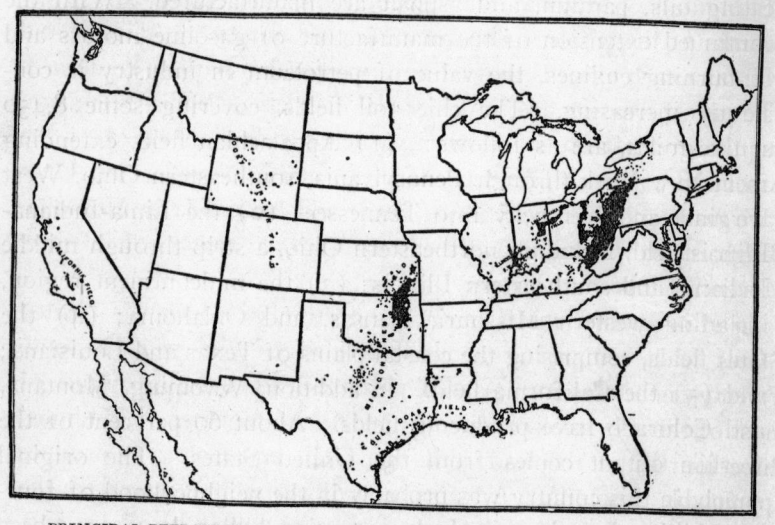

PRINCIPAL PETROLEUM AND NATURAL-GAS FIELDS OF THE UNITED STATES

Superior ores has created a tendency for the manufacturing center of the iron and steel industry to shift slowly to such Lake shore points as Buffalo, Cleveland, Chicago, and Gary. The unmined ore resources of the country are estimated (in metric tons) at 9,855,000,000 capable of producing 5,154,000,000 tons of iron, while the aggregate of the world is estimated at 11,987,000,000 tons. Thus the known iron resources in the United States are nearly equal to those of all the rest of the world.

From primitive times copper has been, next to iron, the metal most essential in the industries. This metal is so ductile and easily worked that the ancients became exceedingly proficient in turning it to a variety of uses. Since the harnessing of electricity, copper, because of its excellent properties as a conductor, has assumed a new importance. Improvements in extracting the metal, and the greater demands for it, have made possible the increase of the world's annual output from 9,000 tons in 1801 to over 1,000,000 to-day. Of this the United States contributes about three-fourths. In 1920 the production of this country was over 1,200,000,000 pounds, valued at over $222,500,000, a value ranking next to iron among the metals. Of this the three states of Montana, Arizona, and Michigan produced almost three-fourths of the whole. The annual copper output of Arizona and Montana each exceeds that of any two foreign nations combined. The purest deposits are in the glacial districts of upper Michigan, but the greatest are in Arizona, which supplies one-third of the American copper. The deposits in Montana, Nevada, and Utah have made possible the cities of Butte and Anaconda in Montana, Bingham in Utah, and the four Arizona towns of Bisbee, Morenci, Globe, and Jerome.

Lead is another metal in the production of which the United States leads, supplying about one-third of the world's total. Twenty-one states and Alaska produce lead, but the output of most of them is small. Missouri and Idaho together yield almost three-fourths of the lead, although Utah and Colorado produce appreciable amounts. Advancing civilization has increased the use of lead both in its pure form and in its numerous alloys, and, like most metals, its production was greatly stimulated by the World War. One-third of the lead goes into white lead for paint.

For the production of zinc, a comparatively new industry in this country (our first records of production are of 1873), the United States and Germany rank as the largest two sources of supply. The metal is used chiefly as an alloy of copper and tin, and in paints. Although twenty states mine zinc-bearing ores, two-fifths come from the region known as the Joplin, comprising southwestern Missouri, southeastern Kansas, and northeastern Oklahoma.

Before 1850 the gold mined in the territory now forming the United States was comparatively small. The discovery of the deposits in California in 1848 quadrupled the world's supply by 1852. The United States held first place for a long time, but in 1898 South Africa took the lead. We are now second, with Australia third. A widely distributed metal, it is found in many parts of the country; its chief sources, however, are in California, Colorado, and Alaska. The output in 1920 was 2,395,017 troy ounces, with a coining value of $49,500,000. Sixty per cent or more of the world's silver comes from the North American continent. Although the richest deposits are undoubtedly in Mexico, the political chaos there during the last few years has enabled the United States to lead the world in the mining of this metal. The big spurt in silver production came after the discovery of the famous Comstock lode in 1859. The chief deposits are in Nevada, Colorado, Montana, Utah, and Idaho. The value of the silver mined in 1920 was $57,400,000, but a small fraction of the value of any of the great staple food products.

Of the remaining metals, aluminum is the most important. It is mined chiefly in Arkansas, Georgia, Alabama, and Tennessee. Numerous nonmetallic minerals might be tabulated—stone, slate, clay, and sand, of primary importance in the building industry; and salt, lime, borax, and others used as food or as chemicals.

The enormous importance of minerals in industry is likely to overemphasize their value as compared with agricultural crops. The grand total value of all crops in 1919 was $15,232,000,000, as against $4,613,000,000 for all minerals. The money value of no single mineral product equals that of corn, cotton, or hay.[1]

Forest Resources.—At the present time the United States with its 350,000,000 acres of forest is the chief wood-producing

---

[1] The following figures for 1919 give basis for comparison:

Mineral products (in millions)		Agricultural products (in millions)	
Coal	$1,515	Corn	$3,508
Iron (ore and pig)	978	Hay and forage	2,523
Petroleum	775	Cotton and cotton seed	2,355
Copper	239	Wheat	1,416
Zinc	66	Tobacco	444
Silver	64	Oats	441
Gold	60	Rye	96
Lead	45	Rice	81
Aluminum	39	Barley	68

Compiled from *Statistical Abstract* for 1921, tables 111 and 145.

nation of the world. A conservative estimate places our present natural consumption of wood in all forms at 100,000,000,000 board feet. The forest belt of the American continent covers roughly the region east of a line drawn from the western shore of Hudson Bay south to Texas, a large area in the highest regions of the Rocky Mountains, and the Pacific forests in the Sierra Nevada and the coast ranges. For convenience in classifying the products, the timbered areas may be divided roughly as follows: (1) the northeast.—The most important species of this group are the conifers, or soft woods, including white pine, spruce, and hemlock; although the various hard woods such as hickory, oak, and maple are found here also. (2) The south. —In the southern states four general types prevail, varying with the altitude—cypress and hard woods in the swamps and lower sections of the river valleys of the Atlantic and Gulf states; yellow pine in the rest of the coastal plain from Virginia to Texas; hard woods on the lower slopes of the Appalachians; and conifers higher up in the mountains. (3) The Great Lake region.—The southern part of this district contains considerable hard wood; in the northern section pine, tamarack, cedar, and hemlock predominate. (4) The Rocky Mountain division is chiefly noted for the western yellow and lodge-pole pines; while (5) the Pacific coast produces soft woods where there is an abundance of Douglas fir, hemlock, pine, cedar, and redwood. The most important, the Douglas fir, attains its best development in the Puget Sound region, where it reaches a height of from two to three hundred feet.

Because of the enormous waste in clearing the land and the destruction of large areas by forest fires, the lumber industry has ceased to be important in most of those regions, and has been continually shifting. The northeastern states, which a half century ago produced more than one-half of all the lumber, now contribute less than one-tenth of the whole. The southern states are now the largest producers, with the Pacific states ranking second. Most of the lumber comes from Washington, Oregon, Louisiana, and Mississippi. Four-fifths of the whole cut comes from the following seven species: yellow pine, Douglas fir, white pine, hemlock, oak, spruce, and western pine. In 1920 the computed total production of lumber was 33,798,800,000 feet, and the value of the lumber cut was not far below that of either the coal mined or the petroleum produced. It is estimated that originally the American forests covered 800,000,000 acres, with a stand of merchantable saw timber of 5,200,000,000,000 square feet, now reduced to 350,000,000 acres, with a stand of 2,200,-000,000,000 square feet "Stated in other fashion," says R. S. Kellogg, "nearly 60 per cent of the merchantable saw timber of the United States has been utilized or destroyed, and the bulk of it has gone in the past fifty years." [1]

Agricultural Resources.—Agriculture continues to be the basic industry of the United States, one-third of the labor force of the country being employed on it. The leading crops are corn, cotton, hay, wheat, oats, potatoes, barley, tobacco, sugar beets and sugar cane, flaxseed, rye, rice, buckwheat, and hops. Of vegetable foodstuffs, cereals are the most important, and of these maize, or Indian corn, takes first place. In 1920 the yield was approximately 3,000,000,000 bushels, exceeding in that year 90 per cent of the world's crop. Production increases continually, and with improved methods of tillage there is reason to believe that the limit is far from reached. Corn is grown extensively in the cotton belt of the south, but three-fourths of our supply is raised in the so-called "corn belt" of the upper Mississippi Valley, which includes the states of Kansas, Nebraska, Illinois, Iowa, Ohio, Indiana, and Missouri. Here are to be found the ideal conditions for its growth, namely heavy rains alternating with much sunshine, and a soil which is drained easily and does not cake.

Although wheat has in the United States only about half the acreage of corn, with a corresponding valuation, its position as the chief food of mankind makes it in some respects more important than corn. Introduced from the Old World by the earliest settlers, it has spread so widely that we now lead in its production. In 1915, under the impetus of war needs, the

---

[1] Kellogg, Royal S., *Pulpwood and Wood Pulp in North America* (1923), p. 148.

PHYSIOGRAPHIC FACTORS

production mounted to over a billion bushels, the largest yield in our history up to that time. Although about four-fifths of the corn produced is consumed on the farm and only about one-fifth offered for sale, almost all wheat is sold and converted into flour. The increase in wheat production was over 550 per cent between 1851 and 1900. The chief producing states are Minnesota and the Dakotas. These states, with Kansas, Nebraska, Illinois, Washington, Indiana, and Ohio, grew nearly 60 per cent of the whole yield.

The United States also leads the world in the production of the other chief cereals—oats, rye, and barley. In 1920 the yield of oats in this country was about 70 per cent of the world's crop, that of rye about 44 per cent, and that of barley about 40 per cent. Oats are grown best in the northern states; about 90 per cent come from north of the thirty-sixth parallel. The middle western and northwestern states are the chief producers of barley and rye. These crops largely enter commerce, though not to such an extent as wheat.

About 15 per cent of the improved land in this country is given over to the production of hay and forage. Of this at least 5,000,000 acres are devoted to hay, which ranks in value third among agricultural crops, being surpassed only by corn and cotton. New York and the middle western states lead in this crop. Hay is shipped extensively in domestic commerce, but not to foreign countries, on account of its bulk.

Since rice is not so important in American diet as in that of some other countries, especially in the Orient, the United States does not rank so high in its production as in the case of the other cereals. The center of American rice production is Louisiana, Texas, and Arkansas, where it is grown mainly by irrigation, though a considerable amount is still produced in the Carolinas and Georgia.

Sugar cane is grown chiefly in Louisiana, with Texas as the only other state producing an appreciable amount. These furnish but a fraction of the sugar used in the United States, most of which is imported from Cuba and our insular dependencies—Porto Rico, Hawaii, and the Philippines. The cultivation of sugar beets, however, has increased rapidly during the last few years.

Ranking second in value among American crops, and the greatest of all crops in the south, is cotton. In 1920 the cotton yield was 12,987,000 bales, valued at over $900,000,000, approximating roughly two-thirds of the world's crop, British India and Egypt ranking second and third, respectively. The cotton belt lies in a strip 1,450 miles long and 500 miles wide, south of the thirty-seventh parallel and east of the hundredth meridian, including all the southern states from North Carolina to Texas. About twenty-seven million acres are given over to this staple.[1]

A superior type, known as Sea Island cotton, but representing in amount only about 1 per cent of the total product, is grown on the sea islands of South Carolina and the interior counties of Georgia and Florida. Egyptian cotton is now being experimented with in California. Before the Great War almost two-thirds of the cotton crop was exported, but now we are using over half of it at home.

Tobacco is grown east of the Mississippi in twenty-five or more states, mainly in the eastern coastal plain, the Appalachian region, and the Mississippi Valley plains. The total product in 1920 was over one and a half billion pounds, with the chief centers of production in North Carolina, Kentucky, and Virginia. Tobacco now ranks only eighth in value among the agricultural crops of the United States, although this country leads the world in its production, and (including the crops of the insular possessions) furnishes about one-third of the world's crop. About three-fifths of the domestic growth is consumed here.

Cultivation of white potatoes is carried on in every state of

[1] The figures of 1920 show the relative importance in production of the cotton-belt states:

State	Amount of production (in 500-lb. bales)	State	Amount of production (in 500-lb. bales)
Texas	4,200,000	Alabama	660,000
South Carolina	1,530,000	Louisiana	380,000
Georgia	1,400,000	Tennessee	310,000
Oklahoma	1,300,000	Missouri	285,000
Arkansas	1,160,000	Virginia	19,000
Mississippi	885,000	Florida	18,000
North Carolina	840,000		

REGIONAL DISTRIBUTION OF PRODUCTS IN THE UNITED STATES

Showing approximately the Productive Areas of principal Agricultural and other Staples, 1912.

Reprinted from a report of the U. S. Bureau of Statistics.

the Union, but in the northern and northeastern sections conditions are most favorable. They are mostly consumed near the place of production. Sweet potatoes are grown extensively in the southern states.

The great importance of the native animal life on this continent to the early settlers and pioneers has been mentioned. Possessing as it does the necessary requisites of temperate climate and immense pasturage areas, this country was destined to take its place as a leading source of animal products. The United States ranks fourth among the nations in the number of sheep produced, second in cattle, and first in swine.[1] The best grazing lands for cattle are in the great plains, the Rocky Mountain table-lands, and their eastern slopes. The centers of dairying, however, are near the great cities, with New York as the leading state. The center of the swine country is, with the exception of Texas, identical with the area of the greatest corn production. While sheep are to be found in every state in the Union, the chief wool states are Montana, Wyoming, Idaho, and Oregon. The two most important horse markets are Omaha and Kansas City, both in the center of the horse-raising area. Mules are more numerous than horses in the southern states, because they can better endure hot climate and hard usage. Owing to the fact that corn is the best cereal for fattening poultry, the center of the poultry industry is in the north central states, from which comes about one-half of the billion-dollar annual yield of eggs and chickens.

The importance of the fisheries in the economic life of the colonies has been touched upon. The United States and Alaska together still lead the world in this industry. The western Atlantic from Newfoundland to the Chesapeake is one of the two most important fishing areas in the world. The shad, mackerel, cod, herring, halibut, bluefish, and oysters on the Atlantic coast supply over forty-one million of the hundred-million-dollar annual haul of the United States, including Alaska. The salmon fishing of Alaska and the Columbia River is the most important phase of this industry on the Pacific coast, the fisheries of which yield

---

[1] For the estimated number of farm animals in the United States on January 1, 1921, see *Statistical Abstract* for 1920, p. 171.

a catch worth fifty million. The Great Lakes supply herring, whitefish, trout, yellow perch, and many other varieties valued annually at over six million.

Water Resources.—The importance of water in our economic life can scarcely be overestimated. Not only is water valuable as the source of fish and certain fur-bearing animals, and as a cheap and convenient highway of commerce, but also as an inexhaustible supply of power, which can be harnessed so as to fill an infinite variety of human needs. More and more the attention of engineers is being given to the development of water power as a source of energy for manufactures. Hydro-electric development has already gone far, notably in three sections: the North Atlantic on the Merrimac, Connecticut, and Hudson rivers; the St. Lawrence region, on the Niagara and other rivers running into the Great Lakes; and on the rivers of the North Pacific. Some 5,000,000 horse power has already been produced, but the possibilities of water power are still in their infancy. Enormous water-power dams have been built at Keokuk, Iowa, on the Big Creek, California, on the Tallulah River, Georgia, and at many other places; but the water resources of the nation have scarcely been tapped. With the increasing need of power and the increasing cost and eventual exhaustion of the supply of fuel, the use of water power through electric transmission from central plants, as in South Africa, must be greatly enlarged in the future.

Summary.—To sum up, it may be said that the United States, with a few minor exceptions, furnishes every mineral necessary to a great manufacturing nation. In her fertile valleys and broad prairies she can produce foods, both of the temperate and semitropical varieties, sufficient to support a population much larger than at present. She can supply enough leather, wool, and cotton for her own needs and a considerable amount for exporting. In addition to her sources of power from coal and petroleum, she has vast reserves of water power, the potentialities of which stagger the imagination. The United States comprises a world in itself. Not only is it the most wealthy and prosperous of the nations, displaying a higher standard of living than any other, but it is a country in which the real wages of work—food, fuel, shelter, and clothing—are higher and in greater abundance than elsewhere. Our resource area is at present second to none, and probably surpasses all others in the world. We have the bases for all the necessities and comforts of life for all our people. The possession of these apparently inexhaustible resources has helped to endow the American people with a hopeful and buoyant confidence, but at the same time has encouraged wasteful exploitation. What we still have to learn is how better to conserve and utilize our resources.

NOTE ON STATISTICS

In any work dealing with economics, wide use must be made of statistics, the nature of which should be better understood. "One of the prime objects of statistics," says W. I. King in his *Elements of Statistical Method* (p. 22), "is to give us a bird's-eye view of a large mass of facts, to simplify this extensive and complex array of isolated instances and reduce it to a form which will be comprehensible to the ordinary mind." To accomplish this, involved mathematics are used to develop economic formulæ and scientific statistics, which then interpret the descriptive statistics compiled by the census and other agencies.

A word should be added as to the accuracy of statistical tables. Absolute accuracy in the material with which we are dealing is probably never possible, but, since the aim is better comprehension of an entire field, relative and not absolute accuracy is the main desideratum. Statistics, as King says (*ibid.*, p. 64), are estimates rather than exact enumerations. It is impossible, for example, to obtain more than an approximate estimate of the number of bushels of wheat produced. Furthermore, statistics of varying kinds differ greatly in relative reliability. For instance, the number of deaths reported is relatively accurate, as the returns are required by law, but a tabulation of the causes of deaths may be far from correct. It should also be kept in mind that the use of round numbers sometimes gives a more accurate impression than the figures carried out, because when attention is directed to the digits the main point being demonstrated may be lost, especially in comparisons.

For a concise account of the manner in which facts for statistical study are collected, see *The Review of Economic Statistics,* Sept., 1920-Jan., 1921. Prefatory sections to each chapter contain valuable material on methods employed and sources of information. For a clear and not too technical account of the nature of statistics, read Willford I. King, *Elements of Statistical Method* (1913), Parts I and II.

Most of the statistical information in this book is obtained from government sources. The government departments at Washington have statistical divisions, compiling elaborate reports, and the Bureau of the Census now maintains a permanent staff collecting information. A large amount of data is conveniently reprinted in *The Statistical Abstract of the United States,* published by the Bureau of Foreign and Domestic Commerce of the Department of Commerce. *The Abstract of the Census of Manufactures* and *The Abstract of the Census* are less bulky than the *Census* to use and condense the most valuable material of each census.

NOTES FOR FURTHER REFERENCE

The serious study of the influence of geography upon history was opened by H. T. Buckle, *History of Civilization in England* (1884), in his epoch-making introduction. The subject has been brilliantly followed by H. B. George, *The Relations of Geography and History* (1901). An economist's point of view may be found in Achille Loria, *Economic Foundations of Society* (1898). The best work on American history has been done by Ellen C. Semple, *American History and Its Geographic Conditions* (1903), and *Influence of Geographic Environment* (1911). An attempt of a geologist to interpret American history is that of A. P. Brigham, *Geographic Influences in American History* (1903). Two short but enlightening sketches are the introduction of D. R. Fox to *Harper's Atlas of American History* (1920) and Chap. I of A. M. Schlesinger, *New Viewpoints in American History* (1922). The relations between geography and history have been emphasized by N. S. Shaler, *Nature and Man in America* (1899) and *The United States of America* (2 vols., 1897).

The influence of frontier life and conditions can be best studied in the articles of F. J. Turner, some of the best of which have been collected in *The Frontier in American History* (1920). Brief studies of special phases are E. R. Johnson *et al., History of Foreign and Domestic Commerce of the United States* (1915), Chap. I, and C. E. MacGill *et al., History of Transportation in the United States Before 1860* (1917). A brief but comprehensive description of physiographic regions, with an excellent map, is J. W. Powell, *Physiographic Regions of the United States,* in National Geographic Monographs, I, No. 3 (1898).

A preliminary discussion of products and resources is to be found in Livingston Farrand, *Basis of American History* (1904), Vol. II of the American Nation Series, and in Ellsworth Huntington, *The Red Man's Continent* (1919) in the Chronicles of America Series. Further figures and facts are obtainable in C. R. Van Hise, *Conservation of Natural Resources in the United States* (1910); in C. G. Gilbert and J. E. Pogue, *America's Power Resources* (1921); in R. S. Kellogg, *The Timber Supply of the United States* (1909); and in such standard commercial geographies as R. S. Tarr, *Economic Geology of the United States* (1895); H. R. Mill (ed.), *International Geography* (1899); C. C. Adams, *A Textbook of Commercial Geography* (1902); E. Van Dyke Robinson, *Commercial Geography* (1910); G. G. Chisholm *Handbook of Commercial Geography* (8th ed., 1911); J. R. Smith, *Industrial and Commercial Geography* (1913); and J. E. Spurr (ed.), *Political and Commercial Geology and the World's Mineral Resources* (1920). Recent volumes designed for classroom use containing much that is pertinent to North America and its resources are: E. Huntington and S. W.

Cushing, *Principles of Human Geography* (2d ed., 1922); E. Huntington and F. W. Williams, *Business Geography* (1922); William H. Emmons, *Geology of Petroleum* (1921); and C. C. Colby, *Source Book for the Economic Geography of North America* (1921).

SELECTED READINGS

Harper's Atlas of American History, introduction by D. R. Fox.

SCHLESINGER, A. M., *New Viewpoints in American History,* Chaps. I, II and III.

SEMPLE, E. C., *American History and Its Geographic Conditions,* Chaps. I–III.

BRIGHAM, A. P., *Geographic Influences in American History,* Chaps. I–III.

COLBY, C. C., *Source Book for the Economic Geography of North America,* Chaps. VIII–XII.

CHAPTER II

SOCIAL AND ECONOMIC ASPECTS OF COLONIZATION

Medieval Commerce.—The discovery of America was brought about by a train of circumstances extending back through centuries of European history and culminating at the end of the fifteenth century. Intellectual, political, and, above all, economic factors contributed to make this a turning point in world history. The fifteenth century and beginning of the sixteenth marked the height of the Renaissance, a period of inquiry and dissatisfaction with the old order. In political life the modern state was being erected on the ruins of feudalism; with the national state came a cessation of the private warfare of the Middle Ages, greater protection to travelers and merchants, and fewer tolls. More settled conditions encouraged the extension of trade and commerce, and the revived economic life led naturally to exploration and discovery. The latter was aided by the compass and astrolabe, by that time in general use; while the news of scientific and commercial progress was disseminated by means of the printing press, invented about the middle of the fifteenth century.

But more important than any motive yet mentioned was the desire on the part of the Europeans to find a quicker and cheaper route to the East. From immemorial antiquity Europe had been dependent upon Asia for most of her luxuries and many of her necessities. The importance of spices in the Middle Ages is difficult to appreciate to-day, when meat is kept fresh by cold storage or curing, but the monotonous diet and coarse food of those times made spices and condiments so desirable that they were frequently sent as gifts of honor from one sovereign to another. Pepper from the Malabar coast of India was a staple import during the Middle Ages and used by all who could afford it. Cloves, cinnamon, nutmegs from the Moluccas and sugar from Arabia and Persia were more expensive and less commonly used, but in great demand. Apothecaries obtained many of their drugs from Asia; among them rhubarb, balsam, gums, aloes, cubebs, and camphor. The precious stones which adorned the persons of the upper classes in Europe came almost exclusively from the East.

Trade with the East, however, was not confined exclusively to such luxuries as spices, drugs, and precious stones. An important class of wares which served manufacturing industries, namely dyestuffs, found their source here. Indigo was the chief staple of Bagdad, while Brazil-wood, producing a red dye, came from India. Other dyes, when the best quality was demanded, were imported from the East. Alum, considered indispensable for fixing colors in dyeing and one of the most desirable products of the Levant trade, was procured mainly in Asia Minor. Manufactured products, superior in workmanship, material, and design to anything known in Europe, came also from the East: glass and cutlery from Damascus, Samarkand, and Bagdad; porcelain from China; a great variety of cottons and silks from India, China, Persia, and Asia Minor. Persian rugs, Cashmere shawls, taffeta silk, damask linen, and japanned ware, all testify to the Eastern origin of the most sought after textiles, rugs, tapestries, and household luxuries.

In return for these products Europe could offer only woolen fabrics and such metals and minerals as arsenic, antimony, quicksilver, tin, copper, and lead. Although these products were much valued in the East, their weight and bulk made transportation on the long overland routes an arduous and unrewarded task. The balance of trade was consequently always in favor of Asia, and Europe was gradually being drained of gold and silver, a situation from which she was eventually saved only through the mines of Mexico and Peru.

While trade with the Near and the Far East was a leading factor in the economic life of the Middle Ages, little was known by Europeans of Asia or the routes thereto. Trade had flourished in ancient times, but during the barbaric invasions of the fifth century and the succeeding conflicts this commerce had been largely broken up. A general awakening of economic life in the eleventh century set in motion with renewed vigor the intercourse with the East, a movement greatly aided by the Crusades (1095-1270). Not alone did the Crusades enlarge the vision and knowledge of Europeans by introducing them again to the learning and products of Asia, but it laid the foundations for the prosperity of the Italians in this trade. Towns at the extreme south of the Italian peninsula, such as Bari, Trani, Brindisi, and Taranto, sprang into importance, as did Amalfi on the bay of Naples, Genoa at the head of the Tyrrhenian Sea, and Venice at the head of the Adriatic.

Medieval Trade Routes.—The products of the East reached Europe over three main trade routes, of which the oldest and most important during the greater part of the Middle Ages was the central passage. Merchandise was gathered from India and the Far East and brought to Ormuz at the north of the Persian Gulf, and thence to Bossorah at the mouth of the Tigris and up the valley to Bagdad. From Bagdad the routes spread like a fan either north to Tabriz, westward to Antioch, Damascus, or Jaffa, or southwest to Alexandria. From here and other ports on the eastern Mediterranean the products found their way to Europe. The southern route was chiefly a sea route leading from India (usually Calicut on the southwestern coast) across the Arabian Sea to the Red Sea. The cargoes in most cases were landed on the western coast and transferred to caravans, which carried them to the Nile, upon which they were floated down to Cairo. This route, although attended by difficulties of navigation, was undoubtedly the cheapest and quickest, and at the close of the Middle Ages the most important. The northern route, which was entirely overland, was in reality a system of routes leading from the inland provinces of China and India to Samarkand and Bokhara on the western slope of the Tian-Shan Mountains. At Bokhara the routes branched, one leading into the Caspian and up the Volga to central Russia and the Baltic, while the other continued west through Tabriz and Armenia to Trebizond on the Black Sea and thence to Europe.

The terminal points, then, of the eastern trade were such cities as Trebizond on the Black Sea, Constantinople on the Bosphorus,

Acre, Beirut, Tripoli, Laodicæa, and Jaffa on the Syrian coast, and Alexandria at the mouth of the Nile. At these cities, and even in such inland points as Damascus, Aleppo, and Antioch, merchants from Spain, France, and Italy met the caravans and

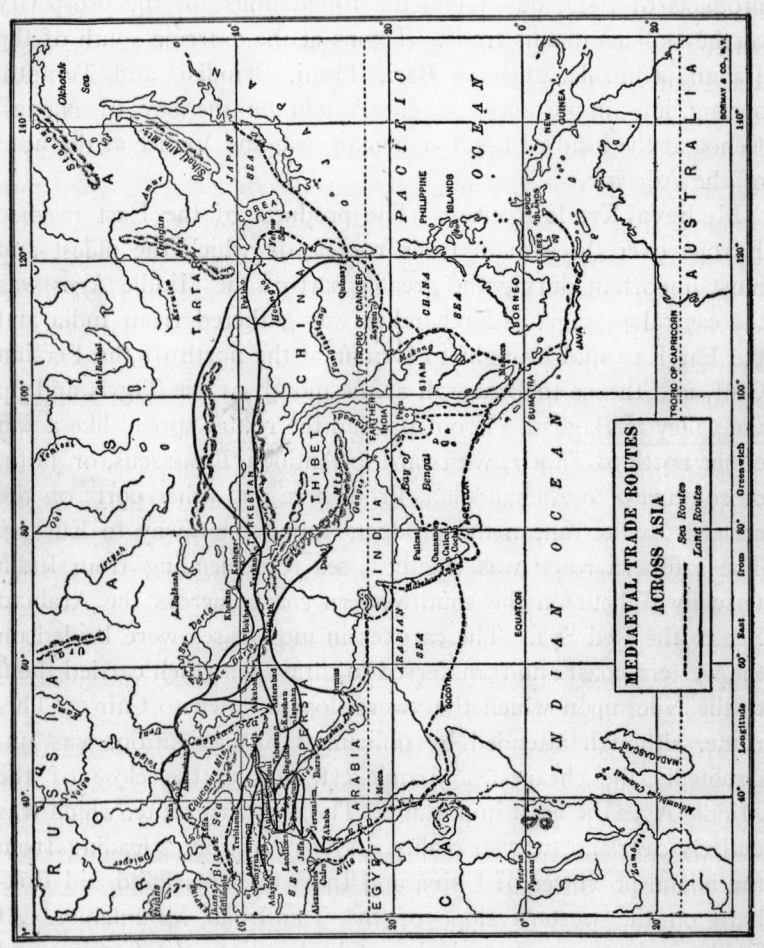

purchased what goods had escaped destruction from the elements and confiscation by pirates. The bulk of the Levant trade during the five hundred years from 1000 to 1500 rested in the hands of the Italians, and the three cities of Venice, Genoa, and Pisa struggled for supremacy in this trade, while Florence became a

ECONOMIC ASPECTS OF COLONIZATION

carry to Europe the bulky commodities hitherto unknown, such as tea, coffee, Indian corn, and tobacco. The growth in business developed better methods of carrying on trade, new industries sprang into existence, manufacturing increased, the capitalist began to appear, all tending to break down the antiquated guild system. The plunder of Mexico and Peru, by providing Europe with a new source of the precious metals, augmented the supply of currency, effecting a revolution in prices. Even agriculture responded to the stimulus of new crops and the necessity of supporting a greater population which came with enlarged commercial activities. The slave trade was revived to provide the labor necessary to work the plantations of the New World.

Upon one class in Europe the Commercial Revolution had its greatest influence. While kings and nobles fought over colonial empires the bourgeoisie or townsfolk, that new middle class just emerging, reaped the benefits. In every country the number of those dependent upon commerce and trade grew rapidly and the long process of the exaltation of the middle class at the expense of the landed aristocracy, the climax of which we see in our own day, was begun. In the arena of imperial politics the Commercial Revolution inaugurated a series of struggles for colonial possessions and commercial power during which the maritime supremacy passed from Portugal to Spain and then, during the Dutch rebellion, to Holland. But England and France were both growing in national spirit and sea power and both made war upon the wealthy but diminutive republic until they had effectually crippled her. With Holland eliminated, England and France engaged in a series of seven great wars extending from 1689 to the overthrow of Napoleon, which left Great Britain supreme upon the sea and the foremost colonial power. All of these nations, however, while relinquishing their maritime supremacy, retained certain parts of their colonial Empire in America.

Motives for the Colonization of America.—As the idea was gradually brought home to Europeans that the new-found land was not the Indies, but two mighty continents, not only did statesmen dream of new empires, and knights and merchants of new sources of riches, but the common man began to think of a new home across the seas where he might escape from the religious, political, and economic tyranny of the Old World. The motives for colonization were varied—religious, political, and economic being inextricably combined.

The age of the Reformation was one in which the religious motive was strong. Prince Henry sent his ships to find not only the Indies, but also the fabled Christian kingdom of Prester John. "We come in search of Christians and spices," said Vasco da Gama. In the breasts of the early Spanish conquerors and explorers the crusading spirit was hot. "He who possesses gold," said Columbus, "does all that he wishes in this world, and succeeds in helping souls into paradise." French Jesuit priests threaded the lakes and rivers in advance of the fur trader, baptizing as they went. The religious impulse moved even the more prosaic English. Drake and Hawkins scoured the Spanish Main to fight Catholics as well as collect booty. Later many felt with the Virginia Company managers that the first object of that plantation was "to preach and baptize into the Christian religion . . . a number of poore and miserable souls wrapt up unto death in almost invincible ignorance." Sir Humphrey Gilbert emphasized the benefits that would accrue to the natives brought by the white man "from falsehood to truth, from darkness to light, from the highway of death to the pathway of life, from the devil to Christ, from hell to heaven." While the work of the priests in opening the routes for settlers should not be underestimated, more important than even the crusading spirit was the influx of settlers who sought freedom from religious persecution at home. Separatists and Puritans founded New England to obtain religious liberty; but Puritan intolerance in turn drove Roger Williams and his followers to Rhode Island, banished Anne Hutchinson, and contributed somewhat to Hooker's settlement of Connecticut. Puritanism drove Cavaliers to Virginia and English Catholics to Maryland. French Protestants found refuge in the Carolinas; while Quakers, Mennonites, Moravians, and other sects found a home in Pennsylvania, New Jersey, and elsewhere.

ECONOMIC ASPECTS OF COLONIZATION

banking and manufacturing center. From the Italian cities the Oriental merchandise was distributed to northern and central Europe through two main channels. German merchants handled the overland reëxport trade from Venice and Genoa. In the former city a building was set aside by the state for their use where they could bargain with the Venetians and be watched by the government. Through the St. Gothard Pass to Basel, Constance, Strassburg, and down the Rhine, or over the Brenner Pass to Frankfort, Nuremburg, or Munich were the usual routes. In addition to the overland passages through the Alps a large amount of trade was carried on with Lisbon, Bruges, and London by sea. At intervals of usually a year a fleet would set out from Venice, the so-called "Flanders galleys," strongly protected by warships, which sailed through the Straits of Gibraltar and up the coast of France. Bruges, the chief objective, was the great market where the merchants of northern Europe met those of the south.

The Commercial Revolution.—Our task up to this point has been to show, first, the importance of the Eastern trade [1] to the Europe of the Middle Ages; and, second, that the center of European medieval commerce lay in the Mediterranean basin. How this trade center shifted from the Mediterranean to the Atlantic, producing a Commercial Revolution, remains to be shown. The story involves the discovery both of new trade routes to the East and of the American continents.

Of the many factors operative in bringing this about, two stand preëminent. In the first place, the eagerness of Europe for the products of Asia, coupled with the difficulty and expense of obtaining them over the existing trade routes, made imperative the discovery of quicker and cheaper means of travel to the East. In the second place stands the decline in sea power of Venice and Genoa, occasioned largely by the conquest of Asia Minor at the

---

[1] To show that the question of Eastern trade routes is still a vital one it is necessary simply to call attention to the history of the Suez Canal and the attempt of Germany to cut in on this route by means of the Berlin-Bagdad Railway. The importance of this was never lost sight of during the World War, as the campaigns in Palestine and the Euphrates Valley demonstrate. See Earle, E. M., *Turkey, the Great Powers, and the Bagdad Railway* (1923).

close of the Middle Ages by the Ottoman Turks. This nomadic and warlike people, imbued with bitter hatred of the Christian and contempt for the merchant, began their conquests about the year 1300. Spreading east and south through Asia Minor, Syria, and Egypt, and northeast into the Balkans, they succeeded in displacing the more peaceful Seljuk Turks and finally in 1453 in capturing Constantinople. By 1500 their empire included practically the whole of what are now the Balkan States, the territory around the Sea of Azov, Asia Minor, Syria, and Egypt, and with the capture of Damascus in 1516 and Cairo in 1517 their control over the three great trade routes to the East was complete. In the process of their conquests they broke the military and colonial power of Genoa and Venice, the great trading cities of the European Mediterranean, and so accentuated the decline of the Italian seaports.

That the effects of the Turkish conquests must have been indirect should be clearly understood. The old theory that the Turks spurred on exploration by imposing crushing burdens upon European trade has been largely discredited by recent investigation. Professor Lybyer, in particular, has shown[1] that there was no serious interference in Eastern trade and no rise in prices in Europe after the Turkish conquests. He contends, on the other hand, that motives "related to religion, crusading, conquest, and adventure probably outweighed the seeking of spices in the minds of the great explorers and their royal supporters," and there is much to be said for this explanation.

Although such Italian travelers as Carpini and the three Polos had added much to the European knowledge of Asia, and the Italian cartographers led in the accuracy of their maps, it was left to the Portuguese to discover the new way to the East. Always a seafaring nation to whom the Atlantic coast of North Africa was not wholly unknown, they were spurred on by the enthusiasm of Prince Henry "the Navigator," a member of the Portuguese ruling family and a man in whom rare business ability was combined with the instinct of the explorer and the zeal of the missionary. One expedition after another was sent by him to

---

[1] *The Influence of the Ottoman Turks upon the Routes of Oriental Trade* in the Annual Report for 1914 of the American Historical Association.

explore the west coast of Africa. In 1434 Eannes rounded Cape Bojador, Cape Blanco was passed in 1441, and in 1445 Denis Diaz reached Cape Verde. It was not, however, until 1487 that Bartholomew Diaz discovered what he called the Cape of Storms and what King John II of Portugal later christened the Cape of Good Hope. Ten years later Vasco da Gama rounded the cape, pushed up the east coast, and in 1498 reached India. Cabral, with a fleet, followed in 1500,[1] and Albuquerque in 1503. The new sea route to India was complete and trade was rapidly established. The strategical position of Portugal, combined with lowered cost of transportation over the new route, threw the Eastern trade into the hands of the Portuguese, who took immediate advantage by laying the foundations of an Eastern empire which they ruled until the crowns of Spain and Portugal were joined. With the opening of the new route to the East the center of the world's trade shifted from the Mediterranean to the Atlantic. Merchants deserted the Rialto and flocked to the quays of Lisbon. The rotting wharves of Venice, in contrast to the now bustling harbor of Lisbon, demonstrated but too clearly how the glory that had for centuries belonged to Venice had passed in a few years to Lisbon.

The Discovery of America.—Portugal was not the only nation where men were dreaming of riches through quicker routes to the Indies. Before da Gama had made his epochal voyage to India, Ferdinand and Isabella of Spain, fresh from their conquests over the Moors, had paused in their building of a great Spanish state to promise aid to the Italian navigator Columbus in his projected westward voyage. Columbus, believing with all educated astronomers and philosophers that the earth was round, thought that by sailing due west he could reach the Indies. "I have always read that the world," said he, "comprising the land and the water, is spherical, as is testified by the investigations of Ptolemy and others, who have proved it by the eclipses of the

---

[1] Blown westward out of his course, he touched Brazil, the point on the American continent nearest the Old World. If Columbus had not discovered America another decade in all probability would have seen Europeans on American soil.

moon and other observations made from east to west as well as by the elevation of the pole from north to south." His thesis was of course correct and not new. His mistake was in conceiving the world to be a smaller sphere than it turned out to be. The greatness of Columbus was not in the originality of his conception, but in his courage in sailing the unknown seas and in his pertinacity in the pursuit of his project. Sailing west in 1492 with his three little ships, he at length ran into what was probably one of the Bahamas. He believed that he had discovered outlying islands of the Indies and returned in 1493, in 1498, and again in 1502, only to meet with disappointment in his efforts to get through to India.

The efforts of Columbus were emulated by John Cabot, the Italian navigator in the service of Henry VII of England, who sailed due west in quest of Cipango, only to land on the barren shore of Labrador. Even after Balboa (1513) had discovered the great western ocean and Magellan's ship *Victoria*[1] had circumnavigated the globe (1519-22) in that greatest feat of navigation of all time, explorers continued for a hundred years to seek for channels leading through or around America to Asia. This quest for a passage to the Indies led to the explorations of Verrazano (1524), Cartier (1534), Frobisher (1576-78), Davis (1585-87), and Hudson (1609). Although no natural opening to Cathay was ever discovered (we now have an artificial cut through the Cordilleras at Panama), these voyages gave to Europeans their first knowledge of what is now the coast of the United States and of the two great rivers of the eastern coast, the St. Lawrence and the Hudson.

The Commercial Revolution, including, as it did, the discovery of America, had incalculable effects upon economic history. But a few of the most important can be designated here. The comparative cheapness of water transportation over the new routes to the Indies reduced the cost of Oriental goods and made possible their more general use. Long ocean voyages developed the construction of stronger and higher ships which could profitably

---

[1] He himself did not complete the journey.

ECONOMIC ASPECTS OF COLONIZATION 35

carry to Europe the bulky commodities hitherto unknown, such as tea, coffee, Indian corn, and tobacco. The growth in business developed better methods of carrying on trade, new industries sprang into existence, manufacturing increased, the capitalist began to appear, all tending to break down the antiquated guild system. The plunder of Mexico and Peru, by providing Europe with a new source of the precious metals, augmented the supply of currency, effecting a revolution in prices. Even agriculture responded to the stimulus of new crops and the necessity of supporting a greater population which came with enlarged commercial activities. The slave trade was revived to provide the labor necessary to work the plantations of the New World.

Upon one class in Europe the Commercial Revolution had its greatest influence. While kings and nobles fought over colonial empires the bourgeoisie or townsfolk, that new middle class just emerging, reaped the benefits. In every country the number of those dependent upon commerce and trade grew rapidly and the long process of the exaltation of the middle class at the expense of the landed aristocracy, the climax of which we see in our own day, was begun. In the arena of imperial politics the Commercial Revolution inaugurated a series of struggles for colonial possessions and commercial power during which the maritime supremacy passed from Portugal to Spain and then, during the Dutch rebellion, to Holland. But England and France were both growing in national spirit and sea power and both made war upon the wealthy but diminutive republic until they had effectually crippled her. With Holland eliminated, England and France engaged in a series of seven great wars extending from 1689 to the overthrow of Napoleon, which left Great Britain supreme upon the sea and the foremost colonial power. All of these nations, however, while relinquishing their maritime supremacy, retained certain parts of their colonial Empire in America.

Motives for the Colonization of America.—As the idea was gradually brought home to Europeans that the new-found land was not the Indies, but two mighty continents, not only did statesmen dream of new empires, and knights and merchants of new sources of riches, but the common man began to think of a new home across the seas where he might escape from the religious, political, and economic tyranny of the Old World. The motives for colonization were varied—religious, political, and economic being inextricably combined.

The age of the Reformation was one in which the religious motive was strong. Prince Henry sent his ships to find not only the Indies, but also the fabled Christian kingdom of Prester John. "We come in search of Christians and spices," said Vasco da Gama. In the breasts of the early Spanish conquerors and explorers the crusading spirit was hot. "He who possesses gold," said Columbus, "does all that he wishes in this world, and succeeds in helping souls into paradise." French Jesuit priests threaded the lakes and rivers in advance of the fur trader, baptizing as they went. The religious impulse moved even the more prosaic English. Drake and Hawkins scoured the Spanish Main to fight Catholics as well as collect booty. Later many felt with the Virginia Company managers that the first object of that plantation was "to preach and baptize into the Christian religion . . . a number of poore and miserable souls wrapt up unto death in almost invincible ignorance." Sir Humphrey Gilbert emphasized the benefits that would accrue to the natives brought by the white man "from falsehood to truth, from darkness to light, from the highway of death to the pathway of life, from the devil to Christ, from hell to heaven." While the work of the priests in opening the routes for settlers should not be underestimated, more important than even the crusading spirit was the influx of settlers who sought freedom from religious persecution at home. Separatists and Puritans founded New England to obtain religious liberty; but Puritan intolerance in turn drove Roger Williams and his followers to Rhode Island, banished Anne Hutchinson, and contributed somewhat to Hooker's settlement of Connecticut. Puritanism drove Cavaliers to Virginia and English Catholics to Maryland. French Protestants found refuge in the Carolinas; while Quakers, Mennonites, Moravians, and other sects found a home in Pennsylvania, New Jersey, and elsewhere.

ECONOMIC ASPECTS OF COLONIZATION 37

Political motives also played their part. Each nation would secure for itself as much of the new land as possible. Settlements in the thirteen colonies were encouraged to check the northward advance of the Spanish and the southward and eastward pressure of the French. The four-cornered struggle for empire between Spain, France, England, and Holland during the sixteenth and seventeenth centuries contributed much in hastening the occupation of America. Divergence in political ideas, often derived from religious tenets, also sent many to the New World.

More important than the religious and political were the economic motives. It was the search for new routes to the Far East that led in the first place to the discovery of America, and during the next century it was the desire to find an opening through the continent that led to the explorations of Cartier, Frobisher, Davis, and others. When gold and silver were discovered in abundance by Cortez (1519) in Mexico, and by Pizarro (1531) in Peru, the dominating impulse of Spain was the exploitation of this source of income. The foundations of New Spain rested during the early years on the precious metals.

In time Europe came to realize that gold was not the only product of value which might be obtained from America. It is believed that even before the discovery of Columbus, fishing vessels from England and France had sailed out to the west until they found fish in plenty. In the fifteenth century the fishing fleets of many nations drew wealth from the Great Banks. The fur trade came soon to rival in value even that of gold. Sugar, tobacco, cocoa, and many other products, including timber and naval stores (tar, pitch, rosin, cordage, masts, etc.), demonstrated the value of the Americas to Europe as a source of raw materials.

As a counterpart to the growth of manufactures in Europe came the appreciation of colonial settlements as a market for the finished products of the looms and workshops of the mother countries. In enumerating the benefits which England would derive from the establishment of colonies beyond the Atlantic, Sir George Peckham wrote that it would revive and promote especially the trades of clothiers, wool men, carders, spinners, weavers, fullers, shearers, dyers, drapers, cappers, and hatters. A pamphleteer writing previous to 1606 on "Reasons for raising a fund for the Support of a Colony at Virginia" speaks of it as a place "fit for the vent of our wares."

The economic motives so far mentioned involve to some extent state as well as private interest and participation. Other economic impulses concerned more directly the individual. The desire to escape the economic restrictions of the European guild system, the hope of bettering his fortunes upon a new soil where land might be acquired easily and the fruits of labor saved from a feudal lord, appealed to the poor but ambitious man. Younger sons of the nobility and impoverished gentlemen saw a chance in the New World to found a fortune or commence life anew. There was undoubtedly a feeling in the sixteenth and seventeenth centuries that England was overpopulated and that the colonies formed a natural outlet for the surplus population. The foundation for this theory was the number of beggars and highwaymen who infested the country, the natural result of the close of the Elizabethan wars and the inclosures of farm land into sheep pastures. To men of this type, especially, America held out hope.

Colonial System of Spain.—Four nations—Spain, France, Holland, and England—strove to dominate the North American continent. Each nation had settlements in what is now the United States, and each nation in its efforts to reproduce on American soil a New Spain, a New France, a New Netherlands or a New England, as the case might be, left here the imprint of its civilization. Although the statesmen of each of these four countries believed thoroughly in the regulation of the economic life of the colonies, nowhere was the idea carried to such extremes as in the treatment of the Spanish colonies by the home government. From 1503 until 1717 all commerce to and from the colonies had to pass through the city of Seville. At the American end the entry of goods during most of this period was limited to the two ports of Vera Cruz on the Mexican coast and Porto Bello on the Isthmus of Panama, the former receiving goods for Mexico and the latter for South America. By 1561 the development of piracy along the Spanish Main [1] led the government to establish the system of yearly fleets which lasted nearly two centuries. Once a year two fleets would form under the protection of warships and sail to the West Indies, where they would separate, one proceeding to Vera Cruz and the other to Porto Bello. The arrival of the fleet at Porto Bello marked the period of the annual fair when the silver wedges from the Peruvian mines which "lay like heaps of stones in the street without any fear or suspicion of being lost" were exchanged for wines, figs, olives, cloth, iron, quicksilver, and luxuries from Spain. Besides gold and silver, which formed the chief item in the return voyage, cochineal, sugar, hides, and drugs were taken back.

This commerce was further restricted by the granting of monopolies on gunpowder, salt, tobacco, and quicksilver, by excises levied on goods sold, by export and import duties averaging perhaps 15 per cent, and by the king's royalty of one-fifth on the yield of the gold and silver mines. In the colonies the culture of olives, vineyards, tobacco, and hemp was forbidden. In addition inter-colonial trade was prohibited. The whole system was highly artificial and was seriously undermined by the wholesale smuggling after the English and Dutch had obtained a foothold in the West Indies and England had through the Asiento, 1713, secured the monopoly of the African slave trade with Spanish America.

A disproportionate emphasis may easily be placed on the part played by gold and silver in the economic life of Spanish America. Although they formed the chief item of exportation, the government was not indifferent to the establishment of agricultural colonies, nor were the colonists wholly occupied with extracting the precious metals. A great majority of the population lived by farming and ranching, the products of which, including hides, corn, the American aloe, or agave, sugar, cocoa, vanilla, and cochineal, were more valuable than those of the mines. Around

---

[1] The Spanish Main properly used means the coasts bordering on the Caribbean Sea; but is sometimes applied to the Caribbean itself.

the two basic industries, mining and cattle ranching, was built up a prosperous and even wealthy civilization, while the English and French to the north were still struggling to maintain a bare existence.

The economic unit of early Spanish America was the *encomienda*, a grant of land carrying with it the authority to command the services of a certain number of Indians. Begun by Columbus in the West Indies, it was later extended and applied almost universally on the continent. Under this system the Indians were forced to till the crops, tend the cattle, and work the mines for their Spanish overlords. While efforts were made to limit the duration of the *encomienda* system and detailed regulations were issued concerning the treatment of the Indians, forbidding their enslavement and advising their conversion to Christianity, it was to be expected that under such a system the natives would degenerate into serfs and often be subjected to the most cruel treatment. The conquest of a large territory with an inferior native population by comparatively few adventurous soldiers bent on rapid accumulation of wealth made inevitable the transplanting of the only system of government known to these men—the feudal system of Europe.

Instead of exterminating or driving away the Indians as did the English settler, the Spaniard made serfs of the majority of them. In 1574, scarcely three generations after the conquest of Mexico, there were in the New World, according to Velasco (historian to the Council of the Indies), two hundred Spanish cities, towns, and mining settlements, containing 160,000 Spaniards, of whom about four thousand were *encomenderos* (lords of Indian serfs), the rest, settlers, miners, traders, and soldiers, controlling an approximate Indian population of 5,000,000 in eight or nine thousand villages. This was a half century before Plymouth was founded. Settling side by side, the Spaniard and Indian have eventually become so intermingled that a large element of the Spanish-American population is "half-breed," a mixture of the two races. In brief, the economic system of Mexico and New Spain was built around the idea of the *encomienda,* where Spanish overlords directed Indian serfs in their work of extracting silver and raising cattle and farm products, all of which were theoretically controlled for the benefit of Spain.

The first permanent settlements in the present United States were made by the Spaniards. They were preceded by the picturesque but unproductive explorations in the south and southwest, of Ponce de Leon (1521), Narvaez (1527), De Soto (1539-42), and Coronado (1540-42). Attempts by the French in 1562 and 1565 to colonize Florida brought the Spaniards under Menendez, who destroyed the French settlement, founded St. Augustine in 1565, and other forts. By 1582 the Spaniards had opened missions on the Rio Grande and Gila rivers, and during the succeeding centuries planted frontier settlements in Texas, New Mexico, and California.

France in America.—Almost a century elapsed after the voyage of Verrazano before the first permanent French settlement was made at Quebec in 1608, but the genius of the French for exploration, and their talent for dominion, were notably demonstrated in the succeeding years. Dominated by patriotism, missionary zeal, and a desire to open up more territory to their traders, priests and explorers pushed their canoes up the St. Lawrence, along the Great Lakes, and down the Mississippi, until by the end of the century French posts extended from New Orleans at the mouth of the Mississippi to Fort Radisson near the western end of Lake Superior and east to Nova Scotia.

The success of the French as colonizers did not measure up to their attainments as explorers and missionaries. This is to be attributed chiefly to the source of economic wealth. The economic backbone of New France was the fur trade. To the Frenchman with initiative the harsh climate and stubborn soil of the St. Lawrence Valley made no appeal. The back country was rich in furs, and in the pursuit of these he penetrated ever farther into the interior. Adaptable in the extreme, the Frenchman would often affect the manners and dress of the Indians, lead them on the war path, live with them, and even intermarry. This won to the French not only the bulk of the fur trade, but the friendship of practically all of the Indian tribes, with the exception of the Iroquois. If wealth was to be gained in New France it must be through furs, and noble and peasant alike engaged in the business. At least a third of the population was occupied in gathering and transporting furs.

As the fur trade was the principal source of wealth, so it proved also to be the chief cause of weakness for the colony. As long as greater profits were to be made in peltries, it was difficult to interest settlers in agriculture. Canada under the French never became primarily a country of homes and farming. The safest basis for a permanent colony was thus lacking. Instead of the 1,300,000 inhabitants which the English colonies boasted in 1754, nine-tenths of whom were engaged in agriculture, compactly settled along a fringe of seacoast and firmly established, the French had only about 80,000 scattered along the rivers and Great Lakes from the Mississippi to Nova Scotia. Beyond the barest necessities of subsistence, agriculture was neglected. There was some attention to fisheries, but practically no manufactures other than household in New France.

From 1600 to 1663 the efforts of the French to colonize and exploit the American mainland were in the hands of commercial companies, the last of which was known as the Hundred Associates. After that date the administration was taken over by the crown, and a government characterized by extreme absolutism and centralization was set up. To make complete the replica of the autocratic system of France in the New World, an order of nobility was created by Richelieu in the charter of the Hundred Associates. To induce members of the lesser nobility to remain in America, seigniories were granted them along the lake and river fronts. When the seigniories were inhabited at all the peasant settlers usually lived on a road perhaps a half mile back of the river or lake, with their fields sloping down to the water on one side and back into the forests on the other. These grants were usually four arpents (768 English linear feet) on the water front and ten arpents (1,920 feet) deep. This long, narrow holding, peculiar to the French, had a twofold *raison d'être*. Comparative freedom from Indian raids made it unnecessary for the French to huddle in fortified villages; but their sociable nature inclined them toward living near one another.

ECONOMIC ASPECTS OF COLONIZATION 43

With the seignories went the rest of the paraphernalia of feudalism. To the seignoir the tenant was expected to pay rent, trifling, to be sure, and generally in kind, to work for the lord a certain number of days a year, patronize his grist mill, present to him one fish out of every eleven caught, and render other feudal dues. In contrast to the situation in Spanish-America, the conditions in New France were not such as feudalism would thrive on. With plenty of vacant land and the fur trade to beckon them on, any attempt to impose a strict feudal system upon the inhabitants was doomed to failure, and the duties of the peasant to the seignor became more nominal than real. The lords themselves, usually poverty stricken and hardly more prosperous than their tenants, were forced to till the fields with their own hands or take to the life of a fur trader. With the seignorial system and despotic government went paternalism. Taught to depend not upon themselves, but upon the home government, the settlers soon lost initiative in economic problems. With their trade shackled by petty restrictions and controlled by government monopoly, it is little wonder that private enterprise in industry was smothered and Canada never prospered under France.

During the century and a half of French colonization in America they planted many trading posts and small settlements in the territory of the present United States. Futile efforts in Florida (1562-1568) were followed after the settlement of Quebec by an extension of French power along the Great Lakes and in the Illinois Valley. The first French went to Louisiana in 1699, where they founded New Orleans in 1718. St. Louis was established in 1764. By 1775 there were several thousand French in the Illinois region and about 14,000 in Louisiana.

The Dutch in America.—The efforts of the Dutch to participate in the profits of the American trade led eventually to the settlement of the Hudson Valley and adjacent region. Henry Hudson in the interests of the Dutch East India Company had explored in 1609 the river which bears his name, and a trading post called New Amsterdam had been established in 1614 by some enterprising merchants of Amsterdam. In 1621 Dutch interests in America were taken over by the Dutch West India Company, a great private corporation to which the States-General of Holland granted a monopoly of the trade not only of the American seaboard, but also of the coast of Africa between the Tropic of Cancer and the Cape of Good Hope. This corporation, interested in trade in gold, slaves, and tropical products, equipping hundreds of privateers, supporting an army and a large navy with which it made war upon Spain and Portugal, found the fur trade of the Hudson Valley but a small item in its numerous enterprises; the Hudson Valley consequently absorbed but a small part of the interest of the directors. This attitude is well expressed in the remonstrance of the company to the States-General against a peace with Spain, when they maintained that their object was not "trifling trade with the Indians nor the tardy cultivation of uninhabited regions," but "acts of hostility against the ships and property of the King of Spain and his subjects." [1]

In spite of the company's lack of interest in the Hudson Valley, much was done here in the way of trade and colonization. Fort Orange upon the site of the present Albany was built in 1622, a village on Manhattan Island was founded in 1623, and settlements were later made not only in the Hudson Valley, but in the Mohawk Valley, on Long Island, and along Delaware Bay. The West India Company, however, intent upon accumulating dividends, was not interested primarily in settling the country. The greatest profits were to be made in furs, and upon the promotion of the fur trade the chief energies of the company and its representatives were bent. Later ship building was carried on to some extent and eventually also prosperous agricultural communities grew up. But during the period of Dutch rule the governor was little more than the business agent of a private corporation whose chief concern was to maintain discipline among the Dutch and prevent friction with the Indians who supplied the furs.

The first farming in New Netherlands seems to have been done not by tenants, but by servants working for the company which owned both the land and the stock upon it. In order to stimulate settlement up the river a scheme of landed proprietors

---

[1] Brodhead, *Collection of Documents*, vol. i, p. 62.

was introduced in 1629. Any member of the company who would bring over fifty families at his own expense should receive a tract of land reaching sixteen miles along the river all on one bank, or half on one bank and half on the other, with no limit as to width. Upon these grantees, or patroons, were bestowed both proprietary rights and subordinate jurisdiction. The patroon could hold manorial courts, with the right reserved for the tenants to appeal to the company; could found townships and appoint officials for them. Upon their estates they had the monopoly of weaving and certain exclusive trading privileges. Here, too, as in New France and New Spain, an attempt was made to graft on the New World the feudal system of Europe. Under this system the most influential members of the company soon gained control of the choicest lands of the Hudson Valley.

The Dutch, who had never accepted the feudal system in its entirety as had other Europeans, chafed under the unaccustomed rule of the company and manor barons. The patroon system was exceedingly unpopular from the start, and in 1640 the company attempted to modify it by reducing the extent of the patroonships and introducing a smaller class of proprietor who was to hold two hundred acres tilled by five men brought over at his expense. Again in 1650 a further effort was made to increase the number of small farmers. A tract of land with implements and stock was granted to the settler, with the understanding that he pay a fixed rent and return the stock or an equivalent at the end of six years. In general the agricultural products and life were not unlike that found in New England, although the big plantations along the Hudson and Delaware Bay, where tobacco was a favorite crop, resembled those of Virginia.

The centralized despotism of the government of New Netherlands in the period before 1629 was modified after the introduction of the patroon system with its almost independent jurisdiction, but the principle of representative government was not recognized in New Netherlands until the closing years of Dutch occupation. At the same time the loss in efficiency and unity of control under a semifeudal patroon system made the colony more susceptible to foreign conquest. Driven like a wedge between the English colonies in New England and the south, it was natural that Great Britain should cast covetous eyes upon the strategic territory of New Netherlands. In fact, English settlers were beginning to filter in from both east and south, many seeking that religious freedom which the Dutch were the first to recognize in America and which Englishmen had sought for in vain in all New England except Rhode Island. This influx of English, combined with the lack of interest on the part of the company at home, the corrupt and despotic government in the colony, and the growing sea power of Great Britain, led to its final conquest in 1664. When the English captured New Netherlands, it contained a population, according to the estimate of Stuyvesant, of about 10,000. Nineteen languages were spoken in New Amsterdam at that time, the city thus early partaking of the cosmopolitan tone which ever since has been a distinguishing feature.

Early English Colonization.—As in the case of New France and New Netherlands, the first colonization by the English was carried on by private corporations. The English crown was unwilling to incur the trouble and expense of founding colonies, but it was glad to grant charters to corporations or individuals who were inclined to risk their own fortunes in such attempts. The expense and difficulties encountered in transporting colonists, equipping them with agricultural appliances, and keeping them alive until the first crops were harvested were so great that only the expectation of future gain and the grant of the broadest powers in the charters could entice stockholders to invest their money in colonizing companies.

The first English charters for the colonization of America were granted to Sir Humphrey Gilbert and to Sir Walter Raleigh, but the failure of both demonstrated that colonization as a private venture was impractical at that time, and as a consequence the next efforts were made by chartered companies. The famous Virginia charter of 1606 created two companies, one consisting of "certain Knights, Gentlemen, Merchants, and other Adventurers, of our city of London and elsewhere"; and the other of "Sundry Knights, Gentlemen, Merchants, and other Adventurers,

ECONOMIC ASPECTS OF COLONIZATION

of our cities of Bristol and Exeter, and of our town of Plymouth." These two groups of stockholders, the big business men of their day, interested in trade expansion in India as well as in America, were known as the London and Plymouth companies. Upon them the king bestowed the coast of the present United States—to the first named company the region between parallels 34 and 41, and to the second that between parallels 38 and 45, with the region between the 38th and 41st open to either on condition that neither settled within one hundred miles of the other. Attempts at colonization were immediately made by both companies, unsuccessfully by the Plymouth group on the Kennebec River in Maine, and permanently by the London Company, whose expedition entered the Chesapeake Bay in 1607 and planted a settlement thirty miles up the James River. A second charter was granted in 1609, to a group including 56 London guilds and 659 individuals, which increased the territorial grant of the London Company and stated very specifically its commercial rights. The latter included the complete control of the natural resources of the country, and levying of export and import duties up to a certain amount. For twenty years it was granted exemption from paying duty on goods imported into Virginia, and for all time was to pay only 5 per cent upon goods brought into England. In return the company was expected to colonize the country and to pay to the king one-tenth of all gold and silver acquired.

From the first arrival until the king had the charter revoked in 1624, the colony was a true plantation. The colonists were servants and employees of the stockholders who resided in England, and the fruits of their labor belonged to the company. For the products of the labor of the settlers the company sent supplies from England of medicines, clothing, furniture, tools, arms, and ammunition, all of which were kept in the common storehouses and allotted by the company's agent to the colonists. But the shiploads of lumber and other forest products gathered and sent to England paid only a small fraction of the expenses incurred by the London Company in its attempt to found the Virginia Plantation. It is important to remember that these commercial companies were more interested in dividends than in colonization, which, as in New Netherlands, was but a means to an end.

Unhealthful environment, infrequent accessions to their number, and mismanagement on the part of the officials contributed to make the first years a period of intense suffering. More businesslike methods introduced by Sir Thomas Dale (1611-16), and the abolition of the system of common industry under Sir George Yeardly in 1619, helped to put the colony on its feet, and the cultivation of tobacco soon assured its future. In 1624 the charter of the London Company was revoked by the king and the settlement became a crown colony. In its efforts to found a plantation the company had expended £200,000 and sent over 14,000 emigrants, but financially the project had been a dead loss. Nevertheless, the London Company secured for England a foothold on the American continent, a service sufficient to win for it undying fame.

The next permanent English settlement in the New World was that of the Separatists at Plymouth. Impelled by the hope that they might find in America an opportunity both for economic betterment and for the worship of God after the dictates of their own hearts, they negotiated with the Virginia Council for patents to settle in their land. A charter was eventually granted by the London Company, now under the control of Sir Edwin Sandys and the Puritan faction, giving the earnest little group the right to found a plantation and govern it by laws of their own in accordance with those of England. The London Company would not finance the settlement and the Separatists had to find assistance elsewhere. Eventually seventy London merchants subscribed £7,000, a sum sufficient for the purpose. Under the terms of the "Articles of Agreement of Plymouth Plantation" each share was to be reckoned at £10. For "adventuring himself" each emigrant was counted as holding one share and was permitted to purchase as many more as he was able. For seven years all wealth produced by the colonists was to go into the common stock, and from this and supplies sent by the London merchants the colonists were furnished food, clothing, and other necessities. At the end of seven years "ye capitall & profits, viz. the houses, lands, goods and chatles, be equally divided betwixte ye adventurers, and planters; wch done, every man shall be free from other of them of any debt or detrimente concerning this adventure."[1] Those who came to the colony before the expiration of the seven years were to share proportionately, according to the time spent. In other words, the plantation was to be a coöperative scheme for the first seven years. Whereas the settlers in Virginia were merely servants of the company who were to receive at the end of the seven years nothing but their freedom, the Plymouth colonists were stockholders in a company for which they all worked and the profits of which they would all share. In addition, their efforts were to be directed by officials of their own choosing and not by representatives sent from England, as in the case of the Virginia Company. The whole plan was a far more generous arrangement than that under which the Virginia Plantation struggled during its first twelve years.

The merchants who had subscribed their money expected immediate and large returns, but the wringing of a mere subsistence from the stubborn soil demanded practically the entire time of the colonists. It is true that some wealth was secured by furs and fisheries, but never enough to return any profits to the stockholders. On their side, the London stockholders were not able to send over supplies in any degree sufficient for the needs of the suffering settlers. As in Virginia, a most serious hindrance to success was the common store and the plan of coöperative industry. When famine threatened in the third year the system was abolished as far as agriculture was concerned, and land was allotted to each man for temporary use only. By 1627, the year in which the agreement with the London merchants ended, the colony was firmly established. Desiring to sever relations with the London merchants in a manner different from that prescribed originally, the colonists made arrangements whereby the interests of the London stockholders were bought out for £1,800, to be paid in nine annual installments of £200 each. In return the merchants surrendered all claims upon the colony. The money was paid chiefly through profits in the fur trade.

[1] Bradford, Wm., *History of the Plymouth Plantation* (Commonwealth of Mass. Ed.), pp. 56-58.

Plymouth was eventually absorbed by the strong Massachusetts Bay Company, most of the stockholders of which were Puritan merchants. Their charter was obtained in 1629, for purely commercial reasons. The pronounced High Church tendency of Charles I and his attempt at tyrannical government, which began in earnest in 1629 with the dissolution of Parliament and the imprisonment of men prominent in opposing his policies, gave to the activities of the company a different turn. To many of the leading Puritans, Massachusetts appeared as an ideal refuge from the hostile policies of the king. Since they belonged to the ruling classes at home, they were unwilling to emigrate as the servants of a plantation company. Consequently they bought up the stock of the Massachusetts Company, pledged themselves to emigrate, and took over their charter with them. Thus we find Massachusetts Bay settled by the controlling members of the company itself. The great migration of 1630-40, which brought to America over twenty thousand settlers, including some of the best stock in England, has left a deep impress upon the whole political, social, and economic fabric of American life.

Other English Colonies.—Virginia and Plymouth were, as we have said, colonized by chartered commercial companies. They bore the brunt of settling a strange land far from the base of supplies. Subsequent English colonization was not attended with the hardships endured by the Pilgrim fathers and the companions of the doughty Captain Smith. Later colonists could profit by the mistakes of their less fortunate predecessors, and it was now possible for private individuals to colonize with success. Later English settlements were effected not only by chartered companies, but also by two other agencies—(1) migrating groups from existing colonies and (2) wealthy proprietors. Examples of the first of these types are Connecticut, Rhode Island, and parts of New Hampshire and Maine.

The little fishing settlements of Maine and New Hampshire were colonized partially by emigrants from England under the protection of Sir Fernando Gorges and Captain John Mason, who had received patents for this region; and partially by inhabitants of Massachusetts Bay, who were successful in extending by 1652 the government of the last named colony over the new country. The colonies of Rhode Island and Connecticut were offshoots of Massachusetts Bay, settled by emigrants who, finding the religious or political attitude of the parent colony unsatisfactory, had moved on farther. Roger Williams established in 1636, on lands purchased from the Indians, a democratic commonwealth where religious toleration was put in practice. Windsor, Connecticut, was founded in 1635, and Hartford in the following year, by dissatisfied groups from Massachusetts. The settlers of neither Rhode Island nor Connecticut had a legal title to the land under English law, being simply squatters on the king's domain; but both succeeded eventually in securing charters confirming their occupation.

Under the proprietary system the king granted a single individual (or a group, as in the Carolinas) estates in America which might be colonized and held by him practically as a feudal lord under the king with very extensive powers and rights, but in most cases with the restraining provision that he must make laws "by and with the consent of the freemen." Land was granted in this way to a number of men, of whom Gorges and Mason have been mentioned. The most important experiments, however, were those of William Penn in Pennsylvania and the Calverts in Maryland. New York for a time (1665-85, with the exception of 1673-74, when it was recaptured by the Dutch) was a proprietary colony of the Duke of York, who handed over New Jersey to his two friends Sir George Carteret and Sir John Berkeley. The last named province, most of which came under the control of the Quakers until taken over as a crown colony in 1702, was settled chiefly by men attracted from the surrounding regions by the liberal land offers. The Carolinas were occupied either by Virginia frontiersmen pushing south or directly from England. They were granted (1663) by Charles II to eight proprietors, the most active of whom was Anthony Ashley Cooper, later Earl of Shaftesbury. John Locke (his former tutor) worked out a highly elaborate model state with a feudal hierarchy, but it was not adaptable and was never put in actual operation. Proprietary rule in the Carolinas came to an end in 1743.

Georgia, the last of the thirteen colonies, was founded in 1733, partly as a result of the desire of the British government to set up a buffer state against the Spanish in Florida and partly through the philanthropic desire to help English debtors commence life anew. For these reasons a charter was given to a group of trustees in 1732, who were to be in control for twenty-one years. But few of the class for whom the colony was founded came, and the population grew slowly.

Land Tenure in the Colonies.—Attempts to transplant the feudal system, or to enforce any system of land tenure which would restrict the amount of land held, were bound to fail. There was too much vacant land to be obtained by mere occupation. The country was too sparsely settled to control and the authority of the proprietor or crown was too distant or too weak to enforce obedience. Furthermore, men who braved the dangers of frontier life demanded actual ownership. As a consequence the quit-rents of the proprietors were collected intermittently and with great difficulty, while laws restricting the amount of land which any single individual might own were generally evaded. Attemps at coöperative agriculture failed in both Virginia and Massachusetts and were followed by the parceling out of land. Eventually in Virginia, 100 acres were given in fee simple to each stockholder for each share owned upon the first division and another 100 acres per share when the grant was "seated." A shareholder also received as a "head right" 50 acres for every person he might transport. This privilege was later extended to all residents. After 1705 the crown granted 50 acres for five shillings on condition that a house be built and three acres of land cultivated within three years. Thousands of acres were granted for meritorious service or through favoritism. The Virginia law was so easily evaded that by 1700 the average plantation was 700 acres. In Massachusetts Bay every adventurer who emigrated or paid the passage of an emigrant was to receive 50 acres. The usual system, in New England, however, was the settlement under the group plan, in which a number of prospective settlers would secure from the General Court a grant of 36 square miles, upon which they would lay out the village, assign plots for homes and

ECONOMIC ASPECTS OF COLONIZATION 53

gardens, and later divide the arable and the pasture land. The land outside that owned in common was held ordinarily in fee simple.

The proprietors as a rule were liberal in their grants, but usually demanded small quitrents. The quitrent had originated in Europe as a money commutation of other services, and was looked upon as a boon. In America, small as it might be, it was considered an unjust relic of a hated system. Penn offered 500 acres to anyone who would transport and "seat" his family, and was willing to sell 5,000-acre tracts for £100 and throw in 50 acres for each servant brought, but he reserved a quitrent of one shilling per 100 acres. In Maryland a settler was given 100 acres for himself, 100 more for his wife and for each servant, and 50 for each child. They were freehold grants subject to a rent of ten pounds of wheat per 50 acres. Anyone who would bring over five settlers was granted 1,000 acres, subject to a quitrent of 20 shillings a year. For bringing over more men, larger grants were made which might be divided up and sublet under the manorial usage. A man with a musket and six months' provisions might receive 150 acres in New Jersey, with a like amount for each servant or slave, and 75 additional acres for each woman, conditions so liberal that many came in from the near-by colonies. In Carolina under the proprietors, the settler was granted 100 acres for himself, 100 acres each for his wife, child, or man servant, and 50 acres for each woman servant, with a quitrent of a halfpenny an acre reserved. The trustees of Georgia allotted each man 50 acres and provided tools. Slavery was at first forbidden here and no man could own more than 500 acres of land, which must be entailed to the male heirs. Primogeniture, entail, quitrents, and other appurtenances of the feudal system, which prevailed in many of the colonies, were mostly abolished during the Revolution.

NOTES FOR FURTHER REFERENCE

Brief but good accounts of the Commercial Revolution are to be found in Clive Day, *History of Commerce* (rev. ed., 1923), in W. C. Webster, *A General History of Commerce* (1903), and H. de B. Gibbins, *The History of Commerce in Europe* (1891). The outline of the story in C. J. H. Hayes, *A Political and Social History of Modern Europe,* Vol. I (1916), is also excellent. The most satisfactory presentation for the average student of the medieval trade routes and the early chartered companies is that of E. P. Cheyney, *European Background of American History* (1904), in the American Nation Series. *The Cambridge Modern History,* Vol. I, contains a general treatment of exploration and colonization. A most illuminating interpretation of the old imperialism is that of J. R. Seeley, *The Expansion of Europe* (1888.) A recent article on the subject is W. R. Shepherd, *The Expansion of Europe* (1919), *Political Science Quarterly,* Vol. 34. Joseph Jacobs, *Story of Geographical Discovery* (1904), is a particularly readable volume.

The activity of Portugal in the Commercial Revolution may be studied in G. K. Jayne, *Vasco da Gama and His Successors* (1910), in C. R. Beazley, *Prince Henry the Navigator* (1895), in the Heroes of the Nation Series, in J. P. O. Martins, *The Golden Age of Prince Henry the Navigator,* trans. by J. J. Abraham and W. E. Reynolds (1914). For Holland read J. E. Thorold Rogers, *The Story of Holland* (1889), in the Story of the Nations Series; Clive Day, *The Policy and Administration of the Dutch in Java* (1904), and George Edmundson, *Anglo-Dutch Rivalry, 1600-1653* (1911). The effect upon England can be traced in the monumental work of W. Cunningham, *The Growth of English Industry and Commerce* (5th ed., 3 vols., 1910–12), or in the shorter but excellent volume of H. de B. Gibbins, *Industry in England* (6th ed., 1912); consult also W. J. Ashley, *The Economic Organization of England* (1914), G. T. Warner, *Landmarks in English Industrial History* (11th ed., 1912), and E. P. Cheyney, *An Introduction to the Social and Industrial History of England* (rev. ed., 1920). For English expansion see H. E. Egerton, *A Short History of English Colonial Policy* (2d ed., 1909), or W. H. Woodward, *A Short History of the Expansion of the British Empire, 1510–1911* (3d ed., 1912). An historical novel which successfully conveys the spirit of the early commercial and religious rivalries is Charles Kingsley, *Westward Ho!* (1855).

A good introduction to the Spanish colonial system is the fifth chapter of H. H. Bancroft, *History of Central America,* Vol. I (1882); a more detailed account is W. Roscher, *The Spanish Colonial System* (1904). Excellent summaries of Spain in America are: Bernard Moses, *The Establishment of Spanish Rule in America* (1898); E. G. Bourne, *Spain in America* (1904), in the American Nation Series; William R. Shepherd, *Latin America* (1914), especially good on the transplanting of Spanish culture; and William S. Robertson, *History of the Latin American Nations* (1922). The spectacular features of the early years of Spain in America are covered in the classic volumes of W. H. Prescott, *The Conquest of Mexico* (1843) and *The Conquest of Peru* (1847).

On France, consult R. G. Thwaites, *France in America* (1904), in the American Nation Series; W. B. Munro, *Crusaders of New France* (1918), in the Chronicles of America Series (Chaps. VIII-XI are best on New France); the histories of Francis Parkman, especially *Pioneers of France in the New World* (1865), *The Old Régime in Canada* (1874), *Montcalm and Wolfe* (2 vols., 1884), and *A Half Century of Conflict* (2 vols., 1892). French explorations are emphasized in Justin Winsor, *Cartier and Frontenac* (1894) and *The Mississippi Valley* (1895).

The story of early New York is told by T. A. Janvier, *The Dutch Founding of New York* (1903); J. R. Brodhead, *History of New York* (2 vols., 1871); and John Fiske, *Dutch and Quaker Colonies* (2 vols., 1899). A more recent volume is M. W. Goodwin, *Dutch and English on the Hudson* (1919), in the Chronicles of America Series, Chaps. II, III, V, and X.

The bibliography on the English colonies is extensive, ranging from single-volume introductions like those of C. L. Andrews, *The Colonial Period* (1912), in the Home University Library; L. G. Tyler, *England in America* (1904), in the American Nation Series; E. B. Greene, *The Foundations of American Nationality* (1922); H. E. Bolton and T. M. Marshall, *The Colonization of North America, 1492-1783* (1921); Carl Becker, *Beginnings of the American People* (1915); C. M. Andrews, *The Fathers of New England* (1919), Chronicles of America Series, Chap. IV; and S. G. Fisher, *The Quaker Colonies* (1919), Chronicles of America Series, Chaps. IV and IX—to more detailed works occupying several volumes. Of the latter type are the scholarly volumes based upon the most recent investigations of H. L. Osgood, *The American Colonies in the Seventeenth Century* (3 vols., 1904-07), and Edward Channing, *History of the United States*, Vols. I-III (1905-12). The great coöperative work of Justin Winsor, *Narrative and Critical History of America* (8 vols., 1888-89), is invaluable. The most noteworthy study by an Englishman is that of J. A. Doyle, *English Colonies in America* (5 vols., 1882-1907); consult also his contribution to *The Cambridge Modern History*. A recent ably written volume on early New England is J. T. Adams, *The Founding of New England* (1921).

The more purely economic aspect is touched on in William Weeden, *Economic and Social History of New England* (2 vols., 1891), and Philip A. Bruce, *Economic History of Virginia in the Seventeenth Century* (2 vols., 1896). On land tenure see B. W. Bond, Jr., *Quit Rent System in the American Colonies*, in the *American Historical Review*, October, 1912. Very usable and accessible collections of source material are E. L. Bogart and C. M. Thompson, *Readings in the Economic History of the United States* (1916); A. B. Hart, *American History Told by Contemporaries* (4 vols., 1901); William MacDonald, *Select Charters Illustrative of American History* (1904); and J. F. Jameson (ed.), *Original Narrative of Early American History* (13 vols., 1906-12). A practical and almost indispensable aid to further study is the *Guide to the Study and Reading of American History* (rev. ed., 1912), by Edward Channing, A. B. Hart, and F. J. Turner

SELECTED READINGS

HAYES, C. J. H., *A Political and Social History of Modern Europe,* **Vol. I,** pp. 43-69.
CHEYNEY, E. P., *European Background of American History,* **Chaps. I-IV, VII, VIII.**
JACOBS, JOSEPH, *Story of Geographical Discovery,* Chaps. III-IX.
BOURNE, E. G., *Spain in America,* Chaps. XIV-XX.

MUNRO, W. B., *Crusaders of New France,* Chaps. VIII-XI.
WEEDEN, W. B., *Economic and Social History of New England,* Vol. I, Chaps. I-V.
BRUCE, P. A., *Economic History of Virginia in the Seventeenth Century,* Vol. I, Chaps. I-VII.

CHAPTER III

COLONIAL AGRICULTURE AND LABOR

Agricultural Achievements of the American Indian.— Colonial agriculture was founded primarily on what the colonist discovered growing here, whether wild or cultivated by the Indian. He brought with him familiar plants of the home land, of which some proved unsuited to this country, but many grew and thrived. Infinitely more important at first, however, were the native products, especially those which the aborigines had painstakingly put under cultivation.

It is erroneous to picture the American Indian as a nomad, destitute of permanent home and solely intent upon hunting and fighting. The Indian was the first American farmer, and agriculture played a large part in the economic life of the tribe. Cartier and Champlain saw fields of corn on the banks of the St. Lawrence, De Soto on the Mississippi, and Coronado in the southwest. Henry Hudson observed a bark house filled with corn and beans stored by the Indians for winter use. The first settlers in the Ohio Valley found corn fields extending for miles along the river bank. General Wayne wrote in 1794 that he had never "before beheld such immense fields of corn in any part of America, from Canada to Florida." Wherever the war path or hunting expedition might lead him, the warrior would return in the planting season and the harvest time. Indications point to the fact that the villages built near these fields were often permanent, as in the case of the Pueblo Indians, or semipermanent, as with the Iroquois of New York.

The extent to which the Indians practiced agriculture is difficult to determine with exactness. It varied, of course, with the different tribes and in different parts of the country, ranging from the Pacific-coast Indian who practiced no agriculture at all, through tribes like the Apaches, who practiced it to a limited extent, to the southwestern Indians, who constructed reservoirs, irrigation systems, and permanent cities, and hunted very little. The Atlantic-coast Indians from Maine to Florida, with whom the English first came in contact, were farmers as well as hunters, and during the first years their corn helped keep the colonists alive. It is the testimony of explorers that the Indians in most parts of the United States relied upon corn and other cultivated products for their principal subsistence.

The chief plants cultivated by the Indians for food were the Jerusalem artichoke, many varieties of beans, peas, red peppers, corn, grapes, onions, pumpkin, and squash. Cotton was grown for its fiber, gourds of many kinds for water jugs, mixing bowls, ornaments, and other purposes, and tobacco for smoking. Many other vegetable products were gathered by the Indians but not cultivated—berries and roots for food and medicine and the maple sap for sugar.

Indian agriculture in the main was built up around the cultivation of corn and tobacco. To prepare the ground, he first girdled the trees or scotched the roots until they were dead. The dead trees and stumps, as well as brushwood, were then burned. With his stick or crude implement he then dug shallow holes three or four feet apart, into which he dropped a few grains of corn and beans. Between the hills he planted pumpkin and squash. It is believed that the coast Indians in some cases used fish to fertilize their fields. As the corn came up it was hilled, and as it ripened was protected from the numerous wild birds by the children who watched from platforms erected in the middle of the field. After the harvest much of it was dried and stored away in pits or caves lined with bark or in corn cribs, thus protecting it from rotting bacteria or fungi. Although far from efficient in their methods, the Indians produced at least a million bushels a year.

The tobacco plant was cultivated in separate fields, but it also grew wild along the Atlantic-coast plain. It was dried either in the sun or before an open fire, and crumpled for pipe use. In comparison with the present product it was of decidedly inferior grade. Jacques Cartier, writing of the use of tobacco as he saw it among Canadian Indians, says: "There groweth also a cer-

COLONIAL AGRICULTURE AND LABOR

tain kind of herbe, whereof in Sommer they make great provision for all the yeere, making great account of it, and onely men use of it, and first they cause it to be dried in the Sunne, then weare it about their neckes wrapped in a little beasts skinne made like a little bagge, with a hollow peece of stone or wood like a pipe: then when they please they make pouder of it, and then put it in one of the ends of the said Cornet or pipe, and laying a cole of fire upon it, at the other end sucke so long, that they fill their bodies full of smoke, till that it commeth out of their mouth and nostrils, even as out of the Tonnell of a chimney. They say that this doth keepe them warme and in health: they never goe without some of it about them. We ourselves have tryed the same smoke, and having put it in our mouthes, it seemed almost as hot as Pepper." [1]

The third native American plant, destined in the years to come to be of world importance and to rank with wheat, corn, and rice as the world's greatest four foods, was the white potato. It does not seem possible to locate definitely its original habitat, but it is thought to be Peru or Chile. When introduced in England they were called Virginia potatoes, but the careless nomenclature of those days (as evidenced by calling the great American bird the turkey) is no proof that Virginia was their original habitat. The potato was cultivated in various parts of Europe, but gained popularity first with the Irish, hence the colloquial name "Irish potatoes." Carried back to the New World, the potato found little favor in New England until the Irish settlers of New Hampshire encouraged its culture. Potatoes were, curiously enough, not considered edible by the early settlers, but were thought to contain some sort of poison, and were carefully kept out of the reach of cattle and horses. And yet this despised tuber has been one of America's greatest gifts to the Old World and has kept countless thousands from starvation.

Sweet potatoes were a native product in the south, and were at once adopted by the English settlers and cooked in many ways.

---

[1] A Shorte and Briefe Narration (Cartier's Second Voyage 1535-36), chapter 10, from Hakluyt's Voyages, edition of 1600, given in Original Narratives of Early American History, *Early English and French Voyages*, p. 68.

Though white potatoes were produced in very small amounts before the Revolution, large crops of sweet potatoes were characteristic of southern agriculture practically from the beginning. It was a favorite food with the slaves, as well as with the white settlers.

The tools of the Indians were of the crudest: the shoulder blade of an antler or deer, a flat stone (usually flint) chipped and tied by thongs to a stick, a clean shell, sometimes a mere stick, were their implements of agriculture. Most of the work was done by the old men, women, and children, but it was not uncommon for the younger men to go into the fields at planting or harvesting time. The work at such periods was often coöperative and accompanied by ceremonies and festivals.

The chief contribution of the Indian to the white man was in agriculture. Owing perhaps to the fact that his only domesticated animal was the dog, he had not advanced in civilization as far as the European, but he had made considerable progress. He understood the bringing of wild plants under control and the breeding of plants by seed selection. He had grasped the idea of fertilization and of working the soil. He practiced multiple cropping. He knew how to preserve foods—berries and fruits with syrup or honey, and fruit, vegetables, and meat by artificial or sun drying. The New England farm, cleared by tree girdling, with its rows of corn twined with bean vines, interspersed with squash and pumpkin and protected by scarecrows, was a counterpart of the Indian field. The Indian gave to the white man both his chief food and his principal export and taught him how to cultivate them.

Agriculture in New England.—As the Pilgrims landed on the bleak New England shores in December, no foods of their own production could be expected until the next summer. Except for game, fish, and what corn they could purchase from the Indians, they had to rely on their rotting English supplies. It chanced that near them were cleared fields once cultivated by Indians who had been swept off in a great year of pestilence (1616-17). The friendly Squanto showed the weakened Plymouth men the Indian method of cultivating corn, both "ye manner how to set it, and after, how to dress and tend it." During times of food shortage, five kernels of corn a day was the individual allotment, enough to keep life in their bodies; and innumerable were the ways of preparing and cooking it gained from the Indians, or devised by thrifty matrons. Its condensed form made it suitable for carrying on journeys, and the comparative ease of its cultivation made it, after the communal system was abolished, always plentiful.

The great deficiency of the Indian as an agriculturist was his failure to domesticate animals. But the settlers immediately imported animals which multiplied rapidly. Cattle were brought in as early as 1624, and formed the basis of rapidly increasing herds and successful dairying. Hogs were raised in great numbers in New England, and a considerable export trade was developed in barreled pork. Horses of a very hardy variety were raised and exported in large numbers to the West Indies. Sheep were early introduced into Massachusetts and Rhode Island, where they were successfully developed and exported to the other northern colonies. Though these sheep were the old unimproved types, able to forage for themselves and to withstand hardships, conditions were at first hard for them, and special legislation had to be made to encourage sheep raising. Usually the sheep of the entire settlement grazed in common, under the care of a paid herder.

Besides corn, pumpkins, squash, and beans, the cultivation of which they learned from the Indians, the New Englanders raised peas, parsnips, turnips, and carrots from seeds, which they brought with them. Wheat, introduced from England, was not immediately successful, but with rye and buckwheat they had greater success. Barley, oats, and other European grains were introduced generally and thrived, but other products experimented with were found unsuitable. Most of the nuts now growing in the woods were indigenous to America, and a valuable food for the pioneer. Many fruits and berries grew wild in New England—cranberries, huckleberries, blackberries, and raspberries; and cherry and plum trees. Apple trees were imported at once, and were especially successful in New England and the middle states. Johnson in his *Wonderworking Providence,* writing of 1642, said that the settlers could have "apples, pears, and quince tarts instead of their former Pumpkin Pies." [1] Orchards were a part of every farm and the large apple production caused them to be "reckoned as profitable as any other part of the plantation."

While the New England farmer of colonial times could with hard work obtain a living from the soil and might even become very prosperous, his methods were of the crudest and most primitive type. A harrow, a spade, a fork, all clumsily constructed of wood, were his chief farm tools. In the early days few could afford a plow, and a town often paid a bounty to anyone who would buy a plow and keep it in repair. For twelve years after the landing of the Pilgrims there were no plows in Plymouth, and in the Massachusetts Bay Colony there were only thirty-seven plows in 1637. One plow would do the work for a considerable territory. Such a thing as scientific farming was unknown and even rotation of crops was rarely practiced. The land was used until its fertility was exhausted, and was then allowed to lie fallow or planted with natural grasses until it recuperated. Owing to the small size of the farms and the settlement in villages, "land butchery" was not practiced in New England to the extent that it was in the south, but methods were bad enough. Their tillage, says the author of *American Husbandry,* was "weakly and insufficiently given: worse ploughing is nowhere to be seen, yet the farmers get tolerable crops; this is owing, particularly in the new settlements, to the looseness and fertility of old woodlands, which with very bad tillage, will yield excellent crops." [2]

Bad as were their agricultural methods, their treatment of livestock was worse. The same observer maintained that in all that concerned cattle the farmers in New England were "the most negligent ignorant set of men in the world. Nor do I know any country in which animals are worse treated. Horses are in general, even valuable ones, worked hard, and starved: they plough, cart, and ride them to death, at the same time that they give very little heed to their food; after the hardest day's work, all the

---

[1] *Original Narratives,* ed., p. 210.
[2] *American Husbandry,* vol. i, p. 81.

nourishment they are like to have is to be turned into a wood, where the shoots and weeds form the chief of the pasture; unless it be after the hay is in, when they get a share of the after-grass." [1] During the early days of scarcity of food for the settlers, laws were passed forbidding the feeding of corn to animals. As harvests became larger, however, this restriction was removed.

With the exception of a few dollars' worth of salt and iron, the New England farmer was self-sufficing. From his field he obtained grains, from his orchard fruits, and from his pasture land meat and dairy products. Flax from the field and wool from the sheep were spun and made into clothing by his wife and daughters. From honey and maple sap he obtained ingredients to sweeten his food; corn whisky and cider furnished him with strong drink. Every farmer had to be a jack of all trades, and his wife just as able to turn her hand to anything.

The original settlements in New England were made under agreements whereby every shareholder or settler was entitled to a certain amount of land; 20 acres at Plymouth when the first division was made, 10 acres and rights of pasturage and mowing in the common land at Salem, and 50 acres at Massachusetts Bay for each adventurer who came over himself or paid the passage of an emigrant. Further settlement was usually made in the following manner. As the vacant land near the seacoast grew scarce, groups or congregations would obtain a grant from the General Court, to which they would move in a body and found a town. Those grants, commonly 36 square miles, were owned by these proprietors and eventually divided among them. From the center of the town, where the meeting-house stood, a wide street was laid out, and along it house lots with perhaps six acres of garden land assigned. Eventually, as the community grew strong enough to defend itself from Indian raids, the rest of the land was distributed, each settler receiving a share in the upland, meadow land and marsh land, and rights in the commons. This system, in combination with the glacial soil, the rigorous climate, and the land laws which allowed the division among several

---

[1] *American Husbandry*, vol. i, p. 80.

heirs, was not conducive to the development of great landed estates. The New England farm continued to be a comparatively small affair, while the New Englander lived in villages and tilled the land with his own hands.

The New England villages, with their houses and gardens grouped compactly, with the village commons and the rights in the remaining land, is very reminiscent of the English manor. Much of the land was held in common, although cultivated separately, and the town meeting was the center where plans were worked out and the cowherds, swineherds, and other officers who cared for the village property were elected. The system described was transitory. As the towns grew larger and labor was diversified, the inhabitants were often glad to sell their scattered strips, and the compact farm with its buildings and land together appeared, resembling rural New England as we know it to-day. The ease with which new land might be acquired, and an independent living assured, practically obviated a non-landholding labor class and necessitated much coöperation among the farmers. Houses were raised, fences built, corn husked, and fields plowed by coöperative effort, and the gatherings were made the occasion of revelry as well as hard work.

Agriculture in the Middle Colonies.—With the exception of the Hudson Valley, where the patroon system of large landed proprietors was started by the Dutch and continued by the English, the land system of the middle states resembled that of New England in the sense that the holdings were generally small and held in fee simple. There were large plantations on the Chesapeake shore of Maryland, but even in this colony the normal holding was small. The proprietors, Carteret, Berkeley, Penn, and the Calverts, anxious to encourage colonization, granted lands on the most favorable terms.

In these colonies the farmer found a climate closely approximating his native land and a soil much richer than in New England. In the fertile limestone deposits of the Mohawk, Hudson, Delaware, and Susquehanna river valleys, crops grew abundantly. Plenty of moisture and longer summers helped to make conditions ideal for farming. The soil of New Jersey was so rich, said

COLONIAL AGRICULTURE AND LABOR 65

Peter Kalm, the Swedish traveler, in 1749, that it made the settlers careless husbandmen. "They had nothing to do but cut down the wood, put it into heaps, and clear the dead leaves away. They could then immediately procede to ploughing, which in such loose ground is very easy; and having sown their corn, they get a most plentiful harvest. This easy method of getting a rich crop has spoiled the English and other European inhabitants, and induced them to adopt the same method of agriculture which the Indians make use of, that is, to sow uncultivated grounds, as long as they will produce a crop without manuring, but to turn them into pastures as soon as they can bear no more and to take in hand new spots of ground, covered since time immemorial with woods, which have been spared by the fire or the hatchet ever since the creation."[1] In the fertile soil the wheat and barley often grew so rank that it reached the height of six or seven feet, with little grain in the heads. An observer in the Mohawk Valley in 1665 wrote that he had seen fields in which wheat was raised for eleven years in succession on the same field, and farms which had not been manured for nine years. Although the same traveler mentions the irrigation of meadows in Pennsylvania, such care was rare; for the fertility of the virgin soil was not conducive to intensive farming. As in New England, rotation of crops was not practiced, but the land allowed to lie fallow when it had worn out.

The middle states were even more predominatingly agricultural than New England. With the exception of furs and lumber the exports of the middle colonies seem to have been almost entirely agricultural. Judging from the export statistics,[2] wheat must have been the chief export, the average acre yielding from twenty to thirty bushels, a larger production than was common in England at that time. Corn was raised throughout these provinces, providing the bulk of food for cattle in the winter. Rye, barley, buckwheat, and oats were also generally grown, the latter with great success. Potatoes, not common in New England, throve upon the black loose soil of New York, and other

---
[1] Kalm, Peter, *Travels into North America,* vol. ii, p. 193.
[2] See p. 112 ff.

products flourished in New Jersey. Fruits suitable to a temperate climate grew in great abundance—apples in New York, peaches and melons in the sandy soil of New Jersey and Delaware. "Peaches are of a fine flavor," says the author of *American Husbandry*, "and in such amazing plenty that whole stock of hogs on a farm eat as many as they will, but yet the quantity that rot under the trees is astonishing. . . . Watermelons are in such plenty that there is not a farmer or even a cottager without a piece of ground planted with them."[1] Large herds of cattle grazed on the coastal lowlands of Long Island, New Jersey, and elsewhere; hogs were abundant, running wild in the woods; and sheep were plentiful in Pennsylvania.

The agriculture of the middle colonies was somewhat influenced by the heterogeneous population. English, Dutch, Germans, Swedes, and other races had their settlements, imported their particular strains of livestock, farmed with their own methods, and raised their favorite crops. Even with the unscientific and wasteful methods of the time, a rude abundance was easily obtained. For the European peasant it was a veritable land of promise.

Southern Agriculture.—The first Jamestown settlers were poor farmers, being more intent upon gold hunting and exploration. The soil of Virginia was too rich and the climate too warm for English wheat. Disease and starvation carried off many of the first settlers; the remainder were kept alive only by the corn purchased from the Indians. Until 1612 the colony dragged out a most precarious existence. In that year John Rolfe began the cultivation of tobacco, a plant introduced into Europe a hundred years before and extensively grown in the West Indies. It was then discovered that the Virginia soil was ideal for its growth. There was a great demand for the leaf in Europe and the gold hunters turned from their dreams of sudden riches to this slower but sounder basis of prosperity. Although Governor Dale frowned on the new occupation and in 1616 ordered that no man plant tobacco until he had put down two acres of his three-

---

[1] *American Husbandry*, vol. i, p. 139.

acre farm with corn, it was impossible to prevent the growth of tobacco to the detriment of grains and vegetables. In 1617 even the market place and streets of Jamestown were planted with tobacco; the settlers found that the same amount of time and labor would yield per acre in tobacco six times as much as in grain, for it sold at times as high as twelve dollars a pound measured in present currency. The king (James I), out of his desire for revenue as well as his dislike for the use of tobacco, "tending to a generall and new Corruption both of Men's Bodies and Manners," levied a tax of one shilling a pound (about 20 per cent) in 1619, and in 1620 sought to limit the importation from Virginia and the Bermudas to 55,000 pounds. But no legislature either at home or in England could stop the steady increase in production. It amounted to 20,000 pounds in 1616, to 60,000 in 1621, to 500,000 in 1627, to 23,750,000 in 1662, and to 130,000,000 in 1790. Virginia had discovered a staple and an economic basis of wealth.

Owing to the fact that the great staple crop was raised for exportation, it was necessary that the plantation be located on the river banks where the tiny ships of colonial days could sail up and take a cargo from each farmer's wharf. The land was rapidly taken up along the James, York, Rappahannock, and Potomac rivers, then along the Chesapeake inlets, and then south into the Albemarle and Pamlico districts of North Carolina. When the land along the rivers was entirely occupied, the late comers were forced to set off a tier of farms immediately back of the river plantations and get the tobacco to their neighbors' wharves as best they might.

Tobacco, it was found, quickly exhausted the richest soil and necessitated the continual use of fresh land. Three years under the most favorable circumstances was the age of a tobacco field, after which it was turned over to other crops. White labor was scarce and costly, for each emigrant hoped to set up for himself, and gradually negro labor was substituted. It was believed that fifty acres of arable land per negro were necessary for profitable cultivation, and an overseer was too expensive unless he had twenty negroes under him. In consequence, the great plantations often had a thousand acres under actual tobacco cultivation besides land for other crops, for a cattle range, and for woodland. Many Virginia tobacco plantations were five thousand acres or over in size. These factors, combined with the case with which the title to new land was acquired, are the causes for the large holdings in the south. Mere occupation and the payment of a small fee or quitrent were sufficient to establish ownership. "Head rights," or the granting of land to those who imported settlers, grants for meritorious service to professional men, and purely personal grants all contributed to the swelling of the large estates. This system of settlement prevented the growth of towns, and promoted a distinctly rural life of scattered plantations in contrast to New England, where occupation usually began by founding a town.

Next in importance to the plantation system, which was undoubtedly the basic feature of southern agriculture, was the cattle range. Shifting arable land and large plantations made inclosures impracticable; and the vast unoccupied regions could readily be utilized for cattle ranges. Cattle, horses, and swine roamed in droves, subsisting on roots and herbage, branded when possible, but wild and often hunted as wild beasts. Each settler had his "right in the woods," which gave him a share in the unbranded cattle. The western cattle ranch with its round-up and brandings was a replica on a much larger scale and later date of cattle ranching during the early days in Virginia and North Carolina. Plantations were often laid out with reference to cattle ranges, and the wild herds of livestock usually marked the fringe of civilization. The treatment of such livestock as was domesticated was even more unscientific than in the north. It was believed by many that the housing and milking of cows in the winter would kill them!

Passing south from Maryland and Virginia into the Carolinas, different conditions were to be found. Although cattle ranching was a feature in both North and South Carolina, in the first named colony the farms were likely to be small and the products diversified, while the plantations of South Carolina were given over to the production of rice and indigo on a large scale.

COLONIAL AGRICULTURE AND LABOR

The farmers of North Carolina, most of whom were emigrants from Virginia, poor men, often indentured servants or debtors, raised chiefly tobacco and corn, and their general economy resembled more closely the self-sufficing farms of New England and the middle colonies.

Rice had been brought into Virginia by Sir William Berkeley in 1647, but its culture was soon abandoned. It was probably in 1693 that rice was introduced from Madagascar and experimentation on its cultivation was begun by Governor Smith. Eventually it was found that the low and swampy coast lands of Carolina with the long hot summer produced an abundant crop. The work in the wet, hot fields was impossible for white men, but with the importation of black labor, rice culture became profitable. By 1754 the planters exported annually from Charleston 104,680 barrels, and 125,000 by the opening of the Revolution.

The raising of indigo, the other great export crop of colonial South Carolina, was the result of experiments performed by Miss Eliza Lucas, the daughter of an English army officer, who was left in charge of the family plantation. The colonial legislature had granted a bounty in 1723 for the production of indigo, but no progress was made until the experiments of Miss Lucas demonstrated its practicability. A bounty of sixpence a pound was granted in 1748 by Parliament, and from then until it was eventually displaced by cotton immediately after the Revolution, the production of indigo was a constantly increasing source of wealth. In 1775 the indigo exported amounted to 1,150,662 pounds.

The fact that Virginia and the Carolinas were in the same latitude as Spain, Italy, and North Africa, led to attempts to raise the Mediterranean products there. The mulberry tree grew wild in Virginia, and a law was passed compelling each farmer to plant annually six mulberry trees for seven years. Even a premium of ten thousand pounds of tobacco to anyone who would produce fifty pounds of silk failed to make the industry profitable. Eliza Lucas was more successful with her silk industry than most planters, and produced enough raw silk to make three dresses, one of which she presented in 1753 to the Princess Augusta, mother of George III. Sugar cane was tried out with little success until after 1751, when the Jesuits introduced the plant into Louisiana. Olives, lemons, pineapples, figs, ginger, almonds, and wine grapes were all tried, with little success. Identity of latitude with Mediterranean countries did not carry with it identity of climate and soil, and few of the semitropical products would thrive. In many cases, too, the lack of cheap efficient labor prevented success. But Eliza Lucas, describing South Carolina in a letter to her brother, says they have "peaches, Nectarines and mellons of all sorts extreamly fine and in profusion, and their Oranges exceed any I ever tasted in the West Indies or from Spain or Portugal."

Recapitulation of Colonial Agriculture.—To lay down with exactness the characteristics of a period extending from 1607 to 1781 is difficult, owing to the changing condition of such an extended time. Certain general facts, however, can be stated. First of all, it is evident that colonial economy was predominatingly agricultural. Even in New England, where commercial and industrial life was most developed, not more than one-tenth of the people were engaged in non-agricultural pursuits. In other colonies the proportion was far less. Although fish and naval stores were important export items, they were of far less value than the agricultural products shipped abroad. The economic life of the south rested upon large land grants, slave labor, and the export of such staples as tobacco and indigo. Nevertheless, it is important to remember that, while the plantation system determined the economy of the south and the planters were the dominant influence in the government, they were at all times outnumbered by the small farmer. It has been maintained by a leading southern historian [1] that "nine-tenths of the South's landowners at any period in her history were small proprietors." The dependence of the prosperity of the south on the success or failure of a single crop and upon the fluctuation of a foreign market led to repeated efforts on the part of colonial assemblies to stimulate a greater production of foodstuffs. In this movement the small farmer played a large part, so that by 1736 foodstuffs and flax were pro-

---

[1] Dodd, W. E., *The South in the Building of the Nation*, vol. v, p. 74.

COLONIAL AGRICULTURE AND LABOR

duced in sufficient quantity both for home needs and exportation, and by 1760 sufficient livestock for local consumption. In contrast to the south, northern agriculture was based upon limited land grants, free labor, and food crops designed chiefly for a home market. New England and the middle colonies from the beginning were self-sufficing and depended on Europe for their luxuries only.

The last half of the seventeenth and the whole of the eighteenth century was a period in which the beginnings of many great improvements in agriculture were undertaken in England by Tull, Townshend, and Bakewell. But the American farmer adopted them very slowly. Rich virgin soil, with an inexhaustible supply to the west, was no incentive to scientific farming. The value of manure was hardly appreciated, crop rotation was rarely used, and "land butchery" was the usual practice. One observer said that the colonial farmer seemed to have but one object—the plowing up of fresh land. "The case is," he says, "they exhaust the old as fast as possible till it will bear nothing more, and then, not having manure to replenish it, nothing remains but to take up new land in the same manner." [1] With land butchery and crude methods in both north and south went ignorance in the care of livestock. But with all these defects, the colonial farmer was the most prosperous in the world. He had learned from the Indian the production of the staples corn and tobacco, had experimented with European grains and fruits to learn what was adapted to American conditions, and had put under cultivation much of the land as far west as the fall line.

Labor in the Colonies.—In Europe during the seventeenth and eighteenth centuries there appeared to be an abundance of labor, but a dearth of resources. In America the reverse was true. The Europeans who settled North America found a virgin continent still unexploited, with a wealth of raw materials awaiting the hand of man. Labor alone was scarce and from the earliest times the problem of obtaining a sufficient supply was most difficult. This is perhaps a partial explanation of the Amer-

---

[1] *American Husbandry*, p. 144.

ican philosophy of the glorification of work and of the Puritan hatred of idleness, the latter exemplified in many colonial laws. The supply of labor from the native population was practically negligible, for the Indian preferred to live his old life rather than to subordinate himself to the white man. The European, possessing enough initiative to pioneer in a new land, was not the type that would readily submit to the authority of others when the chance of becoming a free landowner was so easy and the inducement to strike out for oneself so alluring. The need for labor was greater in the south, where the staple crops were raised on large plantations, and less in the north, where the farmer and his family cultivated a small farm. There were consequently more servants of all types in the south, but in all sections the demand for them was keen.

In the north the scarcity of labor was partially met by coöperation. When an extraordinary situation arose, as at the time of a house raising or a ship launching, the neighbors were called upon and the project accomplished by the associated efforts of the group. What laborers there were in the north were of two classes—free and non-free. Although there were always a few of the former, their number was small, for a man with any capacity could with little difficulty become a landowner and attain a degree of prosperity.

The non-free laborers were of two classes—indentured [1] servants and slaves. The indentured servants were also of two classes—voluntary and involuntary. The voluntary indentured servant was one whose servitude was based upon a free contract. Many a European, anxious to start a new life in America, gladly sold himself for a period of from five to seven years to shipmasters or emigration brokers in payment of his passage to America, his length of service depending on his ability to pay part of the passage money. A few Germans and others voluntarily indentured themselves in order to learn the language and obtain funds to start life more advantageously. The voluntary free servant

---

[1] The name indenture comes from the form of the contract, which was written in duplicate on a large sheet and the halves separated by a wavy or jagged cut, called an indent.

was ordinarily entitled to two weeks to find a purchaser, but, as he was not allowed to leave the ship, this right was of little value.

The second class of indentured servants, those suffering involuntary servitude, were usually debtors, vagrants, or criminals deported by the courts. The vagrancy laws since the days of Elizabeth had been extremely harsh in England. Added to these were the various laws prohibiting the free movement of labor from one parish to another, owing to the fear that paupers might be thrown upon the parish for support. It was easy to fall into debt during hard times, and the penalty for debt was imprisonment. Over three hundred crimes in the seventeenth century were punishable in England by death. With the courts and prisons crowded with paupers, vagrants, debtors, and petty criminals, it seemed the most humanitarian as well as practical policy to ship them over to the colonies. In this way England was relieved of a burden and America supplied with much-needed labor. If these prisoners could pay their own passage money they were free to do as they pleased; otherwise (and this was true of almost all), they were sold for from seven to ten years. In addition to these, thousands were kidnapped and brought over against their will.

The rights of servants were to a certain extent protected. They were entitled to food, clothing, shelter, medical attendance when sick, and they might own property. At the end of their service they were usually given an outfit and in some cases fifty acres of land. Although protected by law from unjust cruelty, the age was a hard one and the lot of the indentured servant, especially the involuntary, was exceedingly unenviable. On the one hand, the cost and need of labor were an incentive to considerate treatment, but on the other hand, the desire of obtaining as much labor as possible in the number of years covered by the indenture was a spur to excessive driving. In many cases the lot of the slave was superior to that of the indentured servant, for the loss to the owner of an able-bodied slave was greater, and hence conducive to better treatment. Most of the servants of the north were recruited from these classes, as were those of the south during the seventeenth century, for slaves were at first unpopular and slavery grew slowly. In 1681 there were 6,000 white servants in Virginia and but 2,000 slaves. Although the colonies protested against the exportation of criminals from England and Parliament in 1671 passed an order prohibiting it, the laws were not observed, and the practice continued well into the next century. Of the "free-willers," thousands came each year during the latter part of the seventeenth and early part of the eighteenth century, but the number gradually decreased until by the Revolution their immigration had practically ceased. Most of the increase in population during the colonial period came from natural increase rather than from immigration. In 1640 there were about 25,000 in British North America; in 1660, about 80,000; in 1690, about 200,000; and by 1770, about 2,000,000. Of the immigrants during the colonial period probably half landed as indentured servants, so that the "Redemptioners" and their offspring formed a most important element in the early population of America.

Slavery.—To-day exploitation is carried on by moving capital to the labor supplies, for capital is concentrated in North America and Europe, whereas cheap labor is to be found in Asia, Africa, and South and Central America. Our forefathers, equipped with resources of land and raw materials, needed labor, and a partial solution was found in slavery. During the fifteenth century Portuguese traders began to import into Europe negroes from the "Slave Coast," that part of the west coast of Africa extending from Cape Verde on the north to Cape St. Martha on the south; from the time of the first Portuguese settlement in 1482 the traffic became regular and lasted for about four hundred years. The slaves were purchased from native brokers living in the coast towns, who obtained them from the tribes in the interior. The latter, well supplied with guns and ammunition, turned over their prisoners of war and the fruit of their raiding parties to be imprisoned in the slave pens along the coast until they could be shipped away.

The first negro slaves were brought to America and sold at Jamestown in 1619 by a Dutch privateer; within a few years they were to be found in all of the colonies. Slaves were unpopular at first, notwithstanding the scarcity of labor, and the number of

COLONIAL AGRICULTURE AND LABOR

negroes grew slowly and for half a century composed only a small fraction of the total population. The slave trade was a monopoly of the Royal African Company of England until 1698, when the traffic was thrown open, after which it expanded rapidly. Bogart estimates the number of slaves carried to all of the colonies in 1700 at 25,000; and from 1713 to 1750 about 20,000 annually. In 1771, 47,000 were carried in British ships alone. By 1760 there were 400,000 slaves in the colonies, three-fourths of whom lived in the south. They formed two-fifths of the whole southern population, varying from a small percentage of the whole in North Carolina and Maryland to over twice the white population in South Carolina.

The slave trade to the English colonies was soon monopolized by British and American ships. The latter proved to be especially efficient. The usual voyage for the Yankee slaver was to load up with rum and other commodities in New England, sail for the Slave Coast, and exchange them for negroes, dispose of the latter in the West Indies or the mainland, and take in a cargo of sugar, molasses, and tobacco for the north. The trip between Africa and the West Indies, known as the "Middle Passage," shows slavery in its gloomiest aspect and the slave dealers at their worst. Crowded in the smallest possible space and chained to the ships, the negroes suffered untold agonies during the slow weeks of the Atlantic passage; if they fell sick, they were thrown overboard, lest they contaminate their fellows. But our ancestors were hardened to suffering and had few compunctions about slavery.[1]

Unpopular at first, a slave economy was adopted on the tobacco plantations of the south as the easiest way to fill the need of labor, and on the rice plantations of South Carolina as the only labor that could endure the climatic conditions on the hot, muggy rice fields. Slavery fulfilled an economic need, and as long as this continued it prospered. Toward the time of the Revolution,

---

[1] Peter Faneuil, one of the most honored and public-spirited citizens of Boston, made his fortune in the slave trade. Faneuil Hall, his gift to the city of Boston, was the scene of many patriotic meetings at the time of the Revolution and has been appropriately called the "Cradle of Liberty." These two facts led to the witty observation that the "Cradle of Liberty rocks on the bones of the Middle Passage."

when the tobacco farms were wearing out, slavery fell again into disfavor, only to be revived a few years later by the invention of the cotton gin.

NOTES FOR FURTHER REFERENCE

The best bibliography of American agriculture is that of Louis Bernard Schmidt, *Topical Studies and References on the Economic History of American Agriculture* (rev. ed., 1923). On Indian agriculture the following are useful: P. A. Bruce, *Economic History of Virginia in the Seventeenth Century*, Vol. I (1895); L. Farrand, *Basis of American History* (1904), in the American Nation Series; John Fiske, *Discovery of America* (2 vols., 1891), Chap. I; J. W. Powell, *The North American Indians*, in N. S. Shaler, *The United States of America*, Vol. I (1897); and G. K. Holmes, *Aboriginal Agriculture: The North American Indian*, in Bailey's *Cyclopedia of American Agriculture*, Vol. IV (1909).

Recent, scholarly, and the most useful work on colonial agriculture is Lyman Carrier, *The Beginnings of American Agriculture* (1923). N. S. B. Gras also has a volume in preparation (1924) on the history of agriculture. A general and popular survey is that of A. H. Sanford, *Story of Agriculture in the United States* (1915), and a more condensed but clearly stated account is that of T. N. Carver, *Historical Sketch of American Agriculture*, in Vol. IV of Bailey's *Cyclopedia of American Agriculture*. Old but still usable for its details is A. S. Bolles, *Industrial History of the United States* (1878), which devotes some space to agriculture, and the contribution of C. L. Flint, *Agriculture in the United States, 1607-1860*, in *Eighty Years' Progress* (1861). Virginian agriculture is exhaustively handled by Bruce, while glimpses of northern agriculture are given in Weeden. Short but illuminating chapters by experts are to be found in Vol. V of *The South in the Building of the Nation*. Among the special studies are those of M. Jacobstein, *The Tobacco Industry in the United States*, Vol. XXVI, No. 3 (1907), Columbia University Studies in History, Economics and Public Law, and L. G. Connor, *A Brief History of the Sheep Industry* (1918), in the Annual Report of the American Historical Association. The most valuable and interesting contemporary account is the anonymously published *American Husbandry* (2 vols., 1775). Other source material may be found in E. L. Bogart and C. M. Thompson, *Readings in the Economic History of the United States* (1916).

The results of much detailed earlier investigation have been brought together by U. B. Phillips, *American Negro Slavery* (1921). More detailed studies in special fields are those of J. R. Brackett, *The Negro in Maryland* (1889); J. C. Ballagh, *Slavery in Virginia* (1902); and E. R. Turner, *The Negro in Pennsylvania, Slavery—Servitude—Freedom, 1639-1861* (1910). On the conditions of white servitude are the investigations of E. I. McCormac, *White Servitude in Maryland* (1904), Johns Hopkins Studies; J. C. Ballagh, *White Servitude in Virginia* (1895), *ibid.*; J. S. Bassett, *Servitude and Slavery in the Colony of North Carolina* (1896), *ibid.*; and K. F. Geiser, *Redemptioners and Indented Servants in Pennsylvania* (Supplement to the *Yale Review*,

COLONIAL AGRICULTURE AND LABOR

1901). C. M. Andrews, *Colonial Folkways* (1919), Chronicles of America, in Chap. VIII deals with labor in the colonies.

SELECTED READINGS

HOLMES, G. K., *Aboriginal Agriculture: The North American Indian,* in Bailey's *Cyclopedia of American Agriculture,* Vol. IV.
SANFORD, A. H., *Story of Agriculture in the United States,* pp. 1-91.
CARRIER, LYMAN, *The Beginnings of American Agriculture,* Chaps. III-VII, IX, X-XXV.
PHILLIPS, U. B., *American Negro Slavery,* Chaps. I-VII.
BOGART, E. L., and THOMPSON, C. M., *Readings in the Economic History of the United States,* pp. 28-40; 82-95.

CHAPTER IV

COLONIAL INDUSTRY

Resources and Colonial Beginnings.—The British colonies in America were destined to be primarily agricultural. Nevertheless, from the beginning, there was some industrial life, and the history of the development of industries starts with the first settlements. The very economic philosophy which led the statesmen of the seventeenth and eighteenth centuries to encourage colonization necessitated the immediate prosecution of industry. The dependence of England upon the Baltic countries for such naval supplies as pitch, tar, flax, hemp, cordage, and masts was an incentive to colonization, for it was expected that these could be obtained in unlimited quantities in America. The vast forests could supply fuel for glass manufacturing and iron smelting, industries which were rapidly denuding England of forests. Iron, copper, potash, potter's clay, and many other products might be obtained from the New World, it was hoped, as well as many tropical commodities which during the Middle Ages were secured from Asia Minor and the East. With the wealth obtained from selling these products to the home country the colonists were expected to purchase British manufactured goods. Neither the landowning aristocracy, which controlled the British Parliament, nor the gradually emerging bourgeoisie foresaw competition from the colonists. The expectation was that the Englishman beyond the seas would occupy himself either in raising raw materials or in manufacturing such products as could not be readily made in England.

In consequence some form of manufacturing seems to have been contemplated from the start. Captain Newport on his second voyage to Jamestown in 1608 carried with him eight Poles and Germans to make pitch, tar, glass, metals, and soap ashes. When the London Council complained that gold and silver were not forthcoming from Virginia, Captain Smith set his men to work and sent back in 1608 a cargo of pitch, tar, iron ore, soap ashes, wainscoting, and clapboards, but at the same time he maintained that "it were better to give five hundred pound a tun for those grosse Commodities in Denmarke, than send for them hither, till more necessary things be provided. For in overtoyling our weake and unskillful bodies, to satisfie their desire of present profit, we can scarce ever recover ourselvs from one Supply to another."[1] With the exception of a load of sassafras gathered near Cape Cod in 1608, this cargo appears to have been the first export from the British colonies to a European country, and was composed almost exclusively of manufactured products.

Further exports from Virginia were largely a duplicate of the first cargo until about 1617, when the colonists turned their attention to the raising of tobacco. The success of the tobacco plant was not in accord with the desires of the London Company and a determined effort was made to divert colonial enterprise again toward manufacturing. Sir Edwin Sandys said at the meeting of the company in 1620 that already one hundred and fifty persons had been sent to set up three ironworks and that the settlers were advised to make pitch, tar, pot and soap ashes, timber for shipping, masts, planks, boards, and cordage. For the manufacture of the last named each family was ordered to set out one hundred plants of silk grass and the governor five thousand. Materials for erecting sawmills were sent out and efforts were made to induce artisans to emigrate to Virginia. But these attempts to substitute manufacturing for tobacco were largely in vain and Virginia soon came to depend for her wealth upon agricultural staples. Beverley in his *History of Virginia* (1705) laments the fact that whereas the colony abounded in the raw materials for essential industries, the plantations depended for their manufactured products almost entirely upon imports from England.

In New England the situation was reversed. Although the settlers were forced during the first few years to exert their every effort in the production of foodstuffs, the sterile soil and rigorous climate were not conducive to large-scale agriculture. The abundance of timber supplied a marketable commodity and in 1621 the

---

[1] Edward Arber's edition (1884) of Captain John Smith's Works, p. 445.

ship *Fortune* was freighted at Plymouth with clapboards and beaver skins. This was the beginning of New England commerce and industrial life, which increased continually during the colonial era.

As was to be expected in an age of crude transportation facilities, colonial industry was largely dependent upon the immediate natural resources. The products of the limitless forests provided furs, naval supplies, and containers for the fish, salt meat, flour, rum, pitch, tar, and turpentine, all of which were traded in extensively. The forests provided abundant fuel for making iron, glass, bricks, and pottery. The herds of cattle which roamed on the outskirts of civilization provided the meat for export, and the hides which could be tanned from the always plentiful bark. The center of this industrial life was New England, although the middle states were not devoid of it, and some products— lumber, naval supplies, and furs—were obtained in greater or less amounts from all of the colonies. While the above-mentioned articles were exported, it must be remembered that by far the larger amount of manufacturing was of a purely household nature. Many of the farms were self-sufficing economic units, the farmer doing his own carpentry, blacksmithing, and tanning, while his wife spun, wove, and made soap and candles.[1]

Forest Industries—Lumbering.—Four industries were dependent upon the forest—lumbering, ship building, manufacturing of naval supplies, and the making of potash. The first exports from Jamestown and Plymouth contained lumber, and some of the earliest settlements in Maine, New Hampshire, and Georgia were logging camps. The demand for timber of all kinds was great and the profits large. When a man by hand in 1650 could manufacture 15,000 clapboards or pipe staves a year, worth £20 a thousand in the Canaries, the desirability of increasing the output by the aid of sawmills was obvious. Artisans were early sent to Virginia to erect sawmills, but the first one in that

---

[1] That colonial industrial life was influenced by the Indians is undeniable. Such influence, however, in comparison to that of the aborigines upon early agriculture, was trivial, and space forbids further treatment here. See Clark Wissler, *The American Indian* (2d ed., 1922).

province was not built until 1652. The first sawmill in New England, however, was probably built prior to 1635, near Portsmouth, New Hampshire. During the succeeding years sawmills and grist mills were established at many convenient points in all the colonies —small mills to supply local demands at remote interior points, and commercial mills to provide export materials along the fall line of New England and the middle colonies, but especially in New Hampshire and Maine, where the fall line approached the coast and the lumber was easily marketable. Soft woods like the white pine, cedar, and spruce provided lumber adapted particularly to ship building. The hard woods, red and white oak, were of chief industrial importance on account of their use in the manufacture of cooperage stock. Cedar in the north and cypress in the south provided house frames and shingles; cherry, birch, maple, and walnut were used for furniture and gunstocks; red maple in the middle colonies for spinning-wheels, plates, and other implements; laurel for pulley axles and weaver's shuttles. The value of the boards, scantlings, masts, spars, staves, headings, hoops, and poles exported in 1770 was over $160,000.

Forest Industries—Ship Building.—With an abundance of white pine, fir, and oak close to the water's edge and a ready supply of pitch pine for tar and turpentine, the colonists had immediately at hand the raw products for ship building. The scarcity of lumber in western Europe and the high prices which the finished vessels could command enticed the colonial settler from agriculture, his natural occupation, to ship building, and made of it a leading industry. The first ship ever constructed in this country by Europeans was a Dutch vessel, the *Onrest,* built by Captain Adrian Block in 1614 on the Manhattan River and destined to an adventurous career of exploration. In the same year Captain John Smith constructed seven boats on the coast of Maine, in which thirty-seven men made a successful fishing trip. The Plymouth colony was joined by a carpenter in 1624, who commenced immediately building vessels; but it was not until 1641 that a boat of any size, a bark of fifty tons, was launched. *The Blessing of the Bay,* constructed at Mystic in 1631 and owned by Governor Winthrop, was the first vessel built in Massachusetts

Bay. The first vessels were built in Connecticut in 1640, in Rhode Island in 1646, in Delaware in 1642, and in Philadelphia in 1683, all of which dates followed closely the founding of the respective settlements. The English Civil War threw the colonists upon their own resources and gave the first impetus to ship building in New England, and in the succeeding years the industry grew rapidly. A book published in London in 1643, entitled *New Englands First Fruits,* says of ship building there: "(Besides many Boates, Shallops, Hows, Lighters, Pinnaces,) we are in a way of building shippes of an 100, 200, 300, 400 tunne, five of them are already at Sea; many more in hand, at this present, we being much incouraged herein by reason of the plenty and excellencie of our Timber for that purpose, and seeing all the materialls will be had there in short time." [1] The Court of Massachusetts in 1641 declared ship building "is a great importance for the common good," ordered a surveyor "be appointed to examine any ship built, and her work, to see that it be performed and carried on according to the rules of the art." In 1667 it ordered a committee of five to draw up and present suitable laws for the regulation of the business. Wherever harbors, labor, and supplies were contiguous there would be seen the keel and framework of a vessel, and many a little river and coast town in New England flourished in consequence. Small progress was attained in the industry in Virginia and the Carolinas until the last half of the eighteenth century.

The home demand for ships for the fisheries and the coasting trade must have exceeded the demand from abroad, but the prosperity of the industry was largely dependent upon the foreign market. Ships were constructed in New England at the close of the colonial period for $34 a ton, fully 20 to 50 per cent cheaper than the cost of construction in Europe. As early as 1676 New England builders were turning out thirty ships a year for the English market, and 300 or 400 commercial boats were built yearly by 1760. By this time one-third of the tonnage (398,000 tons) sailing under the British flag was American built. Another

---

[1] *New Englands First Fruits,* in Sabin's Reprint, No. vii, p. 40.

factor in addition to that of cheapness which influenced the growth of colonial shipping was that the balance of trade with England continued unfavorable to the northern and middle colonies. On this account and also because of the scarcity of money in the colonies, English merchants from the sale of their cargoes would build ships and load them with lumber for Europe.

Exclusive of fishing craft New England owned in 1745 about 1,000 vessels, and in 1775 about 2,000. The tonnage built in the American mainland colonies in 1770 was about 20,000 tons, in 1771 about 24,000, and in 1772 around 26,000, of which over 18,000 were constructed in New England. At the opening of the Revolution Massachusetts was estimated to own one ship for every one hundred inhabitants.

Forest Industries—Naval Stores.—In addition to the actual building of ships and exportation of lumber the American colonies provided a valuable source for naval supplies, as tar, pitch, rosin, and turpentine. These commodities, needed by the navy and merchant marine, were imported chiefly from Sweden, Russia, and Poland. They were considered extremely important as colonial products in the days of mercantilism, the *summum bonum* of which was that a nation be self-sufficing. To encourage the manufacture of these commodities Parliament in 1706 placed a bounty of £4 per ton on tar and pitch imported from the colonies, £3 per ton on rosin and turpentine, and £6 a ton upon water-rotted hemp, all of which bounties were later decreased. Colonial legislatures in some cases added to the bounties, but it was only in the Carolinas that the policy was successful in greatly stimulating the industry. Only 82,000 barrels of tar, 9,000 of pitch, and 17,000 of turpentine, altogether valued at £175,000, were exported in 1770.

Fur Trade.—The trade in furs was a valuable source of income for the colonies north and south and in the early years was more profitable in New England than either fish or lumber. The variations of climate from Florida to Hudson Bay made it possible for many types of fur-bearing animals to thrive, and the needs of the two hundred thousand Indians who lived on the continent previous to its discovery were not sufficient to prevent a constant increase. The first cargo sent from Plymouth to England contained an assignment of furs. "Of Bevers, Otters, Martins, Blacke Foxes, and Furres of price," wrote Captain John Smith of the possibilities of New England, "may yearely be had 6 or 7,000 and if the trade with the French were prevented, many more."[1] The abundance of the fur-bearing animals and the skill of the Indians as hunters and trappers, augmented by the apparently insatiable demand of Europe, made it a prosperous business from the start.

The profits were enhanced, of course, by the unequal bargains made between the Europeans and the Indians. In the early years trinkets or beads valued at a few shillings could be traded for furs worth hundreds of dollars. John Smith obtained for a copper kettle fifty skins valued at about two hundred and fifty dollars. The New Englanders regularly traded a bushel of corn, varying in value from three to six shillings, for a pound of beaver worth, on an average, twenty shillings. Later, indeed, the Indian grew wiser, but he was a poor business man and was almost invariably the victim of the unscrupulous and often dishonest trader. The Indian at the end of the hunting season, after he had stocked up with furs and traveled many miles, was in no position to bargain equally, especially after his appetite had been whetted with the white man's firewater. The sale of firearms and liquor to the Indians was illegal in most of the colonies, but the temptation to the trader was too great. A half-drunk Indian would sell his soul for more liquor, and it was to the advantage of the trader to stock him up with arms and ammunition with which he could increase his season's harvest of furs. Between 1631 and 1636 the Plymouth colony sent shiploads to England valued at $20,000, chiefly of beaver, otter, and black fox. The first shipment of furs from New Netherlands is reported to have yielded a profit of $10,000, and the business during the first eight years, $56,000. The Champlain country of New York, when the traffic was at its height in the middle of the seventeenth century, exported forty thousand skins annually to England.

---

[1] Arber edition, p. 200.

COLONIAL INDUSTRY

The fur trade, however, has a much deeper significance in American history than its commercial and industrial phase. The fur trader pressing after the retreating game supply blazed the trail for the missionary and settler and pointed the way to the west. But though the fur trader opened the western routes and brought material prosperity, his unscrupulous treatment of the red men often resulted in Indian wars with their horrors and devastations. The most serious Indian troubles in the north came from French influence, but the Indian wars of the Carolinas and Virginia were usually attributable to disputes between Indians and fur traders.

The colonial rivalries of the seventeenth and eighteenth centuries went deeper than mere friction over the fur trade, though the latter was likely to be the immediate cause of irritation on this side of the water. New Netherlands was founded originally not as a colonization project, but as a fur-trading post, and as such the Dutch East India Company was chiefly interested in it. The incessant slaughter of animals fast depleted the New England region, and by 1699 the annual export of New York had decreased to 15,000 skins. With the falling off of the fur supply east of the Alleghanies, the English pushed westward, only to find the French firmly ensconced in the St. Lawrence Valley and already tapping the rich supplies of the Mississippi region. The decline of the fur trade east of the Alleghanies was coincident with the outbreak of King Philip's War (1675-76).

New France, even more than New Netherlands, was interested chiefly in fur trading, its prosperity resting almost entirely upon that business. The enterprising French trader believed in going out and meeting the Indian in his own haunts. This meant carrying the French beads, hatchets, firearms, brandy, and other commodities many miles along streams and through forests, a process which so increased their value that he must bargain closely. The Englishman, who did not penetrate so far and whose goods were likely to be manufactured in the colony, could give the Indian more for his furs. In consequence the Indian would travel long distances to trade with the Englishman in preference to the Frenchman, although as a rule his intercourse with the latter was more amicable. In this way the English drew considerable business from their rivals.

The key to the situation rested in the friendship of the Iroquois, the most highly civilized and most powerful of the Indians east of the Mississippi. Controlling as they did the Mohawk Valley, the gateway to the west, they were bound to play a leading rôle in the rivalries between the French and the English in the fight for the western fur trade. The unfortunate act of Champlain in 1609 in aiding Canadian Indians against the Iroquois, the subsequent mistakes of the French governors, and on the other hand the skillful diplomacy of Governor Dongan of New York (1684), won for the English the friendship of the Iroquois and the furs of the west and, in fact, determined more than anything else which of the two nations should control the continent. The founding of the Hudson Bay Company in 1670, headed by Prince Rupert, to which the English king gave a grant to trade in the regions draining into the Hudson Bay, was a direct stroke at the political and commercial interests of France and contributed, along with the friction on the southern border, to the succession of wars. A colonial writer in 1755 estimated the value of furs and pelts imported into England at about £90,000 per annum, and into France £135,000, of which a part came from the north of Europe, a stake well worth striving for. In fact, the French were ever ready to dispute with the English the control of the trade in the St. Lawrence, the Mohawk, and the Ohio valleys, and with the Spanish the control of the southern Mississippi region; while both Spanish and French competed with the settlers of the Carolinas.

Fishing and Whaling.—Whether it was the chronic state of hunger under which Europe suffered during the Middle Ages or the desertion of the western coast of Europe by the food fish that drove the fishermen to push westward for their supply is uncertain. By 1300 they had reached Iceland, and at least a hundred years before there was an English settlement in New England European fishermen were sailing regularly to the "Banks." By 1500 the deep-sea fisheries were in full operation and the Newfoundland waters were frequented by ships of England, France,

Spain, and Portugal. To preserve the fish on the long trip home they must be cured, and for this it was necessary to land and spread them out in the sun where the moisture could evaporate and the salt "strike in." The mystery which hangs over these early trips to the fishing banks, possibly because a great deal of the product was smuggled in on the return voyage, makes it impossible to determine at what date Europeans first began to land upon various parts of the coast to cure their fish. The Portuguese alone in 1550 had four hundred fishing vessels in the American waters. In 1610 England was said to have derived an income of ten millions of dollars from the sale of surplus fish. In consequence, European governments early in the seventeenth century woke to the importance of possessing lands close to the fishing banks. This necessity provided one impetus to North American exploration, and resulted in disputes over lands close to the fishing grounds which have continued almost to the present day.

The New England settlers soon realized that wealth was to be found in the deep sea rather than upon the rocky soil. Said Captain John Smith, "The maine Staple from hence to bee extracted for the present, to produce the rest, is fish; which however it may seeme a base commoditie: yet who will but truely take the pains and consider the sequell, I think will allow it well worth the labour." [1] An apparently exhaustless supply of cod and halibut was to be found off the Newfoundland banks, while closer to the coast mackerel, herring, bluefish, shad, and other varieties were plentiful. The demand for fish was widening in Europe, especially in Catholic countries. New England, close to the fishing grounds and with an abundance of ship-building materials at hand, was in a strategic position to profit, and after 1650 her prosperity was closely connected with fishing. Three hundred thousand cod, the great staple of the industry, were sent abroad in 1641, and by 1675 over six hundred vessels and four thousand men were engaged in cod fishing. By the end of the colonial period the industry was worth $225,000 a year. Eventually the

---

[1] Arber edition, p. 194.

New Englanders divided their fish in three classes. The largest and fattest, as they were the most difficult to cure thoroughly, were consumed locally. The second class, smaller and more easily cured, were exported to the Continent. The third class, tainted, damaged, or too small for the European or American market, were sold in the West Indies as food for slaves, usually in exchange for molasses. The latter, a by-product of sugar manufacture, was brought back and converted into rum. In addition to the manufacture of rum, which was a result of the fishing industry, the demand for salt was stimulated, and salt vats were erected at various points along the shore, where sea water could be evaporated. Fish, molasses, rum, and salt all contributed to make the cooperage industry one of the liveliest in the colonies.

Almost as important as fishing during the last hundred years of the colonial era was the whaling industry. Spermaceti, sperm oil, whale bone, and ambergris were in great demand. Whales were abundant off the New England coast and after 1700 New England seamen began to put off and harpoon the unwieldy monsters when they came up to breathe. When the whales were driven off the coast the whalers followed them to the arctic and antarctic regions. After 1732 an annual bounty of twenty shillings a ton (raised to forty shillings in 1747) was paid on vessels of 200 tons or upward engaged in whaling, and the consequent increasing values of the products spurred on the hunters. The most skillful whalers in the world came from New England and that province practically monopolized the business. Over three hundred vessels and four thousand sailors were engaged in it at the outbreak of the Revolution, most of them hailing from Nantucket, New Bedford, Marblehead, and Provincetown. Upon the basis of spermaceti a candle-making business of some importance grew up.

Textiles—Wool.—Although the colonists depended for some time upon the home country for their textiles, the earliest ships undoubtedly brought over the spinning-wheel and the hand loom. Imported textiles in those days of long voyages and high freight costs were necessarily expensive, and the average farmer

COLONIAL INDUSTRY

was forced by pecuniary reasons to wear homespun or locally manufactured cloth. In consequence, spinning and weaving commenced very soon in the colonies, largely as a household industry. A further impetus to colonial woolen manufacture was given by English export duties on woolen broadcloth, by a law forbidding the export of raw wool from England, and finally by the act of 1660 which prohibited the Dutch trading in the British colonies, thus cutting off the supply of the cheapest woolens.

Sheep were first introduced into the colonies by the London Company at Jamestown in 1609; they increased so slowly that in 1649 there were in Virginia only about 3,000. The Dutch West India Company introduced them in New Netherlands in 1625. The first sheep were brought to New England about 1633, where they were first kept on an island in Boston Harbor as a protection against wolves. The danger from wild animals hindered the growth of the flocks, and under frontier conditions the breeds were inclined to deteriorate. As each successive colony was established sheep were immediately introduced, usually from Malaga, so that by 1661 100,000 were reported in New England alone, and by the end of the colonial period there were perhaps eight million in all of the colonies. Yet this number did not supply all of the wool needed and woolen cloth was imported continually.

The expulsion of a congregation from Yorkshire, the home of the English woolen industry, led to the founding of the town of Rowley, Massachusetts, and the building in 1643 of a woolen and fulling mill, the latter being the first one erected in the American colonies. The success of this enterprise was assured by the Civil Wars (1640-60), which checked immigration and the inflow of gold, brought down prices of agricultural products, and forced the settler to seek cheaper textiles than he had previously bought from England. Eventually every household kitchen became a workshop where the women spun and wove, turning out the rough serges and linsey-woolseys of the time. By the end of the colonial period in New England some faint beginnings of the factory system could be seen in the grouping of several weaving machines under the same roof. Virginia erected her first fulling mill in

1692, and a number were in operation at the opening of the Revolution. Many fulling mills had been established in Pennsylvania by the middle of the eighteenth century, especially in Philadelphia and Lancaster. The latter town was the largest inland community in the country in 1789, and one-third of its seven hundred families were engaged in the making of woolen, linen and cotton cloth. Excellent broadcloth was manufactured in Philadelphia, where also was located a prosperous concern manufacturing spinning-wheel irons. In the other colonies spinning and weaving were carried on almost from the start.

Textiles—Linen.—More common than wool was the use of linen in the colonial period. Linen served nearly all the purposes for which cotton is used to-day. Hemp cloth or linen of varying degrees of fineness was the chief textile used, and was ordinarily mixed with wool to make serge, kersey, or linsey-woolsey. Because of its importance the raising of flax and hemp and the spinning and weaving of it were constant subjects of government legislation and encouragement. The cultivation of flax and hemp was enforced by severe penalties in Virginia. The General Court of Massachusetts in 1640 was directed to further the growing and preparation of flax, and in 1655 the selectmen of the towns were ordered to ascertain the number of spinners in each family and to require each spinner to spin three pounds of linen, cotton, or woolen yarn each week for thirty weeks during the year. In 1737 a spinning school was established in Boston and supported by a tax on carriages and other luxuries. Other colonial assemblies, including New Jersey, Maryland, and South Carolina, granted bounties on hemp and flax. The Scotch-Irish, who were the first settlers in many of the frontier towns in New Hampshire and Massachusetts, brought with them their skill and founded flourishing centers for the making of linens. Hardly less skillful than the Scotch-Irish were the Dutch of New York and the thrifty Germans of Pennsylvania.

Textiles—Cotton.—A certain species of cotton is indigenous to America and was grown to some extent from the earliest European occupation, but its cultivation never became widespread during the colonial period. Although it was regarded until the middle of the eighteenth century rather as an ornamental plant for the garden than as a commercial crop, it was grown in Virginia as early as 1621 and in South Carolina in 1664. Nevertheless, it was widely used in the colonies and imported extensively from the British West Indies. The first cotton received in Massachusetts came from the Barbadoes in 1633; and in the succeeding years it became an important item of import. Being a difficult fiber to work, it was mixed with wool before spinning.

Cotton raising in a serious way and on a large scale apparently began on the eastern shore of Maryland and in Delaware, and from there shifted south. Its cultivation was held back, partly by the difficulty of separating by hand the seed from the cotton fiber, a factor which made the labor cost excessive, and partly by the greater profits from tobacco. The great development of cotton did not come until after the invention in 1793 of the cotton gin, but the increase in production had commenced with the Revolution, when the importation of English textiles declined.

Minerals.—The first settlers in Virginia and Massachusetts were not long in discovering bog iron. The cost of importing such necessities as nails, agricultural implements, kitchen utensils, and firearms, and the presence of bog iron with an abundance of fuel for smelting, turned the attention of the colonists almost immediately to the possibility of mining and manufacturing iron. Iron ore was early shipped from Virginia, and in 1619 one hundred and fifty workmen familiar with its manufacture were sent out to erect ironworks at Falling Creek near Jamestown. A furnace was built, but in 1622 the works were destroyed and the men massacred by the Indians. No further attempt was made here for a hundred years until Governor Spotswood commenced to mine and work furnaces on the Rappahannock.

The first ironworks in Massachusetts seem to have been erected in 1643 near Lynn on the Saugus River, and were in operation over a hundred years, producing at the beginning about seven tons a week. The bogs on the Monatiquot River at Braintree were tapped almost simultaneously with those of Lynn, and after they were worked out furnaces were built at Taunton. Later, ironworks were erected in 1731 at Great Barrington and in 1765 at Lenox. Forges were erected by Joseph Jenks in Rhode Island before 1675 and smelting was commenced in the Connecticut Valley by John Winthrop at New London and at New Haven (1658).

In the eighteenth century rock ores were mined in the uplands of Connecticut, in New York, northern New Jersey, and western Massachusetts. The first ironworks in New York were erected by Philip Livingston at Ancram, where ore brought from Salisbury, Connecticut, was smelted. German miners who settled in Orange County, New York, after 1730 opened up rich deposits. New Jersey, where the first iron was smelted in 1655 by Henry Leonard, became a thriving center. At the end of the century there were ten mines in Morris County alone, besides two furnaces, three rolling and slitting mills, and about forty forges. Some of these deposits are still worked. The first records of iron mining in Pennsylvania tell of the erection in 1716 by Thomas Rutter of a forge on the Manatawny Creek, Berks County. After 1728 forges sprang up in the Schuylkill Valley and the industry commenced to move west. A bloomery was constructed in York County in 1756 and at Boiling Springs in 1762, the latter forming the nucleus of the Carlisle Ironworks. Iron was smelted in Maryland from the early years of the century, but in the Carolinas and Georgia practically nothing was done until after the Revolution.

The purpose of these iron mines was chiefly to supply the immediate needs of the colonists for wagon and sleigh tires, mill spindles, anvils, pots, kettles, forged plates, weights, bells, chains, guns, and cannon. In conjunction with these mines and smelting establishments usually were to be found some variety of casting works. Slitting mills furnished iron rods from which the farmers on winter evenings, by means of a small furnace in the chimney corner, manufactured nails. The colonial smelting furnaces generally were small and crude, producing from a dozen to twenty tons a week, but the principles upon which they worked were not unlike those employed in the modern establishment.

The development of iron foundries was not what the British wanted. Pig iron might be produced in the colonies, but the colonials must not compete with the home manufacturers. By 1750 the casting of iron had reached such proportions that Parliament prohibited the erection of any slitting or rolling mill, plating, forge, or steel furnace, under a penalty of £200. The production of pig and bar iron at the same time was encouraged by permitting, after 1757, admission into the port of London free of duty. Exports which amounted to 1,127 tons in 1728 grew under this stimulus to 7,525 tons in 1771, valued at £20 a ton.

Copper veins were discovered and a mine opened at Symesburg, Connecticut, in 1721. The Schuylers in New York operated a mine and a considerable amount of ore was taken from Hudson County, New Jersey. Valued at £40 a ton, the export was said to have amounted in 1731 to 1,386 tons. Small quantities of lead were mined after 1720 in southeastern Missouri, most of which was supplied to hunters and trappers, who molded it into bullets.

Other Industries.—In addition to the industries mentioned there were, of course, many others, pursued on a small scale and usually for local needs. The tanning of leather and its manufacture into clothing and harness were as important as any. Because of the cost of woolens it was customary to make clothing of animal skins, especially the waistcoat and breeches of the men, and the jerkins and petticoats of the women. Even the bed clothing was made almost entirely of leather. Beaver skins were made into hats in New England and New York, the Board of Trade and Plantations reporting in 1731 that 10,000 were exported annually. Furniture, wagons, carriages, and tools were likely to be made by the farmer himself or by the village blacksmith or carpenter.

The transition from the log cabin to the more elaborate house developed certain building industries such as brick making and quarrying. The first bricks in Massachusetts were imported, but a brick kiln was set up at Salem in 1629, the product of which was used for fireplaces and chimneys. The first brick house in Massachusetts was built in Boston in 1638, and within a few years after that date lime, brick, and tile making were among independent trades pursued in New England. By the end of the colonial period the various colonies were manufacturing sufficient bricks for home consumption and exporting small quantities to the West Indies. Manufacture of glass was stimulated not alone by the need of it for building (glass windows were scarce in the seventeenth century), but in order to provide trinkets and beads for the Indian trade. A glass furnace was erected at Jamestown in 1609 and a number of works had been erected prior to 1775, but the industry was conducted on a small scale. This was due to the fact that the chief cost of glass is labor, and in a new country labor is exceedingly scarce. The most interesting attempt at glass making was that of Baron Stiegel, who laid out the village of Mannheim about eleven miles southwest of Lancaster, and erected several iron furnaces and glass works. The baron was visionary and impractical and his enterprise was not successful. His idea was to found a European feudal estate in this country. Cannon mounted on his castle fortifications were to announce his arrival; and on hearing the salute the workmen were to leave the furnaces and foundries and assemble to pay homage to their liege lord, or honor his guests with appropriate music and ceremony. Such a system was scarcely likely to prove an efficient basis for glass manufacture.

An almost universal industry, household and otherwise, was the making of many kinds of liquor. It throve especially in the coast towns of New England, where the West Indian molasses was distilled into rum, particularly for the slave trade and the fishing expeditions. At one time twenty distilleries in Newport alone were engaged in the business. Not only rum, but also beer, ale, and cider were exported to the West Indies. It must not be thought, however, that this liquor business was concerned entirely with exports; our forefathers were hard drinkers and a large proportion was consumed at home.

A word might be said of many other industries which existed in colonial times, although on a small scale. In time most of the furniture used in the northern and middle colonies was manufactured here. Furniture was made to order by cabinet makers in their little shops, and their skill and honesty are amply demonstrated by the innumerable pieces still in constant use. Such a thing as large-scale factory production was hardly dreamed of as yet. Paper making, printing, and book binding had made considerable progress. This is remarkable when such factors are taken into consideration as the cost of labor and the supposed lack of interest in reading among a thinly scattered population engaged in arduous frontier labor. The first printing press was set up in Cambridge, Massachusetts, in 1639, and by the outbreak of the Revolution there were thirty-six newspapers published in the colonies, fourteen of which were in New England. Most of the printing was done in Boston and Philadelphia.

Government Aid to Industry.—The development of colonial industries was not wholly spontaneous. The era of *laissez-faire* had not yet arrived, and it was natural that colonial legislatures should follow in the footsteps of the mercantile policy of the British Parliament. Based on British precedent, laws were passed to encourage manufacturing and the production of raw materials by bounties and other forms of aid, and to maintain standards of quality workmanship. In some cases laws were actually passed to force certain types of production, as in the case of the Massachusetts law of 1655 already mentioned, which required each spinner in a family to turn out a certain amount of yarn. Mechanics by an early Virginia law were forbidden to plant tobacco or corn. Virginia and Connecticut passed laws requiring the inhabitants to plant flax and hemp, while the latter state forced its towns to purchase cotton in proportion to their population.

More common than compulsion was encouragement by rewards. Maryland in 1671 granted a bounty of a pound of tobacco for every pound of hemp, and two pounds of tobacco for every pound of flax raised. In coöperation with the British Acts of 1722 and 1764, which subsidized American naval stores, bounties on hemp were granted in 1722 by Virginia, South Carolina, and Pennsylvania (removed in the latter state in 1732); in 1763 by New York; in 1764 by North Carolina; and in 1765 by New Jersey. Bounties on flax were also granted by New Jersey and South Carolina, and on linen at different times by several of the states. South Carolina in 1770 offered a bounty of 30 per cent of the value upon all linen and linen thread made in the province. New England sought particularly to promote the manufacture of duck for sail cloth, Massachusetts and Connecticut both granting a bounty of twenty shillings per bolt of specified dimensions and quality. Owing to English jealousy, the same encouragement was not given to woolen cloth. Substantial bounties were offered by Massachusetts, Virginia, and Maryland in the early years, but after the British Woolen Act of 1699, which forbade the exportation of raw wool, colonial legislatures truly felt that it would only "draw the displeasure of Great Britain upon us, as it will interfere with their most favorite manufactury."

The bounties on textiles and naval stores were important, but similar encouragement was given other industries, as, for instance, on ship building in Virginia in 1661 and South Carolina in 1711. Tar and potash were favored in certain states. Pennsylvania established a bounty of twopence a gallon on spirits distilled and exported. Salt, indigo, and wine were aided in South Carolina. During the Revolution various colonies granted bounties for saltpeter, gunpowder, and firearms.

Government aid to manufactures by means of land grants and loans was common. Large tracts of land were frequently given in New England to recompense citizens who would erect iron, salt, potash, or fulling mills. This was duplicated by the towns on a smaller scale for grist mills and sawmills. Money was frequently loaned and sometimes actually given to private individuals to help them set up a business considered essential to the welfare of the colony. In 1770, for example, Pennsylvania donated £1,000 to the Philosophical Society to encourage silk culture. Permission to hold lotteries to obtain money to establish, foster, or rebuild industries was often granted by legislatures. In certain instances manufacturing concerns or materials for manufacture were exempted from taxation. Examples of this were ships on the stocks in Connecticut, and the Evesham and Hibernia ironworks in New Jersey. Mechanics and factory laborers were also frequently exempted from taxation. Encouragement was given by the granting of monopolies and patents, although the colonial legislatures, remembering the evils of monopolies in England, were chary of conferring this right. In some cases the export of certain commodities was prohibited or heavily taxed in order to promote home industries dependent upon them. Import duties were chiefly for revenue and had little bearing on manufactures except that such raw materials as wool and cotton, destined for later manufacture, were usually on the free list. Laws in various states making tobacco, flax, hemp, wool, lumber, tar, turpentine, leather, and oil either legal tender or receivable for taxes were not without effect. The sum total of these various methods of encouragement, extending as they did for 125 years, was decidedly stimulating; but it is doubtful if any industry not suitable to our economy was thereby permanently established.

Government Restrictions.—Holding to the mercantile theory of colonies as producers of raw materials not obtainable in England and as consumers of manufactured goods made by the home country, the British Parliament during the colonial period felt free to pass legislation in support of this theory and thus attempt to stifle competition. Complaints of British manufacturers that the colonists were exporting woolens led to the act in 1699 forbidding the shipment of wool, woolen yarn, or cloth produced in the colonies to any other plantation or country. In 1732 the exportation of hats was similarly prohibited. A law of 1750, while it permitted the entry of bar iron in England, prohibited in the colonies the erection of slitting or rolling mills, plate, forge, or steel furnaces. Since, however, it did not prohibit casting furnaces, the colonists could still make cannon, kettles, salt pans, and other utensils.

In the various Navigation Acts certain "enumerated" products could be sent only to England. Among these commodities were sugar, tobacco, cotton, wool, indigo, ginger, pitch, turpentine, hemp, masts, yards, rice, copper, bog and pig iron, pot and pearl ashes, beaver skins, whale fins, hides, and molasses. While exceedingly irritating, it cannot be said that these laws greatly affected colonial industry, as they were generally evaded. On the other hand, it should be remembered that Great Britain encouraged as well as restricted. Where British and colonial interests did not conflict the mother country was generous, remitting import duties and granting bounties on the products she needed.

Extent of Colonial Industry.—It would be the merest guesswork to attempt even an approximate statement of the extent of colonial industry at various periods. Contemporary accounts are either inadequate, or else exaggerated for the purpose of influencing prospective settlers or setting at rest the fears of the home government. The attempt made by the colonists at the time of the Stamp Act to free themselves from dependence upon British industry brought in 1766 and in 1768 letters from the Lords of Trade to the colonial governors demanding an annual report on manufactures in their provinces. Though inadequate, these reports are a valuable source of information.[1] The general tenor of the replies was to depreciate the extent of colonial industry and to emphasize the dependence of colonists upon Great Britain. While specific industries were mentioned, it was maintained that the wealthier classes bought imported goods and that the lure of the land turned the mechanic into a farmer, to the detriment of manufacturing development. Governor Bernard of Massachusetts went so far as to declare, "I do not think it necessary to send an annual account where I have nothing to inform of."

These reports, apparently aimed to reassure the British government that the colonists were in no position to become economically independent, depreciated unduly the real condition of colonial manufacturing. A truer impression is given in a letter from Comptroller Weare to the president of the Board of Trade; he says, "Upon actual knowledge therefore of these northern Colonies, one is surprised to find out that, notwithstanding the indifference of their wool and the extravagant price of labour the planters throughout all New England, New York, the Jersies, Pennsylvania and Maryland (for south of that province no knowledge is here pretended) almost entirely clothe themselves in their own woolens, and that generally the people are sliding into the manufactures proper to the mother country, and this not through any

---

[1] A summary of these letters is given in Clark, Victor S., *History of Manufactures in the United States, 1607-1860*, p. 208 ff.

COLONIAL INDUSTRY

spirit of industry or economy, but plainly for want of some returns to make to the shops."[1] An English writer stated in 1774 that "the inhabitants in the Colonies . . . do make many things, and export several manufactures, to the exclusion of English manufactures of the same kinds. The New England people import from the foreign and the British Islands very large quantities of cotton, which they spin and work up with linen yarn into a stuff, like that made in Manchester, wherewith they clothe themselves and their neighbours. Hats are manufactured in Carolina, Pennsylvania and in other Colonies. Soap and candles, and all kinds of wood-work, are made in the Northern Colonies and exported to the Southern. Coaches, chariots, chaises, and chairs, are also made in the Northern Colonies and sent down to the Southern. Coach harness, and many other kinds of leather manufactures, are likewise made in the Northern Colonies, and sent down to the Southern; and large quantities of shoes have lately been exported from thence to the West India Islands. Linens are made to a great amount in Pennsylvania and cordage and other hemp manufactures are carried on in many places with great success: and foundery ware, axes, and other iron tools and utensils are also become articles of commerce, with which the Southern Colonies are supplied from the Northern."[2]

When the colonist could afford it, he undoubtedly preferred to buy the finest grade of manufactures from abroad; but the essentials could be obtained here, and the non-intercourse agreements previous to the Revolution demonstrated Franklin's contention that the colonist could supply himself with what was absolutely needed without recourse to Great Britain. To do so, however, at that time was abnormal, for colonial economy was primarily agricultural. Tench Coxe estimated the annual value of American manufactures in 1790 at $20,000,000.

Recapitulation.—In reviewing the subject of colonial industries it is well to recall three things. First, that industry as

---

[1] Massachusetts Historical Society, *Collections,* 1st series, i, p. 74.
[2] *Interest of the Merchants and Manufacturers of Great Britain in the Present Contest with the Colonies Stated and Considered,* p. 11 (London, 1774, reprinted in Boston).

separated from agriculture was the business of considerably less then one-tenth of the population. Second, the business of the colonist, whether agricultural or manufacturing, was chiefly extractive: in New England, lumbering and fishing; in the middle colonies, the raising of livestock and cereals; in Virginia and part of Maryland, the cultivation of tobacco; in South Carolina, rice and indigo; in Georgia and North Carolina, naval stores and lumber; in Louisiana, indigo, sugar, and tobacco. Third, only the faintest origins of the modern factory system can be found. Industries were migratory and the mills likely to be flimsily built. Diversification and specialization of industry had not progressed far. A sawmill and a grist mill under the same roof were usual, while in some places sawing lumber, grinding flour, forging iron, and fulling cloth were all carried on in the same establishment. Furthermore, the development of manufacturing was distinctly handicapped by the disadvantages of a new country—the primary importance of agriculture, the thinly scattered population and scarcity of labor and capital. Added to this was the hindrance of a British commercial law definitely opposed to the development of competing industries. On the other hand, favorable factors for industrial growth were present. British tariffs on foodstuffs forced the colonists to find new markets or to turn to other occupations than agriculture. Lack of money made it difficult to buy from abroad unless the foreign merchant would take commodities in exchange. Freight costs were so high as to make European goods prohibitive to the average colonist. It was well on into the nineteenth century, however, before we became economically independent of Europe.

NOTES FOR FURTHER REFERENCE

The best single volume available is that of V. S. Clark, *History of Manufactures in the United States, 1607-1860* (1916), published by the Carnegie Institution in a valuable series on American economic history. It contains an extensive bibliography. Two books, now out of date, but containing a mass of interesting detail not easily accessible elsewhere, are J. L. Bishop, *A History of American Manufactures from 1608-1860* (3 vols. 1866), and A. S. Bolles, *Industrial History of the United States* (1887). A briefer and later review is C. D. Wright, *The Industrial Evolution of the United States* (1897), which rests heavily upon Bishop. Brief accounts are contained in *Eighty Years'*

COLONIAL INDUSTRY

Progress (1869) under specific industries and in the recent text-books of E. L. Bogart, Katherine Coman, and Isaac Lippincott. Consult also E. L. Lord, *Industrial Experiments in the British Colonies of North America* (1898) in Johns Hopkins Studies in Historical and Political Science, Vol. XVII, and R. M. Tryon, *Household Manufactures in the United States 1640-1860* (1917). Two standard studies on particular sections are W. B. Weeden, *Economic and Social History of New England* (2 vols., 1890), and P. A. Bruce, *Economic History of Virginia in the Seventeenth Century* (2 vols., 1895). Other material on the South is contained in *The South in the Building of the Nation*, Vol. V.

A good chapter on the fur trade is in the high-school text of J. R. H. Moore, *Industrial History of the American People* (1913). See also Katherine Coman, *Economic Beginnings of the Far West* (1912). The best studies on whaling and the fisheries are those of R. McFarland, *A History of the New England Fisheries* (1911), and W. S. Tower, *A History of American Whale Fishing* (1920), University of Pennsylvania Studies, No. 23. Other special studies include M. T. Copeland, *The Cotton Manufacturing Industry of the United States* (1912), and B. E. Hazard, *The Organization of the Boot and Shoe Industry in Massachusetts Before 1875* (1921), both in the Harvard Economic Studies.

A comparison of European industrial life during the same period is extremely valuable. Material with special reference to England may be found in George Unwin, *Industrial Organization in the Sixteenth and Seventeenth Centuries* (1904).

For accessible source material see G. S. Callender, *Selected Readings in the Economic History of the United States* (1909), and E. L. Bogart and C. M. Thompson, *Readings in the Economic History of the United States* (1916).

SELECTED READINGS

CLARK, V. S., *History of Manufactures in the United States, 1607-1860*, Chaps. I-IX.

TRYON, R. M., *Household Manufacture in the United States, 1640-1860*, Chaps. I-VI.

WEEDEN, W., *Economic and Social History of New England*, Vol. I, pp. 165-204; 379-410.

BRUCE, P. A., *Economic History of Virginia in the Seventeenth Century*, Vol. II, Chaps. XII, XIII, XV, XVIII.

McFARLAND, R., *A History of the New England Fisheries*, Chaps. I-VII.

BOGART, E. L., and THOMPSON, C. M., *Readings in the Economic History of the United States*, pp. 42-68.

CHAPTER V

COLONIAL COMMERCE

Physiographic Background.—Physiographic conditions were the most potent influence in the development and direction taken by colonial commerce. The narrowness of the North Atlantic aided the growth of a large commerce with Europe. The continental shelf stretching into the ocean formed an ideal feeding ground for fish, and European fishermen were drawn in large numbers to our shores before actual settlements were made. The single-crop agriculture of the south made commerce imperative. In New England the sterile soil discouraged agriculture, and what products were raised brought the settlers in direct competition with Great Britain. When the mother country put up tariff walls against colonial foodstuffs the New Englander became more and more interested in ship building and commercial activity. In this he was aided by the frequent coastal indentations, which gave to the region an unsurpassed series of excellent harbors upon which grew up the cities of Portland, Maine; Portsmouth, New Hampshire; Gloucester, Salem, Boston, New Bedford, and Fall River in Massachusetts; Providence, Rhode Island; and New London, New Haven, and Bridgeport in Connecticut. The middle colonies, though lacking such a jagged coast line, were favored with two large rivers, the Hudson and the Delaware, which permitted early settlement to a considerable distance inland and the early development of commerce along these streams. In the south numerous rivers ran a sufficient distance inland to allow the small vessels of the period to sail directly to the private wharves of the tobacco planters and to handle with comparative ease the exports as long as the settlements remained in the coastal plain.

Mediums of Exchange.—Although certain economic and geographic factors were conducive to an early and healthy development of commerce, other influences tended to retard it. The New Englander and the inhabitant of the middle colonies must first discover products to export and markets in which to sell them, not an easy task in an essentially agricultural community when the chief potential market was largely closed and the attitude of the home parliament distinctly hostile to the development of industry.

While these hindrances were to a great extent overcome, another retarding influence, namely the scarcity of a convenient medium of exchange, plagued colonial merchants down to the founding of the Republic. The balance of trade with England was against the colonists and thus they were continually drained of metallic currency. A contemporary writer asserts that money from the West Indies "seldom continues six months in the province before it is remitted to Europe." Such coin as was circulated was obtained chiefly from the Spanish, Portuguese, and French colonies. English coins were rarely seen, yet the colonist attempted to carry on business through the terminology of pounds, shillings, and pence. Barter was further complicated by the fact that the English values were not maintained, various colonies placing different standards of value upon the foreign coins. Further difficulties to placing a true valuation upon metallic money were the habit of "clipping" and "sweating," and the fact that the same coin made at different mints might contain different quantities of the precious metal.

One state, Massachusetts, from 1652 to 1684, conducted a minting establishment from which were turned out the "pine-tree shillings" worth 75 per cent of the English coins of the same denomination. In 1690 Massachusetts issued bills of credit to pay the soldiers who took part in the expedition against Port Royal and Quebec, and this experiment in paper money was followed by all the colonies in the hope that such currency would fill a very evident need in commercial life. In most cases the result was disastrous, since depreciation followed and the numerous issues with their uncertain values hindered as much as they helped business. The Act of Parliament of 1751 which forbade the further issuance of paper money in New England, extended in

1764 to the other colonies, although bitterly resented and not wholly effective, put a check on this expedient.

The natural result of the scarcity of a medium of exchange was the attempt, as in Virginia and Maryland, to make the staple crop tobacco, or its paper equivalent, pass as money. Warehouse receipts were issued against tobacco placed in public warehouses, and these circulated readily as money. In New England barter was carried on by means of staples—beaver skins, corn, wheat, and other commodities. This exchange by "country pay," which was simply bartering one commodity for another, was the only practical method under the circumstances.

Piracy and Privateering.—Another hindrance to colonial commerce was piracy, which continued actively until far into the eighteenth century, and privateering, which in many cases could hardly be distinguished from it. Piracy flourished along the American coast, but especially in the West Indies and Caribbean Sea, where the booty was greater, the hand of the law weaker, and the chances of hiding among the numerous islands excellent. Here in the sixteenth century Drake and other English seamen scoured the Spanish Main for treasure ships bound for Europe while England was at peace with Spain. Their successors, however, the pirates of the seventeenth and eighteenth centuries, with their nests in the West Indies and on the mainland, did not limit their activities to the vessels of any single country. Notwithstanding their depredations, the visits of pirate ships to colonial ports were winked at by the authorities, for they disposed cheaply of their ill-gotten gains and bought heavily of ship stores and provisions. Even the stringent law of Parliament (1698) against piracy did not put an end to the practice. Colonial governors were continually accused, and apparently with reason, of harboring these enemies of commerce.

The Second Hundred Years' War between France and Great Britain, involving Spain at various times, provided ideal conditions in which piracy could thrive. The navies of the warring countries were too busily engaged with one another to police the high seas, and piracy was able easily to disguise itself as legitimate privateering. It is probable that the losses sustained by commerce from privateering during the century 1670-1770 were far larger than from piracy. Smuggling was undoubtedly stimulated by both piracy and privateering, and these practices in turn were encouraged by the ease of evading the customs officials, for as long as it was possible to dispose of captured goods readily and at a fair price there were always those willing to risk the gallows to obtain them. Although the sum total of losses from piracy was large, the pirate carried on his activities on a small scale. He has been much overdone in fiction. "When he is seen in authentic evidence he is found to have been for the most part a pitiful rogue. His gains were but small. A share of £200 was wealth to a mere sailor, and one of £1,000 beyond the dreams of avarice. He rarely fought a battleship if he could help it, and indeed nothing is more surprising than his readiness to surrender when the fate before him was the gallows." [1]

A dominating influence upon American commerce during the colonial era was the long period of warfare between France and England. On the one hand it unloosed navies, privateers, and pirates alike to prey upon sea-borne commerce; on the other it created temporary markets. Although the dangers of capture were great, the small sailing vessels of that time—their movements not reported by cables and wireless—were able to make their way quietly from one port to another and transact business even with the enemy.

Commercial Policy of Great Britain and the Colonies.—Reserving for a subsequent chapter a more detailed discussion of the colonial policy of Great Britain, we shall here merely point out again the fact that the European nations during the seventeenth and eighteenth centuries looked upon their colonies chiefly as subsidiary economic units; their function was to produce those raw materials not obtainable in the mother country and to refrain from competition with home industry. As a consequence, a long series of acts was passed by the British government aiming to promote the production of those raw materials not obtainable in England, and to discourage the production or manufacture of

---

[1] Hannay, David, Article "Pirates and Piracy," Enc. Brit., 11th ed.

commodities grown or produced in England.[1] Although this commercial legislation was not disastrous, owing (1) to the constant evasion of the law, (2) to the fact that the colonies were naturally producers of raw materials, and (3) to the discovery of other outlets when colonial foodstuffs were denied the English market, nevertheless these acts did color strongly and affect considerably the commercial life of the colonies.

In theory the American colonies were founded on crown land, and therefore acts of colonial legislatures were subject to review by the king and privy council. In actual practice, however, the legislation of the colonial government was practically uncontrolled, especially in the chartered and proprietary provinces. The interests of the various colonies in commercial matters differed and rivalries sprang up between them, resulting in discriminatory tariffs and divergent commercial laws. Believing thoroughly in the efficacy of governmental control of commerce, the colonial governments levied tariffs on both imports and exports, and resorted to embargoes, tonnage taxes, and other expedients. This legislation was designed not only to stimulate the production and transportation of certain commodities and discourage that of others, but also to raise needed revenue. Nevertheless, each colony, situated as it was on the seacoast, realized that its prosperity rested in no small degree upon active maritime commerce; and the regulations in general were designed to hinder commercial activity as little as possible.

Colonial Commerce to 1700.—The commercial interests of Europeans in the New World were centered first on the transportation of the booty of the Spanish conquests, and on the less romantic but lucrative products of the North Atlantic fisheries. Within thirty years after the landing of Columbus came the conquest of Mexico, followed by that of Peru. The Spaniards gathered in the accumulated wealth of centuries, and when this was exhausted put the natives at forced work in the silver mines and on the ranches and plantations. Under the exceedingly strict Navigation Acts bullion and surplus agricultural products were

---

[1] For a fuller discussion of British commercial policy, see chap. vii

shipped to the home country in exchange for the luxuries of civilized life. While the Spaniards were concentrating on the rich lands of South and Central America, the fishermen of many nations were prosecuting an active industry off the Newfoundland banks. For a hundred years before Jamestown was settled fishermen regularly worked the North Atlantic waters and dried their catch on the near-by shores.

It has been pointed out in an earlier chapter that after the failure of individual initiative, colonization and commerce in the original thirteen colonies were successfully undertaken by chartered companies. Economic motives were undoubtedly uppermost with the majority of stockholders in these plantation enterprises, and they obtained commercial monopolies and favorable tariff concessions from the crown, as did the later proprietors. Early expectation that vast and sudden riches were easily to be obtained soon proved unfounded, and the company agents, such as Captain John Smith, turned the energies of the settlers to the natural resources of the country. The first cargoes sent from the colonies to Europe were on the ships of the commercial companies and were composed chiefly of lumber and furs. The activities of the chartered companies extended over only a few years, when the charters were revoked or bought out; but they were responsible for the first settlement in and for the first successful commerce with the original colonies.

After the decline of the chartered companies and until the Navigation Act of 1660 commerce developed with comparative freedom. Within twenty-five years after the founding of Plymouth, New Englanders were carrying on trade with England, the West Indies, and the near-by Dutch at New Amsterdam. Timber, furs, fish, and grain, and even the ships in which they were carried, were exchanged for wine, sugar, iron, wool, and currency. The commerce of Virginia, which was the largest of any of the colonies during the seventeenth century, was chiefly with England or Holland, the latter often *via* New Netherlands. To these countries she shipped her staple, tobacco, receiving in return wines, iron, clothing, and manufactured goods of every description. A considerable amount of trade was carried on with New

England and an intermittent traffic with the West Indies. The principal commerce of New Netherlands was with the home country and consisted chiefly of furs. Maryland carried on a direct commerce with England, exporting tobacco, while the agricultural products and furs of the Delaware and Jersey settlers found an outlet through Philadelphia.

In 1700 the total population of the colonies was hardly 300,000 people: Pennsylvania and South Carolina had but recently been settled and Georgia was not yet founded. But even thus early the economic tendencies as they developed during the next hundred years were well marked. The relative importance of the fur trade had already declined. New Englanders had found a market for timber and ships in Europe, but their grain and fish were meeting the competition of English farmers and fishermen and were encountering tariff walls. Their problem of finding a market elsewhere for these commodities was being solved by developing the West Indies as an outlet for the sale of grain, salt fish, meat, and timber (particularly barrel staves), in exchange for coin and molasses. The molasses could be distilled into rum and the rum taken to Africa to be exchanged for slaves, the latter in turn brought back and sold to the West Indian and Virginian planters, thus developing a lucrative three-cornered traffic. Household manufacturing was carried on to a greater extent in New England than in the southern agricultural colonies.

The commerce of the more southern states, Maryland and Virginia, as in earlier and later years, was largely in tobacco, although furs and some grain were shipped. Most of this was picked up at the plantation wharves by the English or New England skipper in exchange for manufactured goods or slaves. Some direct trade was carried on between the southern mainland colonies and the West Indies, especially the Barbadoes, whence were obtained sugar, molasses, rum, ginger, and slaves. The isolated position of South Carolina and its nearness to the West Indies identified its interests with the English island plantations to which corn, cattle, pork, and forest products were sent.

In the middle colonies two economic centers were to be found; first, the region of the Hudson, lower Mohawk, Long Island,

and East Jersey, which had an outlet through New York; and second, West Jersey, Pennsylvania, and Delaware, the two principal shipping points of which were Burlington and Philadelphia on the Delaware River. The exports from New York were much the same as from New England, and the commerce in a similar manner was divided between Europe and the West Indies. In volume and value it was less than that of either New England or Virginia. Philadelphia was founded in 1681, and her commerce, while not large, was rapidly growing by 1700. The friendly relations which Penn's representatives maintained with the Indians insured a large supply of furs. Tobacco was raised in Pennsylvania extensively at first, but later gave way to grain.

Statistics showing the total trade of the colonies in 1700 are not available. The figures of the Board of Trade are for the commerce with England alone and show British imports from the mainland colonies to be £395,000 and the exports £344,000. To this should be added the trade with the West Indies and other nations, which would undoubtedly double these figures.

Commerce During the Eighteenth Century.—The commerce of the eighteenth century followed to a great extent the routes of the previous years and was concerned with similar commodities. The exportation of furs declined, while the commerce in grains and lumber increased. The southern provinces continued to devote their chief energies to the growth of staples, producing only the crudest manufactured articles for plantation consumption, since they depended upon Europe for the great part of their manufactured goods and upon others to do the carrying trade. The production of cereals increased rapidly in the middle states, where Philadelphia became the leading port for grain exportation, with New York ranking second. New England continued to draw great wealth from her fisheries, the construction of vessels, the manufacture of rum, the carrying trade of a good portion of the colonial products, and especially the trade with the West Indies and Africa.

As the eighteenth century progressed, the West Indian trade became increasingly important. Not only the British West Indies (the most important of which were Jamaica and the Barbadoes), but also the French and Dutch West Indies, could use the fish, cereals, and lumber of the mainland colonies. In the French West Indies, where up to 1717 the French West India Company had an official monopoly, illegal trading was profitable, and after 1717 competition between the French and English islands became keen. The New Englanders came to depend upon the West Indian trade for (1) the currency with which they bought English manufactured goods, (2) the molasses and sugar with which they manufactured rum for the African trade, and (3) the employment of men and capital in the carrying trade. Their economic life was so tied up with this trade that the Molasses Act of 1733, which attempted to prevent by high duties the importation of rum, molasses, and sugar from the non-British West Indies, was a dead letter. Even during the wars with France, which were waged intermittently during the century, commerce went on much as in time of peace. The supplying of food and building materials to the enemy in time of war in exchange for the tropical commodities apparently did not trouble the conscience of the colonial merchant. Much of this trade was carried on through the neutral Spanish and Dutch ports and had reached such proportions in the French and Indian War that Amherst, the royal governor, actually placed an embargo in 1762 on the trade of New England and the middle colonies.

A special phase of the West Indian commerce, the African slave trade, grew enormously during this period. The causes were: (1) the demand for cheap labor on the plantations of the islands and mainland, and for household servants in the northern colonies; and (2) the large profits which accrued from the trade to English interests and colonial shipowners. Not only were the shipping interests in New England concerned in the slave trade, but also the rum manufacturers who supplied the chief commodity used in the purchase of the negroes. Up until the Revolution almost all of the negroes were taken to the West Indies, and from there a certain portion later brought to the mainland. From 25,000 to 30,000 slaves a year were transported to the West Indies and of these perhaps 10,000 a year were taken to the continental colonies, the negro population of which in 1760 is estimated by Professor Channing at 386,000. Although the increase in negroes was so rapid that certain colonial legislatures took alarm and attempted to place prohibitive taxes on their importation, the profits were so great that the commercial interests of Great Britain succeeded in having the acts vetoed by the home government.

The eighteenth century as a whole showed a continued and healthy increase in commerce commensurate with the growth in population. The annual average of commerce between England and the mainland colonies follows:

Decades	Exports to England			Imports from England		
	£	s.	d.	£	s.	d.
1700–1710	265,783	0	10	267,205	3	4
1710–1720	392,653	17	1	365,645	7	11
1720–1730	518,830	16	6	471,342	12	10
1730–1740	670,128	16	0	660,136	11	1
1740–1750	708,943	9	6	812,647	13	0
1750–1760	802,691	6	10	1,577,419	16	2
1760–1770	1,044,591	17	0	1,763,409	10	3
1770–1780	743,560	10	10	1,331,206	1	5

The period immediately after the close of the French and Indian Wars was commercially prosperous and hopeful. The fisheries, now freed from the danger of privateers, throve. Whaling was aided by the removal in 1764 of most of the duties on whalebone imported by British subjects, and in 1765 lumber was added to the list of articles upon which bounties were paid. With the exception of the ill-fated Molasses Act of 1733, the British government had made, up to this time, comparatively little effort to enforce the Navigation Acts.

With the Sugar Act of 1764, the policy of "salutary neglect" was reversed, and in its place came an attempt to tax and more strictly regulate colonial commerce. This attempt, although bitterly opposed on political and economic grounds, did not immediately effect much injury on colonial commerce. The continued refusal of either side to back down on these principles led, however, in the 'seventies to boycotts on British goods, the closing of the port of Boston, commercial depression, and eventually to war.

Some idea of the variety and extent of colonial exports can be gained from the rough estimates of exports during the decade preceding the Revolution, furnished by the author of *American Husbandry*. In these tables the importance of fisheries in New England and of agricultural products in the middle colonies is clearly demonstrated.

AVERAGE ANNUAL EXPORTS FROM NEW ENGLAND, NEW YORK, AND PENNSYLVANIA, 1763-1766

NEW ENGLAND [1]

Codfish, dried, 10,000 tons, at £10	£100,000
Whale and cod oil, 8,500 tons, at £15	127,500
Whalebone, 28 tons, at £300	8,400
Pickled mackerel and shads, 15,000 barrels, at 20s.	15,000
Masts, boards, staves, shingles, etc.	75,000
Ships about 70 sail, at £700	49,000
Turpentine, tar, and pitch, 1,500 barrels, at 8s.	600
Horses and livestock	37,000
Potash, 14,000 barrels, at 50s.	35,000
Pickled beef and pork, 19,000 barrels, at 30s.	28,500
Beeswax, and sundries	9,000
Total	£485,000

NEW YORK [2]

Flour and biscuit, 250,000 barrels, at 20s.	£250,000
Wheat, 70,000 qrs.	70,000
Beans, peas, oats, Indian corn, and other grains	40,000
Salt beef, pork, hams, bacon, and venison	18,000
Beeswax, 30,000 lbs., at 1s.	1,500
Tongues, butter, and cheese	8,000
Flax seed, 7,000 hhds., at 40s.	14,000
Horses and livestock	17,000
Product of cultivated lands	418,500
Timber planks, masts, boards, staves, and shingles	25,000
Potash, 7,000 hhds.	14,000
Ships built for sale, 20, at £700	14,000
Copper ore, and iron in bars and pigs	20,000
Total	£526,000

[1] *American Husbandry*, vol. i, p. 59.
[2] *Ibid.*, p. 124. Total incorrect as in source.

PENNSYLVANIA [1]

Biscuit flour, 350,000 barrels, at 20s.	£350,000
Wheat, 100,000 qrs., at 20s.	100,000
Beans, peas, oats, Indian corn, and other grain	12,000
Salt beef, pork, hams, bacon, and venison	45,000
Beeswax, 20,000 lbs., at 1s.	1,000
Tongues, butter, and cheese	10,000
Deer, and sundry other sorts of skins	50,000
Livestock and horses	20,000
Flax seed, 15,000 hhds., at 40s.	30,000
Timber planks, masts, boards, staves, and shingles	35,000
Ships built for sale, 25, at £700	17,500
Copper ore, and iron in pigs and bars	35,000
Total	£705,500

Internal and Coastwise Commerce.—Internal and intercolonial trade was more fully developed in the northern and middle colonies than in the southern. The tobacco of the southern plantations could usually be shipped directly to England from the plantation wharves, and the exchange for English commodities could be carried on directly with a minimum amount of preliminary collecting. Likewise, certain commodities, such as rum and slaves, could be purchased directly from the West Indies. The situation in the northern colonies, on the other hand, favored the development of this trade. Furs, the first commodity shipped in any large quantity to England, had to be collected from trading posts established at various points along the coast and inland. Agricultural products, though they were not welcomed in England, found a ready market elsewhere, but, unlike the southern plantations, the New England farms were so situated that their produce had to be conveyed to shipping towns. They were bulky and hard to handle, but these discouraging features were counterbalanced by the cheapness with which ships could be built, the skill of New England mariners, and the necessity of selling the surplus products in order to obtain money to purchase manufactured goods from England. These influences contributed from the start to the development of coastwise commerce in the northern

---
[1] *American Husbandry,* p. 181.

colonies, and to the growth of such shipping centers as Salem, Boston, Providence, New York, and Philadelphia. The New England settlements were hardly founded before the Dutch at New Amsterdam were exchanging with the English gunpowder and European manufactures for furs, fish, tobacco, and grain. The shipping of products to Europe *via* New Amsterdam, however, came to an end with the fall of the Dutch, and the coastwise traffic fell largely into the hands of the New Englanders. While the trade of New England with the south never attained the large proportions of West Indian or European trade, there was still a growing tendency for the southerners to purchase supplies produced in New England and European commodities imported by Yankee skippers. The trade of such colonies as New Hampshire, Connecticut, New Jersey, and Delaware was, in fact, largely coastwise. Their products were transported to Boston, New York, or Philadelphia, and in return foreign merchandise was secured from these cities. Consequently, most of the overseas commerce was carried on by the ships of Massachusetts, New York, and Pennsylvania, or by British ships—although the merchantmen of these colonies also engaged largely in coastwise commerce. Intercolonial commerce was augmented to some extent in the winter, when in the dull fishing season the fishing schooners were loaded with the products of New England homes, which were peddled along the southern coasts and plantation wharves in exchange for tobacco, pitch, tar, and pork. Coastwise commerce was also increased during the war periods, when enemy privateers drove the less venturesome close to shore. From the existing records it seems evident that the coastwise commerce in its entirety surpassed in volume the trade with Great Britain, with southern Europe, and even with the West Indies.

Summary.—The story of American colonial shipping may be divided into the following periods: first, the sixteenth century, when European fishermen frequented the Newfoundland banks and Spanish galleons transported the riches of Mexico and South America to the mother country; second, the early seventeenth century, when chartered companies with monopolistic privileges settled Virginia, New England, and New Netherlands and sought trade for the benefit of investing stockholders; and third, the greater part of the seventeenth and eighteenth centuries, when British, European, and colonial ships developed an overseas and coastwise traffic of an exceedingly high per capita ratio. Hindered by piracy, privateering, long periods of warfare, European tariffs, and navigation acts, nevertheless colonial products found a market and commerce increased. Furthermore, much of the carrying trade was in the hands of colonials, who participated heavily in the European, African, and West Indian trade, and almost exclusively in the coastwise traffic. Built up under many adverse circumstances, it was a monument to the ingenuity and energy of the colonial settler. Since colonial commerce was fostered to some extent by the policy of "salutary neglect" which failed to enforce the Navigation Acts, it is not to be wondered at that the reversal of this policy, which led to tampering with a structure laboriously built up by generations of sturdy pioneers, should have been the chief cause of the American Revolution.

NOTES FOR FURTHER REFERENCE

For brief running accounts of colonial commerce, consult W. C. Webster, *A General History of Commerce* (1903), and Clive Day, *History of Commerce* (rev. ed., 1922). The standard work and the one most satisfactory for student consultation is the coöperative work of Emory R. Johnson, T. W. Van Metre, G. G. Huebner, and D. S. Hanchitt, *History of Domestic and Foreign Commerce of the United States* (2 vols., 1915), based on monographs by other collaborators and published in the Contributions to American Economic History by the Department of Economics and Sociology of the Carnegie Institution of Washington. Most of that part of Volume I devoted to colonial commerce is the work of Professor Johnson. A useful bibliography will be found on pages 112 to 117 of Vol. I.

English colonial policy can be best studied in the works of G. L. Beer, *Origins of British Colonial Policy, 1578-1660* (1908); *British Colonial Policy, 1754-1765* (1907), and *The Old Colonial System* (1913); these are among the most notable contributions made by an American scholar in economic history. A handy introduction to these volumes is the same author's *Commercial Policy of England toward the American Colonies* (1893), No. II, Vol. XI, in Columbia University Studies in Economics, History and Public Law. See also H. E. Egerton, *British Colonial Policy* (2d ed., 1909); and for a brief but most enlightening survey, C. M. Andrews, *The Colonial Period* (1912), in the Home University Library.

On mercantilism consult Gustav Schmoller, *The Mercantile System and Its*

Historical Significance (1896). Edward Channing in his *History of the United States* (Vol. II, 1908) devotes Chap. XVII to Colonial Industry and Commerce. The somewhat desultory but valuable *Economic and Social History of New England* by William B. Weeden (2 vols., 1890) contains much information.

On colonial finances consult the popular account in J. R. H. Moore's *An Industrial History of the American People* (1913) or a more technical treatment in D. R. Dewey, *Financial History of the United States* (rev. ed., 1922). On the medium of exchange in Virginia, see P. A. Bruce, *Economic History of Virginia in the Seventeenth Century* (2 vols., 1907), Vol. II, Chap. XIX.

Treatments of the slave trade are to be found in W. W. Claridge, *History of the Gold Coast* (1915); U. B. Phillips, *American Negro Slavery* (1921); and W. E. B. DuBois, *The Suppression of the American Slave Trade* (1896). On piracy see the Encyclopedia Britannica and J. F. Jameson (ed.), *Privateering and Piracy in the Colonial Period* (1924), a source book.

Various American historians have devoted special chapters to colonial commerce; for example, Edward Channing, *History of the United States*, Vol. II (1908), Chap. XVII; Vol. III (1912), Chap. XIII; C. M. Andrews, *Colonial Self-Government* (1904), in the American Nation Series, Chap. XIX; and E. B. Greene, *Provincial America* (1904), in the same series, Chap. XVII. See also the latter's *Foundations of American Nationality* (1922) for pertinent sections.

Extracts from sources dealing with colonial commerce are included in the *Readings* of Bogart and Thompson and those of G. S. Callender.

SELECTED READINGS

JOHNSON, E. R., et al., *History of Domestic and Foreign Commerce of the United States*, Chaps. I-XI.

CHANNING, EDWARD, *History of the United States*, Vol. II, Chap. XVII; Vol. III, Chap. XIII.

WEEDEN, W., *Economic History of New England*, Vol. I, Chaps. II, IV, V, VII, IX; Vol. II, Chaps. XII-XIX.

DEWEY, D. R., *Financial History of the United States*, Chap. I.

BOGART, E. L., and THOMPSON, C. M., *Readings in the Economic History of the United States*, pp. 69-80; 96-105.

CHAPTER VI

THE WESTWARD MOVEMENT BEFORE THE REVOLUTION

Significance of the Westward Movement.—Conflicting influences have contributed to the formation of American character and ideals. On the one hand, proximity to the ocean and intercourse with Europe retarded the development of a distinctly American civilization, while on the other the westward movement and frontier life have continually worked to efface the European influence and to stimulate the growth of a new nation. The Appalachian barrier and the lack of transportation facilities, until the opening of the nineteenth century, kept the great majority of the people east of the mountains. As late as 1830 the center of population was still on the Atlantic coast. Nevertheless, even before this date, as afterward, the most important factor in the life and history of the people was the continued advance to the west. The movement inland commenced almost immediately after the first settlements were made and lasted until about 1890, when the frontier lines moving west and east joined. "Up to our own day," says Professor Turner, "American history has been in a large degree the history of the colonization of the Great West. The existence of an area of free land, its continuous recession, and the advance of American settlement westward, explain American development."[1]

Leaving until a later time a more detailed discussion of the effects which the westward movement has had, it is necessary here simply to point out by way of introduction some of the most important tendencies resulting therefrom. As the frontier line advanced, environment mastered the settler and as he perforce adopted a different manner of life, the influence of Europe became less. As settlers of various races were thrown together under a

---

[1] Turner, F. J., *The Frontier in American History*, p. 1.

similar environment, there gradually evolved a composite type of American. The never-ending struggle against the forces of nature, against hostile Indians and wild beasts, developed a self-reliant, keen, aggressive, individual type, characterized by antipathy to control and to any attempt to abridge his independence—what Burke called "a fierce spirit of liberty." The comparative equality of wealth in a new community, where each man stood upon his own feet, made it easy to forget the artificial customs of an old world and developed a distinctly democratic outlook. The west has been the most democratic part of America, and our history has been full of the struggles between the democratic frontiersman and the more conservative easterner. The elements which have gone to make up the life of the frontiersman have produced intellectually a type that is restless, energetic, masterful, original, and practical, and at the same time buoyant and optimistic.

On our political life the growth of the west with its different interests has meant the emergence of sectionalism and the demands of the westerner for legislation promoting his own interests—internal improvements, free land, and an inflated currency. At the same time, however, his need for many things which only the national government could give him has been a most potent factor in the growth of nationalism. In the technique of government the drive of the west has been toward such democratic innovations as the direct election of the United States Senators, woman suffrage, the initiative, referendum, and recall. Economically, the growth of the west has brought industrial independence for the nation along with sectional specialization.

Stages of Westward Advance.—Even during colonial times, rather clearly marked stages of westward advance were in evidence. The first stage was usually marked by the activities of the hunter, trader, or missionary. Traders and trappers like John Smith and Daniel Boone, and missionaries like Father Marquette and Marcus Whitman, are typical of the pathfinders who blazed the way. The trail of the hunter followed that of the buffalo and the Indian, and eventually became a highway of civilization. Trading posts erected at convenient points on the western trails

WESTWARD MOVEMENT BEFORE 1775

grew into such cities as Albany, Pittsburgh, Chicago, and St. Louis. Following the trapper and the trader came the rancher, who occupied the land to exploit the grasses. Of all farm products, livestock was in those days the easiest moved, and from the "cow pens" of seventeenth-century Virginia and the Carolinas to the great modern ranches of the western prairies, the frontier ranchman has marked the farthest westward advance.

Close on the heels of the rancher came the farmer, the first wave dispersing in sparsely settled communities and wastefully exploiting the soil. This preliminary farmer stage was in turn succeeded by a more or less intensive farming in denser settlements. Where conditions were favorable, the farming stage has given way to the final, that of city life with its manufacturing and commercial activities.

These are the regular stages through which most of our country has passed, and hunter, rancher, farmer, and capitalist have all played their part in its development. For almost three hundred years this same process has been repeated and the same drama reënacted over and over again, coloring our history and determining our civilization. The recurring stages have been vividly portrayed by Professor Turner: "The Atlantic frontier was compounded of fisherman, fur trader, miner, cattle-raiser, and farmer. Excepting the fisherman, each type of industry was on the march toward the West, impelled by an irresistible attraction. Each passed in successive waves across the continent. Stand at Cumberland Gap and watch the procession of civilization, marching single file—the buffalo following the trail to the salt springs, the Indian, the fur trader and hunter, the cattle-raiser, the pioneer farmer—and the frontier has passed by. Stand at South Pass in the Rockies a century later and see the same procession with wider intervals between. The unequal rate of advance compels us to distinguish the frontier into the trader's frontier, the rancher's frontier, or the miner's frontier, and the farmer's frontier. When the mines and cow pens were still near the fall line, the traders' pack trains were tinkling across the Alleghanies, and the French on the Great Lakes were fortifying their posts, alarmed by the British trader's birch canoe.

When the trappers scaled the Rockies, the farmer was still near the mouth of the Missouri."[1]

Routes of Westward Migration.—Navigable streams marked the first routes of westward migration. The hope of finding a passage through the newly discovered lands to the riches of Cathay led the earliest explorers to probe with their tiny boats the innumerable rivers and estuaries. Later to the fur trader the rivers and lakes offered the readiest access into the interior, and decades, even centuries, before the settler followed him he had clearly pointed out the routes of travel. Nine years before the Dutch settled at the mouth of the Hudson they had established a trading post at Albany; by 1627 merchants of Jamestown were trading with Indians of the upper Potomac and Susquehanna. The French, who had stumbled upon the best route into the interior and had been forced by circumstances to the development of the fur trade, had discovered long before the opening of the eighteenth century the best routes between the basin of the Great Lakes and the Mississippi, and had used the key portages which the Indians made known to them. A principal portage was the one between the upper Ottawa and Lake Nipissing, which allowed the trader to reach Lake Huron *via* the French River and the Georgian Bay. Connections between Lake Michigan and the Mississippi were obtained by the portage between the Fox River and the Wisconsin River, that between the Calumet and the Des Plaines, and that from the St. Joseph to the Kankakee. The last two carrying places named led into the Illinois and thence to the Mississippi. The route from Lake Erie to the "Father of Waters" led by portages from the Maumee to the Wabash, from the Maumee to branches of the Great Miami, from the Sandusky to the Scioto, from the Cuyahoga to the Tuscarawas, from Lake Erie to Chautauqua Lake, and thence to the Allegheny. A strategic portage connecting New York with the Great Lakes was that which led from the Mohawk Valley to Wood Creek, a tributary of the Oswego, and hence to Lake Ontario. The usual route from the Hudson to Canada was

---

[1] Turner, F. J., *The Frontier in American History*, p. 12.

WESTWARD MOVEMENT BEFORE 1775

by portages leading from that river to Lake George, from Lake George to Lake Champlain, and then by the Richelieu to the St. Lawrence. Farther south, portages connecting the east with the Ohio Valley were made between the Susquehanna and Allegheny rivers near Kittanning, between the Juniata and the Allegheny, and from the Potomac to the Monongahela along Wills Creek.

There were four leading routes through the Appalachian barrier. The most northerly and the best, that by way of the Hudson and Mohawk to the Lakes, was closed to early settlers by the Iroquois. To the south was a second route leading from the headwaters of the Mohawk to the upper Allegheny. The third route led across southern Pennsylvania to the Monongahela and thence to the Ohio, a line later followed by the Cumberland road. The fourth and most important to the pre-Revolutionary settlers was southward down the great Appalachian Valley and out through the Cumberland Gap or the Tennessee Valley. A possible route around the south of the Appalachians was closed by the Cherokees. While these routes were known to fur traders long before the Revolution, it was not until the latter part of the eighteenth century that settlers in any numbers followed them. Before that time the need of keeping in close touch with the European market had kept the white man near the rivers, and hostile French and Indians as well as the natural mountain barrier had all contributed to limit settlement east of the Appalachians.

The First Frontier.—The first permanent English settlement, that of Jamestown in 1607, was largely a business venture, engineered by the Virginia Company. A few years only were necessary, however, to demonstrate the impracticability of the original plans, and it was not long before the land was broken up among individual owners, who, after about 1617, devoted themselves largely to raising the staple crop of tobacco. As the market for the crop was beyond the seas, it was necessary to remain close to the river. The early settlers took up land along the James, and from there spread north and south to other rivers, staking off their plantations on the bank and shipping their tobacco to Europe from their own wharves. As the land along the larger rivers was preëmpted, the region inland was tapped and the tobacco floated down in canoes and on rafts to the bigger streams. This tendency to remain near the rivers retarded the building of roads inland and the growth of urban life. Early Virginia was not unlike a federation of peninsulas.

The Virginian lowlands cut into by many little rivers were particularly adapted to a tobacco-plantation economy. But tobacco raising under primitive conditions was wasteful and the movement inland to fresh lands was rapid. New arrivals ascended the rivers to take up new lands as close to the water as possible. This type of westward advance left the frontiers open to Indian attack, and as the years went on the government made repeated efforts to control the frontier advance and to group the frontiersmen into towns at the first falls of the river in the vicinity of Richmond, Petersburg, and other places which might serve as outposts of defense against Indians. The frontier of Virginia pushed south as well as west, and by 1700 the region of the Carolinas north of Albemarle Sound and east of the Chowan River had been occupied.

Whereas the westward migration of the Virginia colonists was more or less haphazard, in New England the colonial government attempted very definitely to superintend the founding of towns and the prescribing of their limits. Upon the request of a group of prospective settlers the colonial governments of New England would grant a tract usually thirty-six miles square. "The settlement of a town normally began," says Professor Osgood, "with the laying out of a village plot and the assignment of home lots. This to an extent determined the location of highways, of the village common, and of some of the outlying fields. On or near the common the church was built, and in not a few cases the site that was chosen for the building went far toward determining the entire lay-out of the town. The idea of a home lot was a plot of ground for a dwelling house and outbuildings, for a dooryard and garden, and usually also an enclosure for feeding cattle and raising corn."[1] Common fields were usual and provisions were ordinarily made in these pioneer towns for reserving lands for the support of a minister and schools. The attempt on the part of the government to control this frontier movement is seen in an order of the General Court of 1636, which directed that none go to the new plantations without the permission of a majority of the magistrates, an order which was probably evaded. On the other hand, the danger from French and Indian attack led the General Court in 1694, after enumerating certain "Frontier Towns," to forbid the inhabitants to desert these outposts on pain of imprisonment and confiscation of their land. The closeness of the New England frontier to the older settlements and the constant danger of enemy attacks, made the position and status of the frontier towns a matter of the most earnest concern.

With few exceptions the expansion of Massachusetts Bay proceeded normally along the lines pointed out. In the case of the migrations to Rhode Island and the Connecticut Valley, new communities of sufficient strength were formed to push out frontier posts of their own. The Puritanism of Roger Williams was a little too radical for those in control at Massachusetts Bay, and, driven out in 1636, he, with his followers, founded Providence. In 1635 and 1636 the migration from Massachusetts to the Connecticut Valley commenced. A congregation from Dorchester settled Windsor in 1635. The next year the Cambridge congregation founded Hartford, the Watertown group settled Wethersfield, and those from Roxbury founded Springfield.

New Amsterdam began in 1612 as a trading post, and in 1614 a similar post was established on an island near the present site of Albany. Schenectady at the rapids of the Mohawk was begun in 1661 and in the succeeding years the intervening land along the two rivers was taken up under the patroon system, and allotted by the patroons or the New Amsterdam officials to tenants. New Jersey was first settled by the Dutch, who emigrated from New York, and by the Swedes, who settled in 1638 on the Delaware. The Dutch conquered the Swedes in 1655 and

---

[1] Osgood, H. L., *The American Colonies in the Seventeenth Century*, vol. i, p. 438.

were in turn (1664) brought under submission by the English, who turned the region over to two proprietors, Carteret and Berkeley. The latter sold his share in 1674 to two Quakers, and eight years later Carteret disposed of his to William Penn and eleven others. Consequently New Jersey was settled by Dutch from New York, Swedes, Quakers, and westward-moving emigrants from New England, all of whom were able to acquire land on advantageous terms. William Penn in 1681 was granted the territory west of the Delaware and between New York and Maryland. Here in the same year Philadelphia was commenced Penn founded the colony primarily as a home for oppressed Quakers, then opened it to the persecuted of all lands, and before the close of the century Quakers from England and Wales, Mennonites from the Rhine Valley, with a sprinkling of Scotch, Swedes, Irish, and French had established a thriving commonwealth.

Maryland to the south had been founded in 1634 by Lord Baltimore as a refuge for Catholics. The first settlement, St. Mary's, was begun about nine miles up the St. George, a tributary of the Potomac Religious freedom brought immigrants of many sects, as did the liberal system of granting land. As the economic life of Maryland strongly resembled that of Virginia in its dependence upon agriculture and its specialization upon a single staple, and as the geographical conditions were similar, the westward advance was like that of her southern neighbor. The Carolinas in 1700 boasted two patches of settlement, the northern around Albemarle Sound, in reality an extension of Virginia, and the southern extending from the Santee River south, with Charleston as its center. The southern colony, founded in 1670 under the patronage of eight proprietors, among whom were some of the most distinguished men in England, including Anthony Ashley Cooper, later Earl of Shaftesbury, was commercially bound to the West Indies, whence most of the inhabitants came. Although settlement was hindered at first by the attempt to impose an artificial system of government—the Grand Model of John Locke—after that plan had been discarded the movement up the river was rapid. Excellent additions to the population were furnished by the Scotch-Irish and by five hundred Huguenots who left France after the revocation of the Edict of Nantes and took up 50,000 acres along the Santee.

Population and Settled Area in 1700.—By 1700 the settled region had advanced approximately to the fall line of the rivers. In New England the seacoast had been taken up almost continuously from the Pemaquid region of Maine to the New York border line, the settlements in Maine running back not more than ten miles and in New Hampshire from fifteen to thirty miles from the coast. Massachusetts had been settled perhaps fifty miles inland. According to an act of the General Court, 1694-95, the "Frontier Towns" were Wells, York, and Kittery in what is now Maine, and Amesbury, Haverhill, Dunstable, Chelmsford, Groton, Lancaster, Marlborough, and Deerfield, while in March, 1699, Brookfield, Mendon, and Woodstock, with a number of others, were added. Rhode Island, with the exception of some tracts in the southern part, was largely occupied by 1700. The General Court of Connecticut in 1704 designated as the frontier towns of that state Symsbury, Waterbury, Danbury, Colchester, Windham, Mansfield, and Plainfield. The entire population of New England amounted at the close of the century to about 80,000 white inhabitants, of which 5,000 were in New Hampshire, 5,000 in Rhode Island, 17,000 in Connecticut, and 55,000 in Massachusetts, including the Plymouth and Maine settlements.

Habitation in New York by this time extended along the Hudson a few miles back of the river on each side to a short distance above Albany, with some outposts on the Mohawk. Most thickly settled were the regions on Manhattan Island, around Kingston, and near Albany. The north and the south shores of Long Island had both been largely preëmpted. The population of New York was probably about 18,000. Settlement in New Jersey covered a broad patch running across the north central portion of the state from New York Bay to the Delaware. In East New Jersey there was a strip of settlement along the Delaware from just below Easton to Cape May and up the coast to Barnegat, extending a little way inland. The population of West New Jersey, more than half of which was Dutch and closely allied to New York in economic interest, was about 10,000, while in East New Jersey there may have been from 5,000 to 8,000 inhabitants. In Pennsylvania the inhabited regions in 1700 extended inland from the Delaware River from forty to fifty miles, and in Delaware, which was then part of Pennsylvania, from ten to fifteen miles Possibly 15,000 were located to the west of the Delaware River and Bay, a group predominatingly English, but including many nationalities.

Of the two provinces along the Chesapeake, Maryland contained about 30,000, and Virginia 60,000. On the eastern bank the peninsula was occupied from Cape Charles northward for fifty miles. The rest of the eastern shore was unoccupied up to the region of Kent Island, where settlements began again, extending to the Pennsylvania boundary line. On the western shore a belt of settlements twenty-five miles broad extended inland from the mouth of the Potomac to the head of the bay. In the Virginia section of the Chesapeake Bay region settlements in some cases ran up the river seventy-five miles In what later (1729) came to be North Carolina, about 3,000 people were to be found north of Albemarle Sound, and south of the Santee River another group of perhaps 5,500 whites were settled

Indians and the Early Westward Movement.—The present-day insignificance of the Indian in numbers and power makes it easy to underestimate his influence in American history. In the first place, he prepared the way in a sense for the European settler. Indian trails have generally marked the routes inland taken by the white men from the earliest days to the era of railroads. The clearings made by the Indian for his crude farming were among the first occupied by the newcomers, who not only used his land, but adopted his methods of agriculture. Secondly, the Indians spurred on the white advance by the temptations which they held out to the fur trader. The latter, returning with tales of the rich western lands, whetted continually the desires of the land-hungry settlers. On the other hand, however, every frontier had an Indian barrier to dispose of, and the problem of removing the red man, by purchase of land or forcible ejection through warfare, was a continual difficulty for almost three hundred years. The Indian held possession of strategic passes and gaps in the mountains, and was able to hold up the immigration into the west. Again should be noted the effect of the Indian barrier upon the life and character of the people. Frontier life meant danger from the Indians, which developed courage and self-reliance. For outside help the frontiersmen must look to a strong government, a decided influence toward nationalism. The Indian danger also developed more community life and less scattered settlement than would have come about otherwise.

By the end of the seventeenth century the lands back to the fall line had been pretty well cleared of Indians, who were driven into the Piedmont region.[1] But the first foothold had not been secured without numerous wars. Virginia's troubles had commenced with the massacre of 1622, in which one-fourth of the whites were killed, followed by the extermination of the Indians on the lower James and York. The encroachment of settlements on the upper James and Rappahannock brought about Opechancanough's War in 1644, and again in 1675 and 1676 the colony experienced considerable difficulty with the Susquehannas. Maryland did not have so much trouble as did Virginia, but between 1639 and 1644 expeditions were sent yearly against the Indians. William Penn's policy of handling the Indians by the golden rule kept that colony remarkably free from molestation. New Netherlands escaped serious disturbance until 1640-46, when by a ruthless war the Dutch managed to tame the Long Island and Hudson River Indians. The Iroquois of the Mohawk Valley were not so easily disposed of, and were able to stem the advancing tide of white men until after the Revolution. Hooker and his congregation had scarcely established themselves in Connecticut before they were forced to defend themselves from the Pequots, and forty years later, in 1675, all New England joined in one of the bloodiest Indian wars in our history. King Philip's War broke the power of the New England Indians, but their

---

[1] The term Piedmont, literally foothills, in American physiography designates that part of the Atlantic coastal plain lying between the low coastal plain proper and the Appalachian highlands. See chap. i.

descendants, driven northward, continued in later years to aid the French in harassing the frontier settlements.

The tendency toward nationalism which the Indian danger fostered was seen most clearly in the colonial period. The need for mutual military defense had in 1643 brought Plymouth, Massachusetts Bay, Connecticut, and New Haven to form the New England Confederation, a real forerunner of the famous Albany Congress of 1754, where an attempt was made to bring together the colonies for united action, chiefly with reference to Indian problems.

The Advance into the Piedmont.—Although the Indian barrier in the coast plains had been largely removed by 1700, there were still outbursts, as in the Tuscarora War (1712-13) in North Carolina and in the Yamassee War of 1715 in South Carolina. Speaking broadly, however, by the end of the century the tidewater settlements to the fall line were fairly well secured. The period from the close of the century until the conclusion of the French and Indian War in 1763 saw the colonists push into the Piedmont region and take up the lands between the fall line and the Alleghanies. New England, following her policy of granting land to approved bodies of men, began in 1713 the plan of locating towns in advance of settlements to protect the boundary claims and ward off Indian attacks. Settlers pushing up the Housatonic Valley into the Berkshires founded Litchfield in 1720, Sheffield in 1725, Great Barrington in 1730, and Williamstown by the middle of the century. In 1735 four contiguous towns were laid out to connect the Housatonic and Connecticut River settlements. During these years most of the land between the Connecticut and the seacoast was also taken up and Connecticut by 1737 had disposed of her unlocated lands. In New Hampshire settlement proceeded up the Merrimac and some distance up the Connecticut. The taking up of lands was also encouraged by the grants of Governor Wentworth of a hundred and twenty towns west of the Connecticut in what was later Vermont. Although the New England expansion of this period was toward the north, it had all of the essential characteristics of the westward movement.

A large and influential element in this frontier advance was furnished by the Scotch-Irish, a race well adapted to the rigors of pioneer life and *"par excellence* the Indian fighters." The Act of 1699 prohibiting the exportation of Irish wool from Ulster, the enforced payment of tithes to the Anglican Church, and the fact that between 1714 and 1718 many of the leases granted to the original settlers expired, all contributed in the early years of the eighteenth century to bring about a great migration of Scotch-Irish to America. So rapid was the influx into New England that the authorities shipped them to the frontier, where they settled in Worcester, founded Pelham, Warren, and Blandford and, following the Connecticut Valley, settled in Windsor, Orange, and Caledonia counties in Vermont, and Grafton County in New Hampshire.

Hemmed in by the Catskills to the west and with the Mohawk pass into the interior blocked by the Iroquois, New York during the period showed very little expansive vitality. Notwithstanding the richness of the soil, the cultivation of the narrow ribbon of land along the two rivers proceeded but slowly, due largely to the fact that the Dutch system of huge manorial grants was continued under British rule. With millions of acres of the choicest lands under the control of a handful of men who wanted to settle tenant farmers upon their lands, it was little wonder that the tide of immigration moved elsewhere. New York caught some of the first wave of German-Swiss immigration, which, commencing in 1683, continued throughout the first half of the next century. This inflow came mostly from the Palatinate, Würtemberg, Baden, and Switzerland, and was caused by religious persecution, political discontent, and economic disorganization following the continental wars. Governor Hunter in 1710 tried to settle 3,000 Germans on the Hudson near Saugerties to produce naval stores, but discontent with their lot led many of them to move on to the Mohawk Valley, where they settled in the country between Fort Hunter and Palatine Bridge. Scotch-Irish moving on from New England mingled with the Dutch in the Mohawk region and entered the Cherry Valley in 1738.

Pennsylvania's reputation as a home for persecuted sects under the magnanimous rule of Penn brought to her shores as permanent settlers, between 1700 and the Revolution, at least 100,000 Germans from the Palatinate and surrounding regions, the ancestors of the present "Pennsylvania Dutch," while 100,000 more were scattered along the frontiers of the other colonies from the head of the Mohawk to Georgia. The Pennsylvania frontier of this period was also the center of the great Scotch-Irish migration which brought to us between 1730 and 1770 close to half a million. Probably one-third of Pennsylvania's population at the time of the Revolution was composed of Germans from the Rhineland and another third of Scotch-Irish. The cost of land in Pennsylvania in 1719 was ten pounds per hundred acres and two shillings quitrent; the price was raised in 1732 to fifteen pounds and a quitrent of a halfpenny an acre. But with the rapid influx of immigrants the management of the lands fell into confusion and a large proportion was occupied by squatters without title. The Germans spread out in eastern Pennsylvania and the Scotch-Irish, coming a little later, planted their outposts in the Cumberland, Juniata, and Susquehanna valleys.

From Maryland to Georgia the story of the occupation of the Piedmont is much the same. As population increased and the rich lowlands were exhausted, more and more land was taken up until the fall line was reached. Then there moved up into the Piedmont a stream of newcomers, mostly of the poorer classes, to claim lands under head rights, or settlers brought in by wealthy speculators to satisfy the requirements for obtaining their vast estates. Efforts were made by each of these colonies, for the purpose of protection, to lure men to the frontier by cheap or free lands and by exemption from taxation. For the same reason attempts were made to prevent the growth of large estates and to stimulate communal life. These efforts were but partially successful, and an aristocratic planter group occupied the Piedmont along with a yeomanry of small farmers. By 1730 settlers from the coast had spread from thirty to fifty miles into the Virginia Piedmont, but in the Carolinas and Georgia the foothills had scarcely been touched.

WESTWARD MOVEMENT BEFORE 1775

After 1730 this westward movement from the coast was augmented by a steady stream of Germans and Scoth-Irish from the northeast. The Blue Ridge mountains of Virginia and the pine barrens of the Carolinas abruptly checked the advance from the coast, but beyond these barriers in the great valleys of the Appalachians lay rich lands to which ready access could be had from the north. Impetus was given also by the fact that the best land in Pennsylvania was already taken up, while land in Maryland could be obtained at a cheaper price and that of Virginia was practically free. Accordingly, a steady stream of pioneers flowed through the Cumberland, Hagerstown, and Shenandoah valleys into the great mountain trough, and finally out through the passes east into North Carolina or west some years later into Kentucky and Tennessee. By 1760 they had reached the uplands of Georgia. In the Piedmont were mingled the settlers of these two converging streams, the vanguard being usually the sturdy and venturesome Scotch-Irish.

Among the moving mass, as it passed along the Valley into the Piedmont, in the middle of the eighteenth century, were Daniel Boone, John Sevier, James Robertson, and the ancestors of John C. Calhoun, Abraham Lincoln, Jefferson Davis, Stonewall Jackson, James K. Polk, Sam Houston, and Davy Crockett, while the father of Andrew Jackson came to the Carolina Piedmont at the same time from the coast. Recalling that Thomas Jefferson's home was on the frontier, at the edge of the Blue Ridge, we perceive that these names represent the militant expansive movement in American life. They foretell the settlement across the Alleghanies in Kentucky and Tennessee; the Louisiana Purchase, and Lewis and Clark's transcontinental exploration; the conquest of the Gulf Plains in the War of 1812-15; the annexation of Texas; the acquisition of California and the Spanish Southwest. They represent, too, frontier democracy in its two aspects personified in Andrew Jackson and Abraham Lincoln. It was a democracy responsive to leadership, susceptible to waves of emotion, of a "high religious voltage"—quick and direct in action.[1]

The Back Country versus the Coast.—Before the Revolution there had developed in the back country a society distinct from the tidewater regions. The men of the Piedmont were

---

[1] Turner, F. J., *The Frontier in American History*, p. 105.

generally small farmers and trappers, destitute of wealth, but well equipped with courage and initiative. Democratic and individualistic, they resented their political and economic subserviency to the minority of the coastal plain. From the beginnings of westward advance a distinct antagonism between the interior and the coast seems to have developed and during this period it can be clearly seen in controversies between the plantation owners of Virginia and the Piedmont settlers; between the backwoodsmen of Pennsylvania and the wealthy Quakers of the east, and between the frontiersmen of New England and the coast-town aristocracy. This antagonism was evident (1) in the contests between the debtor class of the interior and the property-holding class of the coast; (2) in the demands for a more democratic and representative government in which the frontier might be more justly represented; (3) in the dissatisfaction over the defective administration of government and law under which the back country suffered; and (4) in the different moral and intellectual outlook of the two regions.

Absentee landlordism was a curse of the early west. The hope of fortunes in western land soon developed and most of those enjoying means or influence speculated in land. "You may be pleased to know," said a Deerfield petition of 1678, "that the very principle & best of the land; the best for soile; the best for situation; as lying in ye centre & midle of the town: & as to quantity, nere half, belongs unto eight or 9 proprietors each and every of which, are never like to come to a settlement amongst us, which we have formerly found grievous & doe Judge for the future will be found intollerable if not altered."[1] While the actual settlers cleared the land and bore the brunt of Indian wars, the proprietors profited financially in the security of the tidewater country. As the frontiersmen were the debtor class and as specie was difficult to get, they demanded paper money and the payment of taxes in kind—demands which were generally opposed by the older communities. The frontiersman was inclined to feel that his contribution to the defense of the colony

---

[1] Sheldon, George, *History of Deerfield, Massachusetts* (2 vols., 1896), vol. i, pp. 189-190.

exempted him altogether from the burden of taxation. Politically the frontiersman felt that he was discriminated against by means of property qualifications and careful allotment of representation. The aristocracy of the tidewater and coast towns, although outnumbered, managed until the Revolution to keep in their own hands the control of the governments. The counties of Chester, Bucks, and Philadelphia elected twenty-six delegates to the Pennsylvania legislature and the five frontier counties only ten. Jefferson complained that 19,000 men below the falls legislated for more than 30,000 living elsewhere, as well as appointing their chief executive and judicial officers. The desire to escape from eastern control led to the efforts to form such new states as Franklin and Vermont. Dissatisfaction over the administration of the government was keen. Officials were corrupt, justice was expensive and slow, for the counties were large and it was sometimes necessary to travel long distances to court. Aid in time of war was uncertain. Finally the intellectual outlook was different. The frontiersman in religious matters was likely to be a dissenter or neglect the means of grace entirely. The social and economic conditions under which he lived made him democratic and in most cases opposed to slavery. These differences of viewpoint accentuated the more pressing causes of antagonism.

This antagonism led in at least two cases before the Revolution to armed uprisings. When Governor Berkeley of Virginia failed to prosecute vigorously enough operations against the Indians, frontiersmen under Nathaniel Bacon took the matter in their own hands. Thereupon Berkeley (1676) declared Bacon and his followers rebels and attempted to arrest them. The backwoodsmen and small planters rose in rebellion behind Bacon, forced Berkeley to make concessions, gained control of the legislature, and inaugurated numerous democratic reforms. At the high tide of success Bacon died and the rebellion collapsed. Berkeley, with the backing of the tidewater aristocracy, was able to revoke the reform legislation and take such cruel revenge that Charles II declared, "That old fool has taken away more lives in that naked country than I did here for the murder of my father." Almost a century later (1769) rebellion broke out in South Carolina, when the backwoodsmen, under the name of Regulators, demanding reforms, attempted to take the law in their own hands. Although they met the government party in arms on the Saluda, hostilities were averted when their demands were complied with. Two years later the Regulators of North Carolina and the militia of Governor Tryon clashed in the bloody battle of the Alamance. The frontiersmen were defeated, and the reins of government were held by the conservatives until the new constitution of 1776 recognized the rights of the interior. The failure on the Alamance was one influence that drove the first pioneers across the Alleghanies.

The French and Spanish Barrier.—Since 1604 the French had maintained permanent settlements in the New World, and during the seventeenth century had explored and laid claim to the region of the St. Lawrence, the Great Lakes, and the Mississippi. The English fur traders had disputed their possession of the Hudson Bay country, and the English colonists had sent a number of expeditions against Acadia. French ambitions in the Mohawk and Hudson had finally brought them into conflict with the British there. This region was so extensive that as yet there was little real trouble as long as the Stuarts, subservient to France, were on the British throne. When William of Orange, champion of the Protestant cause in Europe and bitter enemy of Louis XIV, became king in 1689, the aspect of affairs changed in America. In that year commenced a series of seven wars between France and England known as the Second Hundred Years' War, which continued with brief intervals until 1815 and comprised sixty years of actual fighting. Rivalry for the commercial and colonial supremacy of the world was the underlying cause of these wars, the first four of which were fought in America as well as in Europe.

It is unnecessary here to enter into the details of these wars. They were characterized by raids against the New England frontier on the part of the French and their Algonquin allies and expeditions against the Canadian fortresses by the English and Iroquois. King William's War (1689-97), Queen Anne's War

(1702-13), and King George's War (1744-48) were but preliminary encounters to the last and greatest of the four conflicts, the French and Indian War (1754-63), which broke the French power in America beyond repair. English frontiersmen had previously reached the mountain passes and were ready to break through into the Ohio and Mississippi valleys, and for a long time English fur traders had been bitter rivals of the French for the products of this region. In the meantime the French authorities, following in the wake of explorers and traders, had erected forts at the strategic points on the Great Lakes, at the portages leading from them to the Mississippi and on the larger rivers—Fort Cahokia, Chartres, Kaskaskia, and New Orleans on the Mississippi, Vincennes on the Wabash, and Duquesne on the Ohio. De Bienville pressed down the Ohio in 1749, nailing up signboards and burying lead plates, proclaiming the land as belonging to the king of France. If the French should be successful in the claims, the English would be clamped in between the Alleghanies and the Atlantic. This attempt to dam back the westward advance made inevitable the struggle for supremacy.

In this great struggle the French were handicapped by lack of numbers, for the 80,000 inhabitants of New France were overwhelmingly outnumbered by the 1,300,000 English. Furthermore, they were not compact, but scattered throughout a vast area. These disadvantages were partially compensated for by their centralized government, which functioned infinitely better in time of war than the disconnected colonial governments of the English, and by the fact that they already held the strategic points in the territories under dispute. However, the stronger economic strength of the English colonies, the superior sea power of Great Britain, the driving force of the elder Pitt, the great empire builder, and the persistence of the British soldiers, Amherst and Wolfe, proved in the end victorious and the great colonial empire of France passed largely to England. By the Treaty of Paris of 1763, France ceded to England all of Canada with the land east of the Mississippi, and of her vast American empire retained only the islands of St. Pierre and Miquelon off the coast of Newfoundland, to be used for drying fish, and the sugar islands of

Martinique, Guadeloupe, and St. Lucia. From Spain England took Florida, but France ceded to that nation New Orleans and the country west of the Mississippi.

The Spanish Border.—Spain remained a factor in the colonization of the present United States long after France had been eliminated, and the gradual advance northward of the Spanish frontier brought Spain into conflict with England, France, and the United States in turn. The first permanent white settlement in the limits of the present United States was made by the Spaniards at St. Augustine, Florida, in 1565, after they had destroyed a French colony fifty miles northward. With the cessation of French attempts to settle the Carolinas no further impetus was given to Spanish advance northward in Florida until the British settled Georgia in 1733. Hostilities commenced during King George's War, when Oglethorpe made an unsuccessful attempt to capture St. Augustine in 1740, and a return attack was made by the Spanish two years later. The Peace of Aix la Chapelle made no change in the Georgia-Florida frontier, but the Peace of Paris at the conclusion of the French and Indian War transferred Florida to England, in whose hands it remained until 1783, when it was returned to Spain.

In the meantime the Spanish frontier was gradually pushing northward from Mexico; the motive was the lure of fabled riches and the desire to win souls to Christianity, and missionaries, soldiers, and fur traders led the way. It was in 1598 that Juan de Oñate entered New Mexico with the definite intent to settle. Santa Fé, founded in 1609, became the capital of the new colony, whose limits were extended during the century as the Franciscans and Jesuits established new missions among the Indians. By 1680 there were more than 2,500 Spaniards in the colony, and by the end of the century they had practically subjected the Indians. Eastern Texas, which had been temporarily occupied in 1693, was reëntered in 1716 by an expedition which founded San Antonio in that year. The expedition into Texas had been undertaken principally as a countermove against the French, who were making trading expeditions westward from Louisiana and establishing relations with the Indians of Texas and Arkansas.

When war broke out in 1719 between France and Spain, the contest spread to the colonies, where it was waged along the whole border from Pensacola to the Platte River. The expedition of the Marquis of Aguayo (1720-22), governor of Coahuila, into Texas clinched the hold of Spain on the new province, although the territory in dispute between France and Spain continued to extend from the Trinity to the Mississippi. By 1700 there were to be seen Spanish ranches in Arizona, but the Spanish advance into California did not occur until later. The Portola expedition founded San Diego in 1769, and the next year a post at Monterey was established. The year of the Declaration of Independence saw the beginnings of San Francisco, while San José was founded in 1777 and Los Angeles in 1779. At the conclusion of the Revolution the land now encompassed by the United States was under the control of two nations, the infant American Republic and Spain, with Great Britain and Russia disputing the northern boundary line.

British Western Policy, 1764-65.—The French understood Indian psychology better than the English, and as they were chiefly fur traders they were not continually encroaching on the land. As a consequence the Indians generally sided with the French in these colonial wars. The irregular practices of the English fur traders and the illegal taking up of land irritated the Indians. The French and Indian War had been accompanied by Indian attacks, the most important being that of the Cherokees in 1760 and 1761. As an aftermath to the war Pontiac, chief of the Ottawas, greatest of Indian warriors and friend of the French, organized a confederation of tribes in the Northwest and during the years 1763 and 1764 attacked the frontiers of Virginia and Pennsylvania and the British forts west of the mountains.

The inability of the colonies properly to handle Indian affairs led the home government to formulate a policy under which a definite boundary should be established between the lands to be settled by white men and those reserved for Indians. In order to carry out this policy the Proclamation of 1763 ordered the colonial governors to grant no warrants of survey or to pass patents for any lands beyond the heads or sources of the rivers which fall into the Atlantic Ocean from the west or northwest, all such territory being reserved for the use of the Indians unless purchased in the king's name at a public meeting of the Indians by the governor or commander-in-chief of the colony in which the land lay. In reality, a series of treaties, the most important of which were those signed at Fort Stanwix (1768) with the Iroquois and at Fort Lochabar (1770) with the southern Indians, established a continuous boundary line (in some cases west of the sources of the rivers) running from the Great Lakes west of the Appalachians to Florida and through the southern part of East and West Florida to the Mississippi, thus opening a large extent of territory to immediate settlement. While in actual practice this proclamation did not delay materially the movement westward, the attempt on the part of the British to control this advance was received in the colonies with dissatisfaction. Many believed, probably without much justification, with the Earl of Hillsborough, that the primary object of the Proclamation of 1763 was to confine the colonists to territory where they could be kept in due subjection to the mother country and where they would be within reach of the trade and commerce of Great Britain. The Proclamation of 1763 was undoubtedly one of the causes which hastened the Revolution.

NOTES FOR FURTHER REFERENCE

Most of the studies of the significance of the westward movement in American history rest upon the pioneer work of Professor F. J. Turner, scattered in various articles published in magazines and the publications of historical societies. The most complete collection of references on the westward movement is that of F. J. Turner and F. Merk, *List of References on the History of the West* (rev. ed., 1922). A few of Professor Turner's essays have been collected in *The Frontier in American History* (1921). Upon these and other contributions of Turner the author has drawn heavily for the three chapters on the westward movement.

Most valuable on the geographical background of the pioneer advance, especially as regards routes of travel, is E. C. Semple, *American History and Its Geographic Conditions* (1903). Consult also Livingston Farrand, *Basis of American History* (1904).

The early volumes of the American Nation Series, by L. G. Tyler, C. M. Andrews, E. B. Greene, R. G. Thwaites, G. E. Howard, and C. H. Van Tyne, give the general political background and touch on the westward movement.

WESTWARD MOVEMENT BEFORE 1775

The same is true of the masterly volumes of H. L. Osgood. The conflict between France and England is recounted by Parkman in a style so brilliant that history and literature may be said to join hands. The story of the frontier advance and Indian conflicts is interestingly told in Theodore Roosevelt, *The Winning of the West* (4 vols., 1889-96), but with little attempt at interpretation. Of especial value are the detailed narratives of Justin Winsor, *Mississippi Basin* (1895) and *Westward Movement* (1897). See also C. W. Alvord, *The Mississippi Valley in British Politics* (1917).

The expansion in Virginia is treated by H. L. Osgood, *The American Colonies in the Seventeenth Century* (3 vols., 1904-07); P. A. Bruce, *Economic History of Virginia* (1896); and John Fiske, *Old Virginia* (1897). On New England consult L. K. Mathews, *Expansion of New England* (1909); H. L. Osgood, *op. cit.;* C. F. Adams, A. C. Goodell, Jr., M. Chamberlain, and E. Channing, *Genesis of the New England Town,* in Massachusetts Historical Society Proceedings, 2d Series, VII, and the second essay in Turner's *Frontier in American History,* entitled "The First Official Frontier of the Massachusetts Bay."

The best introduction to the advance into the Piedmont is Turner's essay on "The Old West," Chap. III, in his *Frontier in American History.* There are several excellent chapters in H. E. Bolton and T. M. Marshall, *The Colonization of North America* (1920), and much information in Roosevelt. More detailed on certain sections are R. G. Thwaites, *Daniel Boone* (1902); Archibald Henderson, *The Conquest of the Old Southwest* (1920); C. L. Skinner, *Pioneers of the Old Southwest* (1919), in the Chronicles of America Series; H. E. Bolton, *The Spanish Borderlands* (1921), in the same series; and F. W. Halsey, *The Old American Frontier* (2 vols., 1901). See also F. J. Turner, *State Making in the Revolutionary Era,* in American Historical Review, Vol. I.

The part which certain nationalities played in this advance may be studied in G. D. Bernheim, *German Settlements in North and South Carolina* (1872); S. H. Cobb, *The Story of the Palatines* (1897); A. B. Faust, *The German Element in the United States* (1909); Oscar Kuhns, *The German and Swiss Settlements of Colonial Pennsylvania* (1901); H. J. Ford, *The Scotch-Irish in America* (1915); and C. A. Hanna, *The Scotch-Irish* (1902). Briefer but still useful are some of the state histories in the American Commonwealth Series.

Maps showing the settlements at various periods may be found at the end of Vol. II of Channing's *History of the United States* and in *Harper's Atlas of American History* (1920). Farrand and Semple contain maps showing portages and routes of travel.

SELECTED READINGS

TURNER, F. J., *The Frontier in American History,* Chaps. I-IV
BOLTON, H. E., and MARSHALL, T. M., *The Colonization of North America,* Chaps. XIII, XV, XVII, XXI.
SEMPLE, E. C., *American History and Its Geographic Conditions,* Chaps. IV-VI.
MATHEWS, L. K., *The Expansion of New England,* Chaps. I-V.
HENDERSON, A., *The Conquest of the Old Southwest.*

CHAPTER VII
THE ECONOMIC CAUSES OF THE REVOLUTION

THE fact that the controversy which preceded the Revolution raged around the question of political rights has obscured the social and economic causes. The principal influences which brought about the break with the mother country may be enumerated as follows: first, the commercial legislation of England and the attempt to enforce it more strictly after 1763; second, the economic depression in nearly all the colonies in the years preceding the war; third, the aversion of a frontier people to any kind of taxation whatsoever; fourth, the attitude and laws of Great Britain in regard to the issuing of paper money in the colonies; and fifth, the gradual all-around weakening of the ties which bound the colonies to the home country and the growth of an independent social and political consciousness. These causes will be discussed in the order named.[1]

Mercantilism.—The European statesmen of the colonial period framed their commercial policies largely in accord with the economic theory of mercantilism, sometimes called Colbertism after its most diligent French exponent. This policy aimed at the building of strong, wealthy, and independent national states, and its supporters believed that this could be achieved only by promoting economic independence in all lines and upon a favorable balance of trade. Specifically, it was hoped to achieve this

---
[1] Although this chapter deals only with the economic and social causes of the Revolution, it is undeniable that there were important political aspects. One of the most significant of these, and one which has received considerable recent emphasis, was the struggle within the colonies between the unenfranchised and the ruling aristocracies. The former, says Professor Carl Becker (American Historical Review, vol. xxix, no. 2, p. 345), long before the breach, were as anxious to diminish the power and prestige of the local ruling aristocracies as to diminish the power of the British government. "The Revolution was not merely a question of 'home rule'; it was also a question of who should rule at home." Even this question, however, had an economic background.

ECONOMIC CAUSES OF THE REVOLUTION

by (1) encouraging native shipping in order not to be dependent upon the ships of foreign countries to carry native products, and by providing an efficient navy with well-trained seamen in time of war; (2) protecting and aiding home agriculture that the nation might produce for itself sufficient foodstuffs and raw materials for manufacture; (3) protecting and stimulating home industries in order to be self-sufficing industrially and to provide employment to citizens; (4) maintaining a favorable balance of trade in order to amass and keep in the home country as large an amount of metallic currency as possible. The nation which had the most gold and silver, it was believed, was the wealthiest.

Mercantilism Applied to the British Colonies.—The mercantile system, as far as England was concerned, had its origin as early as the reign of Richard II (1377-99). The policy was continued and enlarged by Edward II, Henry VII, Elizabeth, James I, and Charles I, but it was not until the Cromwellian régime that the great statutes were passed governing the American trade. The Commercial Revolution had shifted the carrying trade from the Mediterranean to the Atlantic, and the industrious Dutch burghers had been among the first to take advantage of this. The English Navigation Acts under Cromwell may be taken as the first blow in the life-and-death struggle between England and Holland for the world's carrying trade, and mark the beginning of the attempt to apply the mercantile system in a large way to the British colonies. In 1651 the most famous act was passed, the provisions of which were as follows: (1) No goods of the growth or manufacture of Asia, Africa, or America shall be imported into England or the dominions thereof, except in ships of which the proprietor, master, and a major part of the mariners are English; (2) no goods of the growth or manufacture of Europe shall be imported into England or the dominions thereof, except in English ships and in such foreign ships as do belong to that country where the goods are produced and manufactured; (3) no goods of foreign growth or manufacture, that are to be brought into England, shall be brought from any other place than the place of growth and production, or from those ports where alone the goods can be shipped or whence they are usually shipped after transportation. The intent of this act was to give to English or colonial shippers a monopoly of the carrying trade and one of its effects was to stimulate colonial ship building.

The act of Cromwell was strengthened in 1660 by an "Act for the Encouraging and Increasing of Shipping and Navigation" (2 Charles II, Chap. 18) which provided that goods carried to and from England must be transported not only in British-manned ships, but in British-built ships, or ships built in the British colonies. The Act of 1660, besides providing for the protection of shipping and thus the development of the merchant marine, sought to regulate the trade of the colonies so as to add to the monopoly of navigation that of colonial commerce and markets. It was enacted that "no sugars, tobacco, cotton-wool, indigo, ginger, fustick, or other dyeing woods, of the growth, produce, or manufacture of any English plantations in America, Asia or Africa" shall be shipped to any place whatsoever except England. This list was expanded in 1706 by the addition of the naval stores, tar, pitch, turpentine, hemp, masts, and yards; by rice, 1706-30; by copper ore, beaver and other furs, in 1722; by molasses in 1733; by whale fins, hides, iron, lumber, raw silk, and pearl ashes in 1764. The non-enumerated articles, chief of which were fish, grain, and rum, could be exported anywhere until 1766, but after that date exportation was confined to nations south of Cape Finisterre.

Not only did England seek to control colonial exports, but by an act of 1663 she sought to monopolize the handling of imports into the colonies. This act prohibited by high duties the importation into the colonies of any European goods unless brought *via* England and in British (including colonial) built and manned ships, an act which allowed duties and commissions to be collected in England before European goods reached America, and to limit the profits of carrying such commodities to British or colonial merchantmen. Exceptions were made in the case of salt from Spain for the New England fisheries, wine from Madeira and the Azores, and provisions and horses from Ireland and Scotland. These laws were constantly evaded by colonial merchants who shipped such enumerated articles as sugar and tobacco directly to European ports without taking them first to England, under the pretense that the commodities were destined for another colony. In an effort to make this unprofitable, Parliament in 1673 enacted a law (reaffirmed and interpreted in 1696) levying a tax on enumerated articles shipped from one colony to another, equal in amount to the import taxes levied on the same articles in England.

In accordance with this general commercial policy, the English government found it necessary to control the few infant manufactures springing up in the colonies. Colonial manufactures were not to come into competition with the home product. Consequently, the colonial governors were instructed "to discourage all manufactures and to give accurate accounts of any indications of the same." That the colonial governors were only too sympathetic with this point of view is illustrated by the words of Lord Cornbury, governor of New York, 1702-08, who wrote to the Board of Trade:

I am well informed, that upon Long Island and in Connecticut they are setting up a woolen Manufacture, and I myself have seen serge made upon Long Island that any man may wear. Now if they begin to make serge, they will in time make course cloth, and then fine; we have as good fullers earth and tobacco pipe clay in this Province, as any in the world; how farr this will be for the service of England I submit to better judgements; but however I hope I may be pardoned, if I declare my opinion to be, that all of these Colloneys which are but twigs belonging to the main Tree (England) ought to be kept entirely dependent upon & subservient to England, and that can never be if they are suffered to goe on in the notions they have, that as they are Englishmen, so they may set up the same Manufactures here as people doe in England; for the consequences will be that if once they see they can cloath themselves not only comfortably but handsomely too, without the help of England, they who are already not very found of submitting to Government, would soon think of putting into execution designs they had long harbored in their breasts. This will not seem strange when you consider what sort of people this Country is inhabited by . . .[1]

---

[1] Letter to Secretary Hedges, 1705, in O'Callaghan, *Documents Relating to the Colonial History of New York,* iv, p. 1151.

Normally the mainland colonies were primarily agricultural. This was especially true in the south, where efforts to encourage manufactures on the part of the legislatures had been unproductive of results. In the north the situation would have been similar had it not been for the artificial conditions which mercantilism had forced upon the New England and middle colonies. The natural products of these provinces were cereals and other agricultural products, lumber, furs, and fish. If the colonist was to obtain money to buy English manufactured goods he must find a market for his commodities. By a series of laws in the reign of Charles II he was denied a market for his grain and other agricultural supplies in England, and in consequence he was forced either to find a new market or to manufacture for himself. Although greatly restricted, he eventually did both.

Of the manufactured articles, woolen goods were perhaps the first to be started. So hostile were the home manufacturers that as early as 1699 a Woolen Act was passed providing that no woolen goods might be exported from the colonies or sent from one colony to another, and in the following year the duty on woolens imported into the colonies from England was removed. As a result of this legislation the manufacture of cloth for sale was checked and the hold of the English woolen merchants upon the American trade was prolonged for a century. Half of their exports to the colonies were woolen goods. The abundance of beaver gave the colonist a decided advantage in the manufacture of beaver hats. A petition in 1731 from a company of felt makers caused Parliament to institute an inquiry which disclosed the fact that 10,000 hats a year were manufactured in New England and New York and an act containing the following provisions was passed: (1) That after 1732 no hat should be put on board a ship or cart for exportation to England, or for transportation from one colony to another; (2) that no one should make felt hats unless he had served an apprenticeship for seven years. No master should have more than two apprentices, and those could not serve for less than seven years, nor could they be negroes. The penalty for violation was £500. The iron industry, which had commenced in 1643 with John Winthrop's smelting furnace near Lynn, had grown by 1750 to healthy proportions. England was in need of iron, and conflicting interests until 1750 had prevented adverse legislation. To encourage the production of raw material, but discourage the manufacture of iron products, a law was passed in 1750 providing (1) that bar iron might be imported duty free to the port of London, and pig iron to any port in England; and (2) that no mill or other engine for rolling or slitting iron, no plating forge to work with a tilt hammer, nor any furnace for making steel should be erected in the colonies. In 1757 it was provided that bar iron might be imported into any British port free of duty.

While England intended to control colonial imports and exports and to eliminate colonial manufacturing, it should be remembered that the English policy did not work necessarily to the detriment of the colonies. The unfavorable balance of trade between England and the Baltic nations and the dangerous dependence upon them for naval supplies led Parliament in 1706 to place bounties on naval stores imported from the colonies to England. A bounty of £1 a ton on masts was provided; lumber should come in free, and stringent laws were passed to protect the American forests and lumber supply. The bounty on tar was £4 a ton, on pitch £4 a ton, on hemp £6 a ton, and on rosin and turpentine £3 a ton. These were lessened considerably during the reign of George II. In 1748 a bounty of 6d. a pound on indigo imported into England from the British colonies in America was granted. In addition to bounties, aid was given to colonial tobacco growers by the prohibition of the growing of tobacco in England and the high duties imposed on Spanish tobacco, by which the Virginians were given a monopoly of the market. Preferential treatment against foreign commodities similar to that extended to tobacco was allowed on iron, whale oil, silk, pot and pearl ashes, and molasses. The duty on commodities bound for the colonies *via* England was generally refunded so that in some cases the colonists could purchase them cheaper than could the English.

The greatest economic mistake made by England during the colonial period was the passing of the famous Molasses Act of

1733. The squelching of manufactures and the passing of the corn laws left the northern colonies in search of a market where they might dispose of their surplus and obtain specie to send to England for manufactured goods. This market was found in the West Indian Islands, especially those under Spanish, Dutch, and French control, to which the New Englander sent lumber of various sorts, fish, and provisions. In return he took molasses and specie. The molasses was distilled into rum, taken to Africa and traded for slaves, the slaves brought to the West Indies or the mainland and sold to the plantation owners. This was the famous three-cornered trade upon which the prosperity of many New Englanders rested. The middle colonies sent lumber and agricultural products to the West Indies and took back the specie and products in exchange. This trade was beneficial to the West Indian planters, and the Dutch and French who took advantage of it began to undersell the English, so that the latter petitioned Parliament to put a stop to it. It was believed at the time that England's West Indian possessions were far more important than the mainland colonies, and Parliament in 1733 to protect them placed duties on goods imported from foreign plantations into the British colonies of America. If this act had been enforced, it would have effected irreparable damage to the New England and middle colonies. Fortunately for them, it remained a dead letter until 1763.

Effect of the "Old Colonial System."—In surveying the period from 1650 to 1763 to determine the actual effect of the mercantile policy as applied by England to the colonies, it appears that while the policy was essentially selfish and the interests of the colonists sacrificed to those of the home country, the effect was not markedly disastrous. This was due primarily to three reasons: (1) In some cases the interests of the colonists ran parallel to those of the mother country. It was fundamentally sound for them to devote themselves to extractive industries. (2) Bounties offered by England for various products, although in some cases producing artificial conditions, in other instances greatly aided logical development and were a source of wealth for the colonist. Even more beneficial were the preferential rates which gave the colonists a practical monopoly of the British market. (3) More important than all else, the acts, in general, either were not enforced or were evaded. During the first half of the eighteenth century England followed the policy of Robert Walpole, who took for his motto *"Quieta non movere"* ("Let sleeping dogs lie"). It was during this period of "salutary neglect" that the West Indian trade reached such great proportions, and that the evasion of the Act of 1663 requiring European goods to be imported *via* England was so prevalent. Colonial merchants evaded the laws by loading tobacco or other enumerated articles without giving bond that they be delivered in England, by loading or unloading at other than ports of entry, or by collusion with British customs officers.[1] It was estimated that in 1700 one-half of the trade of Boston was in violation of the law. Taking these facts into consideration, it appears that up to 1763 the colonists by evading the laws had not suffered severely from the "Old Colonial System" of England, but, on the contrary, had grown rapidly in population and wealth. But it must still be remembered that the mercantile system embodied procedures that were likely to be detrimental to colonial life. Disadvantages to the colonists were: (1) Monopoly of the carrying trade by English and colonial shippers removed foreign competition and had a tendency to make freight rates higher. (2) A middleman's profit must be paid to the English merchant, since most of the colonial products had to pass through his warehouses. (3) The colonies were regarded as a source of cheap raw material for the English manufacturer, and yet at the same time a market for selling the finished product at his own price. (4) To pay for these manufactured goods the colonists' supply of gold was small

---

[1] David A. Wells, writing on the American Merchant Marine in Lalor's *Cyclopædia of Political Science*, says of the colonial merchants, "Nine-tenths of their merchants were smugglers. One-quarter of all the signers of the Declaration of Independence were bred to commerce, to the command of ships and to contraband trade. Hancock, Trumbull (Brother Jonathan), and Hamilton were all known to be cognizant of contraband transactions and approved of them. John Hancock was the prince of contraband traders, and, with John Adams as his counsel, was appointed for trial before the admiralty court in Boston, at the exact hour of the shedding of blood at Lexington, in a suit for $500,000 penalties alleged to have been incurred by him as a smuggler."

and constantly draining into England; and yet their greatest source of gold, the Spanish, Dutch, and French West Indies, was virtually closed to them.

British Policy After 1763.—The defeat of France and the conquest of Canada under the inspiring leadership of William Pitt stimulated a renewed interest in imperialism on the part of the British government. The retention of Canada and the return of Guadeloupe and St. Lucia to France at the end of the war in 1763 were evidences of a new appreciation of the value of the mainland colonies, and the legislation which followed the conclusion of hostilities showed an earnest intention of welding more firmly the bonds joining the empire together. Commercially, the only change in policy was a stricter enforcement of the older acts and an enlargement of their scope. With the fall of the Bute Ministry in 1763, Grenville, the new Prime Minister, and Townshend, president of the Board of Trade, backed by George III, determined to end the policy of "salutary neglect." Believing that the American colonies should be brought under more direct supervision of the crown and that the colonists should help pay the war debts incurred in their defense, they decided (1) to enforce more strictly the laws of trade, and (2) to raise revenues in the colonies by means of the Molasses Act. It was this attempt to enforce the old commercial policy, along with the new imperialism, that was the greatest of all the causes of the Revolution. The policy of England became now a real grievance and one which to the commercial interests seemed to spell ruin. Their opposition was immediate and strenuous.

The first measure under the new régime was the Sugar Act of 1764, designed to provide for the defense of the colonies. It cut in half the duties of the Molasses Act of 1733 in the hope that the removal of the prohibitive rates might induce merchants to become more honest and that some revenue might be raised. In addition duties were laid on indigo, coffee, wines, silks, and calicoes, while at the same time the number of the enumerated articles was increased. Economic depression was at once felt in New England and the middle colonies, the reaction from which was apparent in the south.

ECONOMIC CAUSES OF THE REVOLUTION

The Sugar Act was supplemented in 1765 by the Stamp Act, which provided that stamps varying in cost from a halfpenny to £10 be affixed to licenses, contracts, deeds, wills, newspapers, pamphlets, almanacs, and other papers. The Stamp Act, following so closely the Sugar Act, created an excitement unparalleled in the colonies. When petitions and remonstrances failed, a boycott of English goods was inaugurated, merchants binding themselves to import no British goods until the act was repealed. In England merchants and manufacturers were affected to such an extent that in 1766 the Stamp Act was repealed and the Sugar Act revised downward, although the concessions were accompanied by a Declaratory Act asserting the legal right of Parliament to legislate for the colonies "in all cases whatsoever."

Rejoicing followed the removal of the Stamp Act, and opposition to the British government might have subsided had not the imperial authority passed in 1765 a Quartering Act, declaring the colonists should provide for the light, lodging, and fuel of the garrisons to be placed in specified districts. In 1767 Charles Townshend, who was now the leading spirit in the Cabinet, forced through Parliament the Townshend Acts imposing duties on glass, paper, painters' colors, red and white lead, and tea. Though not high, these tariffs were on articles of general consumption and raised the cost of living. Immediately following the passing of this act the colonists again resorted to their policy of nonintercourse. The boycott of 1768-69 was more than a voluntary movement; it was backed and encouraged by political bodies, and it was of a much more thorough and universal nature than the non-importation movements following the Stamp Act. The value of English goods imported into New England and the middle colonies dropped from £1,363,000 in 1768 to £504,000 in 1769. While imports from England slightly increased in the southern colonies, the falling off for the whole country was over £500,000, sufficient to cause enough economic unrest in both England and America to bring about in 1770 the repeal of the Townshend Acts.

As a matter of principle, in order to assert the power of the crown over the colonies, a tax of 3d. a pound on tea was retained,

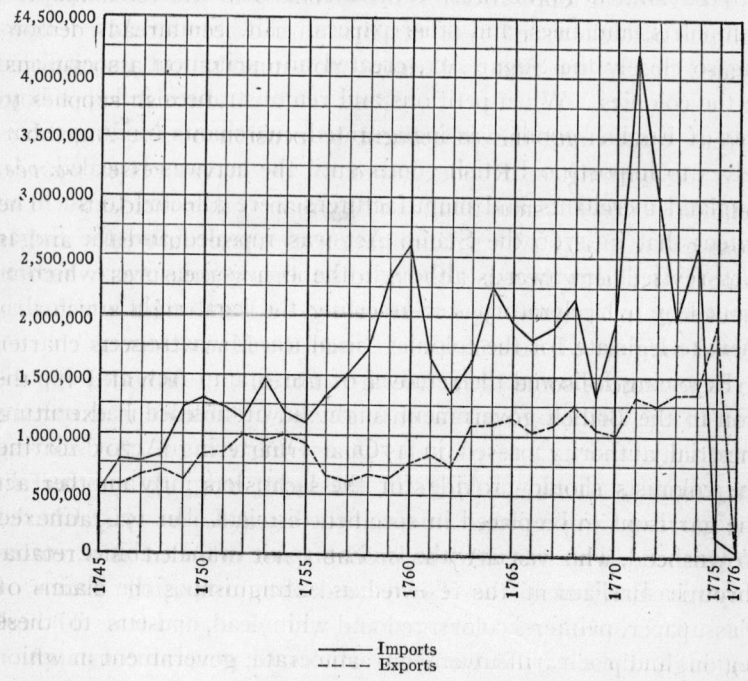

TRADE BETWEEN GREAT BRITAIN AND THE AMERICAN MAINLAND COLONIES, 1745–1776 [1]

——— Imports
– – – Exports

but a refund of 12d. a pound was allowed on tea exported from England to America, thus offering to America cheaper tea than could be purchased in England. The economic effect of the repeal of the Townshend Acts was immediate, the imports of 1769, which had fallen to £1,604,000, jumped to £4,200,000 in 1771, but the bitterness over the retention of the tax on tea continued and the boycott on that commodity was general. In 1773, to ease the financial straits of the East India Company, the British government agreed to refund to the company the entire duty on tea brought into England and subsequently exported to the colonies. This permitted the company to sell tea in America, even with the 3d. tax, cheaper than it could be smuggled in. Oppo-

[1] Johnson, E. R., et al., *History of Domestic and Foreign Commerce of the United States*, vol. i, p. 120.

sition to this act came to a climax in the famous Boston Tea Party (December 16, 1773).

The growing opposition of the colonists to the economic and political policy pursued by Great Britain had been already demonstrated in (1) the first and second non-importation associations, (2) the refusal of the New York Assembly to furnish supplies to British troops and the subsequent suspension of the Assembly, (3) the burning in Rhode Island of the naval vessel *Gaspée*, detailed to prevent smuggling, and by many other incidents. The Boston Tea Party was a culmination of these outbreaks and is important chiefly because of the disciplinary measures which it invoked. In 1774 five acts were passed by Parliament which (1) closed the port of Boston, (2) revised the Massachusetts charter so as to remove some of its liberal features, (3) provided for the trial in England of colonial agents accused of violence in executing their duty, and (4) revived the Quartering Act of 1765 for the purpose of stationing soldiers in Massachusetts. By another act the territory between the Ohio and the Great Lakes was annexed to Quebec. This last act was probably not intended as a retaliatory measure, but it was resented as extinguishing the claims of Virginia, New York, Connecticut, and Massachusetts to these regions and placing them under an autocratic government in which the Roman Catholic church was established by law.

With the passing of these acts, events leading to the Revolution followed one another in quick succession. Immediately a third boycott was organized, encouraged by the different colonial assemblies and by the Continental Congress on December 1, 1774. The colonial merchants, made wiser by their serious losses in the former embargo period, were loath to embark upon this system again, but public opinion carried all before it and the third boycott was more strictly enforced than either of the other two. English imports into the colonies dropped from £2,590,000 in 1774 to £201,000 in 1775. This shrinkage, enormous for the period just at the dawn of the Industrial Revolution, was a stunning blow to the English factory towns and seaports, and Parliament was flooded with petitions. The king and his Ministers would not yield, and in March, 1775, Massachusetts was declared to be in a state of rebellion; the fishermen of New England were forbidden the Grand Banks, and the trade of the New England colonies (extended in April to most of the other colonies) with other countries than Great Britain, Ireland, and the British West Indies was interdicted. Nine months later all intercourse with the colonies was prohibited.

Economic Depression a Cause of the War.—Undoubtedly one potent cause in bringing about the separation was the period of economic depression or "hard times" which preceded the war. It was partially, but not entirely, the result of the commercial and financial legislation of the period 1763-65. Other factors were at work to produce this economic depression and it was not confined to the colonies. It was the period in England of both the Industrial and the Agricultural Revolutions, with the unrest and instability attendant upon these phenomena. With the introduction of the factory system and capitalism on a big scale, elements which are the primary cause of our cycles of good and bad times, came what was perhaps the first of the great cyclical fluctuations which in recent years have become so common.[1] The depression was accentuated by poor crops in England between 1765 and 1774.

The hard times in England were reflected in America; decreased buying power in England combined with the enforcement of the mercantile system was disastrous to the commercial classes of New England and the middle colonies, where commerce was the chief source of private fortunes. The revision and enforcement of the Molasses Act threatened ruin to the prosperity of merchants and shippers, and their misfortune reacted upon the southern colonies. The exports of New England in the ten years 1765-75 reached the total of the first-named year only in 1768 and 1771. In the five years following 1765 the imports into New England failed to reach the figure of that year, although they were higher after 1770. Imports from England into New York reached the figure of 1764 only once (1771) before the end of the Revolution. The exports from Virginia and Maryland did

---

[1] See chap. xxvi.

ECONOMIC CAUSES OF THE REVOLUTION

not reach the figure of 1763 until 1775. The south as a whole did not feel this general movement as did the north, but the fluctuations in trade showed commercial unrest.

This depression was felt not alone by merchants, but by the farmers. The tobacco planters were discovering that a hundred and fifty years of wasteful methods had worn out their lands. Poor crops in 1759 and in later years brought a sharp decline in the amount of tobacco exported. Silk, the exportation of which reached 10,000 pounds in 1759, rapidly decreased in the following years. The exportation of pig and bar iron reached its high point in colonial times in 1771 and decreased rapidly thereafter. Similar reductions in the production of other commodities could be traced, although some products, such as flax, covered by a substantial bounty, showed decided prosperity. The business unrest and depression was accentuated both in England and in America by the three attempts through non-intercourse to bring England to terms. The Townshend duties on necessities of life, the closing of the port of Boston, and the closing of the Grand Banks to New England fishermen, were imperial mistakes leading to colonial rebellion. Hard times have always produced some kind of political unrest in our history and this was a conspicuous example.

Colonial Hatred of Taxation.—As the ultimate battle came to rest on the principle of "no taxation without representation," it has been easy to overemphasize this attitude. As a matter of fact, the colonists bitterly resented being taxed at any time, in any way, or by anybody. The storm over the Stamp and Townshend Acts came as much from general opposition to taxation as from the principle that taxation without representation is tyranny. In a population scattered on isolated farms, enjoying little communication with the outside world, each farm almost self-sufficing, and each farmer dependent for protection largely upon himself and his neighbors, a man got little from the commonwealth except the regulation and enforcement of contracts and the settlement of legal disputes. Government was a matter of small moment to him. Why should he be taxed? He had little ready money anyway and objected strenuously to parting with what he had. Taxes were always light but always extremely unpopular. They were raised with difficulty, the only practical methods being by import duties and taxes on legal proceedings. In the instructions received by Franklin (1778) to present to the French government excuses were made for the neglect of taxation by saying that America had never been much taxed, nor for a continued length of time, "and that the contest, being upon the very question of taxation, the laying of imports, unless from the last necessity, would have been madness." Callender summarizes the situation very well when he says:

It was the fact that social conditions in the colonies were such as to render all taxation except for purely local purposes extremely unpopular. In the unorganized, dispersed society of the colonies it was impossible for men to recognize any connection between most of the governmental expenditures, which occasioned taxation, and their own interests and welfare. Taxes were a burden and did not seem to be justified by necessity, especially after the French had been expelled from the continent. That a great reluctance to pay taxes existed in all the colonies, there can be no doubt. It was one of the marked characteristics of the American people long after their separation from England. Down to the time of the Civil War it constituted one of the difficulties American statesmen always had to face. It was the principal rock upon which the confederation split, and Hamilton recognized it as the chief problem to be solved in the establishment of the new government. Until the Civil War it was strong enough to prevent the establishment of a respectable revenue system in either federal or state finance. It was this unwillingness to bear the burden of taxation that caused nine of the states to default in the payment of interest on their public debts in the early forties, and at least one of them to repudiate the debt altogether. It was fear of this also that caused so long a delay in levying adequate taxes to support the government during the Civil War. Here we have an explanation of that extravagant and, to us now, somewhat incomprehensible opposition to the slight burden of taxation which England proposed to levy upon the colonies.[1]

Financial Legislation.—An important, though little emphasized, cause of the American Revolution was the paper-money legislation of the British government. The question of a medium

---

[1] Callender, G. S., *Selections from the Economic History of the United States*, p. 123. Quoted by permission of the publishers, Ginn and Company.

of exchange had always been an important one in the colonies. No gold or silver mines existed; the colonists were poor and needed other kinds of wealth rather than money; the balance of trade was unfavorable, draining off most of what little currency came into the country, and such as remained circulated slowly. As has been already pointed out, the West Indian trade was the source of specie, most of which was shipped to England in payment of manufactured goods. During their history the colonies adopted many expedients to meet this scarcity of currency. As late as 1634 musket balls were legal tender; wampum was also used as money during these early years. Some attempts were made to coin money, but with little success. The chief coins in the colonies were Spanish and Portuguese, but as each colony put different valuations upon them and merchants attempted to do business in the English phraseology, the confusion was great.[1] Much of the trade was carried on by the bartering of commodities: tobacco in Virginia and Maryland; rice and cotton in the Carolinas and Georgia; corn, cattle, and peltries in the other colonies. Tobacco warehouse receipts were used extensively as money in Virginia. With such a condition of affairs it was only a matter of time when paper money would be issued as essential to an expanding commerce. In addition to the scarcity of specie, the colonies as a whole were in debt to Great Britain and it was natural for a debtor people to look to paper money as a cure.

As a consequence, in 1690 Massachusetts issued fiat money to pay her soldiers returning from the unsuccessful expedition against Quebec. It was the origin of paper money in America, and was kept near to par by making it payable for taxes at 5 per cent advance over coin. Encouraged by the success of the first emission, a second was authorized in 1709. Connecticut, New Hampshire, Rhode Island, New York, and New Jersey followed before 1711, and the remaining colonies later, depreciation generally ensuing. In the meantime in 1704 the so-called "loan banks" were established in Massachusetts, designed to issue paper money on safe mortgages and real estate. Before these

---

[1] See p. 103.

schemes could go far Parliament intervened in 1741 and applied the Bubble Act of 1720, which put an end to these banks. Ten years later (1751) Parliament forbade the issue of bills of credit in New England, and in 1764 extended the prohibition to the remaining colonies. The denial of the right of the colonies to issue paper money was the basis of bitter quarrels between the colonial legislatures and the royal governors. The colonists were in desperate need of a circulating medium of exchange and England's attitude on this matter did as much as anything else to sow the seeds of discontent.

Other Causes of Friction.—The attempts to enforce the mercantile system after 1763, coupled as it was with fresh legislation against paper money and with new taxes, undoubtedly were the chief forces which brought about the Revolution. But they do not tell the whole story. There was an obvious lack in England of an understanding of colonial conditions which was returned in the colonies by a growing distrust of the home country. As an example may be cited the matter of slavery in Virginia. Here the people were beginning to doubt the economic value of the slave trade and to fear a possible negro uprising. In 1761 acts raising the duty on imported slaves were vetoed by the crown, which in 1770 further commanded the governor, upon pain of highest displeasure, to assent to no law by which the importation of slaves should be in any respect prohibited or obstructed. The attitude of the Virginians may be seen in the reply of the House of Burgesses (1772), which said:

> We are sensible that some of your Majesty's subjects of Great Britain may reap emoluments from this sort of traffic; but when we consider that it greatly retards the settlement of the colonies with more useful inhabitants, and may, in time, have the most destructive influence, we presume to hope that the interest of a few will be disregarded when placed in competition with the security and happiness of such numbers of your Majesty's dutiful and loyal subjects.[1]

More potent than the slavery controversy in arousing discontent and in illustrating the lack of understanding was the British

---

[1] *Miscellaneous Papers* in Virginia Historical Society's Collections, New Series, vi, p. 14.

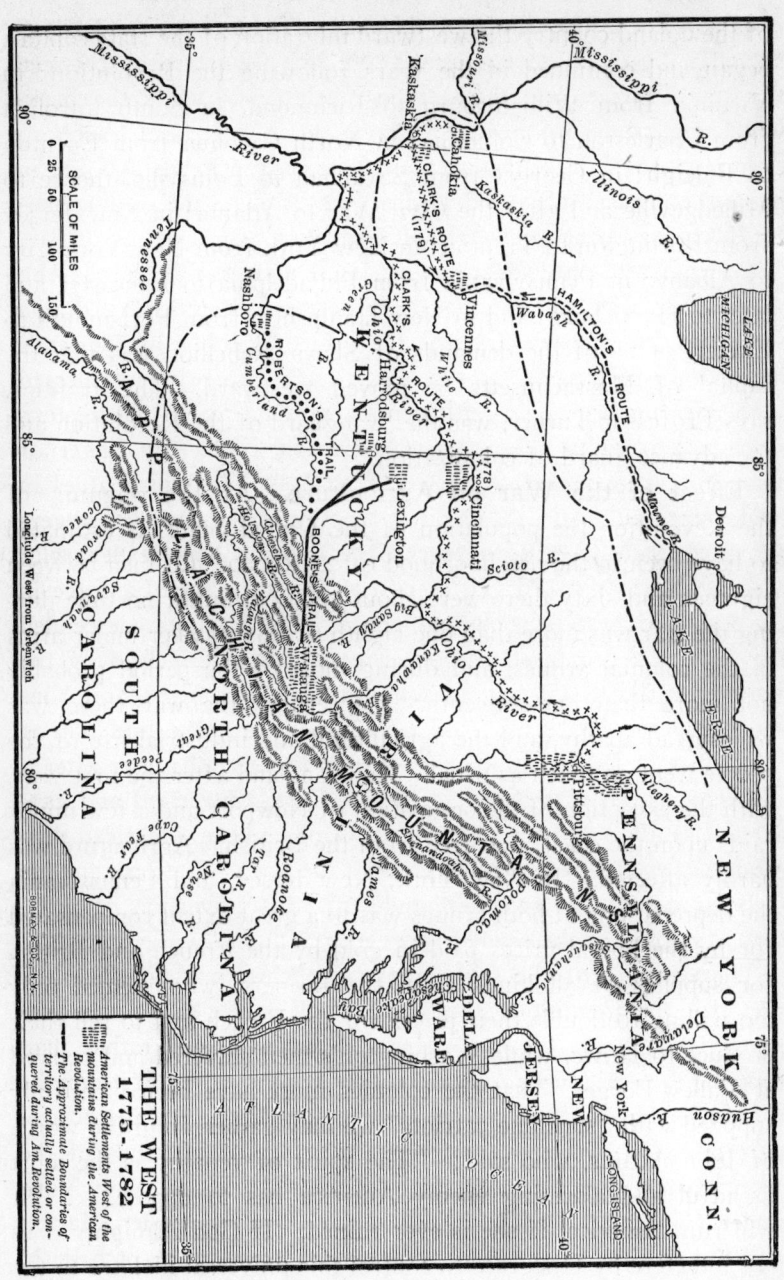

of the upland country the westward migration of the state capitals began and continued in the years following the Revolution: in Virginia from Williamsburg to Richmond, in South Carolina from Charleston to Columbia; in North Carolina from Edenton to Raleigh; in Georgia from Savannah to Louisville, thence to Milledgeville and after the Civil War to Atlanta; in New Jersey from Burlington to Trenton; in New York from New York City to Albany; in Pennsylvania from Philadelphia to Lancaster and later to Harrisburg; and in New Hampshire from Portsmouth to Exeter. One of the demands of Shays' Rebellion was that the capital of Massachusetts be moved westward. The frontier, says Professor Turner, was the "vanguard of the Revolution and the advance guard of colonization."

Effect of the War on Agriculture.—At the opening of the Revolution the population of the United States is estimated to have been in the neighborhood of 2,750,000. Of men between eighteen and sixty there were about 700,000, but at no time during the war was more than one-eighth of this number under arms in the colonial armies, and during most of the period probably not more than one-sixteenth. Concerning the war there was widespread apathy, and the agricultural and industrial life of the people went on much as usual. New England after the first year, with the exception of the occupation of Newport and a few minor raids upon the coast, was free from the British. Agriculture was hardly affected. In New York, New Jersey, and Pennsylvania the depredations of both armies was to a great extent compensated for by the liberal prices paid in gold by the French and British for supplies of all kinds from the farmers, who seemed only too willing to double their prices for the French and to sell their produce to Howe, while Washington's men shivered and starved at Valley Forge. That the colonies must have been plentifully supplied with "sunshine patriots" we may gather from the words of John Adams, who said: "The spirit of venality is the most dreadful and alarming enemy America has to oppose. . . . It will ruin America, if she is ever ruined. If God Almighty does not interfere by His grace to control this universal idolatry to the mammon of unrighteousness, we shall be given up to the chastise-

THE REVOLUTION AND CONSTITUTION 169

ment of His judgements. I am ashamed of the age I live in." [1] "Such a dearth of public spirit," said Washington in 1775, "and want of virtue, such stock-jobbing, and fertility in all the low arts to obtain advantage of one kind or another . . . I never saw before, and I pray God I may never be a witness to again. . . . Such a dirty mercenary spirit pervades the whole that I should not be at all surprised at any disaster that may happen." [2]

Blockade runners were always ready to carry the tobacco of the Virginia plantations to a waiting market in Europe. Comparatively speaking, the last twenty years of the century was the golden age for tobacco for it was still the leading southern product, in 1791 surpassing flour as an export. The production of leaf tobacco rose from 101,800,000 pounds in 1774 to 130,000,000 in 1790, at which time over one-half of the southern population was either engaged in or dependent on its production. In the Carolinas the cultivation and export of rice went on apparently with little interruption. In 1778 the first water mill adapted to cleaning and preparing rice for the market and the model upon which subsequent improvements were based, was erected on the Santee River. The cessation of the British bounties on indigo occasioned by the war marked the beginning of the end of an important industry, a decline rapidly accelerated by the invention of the cotton gin. The cotton plant throve on the same soil and was so much easier to raise and market that by 1796 it had almost entirely supplanted indigo.

The interference in trade caused by non-importation agreements and the first years of the war stimulated throughout the colonies the production of wool. The same was true of cotton in the south. The legislatures of Maryland, Virginia, and South Carolina urged the growing of cotton upon their farmers with such effect, apparently, that Hamilton, writing in 1775, said, "Several of the Southern colonies are so favorable to it that, with due cultivation, in a couple of years they would afford enough to clothe the whole continent." [3] American agriculture with its

---

[1] Adams, John, *Familiar Letters*, p. 232.
[2] Washington, *Writings* (Ford's ed.), iii, pp. 246, 247.
[3] *Works of Alexander Hamilton*, ed. by H. C. Lodge (1885), vol. i, p. 153.

primitive, wasteful methods was stimulated as a whole rather than injured by the war. Knowledge of European improvements was spread by the foreigners whom the war brought into the country. The first society for the promotion of agriculture was founded in Charleston, South Carolina, in 1785, followed by similar societies in Philadelphia, 1785; New York, 1791; and Massachusetts, 1792.

Effect of the Revolution on Industrial Life.—American manufactures were more directly affected by the war than American agriculture. The Revolution enfranchised American industry by doing away with all of the annoying restrictions which the English Parliament under the influence of mercantilism had imposed when it sought to confine the colonies to the production of raw materials. During the boycotts preceding the outbreak of hostilities, the colonists had refused to purchase English goods and great efforts were made to stimulate the manufacture of such necessities as woolens and linens which had formerly been imported in large amounts. The spinning-wheel came into renewed use. Large numbers of people pledged themselves not to eat lamb or mutton or to buy from butchers who sold it, that the wool might be saved for clothing. Women of all classes turned to the production of cloth as a domestic business. The southern planters employed their poorer white neighbors at spinning or weaving or themselves built loom houses and trained their slaves to this work. Homespun was worn by the wealthiest. The necessity of wool cards led Connecticut to loan £300 to Nathaniel Niles, of Norwich, for four years to make wire for card teeth. Massachusetts granted in 1777 a bounty of £100 for the first 1,000 pounds of "good merchantable card wire" made in any water mill in her own territory from iron made in the American states. This activity in spinning and weaving during the early years of the war declined after the cargoes captured by the privateers began to be thrown on the market and importation was resumed.

The manufacture of munitions and necessaries of war was, of course, stimulated. The life in the colonies which made everyone a hunter had developed skilled locksmiths, and small gun factories sprang up at Sutton, Massachusetts; Waterbury, Connecticut;

THE REVOLUTION AND CONSTITUTION

and North Providence. Connecticut in 1775 offered a bounty of 1s. 6d. for each gunlock manufactured, and 5s. for each complete stand of arms to the number of 3,000. Congress founded in 1778 works in Springfield where cannon were cast—the predecessor of the present national armory established there in 1794. The casting and forging of guns and camp kettles was carried on in Pennsylvania and on the Hudson; new furnaces were built in many places in New England and the middle colonies. Rhode Island in 1777 provided for a bounty of £60 per gross ton for "steel of the best quality, or equal in goodness with good German steel,"[1] made in the state during the next three years. In 1778 the legislature of Maine granted to Rev. Daniel Little £450 to aid in erecting at Wells a building for the manufacture of steel. It is claimed that Jeremiah Wilkinson of Cumberland, Rhode Island, turned out in 1777 the first cold cut nail in the world. Massachusetts offered bounties on sulphur extracted from native ores and Rhode Island for powder. Attempts at mining and refining lead were made in Connecticut, and at Cheswell, Virginia, but most of that used was obtained from abroad or from melting down lead roofs, window weights, and other commodities.

A very real shortage of many of the necessities was felt until 1777 in all parts of the country. Salt, molasses, and rum were cut off and attempts were made to supply the deficiency from sea water and cornstalks, but with indifferent results. The increase in newspapers during the war from thirty-seven to over a hundred brought an increase in paper mills. Small establishments were started to manufacture various commodities formerly imported. Ship building, an industry which had been stimulated by the Navigation Acts, was greatly restricted during the war, though resumed afterward. Limited as was the manufacturing, it is remarkable that so much was carried on. Labor, always scarce and expensive in the colonial period, became increasingly so during the war. Enlistments in the army and upon privateers, and the emigration of loyalists with their servants, decreased the supply. Wages of skilled and unskilled labor doubled from 1774 to 1784,

---

[1] Rhode Island Colony Records, vol. viii, p. 240.

giving an indication not only of the profits which were made in manufacturing, but also of the rising cost of living and the increased amount of money in circulation.

Commerce and Privateering.—The Revolution favored maritime commerce in two ways: first in the opening of colonial ports to the world, and second, in stimulating privateering. The non-importation agreements of the years preceding the war had exhausted the country of English goods; as a consequence, the merchants of Spain, Holland, and France eagerly welcomed the new markets, discovering means of evading the British warships and privateersmen to such purpose that by 1777 there was little lack of foreign merchandise. Lists of imports during the war reveal items distinctly in the class of luxuries, such finer textiles as velvets, linens, silks, and broadcloths, as well as teas, coffees, spices, and wines. Later the ports were opened to English goods and considerable quantities were imported through New York. The articles were paid for mainly by exports of flour, tobacco and rice, and by the money which found its way to the colonies through the medium of foreign loans and British quartermasters. These exported staples also had to run the gantlet of the British fleet and privateersmen. Although the British admirals reported the capture of 570 vessels between 1776 and 1779, exportation was sufficiently lucrative to continue with little abatement throughout the war. Twenty-four million pounds of tobacco alone in the years 1777-78 were recorded by the British customs officials, about one-third of the ordinary consumption, received possibly under the pretense that it came from neutral ports, for the Dutch island of St. Eustatia and the French island of Martinique served as ports where cargoes could be transferred and neutralized.

Of almost equal magnitude with legitimate commerce were the operations carried on by privateers. It has been estimated that 2,000 privateers were commissioned, of which 365 came from Massachusetts. Salem had 59 carrying 4,000 men in 1779 and probably 180 during the war. Nearly 200 commissions were issued by Rhode Island, where privateering became so popular that the Assembly found it necessary to check it and pass laws to limit the size of the crews. Newburyport sent out 22 vessels, and

THE REVOLUTION AND CONSTITUTION 173

even the Connecticut towns of New London, Hartford, and New Haven engaged in the lucrative business, although more closely watched by the British fleet. Most of the operations were carried on from the smaller New England towns, for New York, Boston, Philadelphia, Newport, and Charleston were at one time or other under British control. With his usual routine voyage cut off, the American seaman found in privateering a natural outlet. Daring was necessary and the risk great, but the spice of adventure and the lure of profits drew into it the keenest and coolest. It was customary for the owners to split half in half with the crew, according to rank. Captured prizes were either taken to European ports, sold, and the money invested in merchandise to be brought home, or else, if the capture was effected off the American coast, brought in at once. More than 445 prizes were brought in by the Salem fleet. Elias Hasket Derby, the chief ship owner and the enterprising genius of this little town, died in 1799 worth about $1,000,000 realized from privateering profits, a stupendous fortune for those days. In the year 1776 English vessels to the number of 250 were captured, entailing a loss of £1,800,000. Insurance rates from the West Indies to England in this year rose to 23 per cent. This constant heckling on the sea with the continued losses incurred by the English merchants increased the unpopularity of the war in England.

Financing the Revolution.—The most difficult task which Congress had to handle was the providing of funds for the carrying on of the war. No power was given to it to levy taxes, and if such power had been given, it is doubtful whether legislation would have been practicable, owing to the colonists' hatred of any form of taxation. Although the entire cost of the war measured in terms of gold was only about $104,000,000, a sum which should have been easily raised, the Continental Congress was forced to resort to every scheme which ill-regulated patriotism could devise to carry on the struggle. Why should I "consent to load my constituents with taxes," said one delegate, "when we can send to our printer and get a wagon load of money, one quire of which will pay for the whole?" As England's control over currency had been one of the colonists' grievances, they expected that, with the restraining hand of the mother country withdrawn, recourse would be made immediately to fiat money.

The war had scarcely commenced when Congress issued (June 22, 1775) bills of credit for $2,000,000 to be redeemed by the states in proportion to population for Spanish milled dollars at a time and place not specified. From then until November 29, 1779, Congress authorized forty-two emissions of paper money to the amount of $241,552,380. To complicate the situation, the states began issuing competing paper currency, eleven states emitting by 1783 paper to the amount of $209,524,776. This was altogether more than the new nation had any need for, and as its value rested wholly on the success of the struggle and the willingness of the states to redeem the paper, it was natural that depreciation should set in. In ringing proclamations Jay and others urged in the name of patriotism that all good citizens accept it in trade. Patriots were exhorted and Tories forced to receive it. Buoyed up by the French subsidies, the dollar held up close to face value[1] until September, 1777, when it commenced to depreciate steadily until March, 1780, when the Continental dollar sold for 2.45 cents, a value it held until the end of the war. In 1781 it took $100 in paper money to buy a pair of shoes, $40 to purchase a bushel of corn, $90 for a pound of tea, $1,575 for a barrel of flour. "Not worth a continental" became the synonym of worthlessness. People with fixed incomes suffered, but it was the heyday for the speculator and for the debtor, who often with ill-concealed glee dumped his depreciated currency upon his Tory creditor. The soldier, whose pay was seven dollars a month, was hardly rolling in opulence. The evils, however, of fiat money in a primitive self-sufficing community were mitigated because few people needed to use it. The loss was also distributed because the depreciation did not all come at once. The influx of European gold during the war, which brought in more metallic currency than the colonies had ever known, helped to ease the situation.

In addition to the issuing of paper currency, almost every other

---

[1] Channing, Edward, *History of the United States,* vol. iii, p. 393, footnote.

THE REVOLUTION AND CONSTITUTION

means was used to obtain funds. Certificates of indebtedness were issued by quartermasters in payment of supplies which they requisitioned. Domestic loans were floated first at 4 per cent and later at 6 per cent, but without great success. Equally discouraging was the result of the requisitions made upon the states. Lotteries were set up and prize money taken from the sale of captures made by government privateers. Gifts were obtained from private individuals abroad and loans and gifts from foreign governments. In addition were the expenses incurred by the states in maintaining their own militia. Professor Seligman has estimated the cost (in round numbers) of the Revolution in gold as follows:

Paper money	$41,000,000 (approx.)
Certificates of indebtedness	16,708,000
Loan-office certificates	11,585,000
Foreign loans	7,830,000
Taxes (requisitions upon the states)	5,795,000
Gifts from abroad	1,996,000
Miscellaneous receipts	856,000
State debts	18,272,000
	$104,042,000

In 1780 Congress recommended to the states that the notes be taken up at forty to one and $119,400,000 were received and canceled. Under the Funding Act of 1790 some $6,000,000 were taken in at the United States Treasury at the rate of one hundred to one in payment for government bonds. The rest were lost, destroyed, or never redeemed.

Economic Reorganization After the War.—As usually happens after a war, the American Revolution was followed by an economic reorganization which carried in its wake a period of uncertainty and hard times. During the conflict labor and capital had been diverted from agriculture and legitimate trade to manufacturing and privateering. Men had gone into unwonted occupations which ceased with the end of the war. Immediate decline in prices, resulting from the cessation of war demands, in combination with the importation of the cheaper manufactured goods of Europe, were fast ruining such infant manufacturing concerns as had sprung up during the war. The "dumping" of cheap goods upon a market already glutted could have but one result as far as the industries affected were concerned. The reabsorption of the disbanded army into the economic life required time. So also did the replenishment of the stock of slave labor in the south, where thousands of negroes had been taken off by the British and fleeing loyalists. South Carolina and Virginia had felt severely the ravages of war in the later years; in New England the fishing industry and the resulting West Indian trade had been broken up to a considerable extent. Other states found business stagnant and conditions depressing.

Another factor which made the situation even more distressing was the British Navigation Acts. The American Revolution had been fought for freedom of commerce and in repudiation of the whole economic policy of Great Britain as it applied to the colonies. Instead, however, of remedying the situation, the War for Independence made matters worse. The only clause in the treaty of peace (1783) concerning commerce was a stipulation guaranteeing that the navigation of the Mississippi should be forever free to the United States. Jay at this time had tried to secure some reciprocal trade provisions with Great Britain, but without result. Pitt in 1783 introduced a bill into the British Parliament providing for free trade between the United States and the British colonies, but instead of passing this bill Parliament enacted the British Navigation Act of 1783 which admitted only British built and manned ships to the ports of the West Indies and imposed heavy tonnage dues upon American ships in other British ports. This was amplified in 1786 by another act designed to prevent the fraudulent registration of American vessels, and by still another in 1787 which prohibited the importation of American goods by way of foreign islands. The favorable features of the old Navigation Acts which had granted bounties and reserved the English markets in certain cases to colonial products were gone; the unfavorable alone were left and the United States was discriminated against like any other foreign nation. Although the French treaty of 1778 had promised "perfect equality and reciprocity" in commercial relations, it was found impossible to make a commercial treaty upon this basis. Spain demanded as her price for reciprocal trading relations the surrender for twenty-five years of the right of navigating the Mississippi, a price which the New England merchants would have been glad to pay. France (1778) and Holland (1782) made treaties, but not on even terms; Portugal refused our advances. Only Prussia (1785) and Sweden (1783) made treaties guaranteeing reciprocal commercial privileges.

To make the matter more galling, Americans needed European goods, especially the manufactured goods of England, which they were accustomed to from long usage. So necessary were they that in 1784 goods to the approximate value of £3,700,000 were imported and only £750,000 worth of goods sent in return; this meant the paying of the balance in specie or in other credits. John Adams was sent to England in 1785 and remained for three years in a futile effort to negotiate a commercial treaty, arguing unsuccessfully that "it is England's interest to cherish her trade with America, and if a hard policy is adopted America will trade elsewhere or build her own factories." The weakness of Congress under the Articles of Confederation prevented retaliation by the central government. Power was repeatedly asked to regulate commerce, but was refused by the states, upon whom rested the carrying out of such commercial treaties as Congress might negotiate. Eventually the states themselves attempted retaliatory measures, and during the years 1783-88, New Hampshire, Massachusetts, Rhode Island, New York, Pennsylvania, Maryland, Virginia, North Carolina, South Carolina, and Georgia levied tonnage dues upon British vessels or discriminating tariffs upon British goods. Whatever effect these efforts might have had were neutralized by the fact that the duties were not uniform, varying in different states from no tariffs whatever to duties of 100 per cent. This simply drove British ships to the free or cheapest ports and their goods continued to flood the market. Commercial war between the states followed and turned futility into chaos.

Weakness of the Central Government and Dissension

Among the States.—It was the lack of a strong central government that tied the hands of Adams and Jay in their negotiations with foreign nations for reciprocal commercial treaties, and that made it possible for certain states to nullify the retaliatory measures against England of the others. Under the Articles of Confederation (1781-89) which created the so-called "League of Friendship," each state "retained its sovereignty, freedom, and independence," granting to Congress only such rights as could not be easily exercised by the individual states, such as the right to conduct foreign affairs, declare war, raise an army and navy, borrow money, emit bills of credit, etc. The right to levy taxes was not granted, merely the right to make "requisitions" which the states might meet or not. A government without the power to raise taxes was without the power to provide for a standing army to enforce treaties, if they could be made. As a consequence, the government under the Articles of Confederation was one without power at home and without standing abroad. England openly violated the treaty of 1783 by refusing to surrender the northwest trading posts; Spain trafficked with the western frontiersmen in an attempt to instigate a rebellion against the United States; and Barbary pirates levied blackmail on American merchant ships.

At home the union brought about by the Revolution seemed rapidly breaking up. Instead of one nation presenting a united front there were again thirteen bickering states wrapped up in their old selfish provincialism, intent upon their own ambitions and problems. Pennsylvania attacked the Connecticut settlers in the Wyoming Valley as if they had been an intruding war party of Indians, while Connecticut and New York fought over the region of Vermont. These boundary-line disputes were only dramatic examples of hostility which was ever present in commercial relations between the states. A classical example of these commercial wars occurred in 1787 when New York levied import duties and placed other hindrances in the way of New Jersey and Connecticut farm products, which had hitherto largely supplied the New York City market. New Jersey replied by levying a tax of $1,800 a year upon a Sandy Hook lighthouse re-

THE REVOLUTION AND CONSTITUTION

cently purchased by New York and essential to the safety of the harbor; and a mass meeting of business men in New London pledged themselves under a penalty of $250 not to send goods to New York for a period of twelve months.

Chaos in the Currency.—Even more disastrous to economic life than foreign trade restrictions, a weak central government, and interstate rivalries, was the chaos in the currency, a condition left as a legacy from the war period. Congress and the several states during the war had issued $451,077,156 in paper money. The paper money of the states had depreciated in varying amounts, while the money of Congress, the Continental paper, had become practically valueless, simply a commodity in the hands of speculators. Since this money gradually passed from circulation as worthless, business again became dependent upon English, French, Spanish, and Portuguese coins. The innumerable varieties of money complicated barter. Bad as the situation was, it was further disturbed by the draining of much of this specie in order to pay the unequal balance of trade, especially unfavorable in the years 1784 and 1785. With such a scarcity of currency and after two exceedingly trying years, it was to be expected that the old cry for paper money would again be renewed, particularly by the farmer class, the debtors, and the poor generally. The business interests, realizing the effects of more paper money on trade, resisted the demand stubbornly. As their legislatures were under the control of the large planters or wealthy merchants of the coast towns, Massachusetts, New Hampshire, Connecticut, Delaware, Maryland, and Virginia succeeded in escaping further paper money, but only after severe struggles. A mob crying out for paper and threatening the lives of the legislators surrounded the meeting house at Exeter, New Hampshire, and were dispersed by the militia. The farmers of central and western Massachusetts, strong for paper money and hot against the aristocrats of Boston, revolted under Daniel Shays, and were put down only after Governor Bowdoin had sent a good-sized army against them. The other seven states, Rhode Island, New York, Pennsylvania, New Jersey, North and South Carolina, and Georgia, yielded to the demand; but neither laws nor threats of bodily harm could in some cases prevail upon merchants to take the money. The most exciting case in the judicial history of Rhode Island was fought on this question, the case of a certain John Weeden, a butcher, who refused to accept scrip for meat. The judges held the statute unconstitutional, were summoned before the legislature and reproved, but their decision stood.

Important as was the question of paper money at this time from an economic point of view, its social significance was even greater. It served as a tangible question around which the social discontent could rally. The close of the Revolution found the old ruling aristocratic class weakened by the emigration of the Tories. The former middle class had pushed to the front and the small farmer was more of a factor. Dominated by more democratic ideals, they opposed bitterly such projects as the promise of Congress to grant officers half pay for life and the founding of the Society of the Cincinnati. Their fear of an aristocratic class is seen in the abolition of primogeniture and entail, and the seizure of the rights of the proprietors in Pennsylvania and Maryland; their fear of a king in the restrictions built around the executive in the new state constitutions. But the depreciation of paper money, the growing scarcity of specie, and the depression of the three years succeeding the conclusion of peace had borne hard upon them. Oftentimes heavily in debt and with their mortgages being foreclosed, it was but natural that they should look with suspicion upon any moves which seemed to point to the establishment of a new ruling class out of sympathy with the people. Hundreds, disheartened, emigrated to the west, only to receive further evidence of the weakness of the central government. For there the Spaniards had closed the mouth of the Mississippi to their products, thus preventing their reaching a market. It was into this social structure and these conditions that the project of a new constitution and a stronger central government was launched.

The Struggle for the Constitution.—From what has been said it is evident that the years between the close of the Revolution and the adoption of the Constitution were critical economically as well as politically, and the growing demand for a stronger central government was stimulated chiefly, although of course not wholly, by economic needs. As was quite natural, the movement for a Constitution was supported most eagerly by those whose economic interests were most seriously affected by the weakness of the government under the Articles of Confederation, and who were frightened at the growing strength of the democratic element. There was little incentive for owners of capital to invest in manufacturing and shipping during the chaotic years of the post-war period, and those who had invested longed for a government strong enough to protect them against foreign discrimination. Capital was also being continually attacked by the debtor class, who were endeavoring to push stay laws and paper-money acts through the legislatures. Among the most ardent advocates of the Constitution were holders of public securities and speculators in western lands. To the former group it was obvious that a strong central government would be able not only to redeem its own securities, but to tone up the state paper. The land speculators, a group including a large number of the wealthy and prominent men of the time, undoubtedly agreed with a certain member of the Constitutional Convention who had "claims to a considerable Quantity of Western Country" and who was "fully persuaded that the Value of those Lands must be increased by an efficient federal Government." [1]

These interests, led by men of ability, integrity and broad vision, were the chief forces behind the agitation for the Constitution, a small minority, to be sure, but powerful, active, and easily organized, for they were concentrated in the towns and represented in each state. With such a group behind the proposition, it was to be expected that the document would be of a conservative nature and the rights of private property and vested interests carefully safeguarded. It was also to be expected that the opposition to the ratification would be extremely bitter. It came mostly from the agricultural districts and debtor areas. "I believe it to be a fact," said Patrick Henry, "that the great

---

[1] *Documentary History of the Constitution*, vol. iv, p. 678.

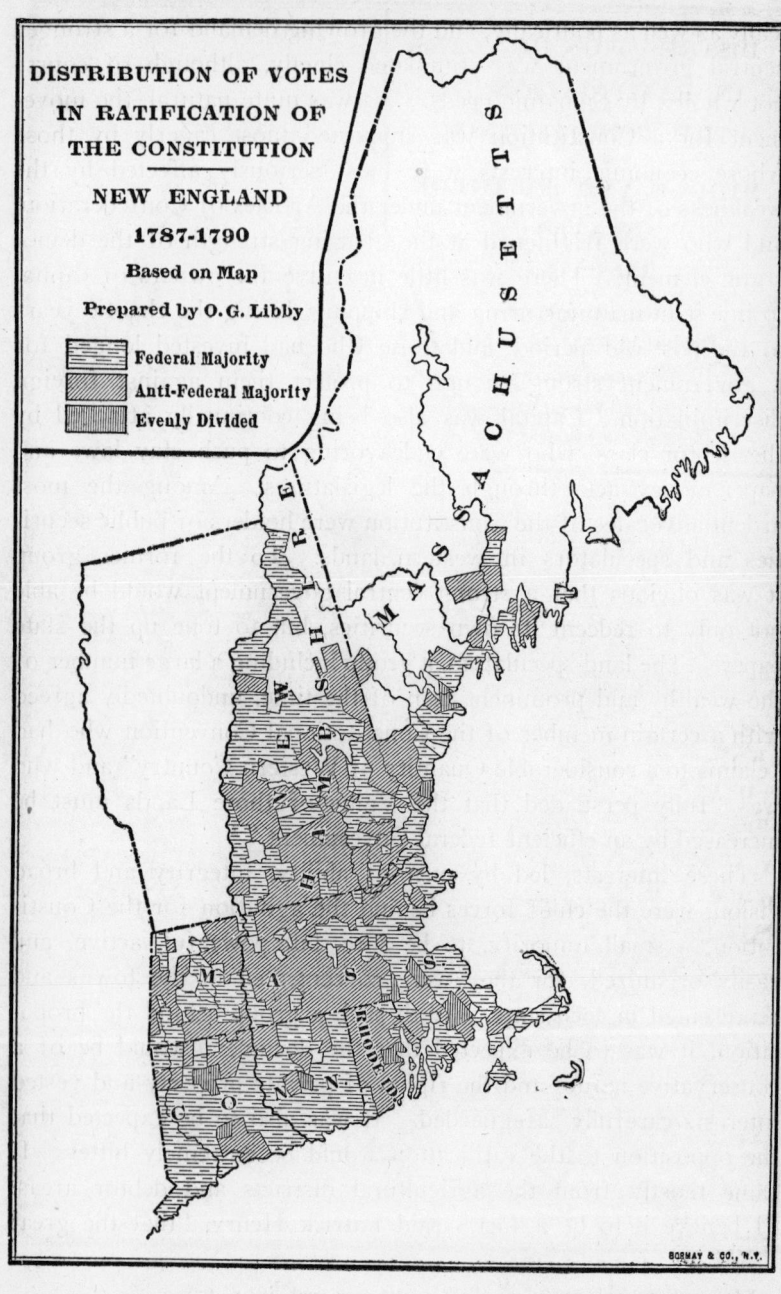

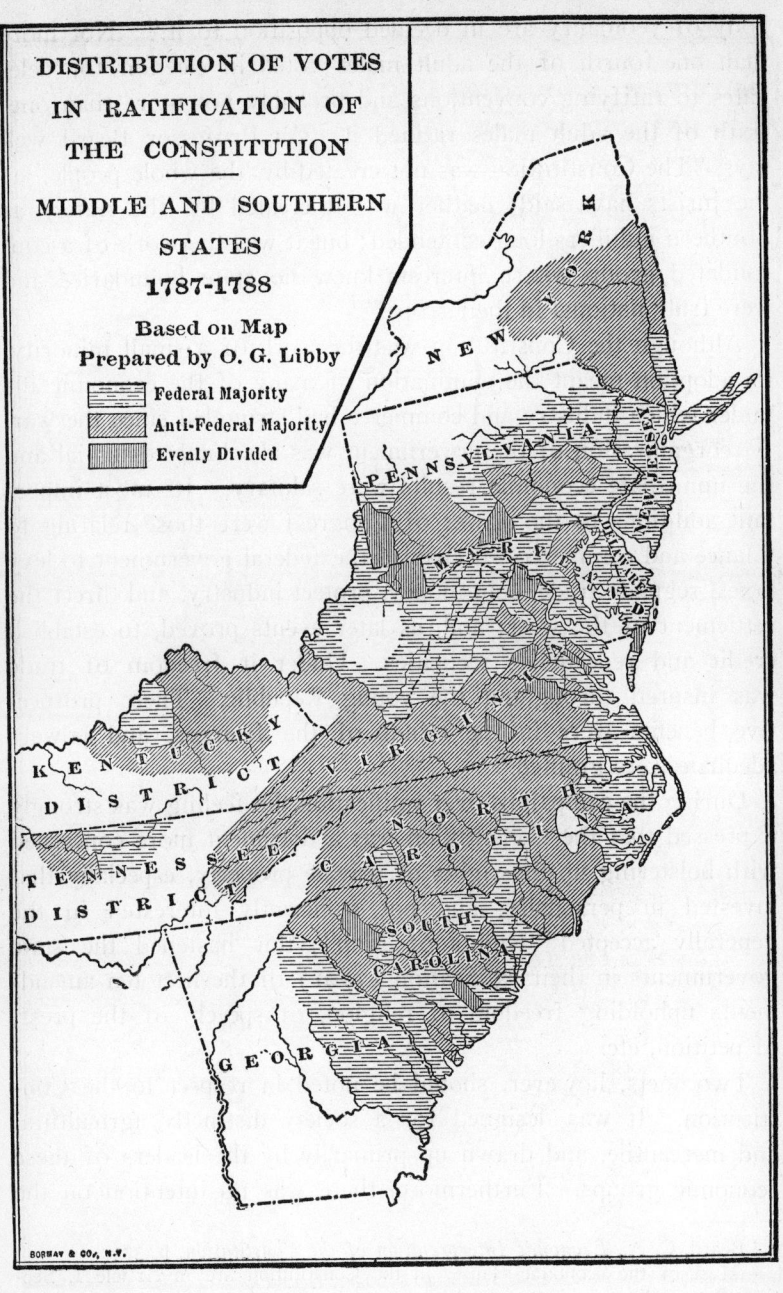

body of yeomanry are in decided opposition to it." Not more than one-fourth of the adult males voted in election for delegates to ratifying conventions and probably not more than one-sixth of the adult males ratified it. As Professor Beard well says, "The Constitution was not created by 'the whole people,' as the jurists have said; neither was it created by 'the states,' as Southern nullifiers long contended; but it was the work of a consolidated group whose interests knew no state boundaries and were truly national in their scope." [1]

Although the Constitution was the work of a small minority, its adoption meant the elimination of many of the economic ills under which industry and commerce had struggled since the war. A reorganization of the government was absolutely essential and the immediate economic results were salutary. Its most important additions to the power of Congress were those relating to finance and commerce—it enabled the federal government to levy taxes, regulate trade, coin money, protect industry, and direct the settlement of the west; and, as later events proved, to establish credit and redeem its securities. Under it freedom of trade was insured throughout the young Republic. These prospective benefits were in the minds of the framers, and powers adequate were granted.[2]

During the campaigns over ratification the feeling was strongly expressed that the Constitution was a document more concerned with bolstering up the rights of private property, especially that invested in personalty, and not sufficiently interested in the generally accepted "Rights of Man"; this hastened the state governments in their ratification (1791) of the first ten amendments upholding freedom of religion, of speech, of the press, of petition, etc.

Two facts, however, should be noted in respect to the Constitution. It was designed for a society distinctly agricultural and mercantile, and drawn up primarily by the leaders of these economic groups. Furthermore, there was no intention on the

---

[1] Beard, C. A., *Economic Interpretation of the Constitution,* p. 325.

[2] Most of the economic clauses in the Constitution are in Article I, Sections 8, 9, and 10.

THE REVOLUTION AND CONSTITUTION 185

part of the framers to set up here unmodified democracy. As a consequence, the readjustment of the Constitution a hundred years later to meet a great industrial development and more democratic conditions has been fraught with difficulty, and further adjustments are subjects of current discussion.

Revival of Prosperity.—While the years between Yorktown and the adoption of the Constitution were indeed critical, it is possible to exaggerate the economic depression. Beginning in 1786, business brightened and trade took an upward swing. Fishing reëstablished itself and again the New England merchants had fish and lumber to take to the West Indies. The slave trade had passed into other hands during the war and the old three-cornered route was a thing of the past. Under the Navigation Act of 1783 the British West Indies were closed to American ships, but the Yankee skipper soon discovered that he could take his products to the Dutch, French, or Spanish islands, from which ports by other hands they reached the British Islands. Methods were found to evade the law so that neither the subsidiary laws of 1786 and 1787 nor the eagle eye of Admiral Nelson could break up illegal trading. The number of ships that cleared for the West Indies was as large as before the war and the amount of exports probably larger. Hurricanes sweeping over the West Indies after 1780 destroyed the crops and other property, increasing the demand for lumber and foodstuffs from the mainland. In place of the African trade, commerce with the Far East and the Baltic sprang up. In 1785 the *Empress of China* entered New York from Canton and in 1787 the *Grand Turk* sailed into Salem from the same port; of forty-six foreign vessels entering Canton in 1789, eighteen were American. Not only had the American people by 1789 effectually regained their footing in shipping, but they were loath to sacrifice such gains as they had made during the war in manufacturing. Many of these industries kept going, and during this period the first cotton factory in the United States was built (1787) at Beverly, Massachusetts, and two years later Slater built his mill at Pawtucket. In 1788 a woolen factory was established at Hartford, with a capital of £1,280, raised by subscription of shares at £10 each. Before the end of the critical period many of the large New England towns had commenced their manufacturing careers. At Philadelphia John Fitch and others were experimenting with the steam vessel, producing one which ran eighty miles a day. It was a period when society was alive not only to political changes, but to the economic possibilities of the new nation.

NOTES FOR FURTHER REFERENCE

As a preliminary to further study A. M. Schlesinger, *New Viewpoints in American History* (1922), Chap. VII, will repay reading. On the financial aspect of the war an excellent résumé is in D. R. Dewey, *Financial History of the United States* (8th ed., 1922). More detailed studies are those of W. G. Sumner, *Finance and Financiers of the American Revolution* (2 vols., 1891); C. J. Bullock, *Finances of the United States 1775-1789,* University of Wisconsin Bulletins (1895); and E. P. Oberholtzer, *Life of Robert Morris* (1903).

Some material on the industrial, agricultural, and commercial background can be found in T. W. Van Metre, *Economic History of the United States* (1921), Chap. VIII; Edward Channing, *A History of the United States* (1912), Vol. III, Chap. XIII; *South in the Building of the Nation,* Vol. V; V. S. Clark, *History of Manufactures in the United States 1607-1860* (1916), Chap. X; W. B. Weeden, *Economic and Social History of New England 1620-1789* (2 vols., 1890), Vol. II, Chaps. XX, XXI; E. R. Johnson et al., *History of Domestic and Foreign Commerce of the United States* (2 vols., 1915), Vol. I, Chaps. VII, VIII; and S. E. Morison, *Maritime History of Massachusetts, 1783-1860* (1921).

On the westward movement read E. L. Sparks, *The Expansion of the American People* (1900); Justin Winsor, *The Westward Movement 1763-1798* (1897); Theodore Roosevelt, *The Winning of the West,* 4 vols. (1889-96); Archibald Henderson, *The Conquest of the Old Southwest* (1920); and E. C. Semple, *American History and Its Geographic Conditions* (1903), Chaps. IV and V. A more extended bibliography is contained in F. J. Turner and F. Merk, *List of References on the History of the West* (rev. ed., 1922).

The part played by the loyalists is set forth in C. H. Van Tyne, *Loyalists in the American Revolution* (1902), and *The American Revolution 1776-1783* (1905), in the American Nation Series. Also see A. C. Flick, *Loyalism in New York* (1901), and Chap. VII of H. E. Egerton, *The Causes and Character of the American Revolution* (1923). On political theories read C. E. Merriam, *A History of American Political Theories* (1903), a lucid and condensed presentation. Material on this period is incorporated in the *Readings* both of Callender and of Bogart and Thompson.

The most authoritative study of the economic phases of the movement for the Constitution is that of C. A. Beard, *An Economic Interpretation of the Constitution of the United States* (1913). His conclusions must be reckoned with by any student of this period. Attention to social and economic conditions, as well as to the political story, is given in A. C. McLaughlin, *The Confederation and the Constitution* (1905), American Nation Series; in Max Farrand, *The Fathers of the Constitution* (1921), Chronicles of America Series; and in J. B. McMaster, *History of the People of the United States*, Vol. I, Chap. I. See also the old and popularly written, but still valuable, volume of John Fiske, *The Critical Period* (1888). A most valuable single chapter containing the results of recent investigation is Chap. VIII of A. M. Schlesinger, *New Viewpoints in American History* (1922). Some economic background is given in T. W. Van Metre, *Economic History of the United States* (1921); in V. S. Clark, *History of Manufactures 1607-1860* (1916); and in E. R. Johnson *et al.*, *History of Domestic and Foreign Commerce* (2 vols., 1915).

SELECTED READINGS

SCHLESINGER, A. M., *New Viewpoints in American History,* Chaps. VII, VIII.
WEEDEN, W., *Economic and Social History of New England,* Vol. II, Chap. XX-XXIII.
CLARK, V. S., *History of Manufactures in the United States,* Chap. X.
DEWEY, D. R., *Financial History of the United States,* Chaps. II, III.
FISKE, JOHN, *The Critical Period,* Chap. IV.
BEARD, C. A., *An Economic Interpretation of the Constitution of the United States,* Chaps. II, III, VI, X, XI.
CALLENDER, G. S., *Selections From the Economic History of the United States,* pp. 122-177.

CHAPTER IX

THE WESTWARD MOVEMENT FROM THE REVOLUTION TO THE CIVIL WAR

The West at the Close of the Revolution.—England and France fought over the lands between the Alleghanies and the Mississippi in the French and Indian War, but it was to the newly founded American Republic that this region was destined to go. The Treaty of 1783 recognized the boundaries of the United States as extending from the Atlantic Ocean and the St. Croix River on the east to the Mississippi on the west, and from the forty-fifth parallel, the St. Lawrence River, the Great Lakes, and the Lake of the Woods on the north to the northern boundary of Florida on the south. The Florida line ran from the Mississippi along the thirty-first parallel to the Chattahoochee, down the River Flint, and in a straight line to the St. Mary's and along that river to the sea. The territory of the thirteen original colonies comprised 341,752 square miles, and the region beyond the mountains 488,248 square miles. The story of how this land was won has already been told. Under the Quebec Act of 1774 the territory between the Ohio and the Great Lakes was annexed to Canada, and as part of Canada George Rogers Clark and his frontiersmen in 1778-79 conquered the Ohio Valley. But fully as important as the conquest of Clark was the actual occupation of the vanguard of settlers led by Boone, Robertson, and Sevier, whose hold on the Ohio and Cumberland valleys was bitterly resented by the Indians.

The opening of the Revolution saw some six or seven thousand French scattered throughout the Mississippi Valley, but of settlers from the Atlantic coast only two or three hundred had broken through the mountain passes and settled on the Monongahela, upon the upper Kanawha, and at Watauga on the upper Holston. The conclusion of peace found twenty-five thousand scattered along the head of the Cumberland, on the Kentucky

River, on the Holston and French Broad, and in groups as far west as the Mississippi. Peace merely accentuated the movement of population, which during the next half century swarmed into the region west of the mountains.

International Background of the Mississippi Valley: Louisiana Purchase and the War of 1812.—The international aspect of the Mississippi Valley from 1781 to 1812 develops a tangled story of the three-cornered attempt on the part of Spain, France, and England to gain control of this region. Each of the three nations intrigued to stir up the Indians against the United States and to detach the settlers west of the Alleghanies from their allegiance to the new government. Although Great Britain had agreed in the Treaty of 1783 to relinquish the forts along the Great Lakes, she continued to hold them in the interest of the fur trade. From these points of vantage she impressed upon the Indians the importance of not ceding land to the oncoming settlers, furnished them with arms and munitions, and encouraged them to oppose the westward advance. Her emissaries tried to induce Kentucky and Vermont to leave the Union. Wayne's victory over the Indians at Fallen Timber, followed by the Treaty of Greenville in 1795, opened most of Ohio to settlers, and nine subsequent treaties up to 1809 gave bit by bit to the white man western Ohio and Indiana. By Jay's Treaty with Great Britain in 1795 the British government promised to evacuate posts in the United States territory, but her active interference in the middle west was not eliminated until the War of 1812. The Old Northwest was not won until Harrison broke the power of the Indians at Tippecanoe (1811), the principal battle of an uprising which merged into the War of 1812 and which was generally believed to be the result of British meddling.

The Spaniards, who had unsuccessfully attempted during the peace negotiations of 1781-83 to prevent the thirteen colonies from obtaining the land west of the Alleghanies, now held Louisiana. From there they intrigued with the Creeks and Cherokees to bind them close to Spain and to use them as instruments to stem the advance into the southwest. Their agents worked among the western leaders, men like George Rogers Clark (who offered his sword both to Spain and to France), Sevier, and Robertson, to foster movements for independence in order to erect buffer states between the thirteen colonies and the Mississippi. To the "Men of the Western Waters" the free navigation of the Mississippi and the right of deposit at New Orleans was essential to their economic prosperity as the only outlet for their produce. Spain held the whip hand, and the indifference of the east made her overtures alluring. Discontent did not subside until Kentucky was admitted to the Union in 1792 and Tennessee in 1796, and until the Treaty of San Lorenzo, negotiated by Pinckney in 1795, had brought about the evacuation of the Spanish posts on the east bank of the Mississippi and the free use of that river with the right of deposit at New Orleans.

The career of France in the Mississippi Valley covered over a century. Her failure in the momentous struggle with England she did not consider as the concluding chapter. A few years later she joined with the American colonies in an effort to disrupt the British Empire, and at the peace negotiations strove strenuously both to limit the boundary of the American Republic to the land east of the Alleghanies and to obtain from Spain the return of Louisiana. After the Revolution French diplomacy aimed in America to keep the colonies disunited, at the same time pushing French interests in the Mississippi. In Europe the effort to gain Louisiana from Spain was never dropped. This dream of resurrecting their lost empire in the New World for the triple purpose of checking England, of rendering the United States subservient, and of provisioning the French West Indies, was pursued under the Bourbons, the government of the Revolution, and the Consulate. Eventually Napoleon succeeded in acquiring Louisiana under the terms of the Treaty of San Ildefonso (1800), but held it only a short time. Influenced undoubtedly by the facts (1) that it was hopeless to try to hold on to an overseas empire while England commanded the sea, (2) that Louisiana was less valuable after the successful revolt in Santo Domingo, (3) that he needed money to carry on war with England, and (4) that a successful war with Great Britain would enable him to get the territory back, anyway, he decided in 1803 to sell the vast region between the Mississippi and the Rocky Mountains to the United States for $15,000,000. Jefferson understood the vital need of the Mississippi to the trans-Alleghany settlements and was a man who could think in terms of a continent. A strict constructionist by principle, he brushed aside in this case his scruples as to constitutionality, and in the face of eastern opposition strongly urged the acquisition.

Notwithstanding the purchase of Louisiana the fate of the Mississippi Valley was not fully determined until 1815, when the outcome at Waterloo laid to rest any further dreams of a French empire in America. In the meantime the War of 1812 had definitely eliminated England. The Second War for Independence was nominally fought over seamen's rights. In reality, however, it was a war of the west. Clay and the western "War Hawks" had pushed it in Congress. Westerners, hungry for a greater empire, had unsuccessfully sought to annex Canada, while the frontiersmen of the southwest cleared the Creeks from southern and western Alabama, and won the only notable land victory of the war at New Orleans. The meager results of the war were chiefly of interest to the west.

The Ordinances of the Confederation and the Land Policy of the United States to 1860.—Among the first problems which confronted Congress upon the cessation of hostilities in 1781 was the disposition of the western territory. Under the vague but inclusive wordings of the original charters Georgia, South Carolina, North Carolina, and Virginia claimed that their boundaries extended west to the Mississippi. The northwest was in dispute between Virginia, Connecticut, Massachusetts, and New York; six states—Maryland, Pennsylvania, Delaware, New Jersey, New Hampshire, and Rhode Island—had no claims on western land. Fear that the greater expansion of the states with western claims might impair their relative importance, and jealousy because the fortunate states could use the western lands to pay off war debts, motivated these six states to repeated demands that the trans-Alleghany regions should be turned over to the national government. The fight was led by Maryland, who had demanded as early as 1779 "that a country unsettled at the commencement of this war, claimed by the British crown, and ceded to it by the Treaty of Paris, if wrested from the common enemy by the blood and treasure of the thirteen States, should be considered as common property, subject to be parcelled out by Congress into free, convenient, and independent governments, in such manner and at such times as the wisdom of the assembly shall hereafter direct."[1] Later she refused to ratify the Articles of Confederation until the states having claims promised to give them up. This they eventually did, though it was not until 1802 that Georgia, the last state, turned over her lands.

In 1784 Thomas Jefferson proposed a plan for dividing the Northwest Territory into a number of states with high-sounding classical names (Sylvania, Assenisipia, Metropotamia, Polypotamia, etc.), whose inhabitants were to enjoy most of the rights of the citizens of the older states and which were eventually to be admitted to the Union on equal terms as soon as their population warranted it. This plan was amplified in the more famous Ordinance of 1787, which provided (1) that not less than three nor more than five states were to be erected out of the territory, but that for the time being it should be administered as one unit; (2) that until the population numbered five thousand male inhabitants the territory should be ruled by a governor and three judges appointed by Congress, who determined the local officers and made the laws, subject to veto by Congress; (3) that after the population reached five thousand the territories could have a two-house legislature, the lower appointed by the people and the upper a legislative council of five men selected by Congress from ten nominated by the lower house. The legislature could send a delegate to Congress with right to debate but not to vote. During this territorial stage the governor had power to veto; political rights were based on graduated ownership of land; a man owning fifty acres had the right to vote for a representative, but to be eligible for the lower house he must own two hundred

---

[1] Instructions of Maryland to her delegates and read in Congress May 21, 1779. Quoted by H. B. Adams, *Maryland's Influence upon Land Cessions to the United States* (1885), Johns Hopkins Studies, vol. iii.

acres, for the upper house five hundred, and for the governorship, a thousand acres; (4) when any of the territories should have sixty thousand inhabitants it might form a permanent constitution and state government and its delegates be admitted into Congress "on an equal footing with the original states in all respects whatever." This ordinance laid down the principles of procedure which have since generally been followed in regard to new territory.[1]

Already (1785) Congress had passed an act providing (1) for a rectangular land survey by the government, (2) for the setting aside of one thirty-sixth of the land for educational purposes, and (3) for the establishment of land offices for the sale of public lands at low prices and in small lots. In general this threefold policy was subsequently followed. After a north and south line, known as the "prime meridian," had been established (the first one set up being the present boundary line between Ohio and Indiana) an east and west base line was then made to intersect it at right angles. From the intersection of the prime meridian and the base line the surveyors ran out perpendicular lines at six-mile intervals. The crossing of these lines divided the land into squares containing 36 square miles. Each of these

---

[1] "The so-called Ohio or Northwest Ordinance of July 13, 1787," said Eduard Fueter, the Swiss historian, "has been called one of the most important laws of the United States (from the point of world history it is perhaps the most important).... Thus the principle was abandoned that the welfare of the colonies ought to be subordinated to that of the mother country; rather was the principle established that colonies which are settled by a people are to be regarded as an extension of the mother country and are to be put on an equal footing in every respect." *World History*, p. 105, trans. by S. B. Fay (1922).

Another famous provision of the Ordinance of 1787 is that which forever prohibited slavery in the entire region. (See chapter on "Economic Causes of the Civil War," p. 370.) In respect to this prohibition, Professor Turner has said (and here he bears out the above quotation from Fueter): "While the importance of the article excluding slavery has often been pointed out, it is probable that the provisions for a federal colonial organization have been at least equally potent in our actual development. The full significance of this feature of the Ordinance is only appreciated when we consider its continuous influence upon the American territorial and State policy in the westward expansion to the Pacific, and the political preconceptions with which Americans approach the problems of government in the new insular possessions" (*The Frontier in American History*, p. 132).

squares was to be a township and subsequently subdivided into 36 squares each one mile square (640 acres), known as sections, with Section 16 reserved for the support of common schools. Most of the states admitted after 1842 reserved also Section 36 for school purposes, thus setting aside for education one-eighteenth of the land surveyed. As the years went by, Congress, under the pressure of the west, provided for the subdivision and sale of the sections into half sections (320 acres), quarter sections (160 acres), and the quarters into sections of 40 acres. An attempt to reserve Section 15 from each township for religious purposes was voted down.

The Settlement of the Old Northwest.—The Land Ordinance of 1785 and the more famous Ordinance of 1787 prepared the way for the opening to settlement of the Old Northwest, the region north of the Ohio and east of the Mississippi. When this land was turned over to Congress, in order to foster religion and education and to reimburse those of her inhabitants whose homes had been burned by British raids during the Revolution, Connecticut reserved a stretch of land 120 miles wide between the forty-first parallel and Lake Erie, which was known as "The Connecticut Western Reserve." Virginia in like manner, to redeem her military bounty certificates, reserved 6,000 square miles, known as the Virginia Military District, between the Scioto and the Little Miami, and Congress for the same reason reserved a block of land between the Scioto River and the Seven Ranges. The rest of the territory, with the exception of the lands sold to the Ohio and Scioto companies and to Judge Symmes, was surveyed and sold by Congress to settlers under the existing laws.

Hardly had the Ordinance of 1787 been passed before the Ohio Company, composed chiefly of Massachusetts speculators under the leadership of Rev. Manasseh Cutler, had purchased 2,000,000 acres of land north of the Ohio with depreciated soldiers' certificates, and in December of that year the first settlers of the Ohio Company left Ipswich, Massachusetts, for the Muskingum River. In the spring of 1788 they founded Marietta, where the Muskingum joins the Ohio and not far from the protecting guns of Fort Harmer. The Marietta settlers were followed in the

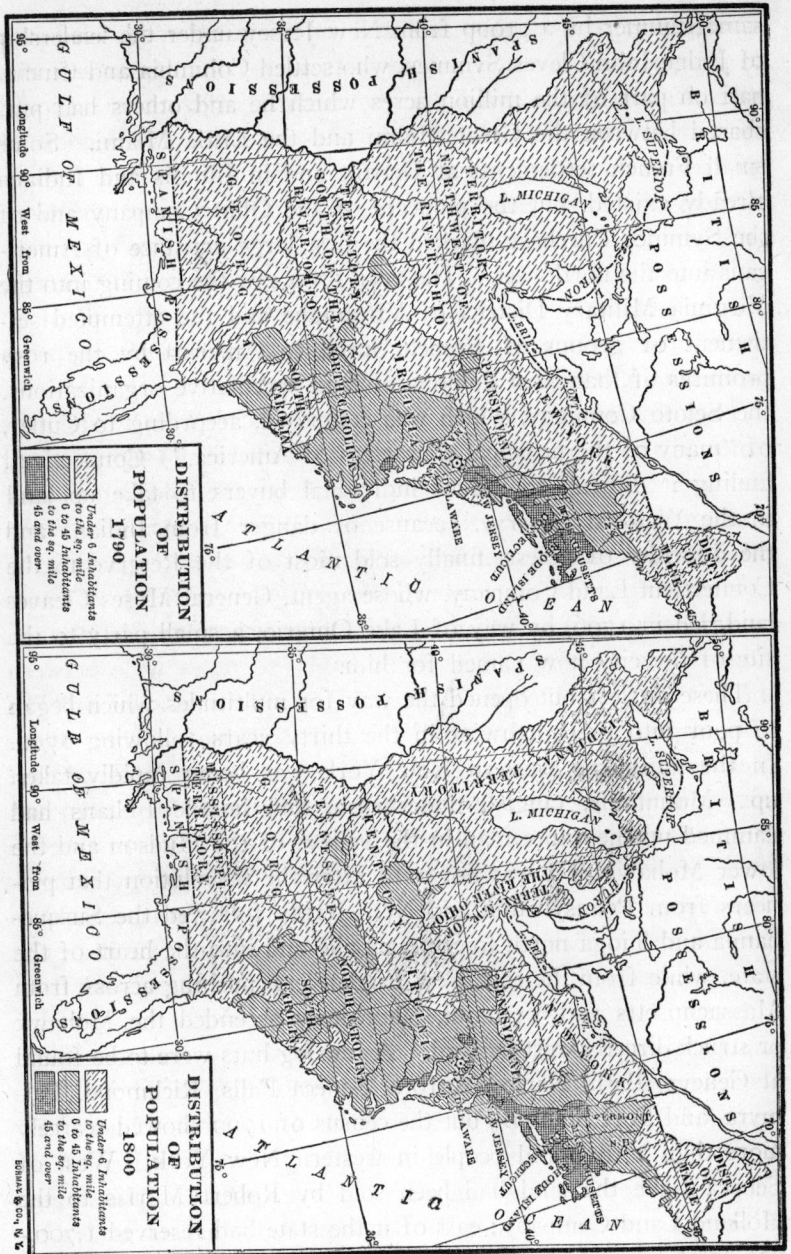

same summer by a group from New Jersey under the leadership of Judge John Cleves Symmes, who settled Columbia and Cincinnati on part of the million acres which he and others had purchased between the Great Miami and the Little Miami. Some small French settlements at Detroit and in Illinois and Indiana already existed, but the pioneers of the Ohio Company and of the Symmes Purchase marked the first large entrance of Americans into the northwest. The year 1790 saw the coming into the Virginia Military District of Virginians and the attempted settlement of groups of French, lured to America by the rosy promises of that most dubious of land speculative organizations, the Scioto Company, which was composed, according to Cutler, of "many of the principal characters of America." Connecticut, finding it difficult to induce individual buyers to take up land in the Western Reserve, because of danger from Indians and the difficulty of access, finally sold most of the Reserve to the Connecticut Land Company, whose agent, General Moses Cleaveland, led in 1796, by way of Lake Ontario, a small party to the site of the city now named for him.

These settlers but opened the way for multitudes which began to pour into the northwest in the thirty years following 1790. In the meantime western New York was being rapidly taken up. Mountains, tangled underbrush, and hostile Indians had dammed up the white man in the valleys of the Hudson and the lower Mohawk. It was not until after the Revolution that pioneers from Pennsylvania and New Jersey followed the Susquehanna and Tioga north to Seneca Lake and into the heart of the state, while from the east New Englanders, pushing across from Massachusetts and Vermont, laboriously ascended the Mohawk or struck directly west by land. A few log huts were to be found at Geneva, Bath, Naples, Aurora, Seneca Falls, Richmond, Palmyra, and Fort Stanwix, but the census of 1790 showed scarcely more than a thousand people in western New York. West of Seneca Lake the land had been sold by Robert Morris to the Holland Land Company; east of it the state had reserved 1,700,000 acres for military bounties. But the plots were soon broken up as the immigrants, chiefly from New England, took up the rich lands on the Tioga, Chenango, Genesee, and Mohawk, streams upon which were shortly to arise cities whose names harked back to classic Greece and Rome.

A combination of causes contributed to the amazingly rapid settlement of the Old Northwest. Immigration of home seekers from Europe, which had amounted to some four or five thousand a year, increased rapidly after the War of 1812, the number entering from 1815 to 1830 amounting to half a million. In the north the economic depression during the period of the Embargo and Non-intercourse Acts, the War of 1812 and immediately after, greatly stimulated the exodus to the west. In the south the planters deserted the worn-out tobacco lands of Virginia and North Carolina for the fresh alluvial soil of the southwest, driving ahead of them the small pioneer farmer of the uplands, who moved on north into the Ohio Valley. Not only was immigration stimulated by economic causes, but the discontent, especially in New England, against the old religious and political oligarchy was potent in the movement. The gradually increasing liberality of the government in its western policy encouraged the taking up of new lands; and the extinction between 1812 and 1830 of the Indian titles opened up much new territory. The victories of William Henry Harrison in the northwest and Andrew Jackson in the southwest over the Indians marked the beginning of the rapid elimination of the red man from these regions. After 1811, when the *New Orleans* was launched on the Ohio at Pittsburgh, the growth of the northwest was aided by steam navigation. Before the advent of railroads the rivers formed the great avenues of travel and traffic, and upon them floated downstream in the flatboats the immigrant or his products. By 1820 sixty steamboats plied on the western waters, and the succeeding years marked the golden age of the river boat. It took the old flatboat months to make the journey downstream from Louisville to New Orleans, but the steamboat could cover the same distance in a few days.

The influx into the northwest was rapid from the start. The very year that Marietta was founded (1788) saw ten thousand float down the river past this point; by 1803 the population of

Ohio was sufficient for its admission as a state. While the prosperity of the east during the early Napoleonic wars held back somewhat the exodus to the west, it was accelerated after 1808. An observer in Robbstown, Westmoreland County, Pennsylvania, a village on the highway to Pittsburgh, claimed that in one month toward the end of 1811, 236 wagons, with men, women, and children, and 600 Merino sheep, passed through to the west.

Old settlers in central New York declared they had never seen so many teams and sleighs loaded with women, children, and household goods traveling westward, bound for Ohio, which was then but another name for the West. One account describes the roads passing through Auburn as thronged all winter long "with flitting families from the Eastern states." Another from Newburg, in New York, declares that during one day in July six wagons with seventy persons, all from Massachusetts, entered and left the village for Ohio, and that scarcely a week passed without its citizens "witnessing more or less immigration of the same kind." [1]

"Old America seems to be breaking up and moving westward," wrote Morris Birkbeck, a European observer in 1817, while journeying on the National Turnpike. "We are seldom out of sight, as we travel this grand track toward the Ohio, of family groups behind and before us." [2] The population of the Old Northwest (Ohio, Indiana, Illinois, Michigan, Wisconsin), which at the opening of the Revolution was composed of but a few thousand French, by 1810 numbered 272,324, by 1830 had increased to 1,470,018, and by 1860 amounted to 6,926,884. Indiana was admitted into the Union in 1816, Illinois in 1818, and Michigan in 1837. By 1830 Ohio had over a million people, more than Massachusetts and Connecticut combined. Indiana in the decade 1810 to 1820 grew from 24,000 to 147,000. That this increase in population seriously drained the east is seen by the fact that Virginia and Massachusetts during the decade 1820 to 1830 remained almost stationary while the western states grew at a rate of one hundred to one hundred and fifty per cent. Chicago, a mere fur trading station in 1830, increased to over a

---

[1] McMaster, J. B., *History of the People of the United States*, iv, p. 383.
[2] *Notes on a Journey in America*, etc., p. 31.

hundred thousand by 1860; Cleveland, with only 6,070 in 1840, increased to 43,000 in 1860. The chief cities of the west about 1830 were Cincinnati, or "Porkopolis," a meat packing center in a rich farming district, with a population of 25,000; Pittsburgh, already an iron city with 12,000 people near the head of navigation of the most popular route (before 1825) to the west;[1] St. Louis with 6,000, the point of exchange between the fur traders of the north and west and the steamboat trade of the Mississippi; and New Orleans, at the mouth, where the inland products were transferred to ocean boats.

The principal route over which this influx took place was the old road that Forbes had cut in the French and Indian War from Philadelphia to Pittsburgh by way of Lancaster and Carlisle. Upon reaching Pittsburgh the immigrant transferred his effects to a flatboat and continued the journey down the Ohio and upon one of its tributaries to his chosen spot.

If the traveller were a settler coming from the East with his family and his goods, he would repair to Pittsburgh, lay in a stock of powder and ball, purchase provisions for a month, and secure two rude structures which passed by the name of boats. In the long keel-boat he would place his wife, his children, and such strangers as had been waiting at Fort Pitt for a chance to travel in company. In the flat-boat or the ark, would be the cattle and the stores. The keel-boat was hastily and clumsily made. The hold was shallow, the cabin was low. Over the stern projected a huge oar which, mounted on a swivel, was called a sweep, and performed all the duties of a rudder. The ark was of rough plank intended to be used for building at some settlement where saw-mills were scarce. . . . In these craft, if the water were high and swift, if they did not become entangled in the branches of overhanging trees, if the current did not drive them on an island or dash them against the bank in a bend, if the sawyers and planters were skilfully avoided, and if no fog compelled the boatmen to lie to and make fast to a tree, it was possible to drift from Pittsburgh to Wheeling in twelve hours.[2]

Another important route was from Albany up the Mohawk to the Genesee turnpike, then to Lake Erie and Ohio. After the

---

[1] The real head of navigation was Old Fort Redstone, above Pittsburgh, from which point many embarked.

[2] McMaster, J. B., *History of the People of the United States*, vol. ii, p. 144.

Erie Canal was completed in 1825 this route became more popular and contributed not only to the settlement of the Ohio Valley but to that of western New York. Another New York route was along the Catskill turnpike to the headwaters of the Allegheny. From Baltimore the traveler followed a turnpike to Cumberland, where began the National Road across the mountains to Wheeling, on the Ohio, with branches leading to Pittsburgh. The wagon road from Virginia into central Kentucky was the chief southern route, while from Kentucky and Tennessee many routes passed to the Ohio in the region of Cincinnati or Louisville.

Although New Englanders founded Marietta and Cleveland, the majority of the population came from elsewhere. New England up to 1820 was still settling her own northern frontier and that of western New York. In Ohio the most numerous groups were from the central states of Pennsylvania and New Jersey, with Cincinnati as their commercial center. Next in importance were immigrants from Virginia, who outnumbered the New Englanders of the Western Reserve. Indiana and Illinois received in their northern counties some accessions from New England, but were mainly settled by the yeomen farmers of the up-country of Virginia and North Carolina and by the restless pioneers of Kentucky and Tennessee, who had been pushed out by the more wealthy planters. From this stock came Abraham Lincoln. Many of these immigrants from the south were Scotch-Irish, the "Hoosier" element of Indiana coming chiefly from North Carolina. The native stock which settled the Mississippi Valley was preponderantly from the south. Nevertheless, the northwest did not take on the tone of southern civilization. The poor whites of the south, with their Presbyterian and Quaker background, mingling with the pioneers from New England and the central states, developed communities of small farms rather than plantations, where slaves were few and democracy was strong. Of direct immigration from Europe into the Old Northwest, the chief strain was German. Over half a million came between 1830 and 1850, and a million more in the next decade took up lands in central Ohio around Cincinnati, in the Wisconsin counties along Lake Michigan, as well as in Indiana, Illinois, Michigan, and other states in the Mississippi Valley.

The Settlement of the Old Southwest.—The first great trans-Alleghany migration, as we have seen, was south of the Ohio into Kentucky and Tennessee. This went on steadily during the Revolution, the first census showing over seventy thousand in Kentucky, and thirty-five thousand in Tennessee. Repeated efforts by the pioneers of Kentucky and Tennessee to free themselves from the parent states of Virginia and North Carolina were successful in 1792 and 1796, when the newly settled regions were respectively admitted to the Union as states. This migration had been undertaken chiefly by the yeomen farmers of the up-country of the south who had originally found their way down the Great Valley from Pennsylvania, or by the small farmers who had been pushed out by the more wealthy planters of the tidewater.

The Industrial Revolution, and especially the invention of the cotton gin in 1793, provided an apparently insatiable market for raw cotton and suddenly turned the eyes of the south to the development of a new staple. The tidewater lands of Virginia and the Carolinas seemed to be wearing out, which drove the southern planters to look westward for new and richer fields. The demand for cotton was the chief determining factor in the second stage of the settlement of the southwest. As the importance of cotton increased, the planter pushed on behind the small farmer, who had first pioneered across the mountains.

By the side of the picture of the advance of the pioneer farmer [says Turner] bearing his household goods in his canvas-covered wagon to his new home across the Ohio, must therefore be placed the picture of the southern planter crossing through forests of western Georgia, Alabama, and Mississippi, or passing over the free state of Illinois to the Missouri Valley, in his family carriage, with servants, packs of hunting dogs, and a train of slaves, their nightly camp fires lighting up the wilderness where so recently the Indian hunter held possession.[1]

Or as Timothy Flint describes it:

---
[1] Turner, F. J., *Rise of the New West*, p. 92.

The southern settlers who immigrate to Missouri and the country southwest of the Mississippi, by their show of wagons, flocks and numbers create observation, and are counted quite as numerous, as they are. Ten wagons are often seen in company. It is a fair allowance, that a hundred cattle, beside swine, horses and sheep, and six negroes accompany each. The train, with the tinkling of an hundred bells, and the negroes, wearing the delighted expression of a holiday suspension from labor in their countenances, forming one group; and the family slowly moving forward, forming another, as the whole is seen advancing along the plains, presents a pleasing and picturesque spectacle.

They make arrangements at night fall, to halt at a spring, where there is wood and water, and a green sward for encampment. The dogs raise their accustomed domestic baying. The teams are unharnessed, and the cattle and horses turned loose into the grass. The blacks are busy spreading the cheerful table in the wilderness, and preparing the supper, to which the appetite of fatigue gives zest. They talk over the incidents of the past day, and anticipate those of the morrow. If wolves and owls are heard in the distance, these desert sounds serve to render the contrast of their society and security more sensible. In this order they plunge deeper and deeper into the forest or prairie, until they have found the place of their rest.[1]

This was a new type of migration.

Just as the eighteenth-century frontier farmer had not been able to withstand the advance of the southern planter, so now the pioneer of the southwest was in turn displaced. Unable to refuse the high prices which the planter offered him for his land, and outbid in the competitive land sales, the small farmer had the option of adopting the slave plantation economy, of retreating to the less desirable soil of the mountains, or of striking again north or west for new lands. Hindered by poverty or religious scruples from adopting slavery, these pioneers left the rich black soil of the southwest to the planters, retreating to the mountains to become the "poor whites" of the south; or pushing north of the Ohio or across the Mississippi to become again the founders of new states. With the elimination of the yeoman farmer the southwest fell under the control of the cotton aristocracy and

---

[1] Flint, Timothy, *The History and Geography of the Mississippi Valley*, vol. i, p. 191.

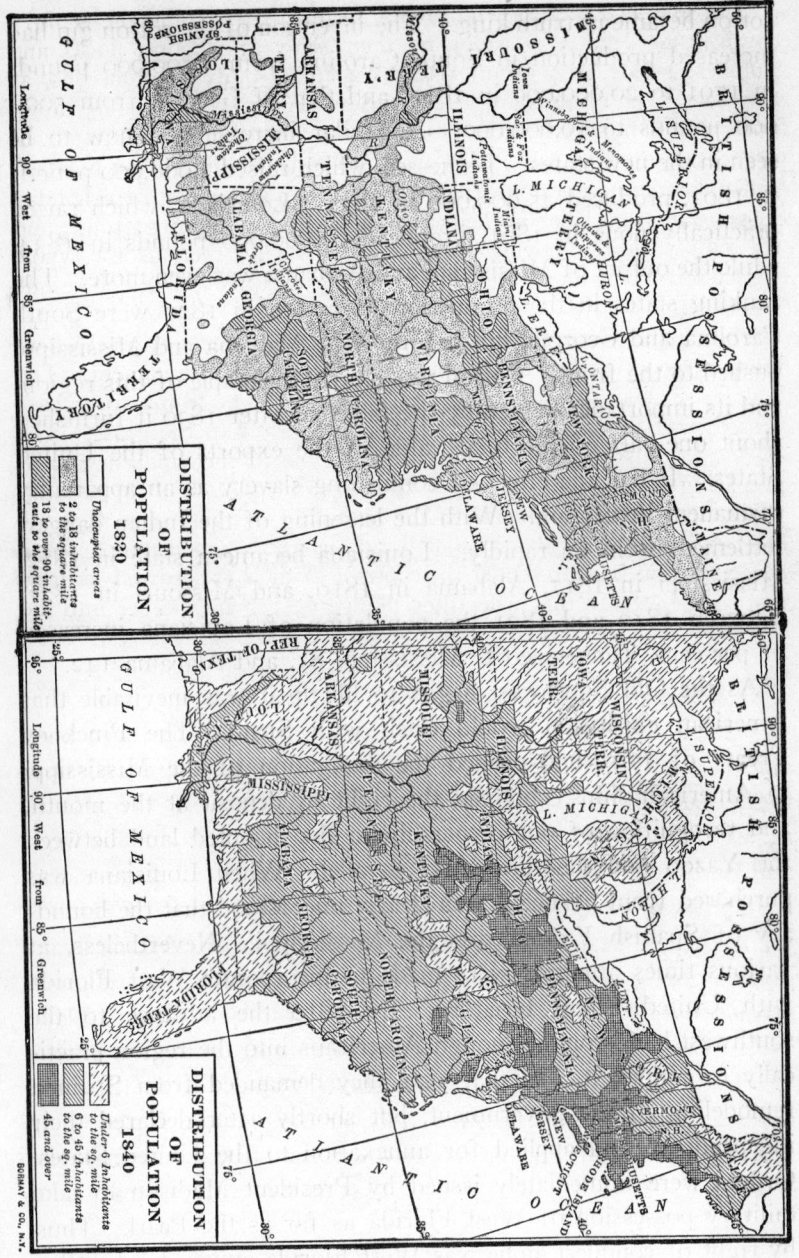

cotton became in truth king. The invention of the cotton gin had increased production in South Carolina from 1,500,000 pounds in 1791 to 20,000,000 in 1801, and that of Georgia from 500,000 pounds to 10,000,000. A similar increase was now to be seen in the new states. Tennessee, which raised 1,000,000 pounds in 1801, produced 45,000,000 in 1834. Louisiana, which raised practically none in 1801, produced 62,000,000 pounds in 1834, while the output of Mississippi and Alabama was even more. The ranking states in the production of cotton in 1820 were South Carolina and Georgia, but before 1834 Alabama and Mississippi pushed to the front. Cotton was clearly the staple of this region, and its importance is seen by the fact that after 1830 it furnished about one-half of the total value of the exports of the United States. It had the effect of confirming slavery as an apparently permanent institution. With the lessening of the Indian danger, settlement went on rapidly. Louisiana became a state in 1812, Mississippi in 1817, Alabama in 1819, and Missouri in 1821. Between 1812 and 1821 the population of Louisiana increased 41 per cent, Tennessee 61, Mississippi 81, and Alabama 142.

As the Old Southwest began to fill up it was inevitable that American expansion would clash with Spain. The Pinckney Treaty of 1795 had opened the navigation of the Mississippi to American ships, obtained the right of deposit at the mouth, and recognition of our ownership to the disputed land between the Yazoo and the thirty-first parallel. When Louisiana was purchased from France it was clearly understood that the boundary of Spanish Florida was the Mississippi. Nevertheless, at various times attempts were made to incorporate West Florida with United States territory. By 1810 the advance to the southwest had brought enough Americans into the region practically to control it. In that year they demanded from Spain a remodeling of the government, but shortly after declared their independence and applied for annexation to the United States. Orders were immediately issued by President Madison to take military possession of West Florida as far as the Pearl. Thus by right of conquest alone was West Florida annexed. In 1819 Spain saw that it was useless to attempt to ward off further

American aggression and agreed to give up East Florida, the United States assuming the claims of the citizens against Spain amounting to $5,000,000 and giving up all claims to Texas.

Frontier Life.—The first task of the immigrant, whether he came by foot, horseback, crude wagon, or by river boat, was to decide upon a place to live. If he was a squatter his chief interest was to find some land distant from settlement where water was abundant, and where some stream might furnish him a chance to reach a market for his produce and to purchase salt and the few necessities he might need during the year. If he intended to comply with the law, he either had purchased before going west or else, providing he had been lucky enough to avoid the land speculators who swarmed the western towns, he filed a claim at the land office upon paying the price of the land. Arriving at the site he had chosen, usually with wood and water as prime elements in his choice, he built a rude log cabin, oftentimes with the help of neighbors. His next task was to clear away the underbrush and girdle the trees upon land sufficient to plant the first year's corn. The fertility of the fresh soil would ordinarily yield fifty to sixty bushels per acre the first year, and further clearing normally produced seventy to a hundred the second year. The cattle, hogs, and horses could easily pick up enough food during most of the year, while with little attention his garden produced sufficient for the table. A rude plenty was thus provided, and by the third or fourth year the settler was in a position to further improve his house and to sell surplus products. If the site was good other settlers would soon appear, and in their wake might come a tanner, or the builder of a sawmill, then possibly a professional innkeeper. The nucleus of a town having been formed, work was to be had for a blacksmith, carpenter, wheelwright, or saddler, and eventually one or more stores would grow up. Later, as the resources of the community grew, there would come the demand for canals and better roads. The distance of the dwellings from one another made social intercourse highly valued, and husking bees, quilting parties, house "raisings," and even revival services contributed a boisterous but stimulating change from the day's drudgery.

Not all of the settlers by any means stayed upon the claims which they first picked out. The large amount of unoccupied land and the ease with which it could be acquired developed a restless, moving people. Markets were usually at a distance and, before canals and railways, almost impossible to reach. In consequence, money was a scarce commodity of which the backwoodsman saw little in the course of his life. The quickest method of acquiring specie was to sell the partly cleared farm to a newcomer and "clear again for the tall timber." Some men repeated this process a half dozen times in the course of their lives—almost professional pioneers, who broke the way for more permanent home builders. As the population grew the latter were in turn followed by the capitalist.

The frontier stages have been well described in a much-quoted passage from J. M. Peck's *A New Guide for Emigrants to the West*, published in Boston in 1837.[1]

Generally, in all the western settlements, three classes, like the waves of the ocean, have rolled one after the other. First comes the pioneer, who depends for the subsistence of his family chiefly upon the natural growth of vegetation, called the "range," and the proceeds of hunting. His implements of agriculture are rude, chiefly of his own make, and his efforts directed mainly to a crop of corn, and a "truck patch." The last is a rude garden for growing cabbage, beans, corn for roasting ears, cucumbers and potatoes; a log cabin, and, occasionally, a stable and corn-crib, and a field of a dozen acres, the timber girdled or "deadened" and fenced, are enough for his occupancy. It is quite immaterial whether he ever becomes the owner of the soil. He is the occupant for the time being, pays no rent, and feels as independent as the "lord of the manor." With a horse, cow, and one or two breeders of swine, he strikes into the woods with his family, and becomes the founder of a new county, or perhaps State. He builds his cabin, gathers around him a few other families of similar taste and habits, and occupies till the range is somewhat subdued, and hunting a little precarious; or, which is more frequently the case, till neighbors crowd around, roads, bridges, and fields annoy him, and he lacks elbow room. The preëmption law enables him to dispose of his cabin and corn-field to the next class of emigrants, and, to employ his own figures, he "breaks for the high timber," "clears out for the

---

[1] Pp. 119-121.

New Purchase," or migrates to Arkansas or Texas, to work the same process over.

The next class of emigrants purchase the lands, add field to field, clear out the roads, throw rough bridges over the streams, put up hewn log houses, with glass windows, and brick or stone chimneys, occasionally plant orchards, build mills, school houses, court houses, &c., and exhibit the picture and forms of plain, frugal, civilized life.

Another wave rolls on. The men of capital and enterprise come. The "settler" is ready to sell out and take the advantage of the rise of property—push farther into the interior, and become, himself, a man of capital and enterprise in turn. The small village rises to a spacious town or city; substantial edifices of brick, extensive fields, orchards, gardens, colleges and churches are seen. Broadcloths, silks, leghorns, crapes, and all the refinements, luxuries, elegancies, frivolities and fashions, are in vogue. Thus wave after wave is rolling westward:—the real *el dorado* is still farther on.

A portion of the two first classes remain stationary amidst the general movement, improve their habits and condition, and rise in the scale of society.

The census of 1820 showed the settled area as including Ohio, southern Indiana, and Illinois, one-half of Louisiana, and a patch in southwestern Missouri. That of 1830 showed Indiana and Illinois practically filled, with the settlements about Detroit now extending well into the interior of Michigan. The unsettled tracts in Maine and western New York were decreasing, the latter due chiefly to the Erie Canal, while western Virginia and Tennessee were almost wholly occupied. In the southwest the Creeks and Cherokees in Georgia and Alabama, and the Choctaws and Chicasaws in Mississippi, had held back the advance, but outside the Indian reservations the white population had filtered into western Georgia, northern Florida, and southern Alabama. West of the Mississippi the population of Louisiana had increased, with sparse settlements in Arkansas, and there was sufficient population along the Missouri River to warrant the admission of Missouri as a state in 1821. By 1840 the state of Michigan had been created, as well as the territories of Wisconsin and Iowa. The southwestern Indians had been removed across the Mississippi, and within a few years the land left vacant by them was occupied. The same fate had been meted out to the Sac and the

Fox, whose lands in Illinois were soon taken up by population stretching into Wisconsin. West of the Mississippi settlers had further ascended the Missouri into Iowa territory.

Trans-Mississippi Advance Before 1860.—When the United States purchased Louisiana in 1803 she divided the country into two parts—the Territory of Orleans with New Orleans as its capital, extending to the present southern boundary of Arkansas, and the District of Louisiana, with St. Louis as its capital. In and around New Orleans there was scattered a heterogeneous population of French, Spanish, Americans, negroes, and Indians, while the upper Louisiana territory contained some ten thousand, of whom half were Americans. Thus by the opening of the century we find that white settlers from the east had crossed the Mississippi, lured by the lenient land laws of the Spaniards and the rich fur trade. St. Louis had already become a market for furs and lead, which were floated down the Missouri and the Mississippi.

The first real knowledge of the size and resources of the newly purchased land came from the famous expedition of Meriwether Lewis and Captain William Clark, the impetus for which was due to Jefferson. Starting from St. Louis in the spring of 1804 with a party of forty-five men, they ascended the Missouri 1,600 miles to a point near the present Bismarck, North Dakota, where they spent the winter. Continuing up the river in the spring, they reached the Rockies and followed western streams until finally in November they floated down the Columbia to the Pacific, having covered over four thousand miles from the point of departure. On the return journey the little party broke into three detachments, uniting again near the confluence of the Yellowstone and Missouri and reached their starting point two years and four months after their departure. Further explorations were carried on by Captain Zebulon Pike in 1805 and thereafter, who traced the Mississippi practically to its source and explored the Arkansas and Red rivers, penetrating into the Rockies, where he discovered the mountain which bears his name.

As the fertility and resources of the new purchase became better known, pioneers crossed the Mississippi and ascended the

Missouri. Toward the south the cotton planters, ever hungry for fresh land, were crossing the river into Louisiana. Ahead of them were the cattle rangers, who from the first settlement of Jamestown had extended just west of the line of permanent settlement. By the 'thirties these cattle rangers had collided with the Spanish frontier cattlemen extending their activities northward. Not to be denied, the American rangers and cotton planters overran Texas, then part of the Mexican Republic, rebelled against Mexico, won their independence in 1836 (recognized by the United States and many European countries, but not by Mexico), and petitioned for annexation to the United States. This was eventually consummated (1845) and led almost immediately to the Mexican War of 1846-48, by which Mexico recognized the independence of Texas to the Rio Grande and was despoiled of the vast region which includes the present states of California and New Mexico, most of Arizona, Nevada, and Utah, and parts of Colorado and Wyoming. Eighteen million was paid for it, and five years later ten million more for the Gadsden Purchase in order to push the boundary line south of the Gila River and in this manner insure the possibility of a southern transcontinental railway in United States territory.[1] This was the conclusion of the long-cherished designs on Texas, New Mexico, and California which extended from the fantastic intrigues of Burr to the actual conquest.

In the meantime—basing our claims upon (1) the discovery of the Columbia by Captain Gray of Boston in 1792, (2) the explorations of Lewis and Clark in 1804-06, (3) the actual occupation by Americans at the trading post of Astoria, founded in 1811, and (4) the later settlements due to the steady influx of Americans—agreement was concluded with England by which the boundary line of the forty-ninth parallel was continued to the Pacific. The settlement of the Oregon line in 1846, the acquisition of New Mexico and California in 1848 and the Gadsden Purchase in 1853, took our territories to the Pacific coast, rounding out the continental boundaries of the United States.

---

[1] On this transaction see the pertinent discussion of H. H. Powers, *America Among the Nations*, p. 79 ff.

Hardly had these acquisitions been obtained before a remarkable impetus was given to the occupation of the west coast by the discovery of gold in 1848 on the mill race of John Sutter on the American River, about sixty miles from Sacramento. Immediately from the four corners of the earth prospective gold diggers flocked to California. By the end of 1848 six thousand had arrived, while in the next year thirty-five thousand came by sea and forty-two thousand by land. The population in 1850 was 92,597, more than that of the state of Delaware. The sea routes were either around the Horn or to the Isthmus of Panama, overland to the Pacific and up the coast. Engineers were sent out in 1849 to plan a railroad across the Isthmus, which was completed with great difficulty five years later, and did an enormous business until the first transcontinental line was completed in 1869. Overland there were two routes. The northern and shorter route, known as the Oregon Trail, led from St. Joseph or Independence near the Missouri along the Platte River to Fort Laramie. From Fort Laramie the trail led through the South Pass to Fort Bridger, where the traveler might turn south by way of the Mormon Trail and the "Hastings cut off" to the Humboldt or proceed to Fort Hall and then bear south along the Snake River and Goose Creek to the American Desert. Proceeding along the Humboldt and the Truckee, he found himself at last in the heart of the gold region. The gold seeker, traveling by way of the Southern or Santa Fé, Trail, might strike southwest from Fort Leavenworth or Independence *via* Fort Dodge to Sante Fé or he might push due west from Fort Smith, Arkansas, along the Canadian or Red River to the Pecos, and thence to Santa Fé. At Santa Fé two routes could be followed, the northern, or "Californian cut off," which connected with the Salt Lake Trail, or the southern, known as Kearney's Route, which crossed the Colorado near its mouth and led north to Monterey. Over these routes for the next twenty years "prairie schooners" by the thousands creaked their way, while multitudes endured every privation and suffering in quest of wealth. Those who escaped trouble with Indians and braved successfully the deserts had still the risks of cholera and typhoid prevalent in the unsanitary surroundings of the boom mining towns.

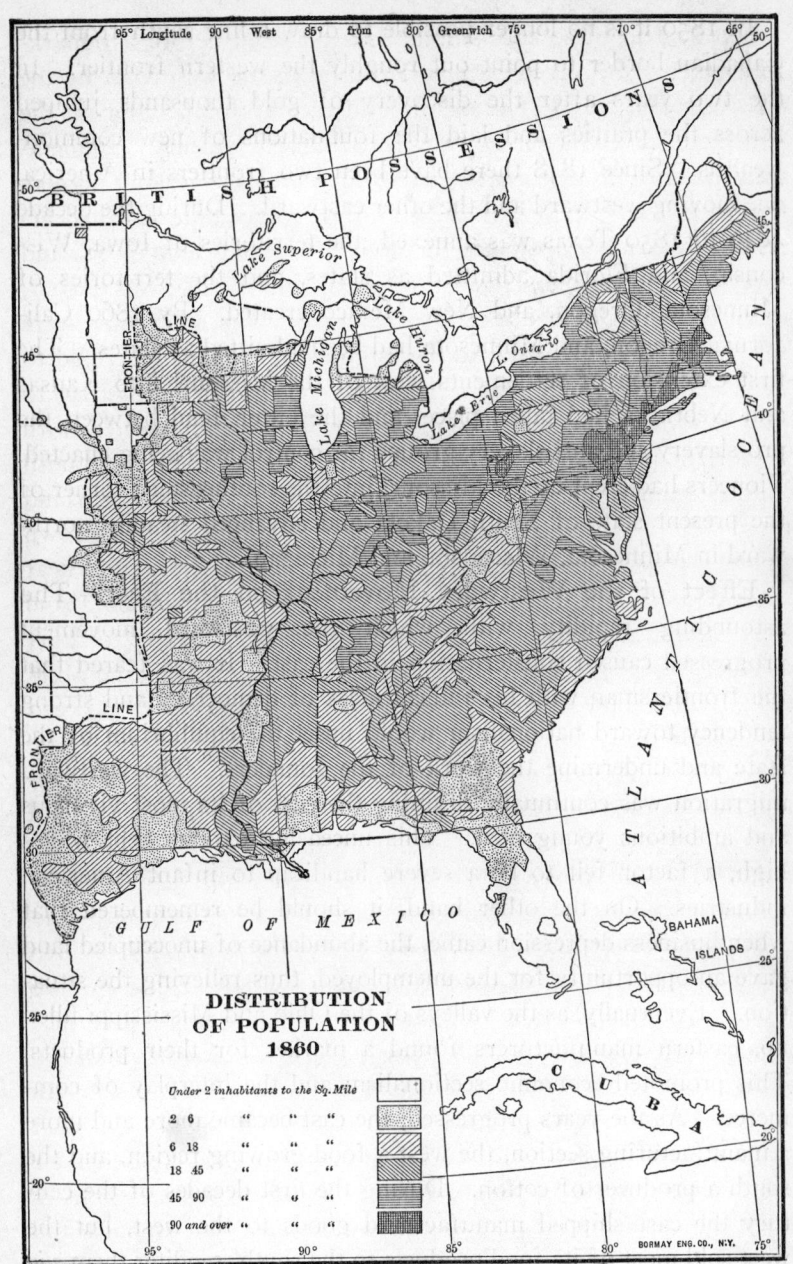

DISTRIBUTION OF POPULATION
1860

Under 2 inhabitants to the Sq. Mile
2 6   "   "   "   "
6 18 "   "   "   "
18 45 "   "   "   "
45 90 "   "   "   "
90 and over "   "   "   "

In 1850 it is no longer possible to draw a line south from the Canadian border to point out roughly the western frontier. In the two years after the discovery of gold thousands jumped across the prairies and laid the foundations of new commonwealths. Since 1848 there have been two frontiers in America, one moving westward and the other eastward. During the decade 1840 to 1850 Texas was annexed, the territories of Iowa, Wisconsin, and Florida admitted as states, and the territories of Minnesota, Oregon, and New Mexico created. By 1860 California, Oregon, and Minnesota had been admitted as states. The first extension of settlements west of the Missouri into Kansas and Nebraska had taken place, and the bitter feud between the pro-slavery and anti-slavery groups for control was being enacted. Pioneers had crept up the Missouri into the southeastern corner of the present state of South Dakota and advanced steadily northward in Minnesota, Wisconsin, and Michigan.

Effect of the Westward Movement on the East.—The astounding rapidity with which the westward movement progressed caused consternation in the east. It was feared that the frontiersman with his radical ideas of democracy and strong tendency toward nationalism would upset the equilibrium of the state and undermine the work of the founders. The westward migration was continually draining the east of its most vigorous and ambitious young men. This unceasing exodus kept wages high, a factor felt to be a severe handicap to infant American industries. On the other hand, it should be remembered that when business depression came, the abundance of unoccupied land gave an opportunity for the unemployed, thus relieving the situation. Eventually, as the valleys of the Ohio and Mississippi filled up, eastern manufacturers found a market for their products. This promoted economic sectionalism and the interplay of commerce. As the years progressed, the east became more and more a manufacturing section, the west a food-growing region, and the south a producer of cotton. During the first decades of the century the east shipped manufactured goods to the west, but the west sold most of its food products to the south, sending them *via* the river routes. The south in turn sold most of its product to

Europe, although some cotton was sent north. After the completion of the Erie and Pennsylvania canals, and especially after the opening of railroads, western products began to move east as well as south, and before the Civil War railroad iron had securely connected the region north of the Ohio with the economic interests of the northeast.

The Influence of the West Upon American Politics During This Period.—The influence of the west upon American politics during this period was exceedingly strong. In state politics the aggressive frontiersmen demanded the rewriting of the state constitutions in the interest of democracy, as, for example, in New York in 1821 and Virginia in 1830. Gerrymandering and other devices were employed to ward off the growing power of the new communities, but without great success. The buoyant and intense spirit of the new west was not to be denied. In national politics the story was the same. The Federalists, representing the eastern aristocracy and what manufacturing and business interests there were at the time, lost control of the executive and legislative branches at the end of the administration of Adams. With the accession of Jefferson, himself a Piedmont farmer, the influence of the western agriculturists was more in evidence, an influence which steadily increased as the representatives of one new state after another, elected by universal manhood suffrage, took their seats at Washington. During the period of the Republican presidents (Jefferson, Madison, Monroe, and John Quincy Adams) the spirit of the west was shown in the growth of strong national policies. The War of 1812 was a western war. The demands of the new communities led to the building of the National Road, notwithstanding the objections of the strict constructionists. The tariffs of 1816, 1824, and 1828 were passed only by the support of Kentucky and the northwest, influenced chiefly by the arguments of Henry Clay, whose "American System" conceived of a manufacturing east as the logical market for the agricultural west. The famous decisions of the arch-Federalist, Chief Justice Marshall, while bitterly resented by the westerners, were generally in line with the nationalism so strong in the west. After 1820 there came a reaction from the nationalism of the previous years, which encompassed the south and west and swept Andrew Jackson into the Presidency in 1828. With his sincere belief in the ability and right of the masses to rule, Jackson, a true representative of the new west, bitterly fought privilege and vested wealth as he saw it in the Second United States Bank. Although strongly grounded in the idea of states' rights, his frontier training gave him a vision which made him stand firmly against the nullification of South Carolina. Even the Whig party, in a sense a sort of resurrected Federalist party, which opposed him, was led by a westerner, Henry Clay.

NOTES FOR FURTHER REFERENCE

The most complete bibliography of the westward movement during this period is that of F. J. Turner and F. Merk, *List of References on the History of the West* (rev. ed., 1922), an absolute essential to further study. Consult also L. B. Schmidt, *Topical Studies and References on the Economic History of American Agriculture* (rev. ed., 1923). Interesting source material is incorporated in Bogart and Thompson, *op. cit.*, and in A. B. Hart, *American History Told by Contemporaries*, Vol. III. Among the most valuable of the contemporary accounts are Harriet Martineau, *Society in America* (1837); J. W. Monette, *History of the Discovery and Settlement of the Valley of the Mississippi* (1846), by an early inhabitant of the valley; Timothy Flint, *Recollections of the Last Ten Years* (1826); and *History and Geography of the Mississippi Valley* (1832); J. M. Peck, *Guide for Emigrants to the West* (1837); R. G. Thwaites (ed.), *Journals of Lewis and Clark* (1904-05); and *Early Western Travels* (32 vols., 1904-07).

For an interpretation of the movement the essays of F. J. Turner are the best, especially those collected in *The Frontier in American History* (1921). In F. J. Turner, *Rise of the New West* (1906), American Nation Series, the westward movement is interwoven with the political history. Excellent chapters are to be found in E. E. Sparks, *The Expansion of the American People* (1900), with valuable illustrations; Edward Channing, *A History of the United States* (1921), Vol. V, Chap. II; K. Coman, *Industrial History of the United States* (rev. ed., 1910), pp. 120-132 and 154-162; E. L. Bogart, *Economic History of the United States* (rev. ed., 1922), Chap. XII; J. B. McMaster, *History of the People of the United States* (6 vols., 1883-96), Vol. II, p. 144 ff.; Vol. III, pp. 100-142, 459-496; Vol. IV, pp. 381-428; Vol. V, p. 160 ff. The work of McMaster, in the opinion of Channing, is "the best bit of writing" on the subject "that has been done." Valuable also are Theodore Roosevelt, *Winning of the West* (4 vols., 1889-96), and Katherine Coman, *Economic Beginnings of the Far West* (1912).

The movement into the northwest is developed in F. W. Hasley, *Old New York Frontier* (1901), and in L. K. Mathews, *Expansion of New England*

(1909) ; and that into the southwest in U. B. Phillips, *Origin and Growth of the Southern Black Belts,* American Historical Review, IX, p. 798, and C. L. Skinner, *Pioneers of the Old Southwest* (1919), Chronicles of America Series.

On the diplomatic background of the Mississippi Valley see F. J. Turner, *Policy of France Toward the Mississippi Valley,* American Historical Review, X, pp. 249-279. The land policy of the United States may be studied in various articles in McLoughlin and Hart, *Cyclopædia,* and in P. J. Treat, *The National Land System, 1785-1820* (1910). For geography and routes of travel see E. C. Semple, *American History and Its Geographic Conditions* (1903). On routes to the interior read A. B. Hulbert, *Paths of Inland Commerce* (1920), Chronicles of America Series.

SELECTED READINGS

TURNER, F. J., *The Frontier in American History,* Chaps. IV-VII, XIII.
CHANNING, E., *A History of the United States,* Vol. V, Chap. II.
MATHEWS, L. K., *The Expansion of New England,* Chaps. VI-X.
MCMASTER, J. B., *History of the People of the United States.* See above.
COMAN, K., *Economic Beginnings of the Far West,* Parts III and IV.
SEMPLE, E. C., *American History and Its Geographic Conditions,* Chaps. X, XI.
CALLENDER, G. S., *Selections from the Economic History of the United States 1765-1860,* Chap. XII.
BOGART, E. L., and THOMPSON, C. M., *Readings in the Economic History of the United States,* Chap. XI.

CHAPTER X

AGRICULTURE FROM THE CLOSE OF THE REVOLUTION TO THE CIVIL WAR

COLONIAL agriculture had been carried on in both north and south by the crudest methods and in the most wasteful manner; as a whole the conditions were not radically disturbed by the Revolution. Nevertheless, the war was not without its effect. On the one hand, a market was provided by the British and the French armies which made the near-by farmers prosperous. On the other hand, certain deleterious effects are to be noted. Many farmers and laborers deserted agriculture for war; British cruisers intercepted the usual trade with the West Indies and many farms fell into decay. The excitements of campaigning were not conducive, at the conclusion of hostilities, to a speedy resumption of former activities. The uncertainty of the days of the "Critical Period" were enervating to ambition and industry, and the soldier with his pockets full of military land scrip was tempted to try his fortune beyond the mountains. Conditions improved gradually during the more prosperous years following the adoption of the Constitution, especially after the Continental wars opened up new markets,[1] but the unscientific methods, which so forcibly struck European travelers, continued for many years.

Effect on Agriculture of Unoccupied Public Lands and the Westward Movement.—For almost three hundred years the greatest single influence on American agriculture has been the existence of unoccupied land readily accessible to the people because of the liberal land policy of the government. As we have seen, a law of 1785 allowed anyone to buy 640 acres at one dollar an acre. This was changed in 1800 to allow the purchase of 320 acres or more at two dollars an acre, one-fourth in cash and the rest in three annual payments. The poor man was further favored in 1820 by an act which allowed him to purchase eighty

---

[1] See chap. xi.

acres (one-eighth of a section) at $1.25 an acre, this liberal tendency continuing until the Homestead Act of 1862 finally granted free land. Anyone endowed with a vestige of ambition and willingness to work might easily acquire land and a start in life. An ordinary laborer in the new country might save enough in a year to purchase his eighty acres, while a skilled mechanic or school-teacher, both in great demand on the frontier, might purchase in less time. The proceeds from the sale of two horses or eight cattle would buy a quarter section. Under this policy and during the years 1783-1860 most of the land between the Alleghanies and the Mississippi was taken up while the advance tide of immigration had swept into Texas, covered Missouri, and was penetrating into Kansas and Minnesota. The Mormons were carrying on agriculture around the Great Salt Lake and farms were springing up on the Pacific slope.

Perhaps the greatest effect of this easily acquired land was to perpetuate the old-fashioned and criminally wasteful methods. In a letter to Arthur Young the treatment of land in Virginia is thus described by Washington:

> The cultivation of tobacco has been almost the sole object with men of landed property, and consequently a regular course of crops have never been in view. The general custom has been, first to raise a crop of Indian corn (maize) which according to the mode of cultivation, is a good preparation for wheat; then a crop of wheat; after which the ground is respited (except from weeds, and every trash that can contribute to its foulness) for about eighteen months; and so on, alternately, without any dressing, till the land is exhausted; when it is turned out, without being sown with grass seeds, or any method taken to restore it; and another piece is ruined in the same manner.[1]

The tobacco grower, when his land wore out, found it cheaper to take up new land than to care for the old, and tobacco culture advanced westward into Kentucky. The same was true a little later of cotton, this factor contributing to the rapid occupation of Alabama, Mississippi, and Texas. Northern agriculture was

---

[1] *The Writings of George Washington*, collected and edited by Worthington C. Ford. New York, 1891, xi, 178 ff.

influenced in a similar way. Why slave to cultivate intensively a small farm in Massachusetts or New Hampshire, when an abundance of richer soil awaited the plow in western New York or the Ohio Valley? After the pioneer had reached the new country the temptation was always present to skim the cream from the fresh land and then sell out and try his fortune farther on. Before the development of rapid water and steam transportation the pioneer farmer was handicapped by a lack of markets, which naturally lessened incentive to improve his holdings. Enough could easily be raised to support his family without calling forth the latent possibilities of the new farm.

Even more harmful to agriculture than the ease of acquiring fresh lands after the old had been ruined was the fever of land speculation that seized the American people and continued decade after decade during this period. Exaggerated, to be sure, but with elements of truth is the picture drawn by an Englishman:

Speculation in real estate has for many years been the ruling idea and occupation of the Western mind. Clerks, labourers, farmers, storekeepers, merely followed their callings for a living, while they were speculating for their fortunes. There are no statistics which show how many Yankees went out West to buy a piece of land and make a farm and home, and live and settle, and die there. I think that not more than one-half per cent of the migration from the East started with that idea: and not even half of these carried out the idea. The German immigrants, indeed, were better entitled to be called settlers; but all classes and people of all kinds became agitated and unsettled, and had their acquisitiveness perpetually excited by land speculations in some shape or other—new railways, roads, proposed villages and towns, gold mines, water-powers, coal mines—some opportunity or other of getting rich all at once by a lucky hit. . . .

In the United States, vast numbers of the population became excited with dreams of sudden wealth, and the idea of a life of labour was scouted as the suitable destiny of mere timid, non-enterprising, weak people, or plodding Dutch or English, but altogether beneath the notice of Young America.

The people of the West became dealers in land, rather than its cultivators. Scorning cheap clocks, wooden nutmegs, and apple-parers, the Yankee, stepping from the almost ridiculous to the decidedly sublime, went out West, and traded in the progress of the country. Every one of any spirit, ambition, and intelligence (cash was not essential) frequented the National Land Exchange, a vast concern, extending from the Mississippi to the Pacific.

By convenient laws, land was made as easily transferable and convertible as any other species of property. It might and did pass through a dozen hands within sixty days, rising in price at each transfer; in the meantime producing buffaloes and Red Indians. Millions of acres were bought and sold without buyer or seller knowing where they were, or whether they were anywhere; the buyer only knowing that he hoped to sell his title to them at a handsome profit.[1]

Chicago between 1830 and 1840 was a center of speculation.

The plats of towns, for a hundred miles around, were carried there to be disposed of at auction. The eastern people caught the mania. Every vessel coming west was loaded with them, their money and means, bound for Chicago, the great fairy land of fortunes. But as enough did not come to satisfy the insatiable greediness of the Chicago sharpers and speculators, they frequently consigned their wares to eastern markets. Thus, a vessel would be freighted with land and town lots, for the New York and Boston markets, at less cost than a barrel of flour. In fact, lands and town lots were the staple of the country, and were the only articles of export.[2]

Speculating on the progress of the country entered into the very soul of the pioneer and was part of his being. In picking out a site for his claim the first consideration was its situation in respect to a possible rise in value. The typical frontiersman of these days was a man who laid out his claim, erected a rude cabin, and worked the land only until he could sell out at a profit.

Upon the eastern farmer the effect of all this was demoralizing. He, too, imbibed the spirit of land speculation and many an eastern farm was for sale by the proprietor who was anxious to unload and try his fortune elsewhere. This trading on the progress of the country gave the American farmer a migratory tendency which was impossible under Old World conditions and unusual in a group which by its very occupation moved slowly. It broke down local attachments and discouraged intensive improvements. The farmer was not building for his descendants, but for the first possible opportunity to sell. Eastern agriculture

---

[1] Mitchell, D. W., *Ten Years in the United States* (1862), p. 325.
[2] Ford, Thomas, *History of Illinois*, p. 181.

was further demoralized by western competition. After canals and railroads had provided an outlet for the bulky agricultural products of the west, the farmer of New England and the middle states found it impossible to compete successfully in the raising of grain and meat and was forced to reorganize his economy to that of truck farming, fruit raising, or dairying. This reorganization took time and was attended with difficulties. Farming in the east was further handicapped by difficulty in obtaining sufficient labor. Higher wages and greater opportunities drained off to the west the best of the farm hands, and tended to keep the labor cost higher in the east for both agricultural and industrial workers.

Southern Agriculture—Rise of Cotton.—Undoubtedly the most striking feature in the agricultural history of the first half century of the Republic was the rise of cotton. During the colonial period little progress was made in cotton culture. Lack of a market and the apparent overshadowing importance of tobacco discouraged its growth, despite the efforts of colonial governments to the contrary. Some cotton was raised to be woven into cloth, but its use was confined to the poorest classes. Interruption of trade with Great Britain during the Revolution, which cut off the importation of foreign fabrics, turned the minds of southerners to the production of cotton as a means of filling the need, and the legislatures of Maryland, Virginia, and South Carolina urged upon their people its possibilities. The chief difficulty with which the cotton grower had to contend was the separation of the seeds from the cotton fiber; even with slave labor it was a costly process.

The years 1790-1830 witnessed a veritable revolution in southern agriculture as far as the product was concerned. By the latter date cotton had become the principal southern crop and the largest single item of export from the country. This rapid development was occasioned first of all by the equally sudden opening of an available market. In England between the years 1767 and 1780 Hargreaves, Arkwright, and Crompton had constructed devices which did away with the old-fashioned spinning by hand and made it possible to spin rapidly by water power and later by steam. In 1785 similar improvements in weaving were inaugurated by Cartwright and an Industrial Revolution effected. Eventually the designs of these machines were smuggled into America[1] and factories established here. Rapid and cheap manufactures reduced prices and created demand. Any amount of cotton could now be manufactured; the problem was to obtain it. In the year 1786 almost by accident "sea-island" cotton was introduced from the Bahamas and was found to thrive along the seacoast. Having a longer fiber than the "short staple" variety, it was possible to separate the seeds by running the fiber between rollers turned in opposite directions. The demand for it abroad was immediate and its cultivation spread rapidly among the seacoast farmers. The planters were the more ready to take up the cultivation of cotton because the tobacco lands were wearing out and the market for indigo and rice had been injured by the separation from Great Britain. The south badly needed a new crop at the same time that the factory owners of England were clamoring for the raw product.

The impetus for cotton culture was undoubtedly present. Only the difficulty of separating the seeds from the cotton held back expansion. Climatic conditions kept the sea-island cotton to the lowlands, and the short-fibered upland cotton, upon which the greater part of the south had to depend, could be cleaned only with painful slowness at an average of about a pound a day per slave. The problem was solved in 1793 by Eli Whitney, a wide-awake Yankee and a mechanical genius, who had gone south immediately after graduation from Yale to teach school. In a letter to his father written in 1793 he tells the story of the invention of the cotton gin, one of the greatest episodes in American history:

I went from N. York with the family of the late Major General Greene to Georgia. I went immediately with the family to their Plantation about twelve miles from Savannah with an expectation of spending four or five days and then proceed into Carolina to take the school as I have mentioned in former letters. During this time I

---

[1] See chap. xii.

heard much said of the extreme difficulty of ginning Cotton, that is, separating it from its seeds. There were a number of very respectable Gentlemen at Mrs. Greene's who all agreed that if a machine could be invented which would clean the cotton with expedition, it would be a great thing both to the Country and to the inventor. I involuntarily happened to be thinking on the subject and struck out a plan of a Machine in my mind, which I communicated to Miller (who is agent to the Executors of Genl. Greene and resides in the family, a man of respectability and property) he was pleased with the Plan and said if I would pursue it and try an experiment to see if it would answer, he would be at the whole expense, I should loose nothing but my time, and if I succeeded we would share the profits. Previous to this I found I was like to be disappointed in my school, that is, instead of a hundred, I found I could get only fifty Guineas a year. I, however, held the refusal of the school untill I tried some experiments. In about ten Days I made a little model, for which I was offered, if I would give up all right and title to it, a Hundred Guineas. I concluded to relinquish my school and turn my attention to perfecting the Machine. I made one before I came away which required the labor of one man to turn it and with which one man will clean ten times as much cotton as he can in any other way before known and also cleanse it much better than in the usual mode. This machine may be turned by water or with a horse, with the greatest ease, and one man and a horse will do more than fifty men with the old machines. It makes the labor fifty times less, without throwing any class of People out of business.[1]

The contraption in its first crude form consisted of a cylinder equipped with teeth projecting through strips of metal which drew in the cotton fiber, leaving the seeds behind, and a second roller, equipped with brushes to free the teeth from the lint, revolving in the opposite direction. Operated by hand, the machine would clean fifty pounds a day, by water a thousand. Whitney's idea was to set up his engines throughout the south and gin the cotton at so much a pound for the planters. But the demand was so great that his machines were stolen and his patents infringed. South Carolina paid Whitney and his partner, Miller, $50,000 for the unrestricted use of his machines and North Caro-

---

[1] *Correspondence of Eli Whitney*, in American Historical Review, iii, pp. 99-101.

lina $12,000, most of which was spent in lawsuits to protect the patent.

Southern Agriculture—Effects of the Invention of the Cotton Gin.—The effects of the invention of the cotton gin were immediate and far-reaching. Cotton became the greatest commercial crop of the south and the largest single export of the United States. With the exception of 1808, the year of the Embargo Act, and 1812-14, the war years, the growth of cotton culture was rapid and steady. The production in 1790 was 4,000 bales of 500 pounds, which jumped, after the invention of the gin, to 73,222 in 1800. Each decade following saw roughly a doubling of production: 177,824 bales in 1810; 334,728 bales in 1820; 732,218 bales in 1830; 1,347,640 bales in 1840; 2,136,003 bales in 1850; 3,841,416 bales in 1860.[1]

Cotton, which constituted in 1810 about 22 per cent of the value of the total export, reached in 1860 over 57 per cent. In value the exports of cotton increased during the same years from $66,757,970 to $333,576,057. The importance of cotton in the

[1] AVERAGE ANNUAL PRODUCTION AND EXPORTS OF AMERICAN COTTON FOR FIVE-YEAR PERIODS, 1791-1865, AND AVERAGE ANNUAL PRICES FOR MIDDLING UPLANDS COTTON IN NEW YORK AND LIVERPOOL.

Years	Average annual production in the United States in pounds.	Average annual exports from the United States in pounds.	Percentage of crops exported.	Average New York prices for middling uplands—cents.	Average Liverpool prices for middling uplands—pence.
1791–1795	5,200,000	1,738,700	33.43	31.7	No data
1796–1800	18,200,000	8,993,200	49.41	36.3	No data
1801–1805	59,600,000	33,603,800	56.38	25.0	15.4
1806–1810	80,400,000	52,507,400	65.38	18.9	18.4
1811–1815	80,000,000	42,269,400	52.83	14.8	20.5
1816–1820	141,200,000	91,144,800	67.38	26.2	16.7
1821–1825	209,000,000	152,420,200	72.93	16.2	9.2
1826–1830	307,244,400	254,548,200	82.84	10.9	6.5
1831–1835	398,521,600	329,077,600	82.57	11.9	8.0
1836–1840	617,306,200	513,315,800	83.15	13.0	6.7
1841–1845	822,953,800	691,517,200	84.03	7.7	4.7
1846–1850	979,690,400	729,524,000	74.46	8.7	5.2
1851–1855	1,294,422,800	990,368,600	76.51	9.6	5.4
1856–1860	1,749,496,500	1,383,711,200	79.51	11.5	6.7
1861–1865	No data	No data	No data	58.9	19.1

Source: *The South in the Building of the Nation,* vol. v, p. 211.

export trade explains the source of southern wealth, their attitude on the tariff, their overconfidence at the beginning of the war, and their eventual failure.

The invention of the cotton gin and the rise of cotton culture were important in the opening up of the southwest. Beginning in Georgia and South Carolina, cotton growing after 1800 spread into North Carolina and southeastern Virginia and across the mountains into Tennessee. After it was realized that the rich alluvial soil of Alabama and Mississippi was better suited to cotton than the uplands, prospective growers entered in a steady stream, pushing before them Indians, Spaniards, and cattle ranchers. Crossing the Mississippi, they drove the cattle men into Texas, brought about the Mexican War with its resulting annexations, and by 1860 had preëmpted the coastal regions of Texas. The exhausting methods of early cotton culture, as with tobacco, wore out the soil and impelled the planter to search for further virgin land. As late as 1820 over half the cotton grown in the country was raised in Georgia and South Carolina. By 1850 Alabama had taken first place, with Georgia second, Mississippi third, and South Carolina fourth. In 1860 Mississippi (1,195,699 bales), Alabama (997,978 bales), and Louisiana (722,218 bales) raised over half the total product, while the yield from Texas (405,100 bales) now surpassed that of South Carolina (353,413 bales). The most pronounced periods of expansion in the southwest were in the flush years preceding the crisis of 1837, when land speculation there reached its highest point, and in the period immediately following the annexation of Texas. With the shifting of the center of cotton culture went a change in the centers of trade, of wealth, and of political power. The importance of Charleston and Savannah declined, while Memphis, Mobile, and New Orleans rose as commercial centers. In the decade 1850-60 New Orleans handled about half the cotton crop.

The effects of the invention of the cotton gin permeated the whole social as well as the economic life of the south by fastening upon it the system of slavery. Slavery as an institution was decidedly under fire in the years immediately following the Revolution, and its desirability was questioned. Men like Washington and Jefferson freed their slaves and with other prominent southerners urged the abolition of the whole iniquitous system. The anti-slavery movement had made considerable progress when the invention of the cotton gin again made slavery profitable. Here was a crop preëminently adaptable to slave labor. Almost the whole negro family could work in the cotton fields during the greater part of the year. A single overseer could supervise a large number of slaves, while cotton culture as a whole was suitable to the crude and wasteful methods of ignorant slave labor in a country where large stretches of fresh land could be taken up as the old wore out. Although the number of slaves increased from 697,890 in 1790 to 3,953,580 in 1860, the growth lagged behind the general rate of population increase, and as larger regions were brought under cultivation the demand for slaves grew. Before the cotton gin was invented a good negro brought $300, but twenty years later the price had doubled. The average value was around $800 in 1830, $1,200 in 1850, and from $1,400 to $2,000 in 1860. Abolition sentiment not only died away in the south, but also in certain of the border states, where it had been strongest. Selling their surplus slaves, the border states, particularly Virginia, actively took up the business of slave raising and slave trading; at the same time along the whole coast the law prohibiting the slave trade was openly evaded. Fearful of the increasing criticism of the system, the slave owners, ably represented in Congress, met attacks boldly and worked unceasingly to extend the slave area.

In the north the effects of the rise of cotton were also felt. The infant textile industries of New England were stimulated; and north of the Ohio the farmers found a market for their corn and pork on the cotton plantations.

Southern Agriculture—Tobacco, Sugar, Rice, and Other Crops.—By far the greatest southern product during the colonial period was tobacco. The last decade of the century saw it at the height of its importance, when over half the population of the tobacco states was engaged in or dependent on its cultivation. It headed the list of exports in 1791 when $4,591,293 worth was shipped abroad. Still the leading export in 1800, it declined rapidly in importance, owing to the disastrous effects of the Embargo Acts and the War of 1812, the competition of Cuba, Columbia, and Sumatra, high import taxes imposed on it by foreign countries, the gradual exhaustion of the land, and the rise of cotton. The effect of the shifting of land and slaves from tobacco to cotton was distinctly noticeable; the industry remained practically stationary until 1840, when the exports again equaled the amount shipped in 1790. The substitution of curing by fire for the old-fashioned open-air process and the introduction of a new yellow-leaf species gave the industry a new life. Between 1850 and 1860 the production more than doubled, so that at the opening of the war the south continued to be the greatest producer of tobacco in the world. About half the tobacco raised at this time was exported to England and Germany.

Virginia continued to be the leading tobacco state, but the center of culture moved steadily west. By 1859 the yield of Kentucky, Ohio, Tennessee, and Missouri was greater than that of the tidewater states—Virginia, Maryland, and North Carolina. Virginia and Kentucky together furnished more than half of the tobacco grown in the United States. Louisville and St. Louis were the western centers for the trade and preparation of the leaf.

PRODUCTION OF LEAF TOBACCO IN 1860 BY STATES [1]

State	Pounds
Virginia	123,968,312
Kentucky	108,126,840
Tennessee	43,448,097
Maryland	38,410,965
North Carolina	32,853,250
All others (including territories)	85,401,997
Total	434,209,461

Sugar cane was brought by the Jesuits from Santo Domingo to Louisiana in 1751, but sugar was not successfully refined until toward the end of the century, when Etienne De Bore made a spectacular success of sugar growing and refining on his planta-

---

[1] Eighth Census, Agriculture, p. xcvi.

tion near New Orleans. His experiments were followed by a rapid increase in the growth and refining of sugar. The average annual production from 1820 to 1830 was 52,000 hogsheads, increasing to an average of 280,000 hogsheads in the decade 1850 to 1860. Although most of the southern states raised some sugar, it was only in Louisiana that the cultivation assumed large proportions. Here the production amounted in 1823 to 15,401 tons (2,240 pounds each), in 1844 to 102,678 tons, and in 1861 to 235,856 tons; in 1844 there were in Louisiana 762 sugar estates cultivated by 51,000 slaves and capitalized at $60,000,000. Most of the sugar consumed in the Mississippi Valley was refined at New Orleans, St. Louis, or Cincinnati.

Rice, which had been a leading colonial crop along the seacoast, continued to be successfully grown, the production more than tripling in the years 1820 to 1850. The high-water mark was reached in 1850, after which a decline set in. Up to that year South Carolina raised more than one-half the crop, with Georgia producing most of the rest. The rice grower as a class was undoubtedly the most scientific and intelligent farmer in the pre-war south. Forced by necessity to use continually the same soil, his attention was early turned to the use of fertilizer, and to the reclamation of swampy land. The fact that only the moist soil was suitable to rice resulted in plantations of diversified products quite different from the cotton and the tobacco plantation. The rice grower was encouraged in his efforts by a steady market and the acknowledged fact that the Carolina product was the best in the world.

Another crop of considerable importance in certain sections of the south was hemp. The beginnings of the hemp industry were almost simultaneous with the settlement of Kentucky. Hemp became as important here in the early years as tobacco had been in early Virginia. It was used principally for bagging and rope for cotton bales and for negro cloth, but the market extended to the north and east. From Kentucky hemp growing spread into Tennessee, Arkansas, and Missouri, where it was the chief factor in winning the latter state to slavery. Henry Clay, the champion of the "American System," was the representative of the Kentucky hemp growers. Kentucky in 1850 produced 17,787 tons of hemp, Missouri 16,028, and Tennessee 595.

Although the economic life of the south was dominated by these commercial crops already described, other products were raised, chiefly for home consumption. In addition to the usual vegetable gardens, cereals were raised in 1859 as follows: Indian corn, 433,067,490 bushels; wheat, 49,157,701 bushels; oats, 32,163,229 bushels; and rye, 4,070,475, with small amounts of barley and buckwheat. The southern soil was more adaptable to corn than the other cereals, and it served as the chief food for the slaves. The amount raised, however, was barely half that of the five states north of the Ohio. In the animal industry the south, particularly Kentucky, was famous for the excellent speed horses bred there. Kentucky was also noted for its breeding of shorthorn cattle and Hampshire hogs. Preëminence of Virginia during these years in sheep raising can be traced back to Washington and the subsequent efforts of agricultural societies. That the interest of the south in livestock was not confined to fancy breeding is evidenced by the estimated value of all livestock, which the census of 1860 placed at $381,778,601.

The Great Lakes and the Opening of the Prairies.—More significant in the history of American agriculture than even the rise of cotton was the opening to agriculture of the region between the Ohio River and the Great Lakes, westward into the prairies of Iowa and Kansas. Its white population in 1790 of a few thousand fur traders and French farmers grew by leaps and bounds; Ohio entered the Union as a state in 1803, Indiana in 1816, Illinois in 1818, Missouri in 1821, Michigan in 1837, and Minnesota in 1858. The lands of western New York and northern Ohio, probably at one time covered by water, are endowed with a fine rich soil, free of bowlders and easy to cultivate, and blessed with a climate tempered by the expanses of water. Nearly 100,000 square miles of water surface furnished the finest inland navigation in the world, facilities augmented by a natural outlet through the Mohawk Valley to the eastern coast. Tributaries of the Ohio and the Missouri permit the carrying off of the produce of the North Central states by way of the Mississippi. Natural

AGRICULTURE FROM 1783 TO 1860

highways were supplemented in the early decades by the Erie Canal, the Pennsylvania Canal, and the Cumberland Road, and after 1840 by the railroads, easily built in this region.

The first settler, pressing into the wooded stretches bordering on the Ohio and its tributaries, found the soil richer, perhaps, but the problems much the same as on his New England or Kentucky farm. It was a case of clearing the forest and duplicating the old crops. As the land was taken up and the advance pushed into Indiana and Illinois, Minnesota and Iowa, the pioneer found himself in a different type of country with new problems to be faced. Here was to be found prairie land without forest, which made building expensive and firewood scarce. Frequent absence of water was a deterrent, while the tough prairie soil was extremely hard to break with the early wooden plows. There was also a mistaken idea that the prairie land was not fertile. When the real richness of the soil was realized, especially after improvements had been made in plows, the land was quickly put under cultivation and the occupation was the more rapid because the land needed little preparation for sowing. Practically no clearing was necessary, the soil was simply broken with a plow or ax, and a crop of corn planted. The first crop usually broke up the sod sufficiently to let it rot, and in the second year a crop of wheat was possible.

The development of the Lake region and prairies depended largely upon markets and transportation. The Industrial Revolution was turning western Europe into an urban manufacturing civilization and was beginning to make itself felt in America. This brought an increased demand for foodstuffs. In the south the planters believed it more profitable to purchase food and turn the entire labor energy of the slaves to raising the staple crop. As a consequence the North Central states developed into a great food-producing region, whence the products in two continuous streams moved eastward by canal and railroad and south by steamboat and rail. Western New York and northern Ohio became interested in fruit growing and dairying, while the rest of the middle west produced meat and cereals.

What little exportation of foodstuffs there had been from this region previous to 1825 had been in the form of beef, pork, and mutton on the hoof. Before the era of canals and railroads and the development of the packing industry, the farmers of the western counties of Pennsylvania, Maryland, Virginia, and Ohio annually took great droves of cattle and hogs along the highways leading into Philadelphia and Baltimore. Livestock raising has always been a leading frontier occupation, for every frontier has provided animal food at low cost. In the middle west, where hog growing was destined to become a great industry, conditions were especially favorable. Hogs at first could find maintenance on the mast [1] and herbage of the forests on the fringe of civilization; later on, corn, the principal food for hogs, was found especially adaptable to the soil and soon became the leading crop. About 1818, meat packing as an industry came into existence west of the Alleghanies. The development of markets, of transportation, of banking facilities, and the growth of a sufficient population to provide a steady supply of animals had progressed far enough by this time to build up a considerable meat-packing industry in Cincinnati. In the early years meat was preserved by salting and smoking and, as this process did not require large capital, packing establishments sprang up in many of the larger towns. Nevertheless, Cincinnati retained the leading position until 1860, although the center moved continually westward. As Chicago grew in size, its superior transportation facilities, combined with its nearness to the new pork states, enabled it to take the lead. Upon the meat industry and the manufacture of such by-products as leather goods, fertilizers, glue, candles, soap, lard, salt, and barrels, the early prosperity of both Cincinnati and Chicago rests. It is still basic in these centers.

The chief stimulation for cereal production came in the early years from the need of food for animals and of ingredients for the manufacture of whisky. Corn was too bulky to transport, but it could be fed to animals, which in turn could be driven perhaps hundreds of miles to the market. Corn and rye could be reduced to the less bulky and more concentrated form of spirituous

---

[1] Fruit of the oak and beech or other forest trees.

AGRICULTURE FROM 1783 TO 1860

liquors. It was this fact that made whisky a characteristic pioneer product, and that explains the opposition of the frontier farmer of Pennsylvania to the excise tax of Hamilton. Corn during all of this period remains far and away the chief agricultural product, although little found its way out except in the form of whisky or meat.

After the opening of the Erie Canal and the growth of railroads it was possible to ship cereals east as well as south and a great impetus to other lines of cereal production was given. Wheat was consigned either to millers at Cincinnati, Louisville, and St. Louis to be prepared for local and southern consumption, or to the east *via* Buffalo and Pittsburgh. Some eventually reached Europe *via* New York or New Orleans. The Middle Atlantic states in 1850 still produced more wheat than the North Central, but in the next decade western production increased 125 per cent to 15.5 for the Middle Atlantic. As the western wheat grew in importance, the milling center shifted from the coast streams to Rochester on the Erie Canal, and then to Chicago and St. Louis, and eventually to the great mills of the northwest.[1]

Beyond the Mississippi.—The extension of agriculture beyond the Mississippi into Missouri, Minnesota, Iowa, Kansas, and

---

[1] Farm Statistics for the regions now occupied by the states of Ohio, Indiana, Michigan, Illinois, Missouri, Minnesota, Iowa, Nebraska, Wisconsin, and Kansas. (Area, 384,510,080 acres.)

	1840	1850	1860	1870
Population	3,352,000	5,404,000	9,092,000	12,967,000
Cash value of farms, including machinery and livestock	....	$914,637,000	$2,523,256,000	$5,132,815,000
Corn, bu.	105,853,000	222,209,000	406,146,000	439,112,000
Wheat, bu.	27,518,000	43,842,000	95,004,000	194,764,000
Oats, bu.	30,335,000	42,329,000	62,951,000	159,690,000
Rye, bu.	1,141,000	840,000	4,105,000	6,473,000
Barley, bu.	472,000	832,000	4,909,000	10,603,000
Buckwheat, bu.	886,000	1,602,000	4,105,000	6,473,000
Tons of hay	1,594,000	3,336,000	7,059,000	12,440,000
Pounds of butter and cheese from dairies	20,880,000	105,110,000	181,308,000	228,367,000
Cheese from factories	....	....	....	23,904,000
Value of animals slaughtered	....	$25,419,000	$62,722,801	$208,586,441

Nebraska was a continuation of the frontier experience of the states north of the Ohio, with essentially the same products and civilization. Farther south, in Louisiana and Arkansas the fringe of settlements was pushed beyond the Mississippi long before the region east had been occupied, and tobacco, cotton, and sugar as well as cereals were grown. American advance into Spanish territory began in 1821, when Moses Austin of Connecticut obtained a grant from the Spanish government upon which to plant a colony of settlers from the United States. The same year his son, Stephen, established the first Anglo-American settlement in the Brazos and Trinity valleys. Probably 30,000 Americans in the next fifteen years followed Austin's band, taking up land mainly along the rivers between San Antonio and Nacogdoches and the coast. Most of them were southerners who either brought their slaves and raised cotton or engaged in ranching. This immigration took place during the revolutionary epoch when Mexico was winning her independence from Spain. American frontiersmen, however, found themselves equally unwelcome to both Spaniards and Mexicans, and friction developed. The Mexicans resented the occupation of the land and passed acts prohibiting slavery. The land-hungry Americans revolted, won their independence from Mexico in 1836, and were annexed to the United States in 1845. By 1860 Texas had a population of 604,215 living on 42,891 farms and raising 405,100 bales of cotton, besides having extensive ranching industries.

New England Agriculture.—In rural New England during the first half of the century changes were enacted which brought about a veritable agricultural revolution. As in colonial days, farm life at the opening of the century was characterized by self-sufficiency. With few markets for his produce the farmer was unable to buy from without; what he needed in the way of food, clothing, and tools was raised or manufactured on the farm. After 1810 this condition began to change gradually as the Industrial Revolution slowly transformed New England into a manufacturing center. Although the early factories were small, they were widely dispersed, and the growth of the factory system was accompanied by a growth of urban population. Only three towns in New England in 1810 boasted over 10,000 inhabitants—Boston, Providence, and New Haven, with a combined population of 56,000—whereas in 1860 there were twenty-six such cities with a population of 682,000. This urban population, which in 1860 amounted to about one-third of the total of southern New England, provided a market for agricultural products. Farming, which had been uniform, began to give way to specialization: intensive farming of root crops in the regions adjacent to the cities, wool growing in the hilly country, and the fattening of beef cattle in the Connecticut Valley. New markets created new interest in farming, which was augmented by agricultural societies and by the use of improved machinery after 1830, particularly the iron plow.

No sooner had the New England farmer accustomed himself to the new conditions than a second readjustment was forced upon him by the growth of railroads. Cheap transportation made it impossible for him to compete with western wool, beef, and pork; in consequence, although corn continued to be the agricultural backbone of New England, the production of beef and pork declined. Attention was turned to dairying and truck gardening. This agricultural upheaval brought striking changes to rural life. The self-sufficient domestic system broke down before the factory system. Markets for agricultural products brought ready money to the rural districts, new comforts to the farmer, and a higher standard of living. Women released from household industries sought work in the factories. At the same time many men, discouraged by the uncertainties of this period of readjustment, followed the lure of new opportunities in the cities or of richer lands in the west. The whole situation was further complicated by a readjustment of land values usually, but not always, detrimental to the farmer.

Transformation of Farming—Labor-saving Machinery.—The period from 1830 to the Civil War witnessed the beginnings of revolutionary changes in American agriculture. Fresh agricultural labor as well as enlarged markets were provided by the rapidly increasing population due (1) to natural growth in a civilization conducive to large families, (2) to immigration which amounted during these years to over 4,500,000, and (3) to the growth of the factory system with its increase of urban population. The rapid building of railroads after 1850 stimulated the farmer by bringing both the products and the markets of the world to his door. The discovery of gold in California and an increased demand for foodstuffs due to the repeal of the English Corn Laws in 1846, both helped to raise prices. Scientific agriculture made rapid strides, but especially to be noted was the invention and adoption by the American farmer of labor-saving machinery.

In a nation where land was plentiful but labor scarce it was to be expected that the first great advance would come in labor saving rather than in land saving devices. American farming implements were designed to increase the yield per man rather than the yield per acre. In the latter we have not kept up with more thickly populated countries. Up to this time a farmer's equipment consisted usually of a crude wooden plow, harrows, hoes, shovels, forks, and rakes, poorly constructed and often homemade. The first great improvement was the metal plow, which came into general use after 1825. Colonial plows had been covered with strips of iron, and as early as 1790 Charles Newbold of New Jersey was working on the idea of a cast-iron plow, which he finally patented in 1797. He spent his entire fortune of $30,000 to introduce his invention, but the farmers would have none of it, many claiming that the iron poisoned the soil and made the weeds grow. The conception of an improved plow was not lost sight of, men like Jefferson and Webster making studies of types and materials. Eventually Newbold's plow of one solid piece of cast iron was improved on by Jethro Wood of New York, who patented in 1819 a plow, the different parts of which interlocked and could be replaced if broken. Manufacturers and inventors infringed his patents, but the farmer profited. Eventually moldboards were designed more adaptable to breaking the matted grasses of the prairies, the sticky soil of which also necessitated a smoother surface; this was provided by the steel mold-board first made in 1833 by John Lane of Chicago, and a few years later by John Deere. The life work of James Oliver of South Bend,

AGRICULTURE FROM 1783 TO 1860

Indiana, resulted in 1869 in the chilled-steel plow, which eliminated blowholes and made the metal less brittle. After the 'thirties the metal plow was adopted rapidly by the farmer. In Massachusetts alone in 1845 there were seventy-three plow-manufacturing concerns with an output of 61,334 plows and other implements. In 1855 the number of establishments had decreased to twenty-two, but the yield had increased to 152,688 plows valued at $707,175. Two factories in Pittsburgh in the 'thirties were turning out plows at the rate of 34,000 a year.

Simultaneously with the improvement in plows came the invention of the mowing and reaping machines, to keep pace with the increased production which the new plows made possible. The grain cradle had come into use about 1800 and had considerably facilitated both the cutting and gathering up of the grain, but harvesting was still a painfully slow process. Many men experimented during the following years on the problem of a reaper, and many minds contributed to the eventual machine. A patent for a mowing machine had been granted to William Manning of New Jersey in 1831, but the two men who succeeded in building a practical reaper were Obed Hussey and Cyrus McCormick, whose patents were dated, respectively, 1833 and 1844. Hussey's poverty prevented the rapid manufacture of his machines and they were not put out in quantity until 1845. Greater success was experienced by Cyrus McCormick, born of Scotch-Irish ancestors, who had emigrated from Pennsylvania into the Shenandoah Valley, and who inherited from his father a mechanical genius and an interest in farm machinery. Turning his attention to the development of a practical reaper, he continued, after securing his first patent, to manufacture them in his workshop on the Virginia farm and to perfect further improvements. Believing his machines would be more practical on the level land of the west, he moved in 1845 to Brockport, New York, on the Erie Canal, and three years later to Chicago, where by 1860 he was turning out 4,000 machines a year.

The principle of these early mowers and reapers was the same— a number of blades or "wipers" swept the grain against the cutting surface, after which it was pushed on to a receiving table and automatically shoved off when enough had been gathered to make a sheaf. Laborers following the machine tied the sheaves. These early machines, clumsy as they were, showed their superiority over hand labor and improvements came rapidly. By 1855 nearly 10,000 were in use. At the International Exposition at Paris in that year an American reaper cut an acre of oats in twenty-one minutes, one-third of the time consumed by the foreign makes. In 1857 the United States Agricultural Society held a national trial at the New York State Fair at Syracuse, where forty mowers and reapers were entered. The results demonstrated that the most serious disadvantages, such as side draught, clogging, and inability to start in standing grain had been practically eliminated. In the succeeding years the machines were widely introduced, a fact which explains the great crops during the labor shortage of the Civil War.

A necessary further improvement was furnished when a satisfactory thresher was added to the mechanical devices upon which the farmer could depend. With the old-fashioned hand flail progress was painfully slow; from eight to sixteen bushels a day was the average production per man. Experiments went on both in Europe and America in an attempt to devise flails which might be attached to cylinders and driven by horse or steam power, but it was not until 1850 that the separator was attached to the thresher and the whole process of threshing and winnowing was carried on in the same machine. At the Paris Exposition the American machine entered by Hiram and John Pitt won first prize, threshing 740 liters in an hour to 410 liters by the English machine, its nearest competitor.

The invention and improvement of other farming implements accompanied the greater inventions. The horse hay-rake, which did the work of from eight to ten men, came into use in the early part of the century, and the curing of hay was aided years later by the invention of the tedder. The production of corn was further increased by the invention of the corn planter and the two-horse cultivator.

Scientific Farming.—In England the work of Arthur Young, Jethro Tull, Viscount Townshend, Robert Bakewell, and

AGRICULTURE FROM 1783 TO 1860

others had in the eighteenth century demonstrated to Englishmen what might be done in the way of scientific farming. Little interest was shown in agricultural improvements in the colonies, and it was not until after the Revolution that the American farmer became interested in better methods. The lead was taken by such wealthy planters as Washington and Jefferson, who were farmers on a big scale and intensely interested in agricultural experiments. Washington, who has been described as "not only the greatest man, but the greatest agriculturist of the period," turned from tobacco raising to an intensive cultivation of other products. He was the founder of the mule-raising industry in the country, the fine Kentucky breed of later years descending directly from the best asses of Spain and France sent as presents to him from Lafayette and the king of Spain. His experiments in sheep raising, continued by George Washington Parke Custis, did much to better the breed of sheep in the south. A few of the excellent merino sheep had been smuggled out of Spain, but it was not until the Napoleonic wars that it was possible to obtain them in large numbers. Jefferson and Livingston were especially interested in the introduction and wide distribution of merinos. A similar improvement also took place in the quality and size of cattle. English shorthorn or Durham cattle were imported into Kentucky in 1817 and in succeeding years great numbers were bought by farmers who desired to better their stock. Henry Clay in 1817 imported the first Herefords, but this breed did not develop rapidly until found adaptable to the Texas ranges in the 'seventies. Other standard breeds were now imported and farmers came to take more of an interest in the bettering of their herds.

Until 1840 the usual mode of travel was by horseback, and this fact undoubtedly explains the early improvement of breeds of horses. The thoroughbred stallion, Messenger, progenitor of the Standard-bred horse so closely connected with the agricultural history of Kentucky, was brought into New Jersey from England in 1788. The American saddle horse, also largely bred in Kentucky, descends from the thoroughbred stallion, Denmark, brought into Kentucky in 1839. The Morgan horses of New England were also excellent types. The best kind of American horse as developed by the Kentucky breeders was a saddle horse with easy motion and a rapid walk.

The knowledge of these superior breeds, of the new inventions and of the improved methods of tillage, was disseminated by five means: (1) agricultural societies, (2) agricultural fairs, (3) farm periodicals and literature, (4) agricultural schools, and (5) government aid. The beginnings of all of these are to be found in this period. The first agricultural society was the Philadelphia Society for Promoting Agriculture, founded in 1785 and including in its membership Washington and Franklin. Similar societies were founded in South Carolina in the same year, at Kennebec, Maine, in 1787, in New York City in 1791, and in Massachusetts in 1800. The first half of the century saw like organizations springing up all over the country, whose purpose was to spread information, lend mutual aid, and stimulate improved methods by holding fairs and offering prizes. These early organizations were neither cultural nor political in their purpose.

A distinguishing feature of American agricultural life is the county fair. The first agricultural fair was held in Washington in 1804, but the idea took root chiefly through the influence of Elkanah Watson, who organized a cattle show at Pittsfield, Massachusetts, in 1810, an exhibition which led to the founding in that year of the Berkshire Agricultural Society, the first permanent fair association in America. Watson pushed his idea in other states and similar societies were founded. The first state aid for agricultural fairs was granted by New York in 1819, when $20,000 was appropriated for two years. The United States Patent Office in 1858 printed a list of over 900 agricultural societies, most of which were state or county organizations existing for the purpose of holding fairs. The interchange of ideas, the new information obtained, and the rivalry promoted by these fairs have made them of great importance, especially during this period in the exhibition of the worth of the new machinery.

Agricultural journalism sprang up along with the associations and fairs. Its real beginning dates from 1819, when John S. Skinner founded *The American Farmer* at Baltimore, a weekly paper which continued until 1862. This paper was followed in the same year by *The Plow Boy,* printed at Albany, by *The New England Farmer,* first printed at Boston in 1822, and *The New York Farmer* at New York in 1827.

Agricultural education in America undoubtedly commenced with special instruction in the established schools. The first institution devoted principally to the teaching of agriculture was the Gardiner Lyceum, established at Gardiner, Maine, in 1821. This school went out of existence after several years, as did a number of other seminaries founded with the same purpose in view. Agricultural education waited on state aid to become a real factor. The state constitution of Michigan, adopted in 1850, provided for a college of agriculture. In accordance with this provision the legislature appropriated $40,000 for buildings, instruction, and maintenance, and in 1857 a state college of agriculture was opened, the first institution of its kind in America. Two years later Maryland and Pennsylvania followed the example of Michigan in establishing state-supported institutions. The great growth of agricultural education, however, came after the passing of the Morrill Act in 1862.

National aid to agriculture commenced in 1839, when Congress appropriated $1,000 to be expended by the Commissioner of Patents for the collection of statistics and investigations for the promotion of agriculture. After 1842, with the exception of one year, gradually increasing appropriations were made for this purpose. Annual agricultural reports were printed after 1854. Agricultural matters were handled by the Patent Department until 1862, when a separate bureau of agriculture was set up. These modest beginnings of federal and state interest in the problems of the farmer hardly presaged the immense government activity along this line in recent years.

NOTES FOR FURTHER REFERENCE

The best bibliography that has been collected on American agriculture is that of L. B. Schmidt, *Topical Studies and References on the Economic History of American Agriculture* (rev. ed., 1923), an essential to further study. There are many articles of exceptional value in Vol. IV of Bailey's *Cyclopedia of American Agriculture*. A recent short and popular account is that of A. H.

Sanford, *The Story of Agriculture in the United States* (1915). There are chapters in both E. L. Bogart, *Economic History of the United States* (rev. ed., 1922), and Isaac Lippincott, *Economic Development of the United States* (1921), which emphasize the outstanding features. Among the best of the early accounts are those of C. L. Flint, *Agriculture in the United States, in Eighty Years' Progress* (1869); also in the *Annual Report*, U. S. Department of Agriculture (1872), and in the *First Annual Report of the Massachusetts Board of Agriculture* (1854). See also the Introduction to the volume on Agriculture in the *Eighth Census of the United States* (1860) and W. N. Brewer, Report on the Cereal Production of the United States in the *Tenth Census* (1880) volume on Agriculture, Part II. Some source material may be found in Bogart and Thompson, *op. cit.*, and in T. N. Carver, *Selected Readings in Rural Economics* (1911).

On New England farming of this period read P. W. Bidwell, *Rural Economy in New England at the Beginning of the Nineteenth Century,* in Transactions of the Connecticut Academy of Arts and Sciences, Vol. 20 (1916), and *The Agricultural Revolution in New England,* in the American Historical Review, Vol. XXVI, No. 4 (1921). On southern agriculture the following are valuable: M. B. Cairnes, *The Slave Power* (1863); M. B. Hammond, *The Cotton Industry* (1897); A. B. Hart, *Slavery and Abolition,* in the American Nation Series; F. L. Olmsted, *Journeys and Explorations in the Cotton Kingdom* (1861); James A. B. Scherer, *Cotton as a World Power* (1916), a study in the economic interpretation of history; M. Jacobstein, *The Tobacco Industry in the United States,* Columbia University Studies, Vol. XXVI, No. 3 (1907). Excellent chapters will be found in McMaster, Vol. VII, Chap. 76, and in Rhodes, Vol. I, Chap. 4. There are a number of very rewarding studies of different phases of southern agriculture in *The South in the Building of the Nation,* Vol. V. See Chap. XVI for a further bibliography of southern agriculture.

On the public domain see I. Donaldson, *The Public Domain* (1884); L. H. Haney, *A Congressional History of Railways in the United States,* Vol. I, to 1850 (1908); Vol. II, 1850-1887 (1918), Bulletin of the University of Wisconsin (1910); R. T. Hill, *The Public Domain and Democracy,* Columbia University Studies, Vol. XXXVIII (1910); the Annual Reports of the Commissioner of the General Land Office from 1860 to 1900; and George M. Stevenson, *The Political History of the Public Lands from 1840-1862* (1917). *The Centennial History of Illinois,* Clarence W. Alvord, editor-in-chief, contains much valuable material. See especially S. J. Buck, *Illinois in 1818* (1917).

Among the literature on this period are: Robert Dudley (pseud.), *In My Youth* (1914), the fictitious autobiography of a Quaker settler of Indiana in the middle of the nineteenth century; and Hamlin Garland, *A Son of the Middle Border* (1922), an autobiographical narrative of middle-western family life in the period after the Civil War, continued in *A Daughter of the Middle Border* (1921). Herbert Quick, *Vandemark's Folly,* deals primarily with Iowa in the 'fifties, but gives a picture of the life of western New York and Wisconsin settlements. A sequel, *The Hawkeye* (1923), pictures Iowa in the middle and later years of the nineteenth century.

SELECTED REFERENCES

CARVER, T. N., in Bailey's *Cyclopedia of American Agriculture,* Vol. IV, pp. 50-69.

SANFORD, A. H., *The Story of Agriculture in the United States,* pp. 92-199.

BIDWELL, P. W., *The Agricultural Revolution in New England,* in the American Historical Review, Vol. XXVI, No. 4 (1921).

HAMMOND, M. B., *The Cotton Industry,* Chaps. I-III.

CHAPTER XI

THE AMERICAN MERCHANT MARINE AND THE DEVELOPMENT OF FOREIGN COMMERCE TO 1860

Colonial Shipping and the Revolution.—Some attention has been devoted to the history of ship building and the merchant marine during the colonial period and the Revolution. We have seen how the colonist, dependent upon the ocean for communication with the home land and with other colonies, commenced the construction of ships almost immediately upon his arrival. The best of material was at hand and the market was steady. The fisheries and the West Indian trade absorbed most of the ships built; although Europeans, after the Yankee ships had demonstrated their superiority, bought heavily. Ship building was a favored and protected industry from the start. At the opening of the Revolution at least one-third of the tonnage under the English flag was American built; Massachusetts was said to own one sea-going vessel for every one hundred inhabitants. So anxious was England to increase her power on the sea that Parliament forbore from passing legislation directly antagonistic toward the building of ships; lax enforcement and easy evasion of the Navigation Acts up to 1763 permitted an extensive commerce and normally developed a ship-building industry.

The immediate effect of the Revolution upon American merchant shipping was disastrous. The fishing industry off the New England coast was for the time being effectually limited and the British navy broke up the intercourse with the West Indies, Spain, and Portugal. As the war progressed, however, merchantmen discovered ways of evading the enemy, and in privateering shipping interests found an outlet for their energies and compensation for their losses. While the slow and heavy ships were driven from the sea, the lighter and more easily managed were mounted with guns and directed against English commerce. In 1781 there were 449 American privateers, whose crews were composed of

MERCHANT MARINE AND COMMERCE 243

almost as many men as were serving in the Continental army. Silas Deane wrote in 1777 to Robert Morris that as American privateers and cruisers "sailed quite around Ireland and took or destroyed seventeen or eighteen sail of vessels, they most effectually alarmed England, prevented the great fair at Chester, occasioned insurance to rise, and even deterred the English merchants from shipping goods in English vessels at any rate, so that in a few weeks forty sail of French ships were loading in the Thames on freight, an instance never before known."[1] He adds that "even the packet boats from Dover to Calais were for some time insured." A witness before a special Parliamentary inquiry in 1778 stated that the losses suffered by British merchants from American privateers "could not be less than two million two hundred thousand pounds." Privateering served the purpose not alone of harassing the enemy, but of keeping alive the maritime spirit and holding capital to the shipping industry. In consequence, the close of the war found the American merchant marine in a fairly prosperous condition.

The history of our merchant marine from the Revolution to the Civil War may be divided for the sake of convenience into the following periods:

1781–1789, Period of uncertainty.
1789–1810, The first period of swift growth and prosperity.
1810–1820, The War of 1812 and the unfavorable reaction from the war.
1820–1830, The second period of rapid growth and prosperity.
1830–1860, Period of overproduction in ship building and the beginning of decline.
1860–1865, Civil War.

Period of Uncertainty, 1781-1789.—The years between the Revolution and the adoption of the Constitution were years of uncertainty for the American shipping interests. The West In-

---

[1] The Deane Papers, vol. ii, p. 108, New York Historical Society Publications. Letter, dated "Paris, 23rd. August, 1777," to Robert Morris, who was a member of the Secret Committee of Congress. This entire letter, pp. 106-111, is well worth reading for its account of the activities of "American ships of war, private as well as public."

dian trade, so important to the colonists, was forbidden by an Order in Council of July, 1783, which proclaimed that all trade with the British West Indies must be carried on in British ships. Although these orders were evaded to a considerable extent, the fact that as many as fifteen thousand slaves in the British West Indies died from starvation between 1780 and 1787 shows but too clearly that trade in the bulky articles of foodstuff was greatly restricted. British ship owners were forbidden the privilege of purchasing vessels built in America, and the list of American products which could lawfully be imported into England was restricted chiefly to naval supplies. Our ambassadors failed in their attempts at relief through treaties, and retaliatory measures were unproductive because of colonial jealousies. Ship owners were ruined, and ship builders and seamen were destitute of work. No group of interests more clearly saw the need of the encouragement and protection of a strong central government and no group more ardently demanded it.[1]

Perhaps it was the discouraging outlook for trade in former channels that led during this period to the opening of new sources of commerce. The *Empress of China* sailed in February, 1784, from New York for Canton, returning in May of the year following. Elias Hasket Derby of Salem in 1785 sent his *Grand Turk* to the Far East. Before five years had elapsed after the return of the first vessel, the ship owners of New England and New York were trading regularly with Asia and had broken the long monopoly of the East India Company.

First Period of Growth and Prosperity, 1789-1810.—Commerce with the Far East ushered in the golden age of American shipping. Sensitive in the extreme as merchant shipping is to outside influences, the favorable reaction occasioned by the adoption of the Constitution, by tonnage taxes imposed on foreign vessels, and by the establishment of public credit, was nevertheless astonishing. The tonnage registered for foreign trade jumped from 123,893 tons in 1789 to 981,017 in 1810. The imports carried in American bottoms during the same period

---

[1] See above, pp. 178, 181.

increased from 17.5 per cent to 93 per cent, and exports in American bottoms from 30 per cent to 90 per cent.

Shipping interests had been among the most ardent in the support of the new Constitution, and the first Congress hastened to their protection. The first act passed by the first Congress (with the exception of a formal statute with reference to the taking of oaths) was the Act of July 4, 1789, which, although designed for the "encouragement and protection of manufactures" and for obtaining revenue, gave very real aid to shipping by allowing a discount of 10 per cent of the tariff duties upon imports brought to this country in ships built and owned by American citizens. In order to encourage the newly developed trade with the Far East, this same act allowed upon tea imported direct from the East a reduction in duty which made the tariff paid by the American ship less than half that of the foreign vessel, and at the same time dealt a blow at the East India Company by placing high duties on tea bought from them in Europe, even if imported in American ships. The next act of the same Congress, that of July 20, 1789, imposed duties of six cents a ton on American-built ships owned by Americans upon entering our ports, but thirty cents a ton was charged on American-built ships owned by foreigners, and fifty cents a ton on foreign-built and owned ships. It was provided at the same time that American ships in the coastwise trade should pay tonnage duty only once a year, while foreign ships must pay it at every entry. This act presaged the early absorption of the coastwise trade by American ships.

An act of 1790 on the government and regulation of seamen provided a code of law in advance of the time. It stipulated that a written contract must be entered into between master and seamen specifying the voyage and rate of wages, without which the master could not have full control of his men. He was also required to pay them the highest current wages, with the ship itself as a guaranty. Masters were liable to severe penalties for abandoning American sailors in a foreign country, while seamen who signed articles and deserted their ship might forfeit their wages and be brought back under compulsion. With but few changes, this has been the basis of the law covering seamen until the salutary modifications of the La Follette Act of 1915. It helped during this period of rapid expansion to maintain on American ships both an excellent personnel and a high standard of discipline. The development of a strong merchant marine was also helped by the creation in 1798 of an American navy, which repeatedly gave a good account of itself in the succeeding years.

Much of the remarkable growth in shipping was absorbed in the first years of this period in the Far Eastern trade, but when the energies of the European nations were taken up with the Napoleonic wars, American merchantmen were to be found wherever business was to be obtained. "The unfolding of the great West," says Marvin, "had scarcely begun. Kentucky was not admitted to the Union until 1792; Tennessee, until 1796. Not only did most of the American people live within reach of the ocean, but the ocean everywhere seemed to be the nearest, the most natural, and the most inviting field of adventure. It was true of many more American towns than tide-encircled Boston that 'Each street leads down to the sea.' Down these streets went most of the young men who had dreams in their heads and iron in their blood, and they always found ships waiting."[1] It was the day when Captain John Kendrick and Captain Robert Gray showed the way to the northwest coast and laid the foundations of the claims of the United States to the region of Oregon and Washington. "At the end of 1793," says Ugo Rabbeno, "the tonnage of the United States exceeded that of every other nation except England; their foreign trade ranked in point of value next to that of England, and, proportionally to the population, the United States were the first commercial nation of the world."[2]

The years 1789 to 1810 were not, however, years of peaceful and uninterrupted development. During practically the entire period we were in difficulties with our chief rivals, England and France, while between the years 1801 and 1805 our tiny navy engaged in war with Tripolitan pirates in defense of our merchant marine. In 1792 war broke out between France and Austria, a war which was eventually to involve all Europe and to

---
[1] Marvin, W. L., *The American Merchant Marine*, p. 43.
[2] Rabbeno, Ugo, *The American Colonial Policy*, p. 141.

MERCHANT MARINE AND COMMERCE   247

continue, with but two short interruptions until the year 1815. The superiority of the British navy was soon apparent and it was not long until the merchantmen of France and her allies had entirely disappeared from the sea. The result was to throw the carrying trade of France into the hands of American merchantmen flying the flag of the only neutral nation of importance on the ocean. While all Europe was engaged in a life-and-death struggle Yankee merchants and ship owners reaped handsome profits from transporting the products of the French, Dutch, and Spanish colonies. Intent upon weakening her foe as well as hindering this new rival, England began to enforce the so-called "Rule of War of 1756," by which she maintained that a neutral might not enjoy in time of war a carrying trade prohibited in time of peace. This rule would have prevented American merchantmen from trading with the French West Indies, as the latter had been theoretically closed before the war. Provisions at that time were considered contraband of war, and the warships and privateersmen of both England and France were ordered to capture any vessels so laden. As the chief exports of the United States were provisions, the effect, if the orders were enforced, would be disastrous. Especially irritating was the claim made by Great Britain that British sailors found on American ships might be taken and forced to serve on British men-of-war. In the later years of the war, as seamen became scarcer, Great Britain became more and more unscrupulous in her use of this alleged right. In the years 1806 and 1807 it was believed by the State Department that as many as six thousand American seamen were serving under compulsion in the British navy.

Although the chief trouble was with England, the first actual fighting occurred with France. Rightly or wrongly, she felt that under the Treaty of 1778 we were in duty bound to aid her in the war with Great Britain; she resented, therefore, Washington's proclamation of neutrality (1793); and she maintained that we had in the Jay Treaty (1794, ratified in 1796), and in several other ways, violated our treaty agreements with her. Her resentment led to repeated violations of our neutrality and insults to our government, of which the "XYZ affair" created most stir in this country. While war was never officially declared, France in 1796 informally stated that the alliance with the United States was at an end; and Congress, on July 7, 1798, voted to abrogate all treaties with France. Organization of a Navy Department (1798) and other preparations for war were pushed forward, Congress suspended commercial intercourse with France and her dependencies, and between 1798 and 1800 naval engagements between the two nations were fought in the West Indies and eighty-four French ships, mostly privateers, were captured. A treaty in 1800, which settled for the time being the chief points in controversy, closed this episode.

Notwithstanding interference from British warships, French privateersmen, and Tripolitan pirates, the years up to 1807 were prosperous for American shipping and agriculture. Registered tonnage in foreign trade had mounted from 123,893 tons in 1789 to 810,163 in 1807; exports from $19,012,041 in 1792 to $108,-343,150 in 1807, and imports from $29,200,000 in 1792 to $246,-843,150 in 1807. The proportion of this trade (import and export) carried in American ships jumped from 23.6 per cent in 1789 to 92 per cent in 1807. Only in two subsequent years (1825 and 1826) was this mark surpassed. The ship-building industry was especially prosperous, taking care not only of this remarkable growth in local tonnage, but actually, between 1789 and 1812, selling to foreigners over 200,000 tons. The rapid increase in exports was made up chiefly of provisions, a market for which had been created by the war. Europe, too busy fighting to raise sufficient foodstuffs, called increasingly upon America to furnish grain and meat, as well as such raw materials as cotton, wool, and leather. While the prices of foodstuffs and raw material soared, farmers reaped a golden harvest, shared by the shippers who transported the products and the merchants who were able to dispose during these prosperous years of increasing amounts of imported goods. Sailors' wages rose from eight to thirty dollars a month and foreigners became naturalized in order to partake of the huge profits of American ship owners. The whole situation was remarkably similar to the first years of the World War, 1914-17, when the United States, as the great neu-

MERCHANT MARINE AND COMMERCE

tral, profited from supplying foodstuffs and other products to the warring nations.

This period, one of prosperity not only for merchant shipping, but for the country at large, was halted in 1807. The year 1805 witnessed the two great battles of Austerlitz and Trafalgar, the former leaving Napoleon master of the Continent, and the latter leaving Great Britain mistress of the sea. By this time both nations had come to the conclusion that only through economic boycotts could eventual victory be gained. Already in 1804 English Orders in Council had declared French ports, from Ostend to the Seine, under blockade; these orders were extended in May, 1806, to include the coast from Elbe to Brest. Napoleon, recently victorious over Prussia, issued from Berlin his answer in the form of a decree declaring the British Islands under blockade. A year later, November, 1807, England issued a second Order in Council declaring that no neutral vessel should trade with the ports of France or her allies, without touching and paying duties at a British port. Napoleon, now thoroughly convinced that the ruin of England could only be brought about by a strict enforcement of the "Continental System," retaliated with the Milan Decree, 1807, declaring that any vessel sailing to or from Great Britain or her colonies was liable to seizure, as was any ship that submitted to the Orders in Council of 1807 and paid duties to England. It is true that these were merely paper blockades, impossible of enforcement on the part of France with her navy driven from the seas, and but inadequately enforced by England, yet they resulted in the capture of about 1,600 American ships and $60,000,000 worth of property. These orders and decrees, if enforced, meant the virtual prohibition of neutral trade, and as such meant the elimination of American ships from the European trade.

This utter disregard of neutral rights and the heavy losses inflicted upon American shipping, accompanied as they were by such virtual acts of war as the firing upon the American frigate *Chesapeake* by the British man-of-war *Leopard,* finally aroused Jefferson to action. Anxious to preserve peace, he believed that Europe could be brought to terms by an embargo. At his advice Congress in December, 1807, passed an Embargo Act which prohibited American ships from sailing to foreign ports and allowed coasting trade only under condition that the owner give bonds double the value of the cargo that the same be relanded in the United States. This applied even to the smallest fishing smack. The navy and revenue cutters in later acts were put at the disposal of the executive. Instead, however, of starving Great Britain into submission, the act bid fair to ruin our own shipping. American exports dropped from $108,343,150 in 1807 to $22,430,960 in 1808; imports from $138,500,000 to $56,990,000. New York, said a British traveler at that time, "was full of shipping, but they were dismantled and laid up. Their decks were cleared, their hatches fastened down, and scarcely a sailor was to be found on board. Not a box, bale, cask, barrel, or package, was to be seen upon the wharves. Many of the counting houses were shut up or advertised to be let; and the few solitary merchants, clerks, porters, and laborers that were to be seen, were walking about with their hands in their pockets. . . . The coffee-houses were almost empty; . . . The streets near the waterside were almost deserted; the grass had begun to grow upon the wharves. . . ."[1] The testimony of this gentleman may not have been unprejudiced, but McMaster estimates that 55,000 sailors and 100,000 mechanics and laborers were thrown out of work, that during the period of the Embargo Act ships lost $12,500,000 in net earnings. The customs revenues sank from $16,000,000 to a few thousand. Thirteen hundred men in New York City alone were thrown into prison as debtors ruined by the Embargo.[2]

The disastrous effects of the Embargo Act were felt in all of the states, but especially in New England, where one hundred Massachusetts towns adopted resolutions against the Embargo. Smuggling was rampant and from certain New England ports

---

[1] Lambert, John, *Travels through Canada and the United States of North America in the Years 1806, 1807, and 1808*, vol. ii, p. 65.

[2] W. P. Trent in his excellent sketch of Jefferson in *Southern Statesmen of the Old Régime*, p. 80, comments as follows on Jefferson's efforts to avoid war: "In short, what chiefly affects me when I study the whole matter is the pathos of it,—a philosopher and a friend of peace struggling with a despot of superhuman genius and a Tory cabinet of superhuman insolence and stolidity."

"loaded vessels literally fought their way to the sea," for New Englanders found that they could carry on a prosperous commerce notwithstanding the many adverse factors. The pressure of the shipping interests was so strong that in March, 1809, the Embargo Act was repealed, and in its place was substituted the Non-intercourse Act, which prohibited trade only with Great Britain and France and their possessions. The Non-intercourse Act was repealed in 1810, and new legislation enacted called the Macon bill, which provided that as soon as either England or France withdrew its decrees against our shipping, the Non-intercourse Act would be revived against the other country. Napoleon immediately (August 5, 1810) announced that the Berlin and Milan decrees were repealed, and Madison issued a proclamation reviving the Non-intercourse Act against England if she did not repeal the Orders in Council before February 2, 1811. England ignored the proclamation and Napoleon, in spite of his announced repeal of the Berlin and Milan decrees, continued, as before, to seize and rob American ships wherever he could lay hands on them. Nevertheless, the years 1808 and 1809 showed increases in both imports and exports and the registered tonnage in foreign trade reached 981,019 tons in 1810, a mark not equaled again until 1847. With all the setbacks and discouragements it had been a period of remarkable growth and prosperity, the worst blow being the Embargo Act dealt by our own government.

The War of 1812 and the Reaction from the War, 1810 to 1820.—The year 1811 found the United States fast drifting into a war with England. Although France and England had both apparently vied with each other in heaping insults upon our government and bringing losses upon our shipping, it was the latter nation that was in a position to enforce her attitude and to cause the most trouble. In the spring of 1812 as a preliminary to war and in order to give merchantmen a chance to reach a safe harbor, Congress imposed a third embargo; and on June 18, 1812, declared war. The causes, as reviewed by Madison in his message of June 1, were violations of our flag on the high seas, confiscation of our ships, illegal impressment of seamen, blockade of our ports, the obnoxious Orders in Council, and the inciting of Indians against our borders. All of these causes but one had to do with the violations of the rights of American merchantmen, yet it was the young "War Hawks" of the south and west, led by Henry Clay, who urged the war in Congress; while New England, which had been most disastrously affected by the British Acts, bitterly opposed the war, not only refusing in certain instances to fight and loan money to carry on the war, but actually rendering aid to the British and making preliminary movements toward secession. The result of the embargoes in England had been to raise the cost of food, and with war in sight the Orders in Council were withdrawn five days after war had been declared. Modern cables might have prevented the war.

The War of 1812 was primarily a war on the sea. With the exception of the battle of New Orleans, which was fought after peace had been signed, the conflicts on land were of minor importance and were usually unsuccessful to American arms. Attempts to invade Canada were discouraging failures. The American navy of but twenty-three vessels of all classes gave an excellent account of itself, capturing 254 naval and merchant ships of the enemy before being destroyed or shut up in American harbors. By the end of the war the British navy, numbering at that time about a thousand ships, had effectually blockaded the American coast and captured some 1,400 merchant vessels and fishing boats. Exports dropped from $61,316,832 in 1811 to $6,927,441 in 1814; imports during the same period from $53,400,000 to $12,965,000. The most effective work on the American side was done by the privateersmen, who took 1,300 prizes valued at $39,000,000. The self-reliant American seaman, realizing that he was fighting his own battle, was at his best aboard a privateer. That his work bore fruit may be gleaned from the words of a resolution of Glasgow merchants passed at a meeting in September, 1814, which said:

That the number of privateers with which our channels have been infested, the audacity with which they have approached our coasts, and the success with which their enterprise has been attended, have proved injurious to our commerce, humbling to our pride, and discreditable to the directors of the naval power of the British nation, whose flag, till of late, waved over every sea and triumphed over every rival. That there is reason to believe that in the short space of less than twenty-four months, above eight hundred vessels have been captured by that power whose maritime strength we have hitherto impolitically held in contempt. That at a time when we are at peace with all the world, when the maintenance of our marine costs so large a sum to the country, when the mercantile and shipping interests pay a tax for protection under the form of convoy duty, and when, in the plenitude of our power, we have declared the whole American coast under blockade, it is equally distressing and mortifying that our ships cannot with safety traverse our own channels, that insurance cannot be effected but at an excessive premium, and that a horde of American cruisers should be allowed, unresisted and unmolested, to take, burn, or sink our own vessels in our own inlets, and almost in sight of our own harbors.[1]

By the spring of 1813 flour was selling in England at $58 a barrel, beef at $38, pork at $36, and lumber at $72 a thousand. Peace, which was signed on December 24, 1814, made no mention of the causes for the war—impressments, right of search, or blockades—but with the removal of the causes and with the record our seamen had made in asserting their rights, it seemed likely that the latter would not soon again be called into question.

The natural reaction after the war was immediately felt. The conclusion of peace released shipping and the registered tonnage for foreign trade went up from 674,633 in 1814 to 854,295 in 1815. Exports increased from $6,927,441 in 1814 to $93,281,133 in 1818; imports from $19,892,441 in 1814 to $147,103,000 in 1816. This sudden prosperity was not entirely healthy. The flood of imports glutted the market and forced manufacturing concerns which had started during the war to suspend operations. This, combined with the disturbed and unsettled financial condition, brought in 1818 a sharp decrease in tonnage registered in foreign trade, and in 1819 a marked decrease in both exports and imports.

The years immediately following the war marked the beginning of legislation to establish commerce on the principle of reciprocity.

---

[1] Quoted by Winthrop L. Marvin, *The American Merchant Marine*, p. 129.

The Act of March 3, 1815, provided that all discriminating duties imposed by former laws on the tonnage of foreign vessels or the goods imported therein would be repealed in the case of any foreign nation abolishing its discriminating and countervailing duties against us. On the other hand, an act was passed in 1817, in imitation of the European navigation laws, forbidding the importation of goods from any foreign port, except in American vessels or vessels of the country from which the goods came, and at the same time absolutely closing the coasting trade to foreign vessels, but providing for repeal in the case of nations removing such restrictions upon our vessels. An act of 1828 provided for reciprocity with foreign nations in the indirect carrying trade. The result of these three acts was the beginning of a long series of reciprocal treaties with foreign nations, commencing with the treaty of July 3, 1815, with Great Britain, which abolished differential duties with respect to direct trade with the two countries. This was followed by a treaty with France in 1822, and Prussia in 1828, guaranteeing reciprocal liberty in commerce, and in the succeeding years with most of the countries of Europe and Central and South America. It was not until 1830 that England opened the West Indian ports to United States commerce, and the restrictions of the Act of 1817 against her were removed.

Second Period of Rapid Growth and Prosperity, 1820-1830. —The years 1820 to 1830 witnessed a second period of remarkable growth and prosperity. Although the amount of tonnage registered in foreign trade did not equal that of the years 1815-17 or the figures of the next two decades, the proportion of American carriage in the foreign trade reached 92.5 per cent in 1826, a larger percentage than has been attained before or since. J. R. Soley maintains that "In every respect we may say that this period represents the most flourishing condition of shipping in American history."[1] Not only were we carrying practically all of our own goods, but the reputation of Yankee ship builders for turning out models which surpassed in speed, strength, and durability any vessels to be found, brought about the sale between

---

[1] In N. S. Shaler, *The United States of America,* vol i, article, "The Maritime Industries of America," p. 539.

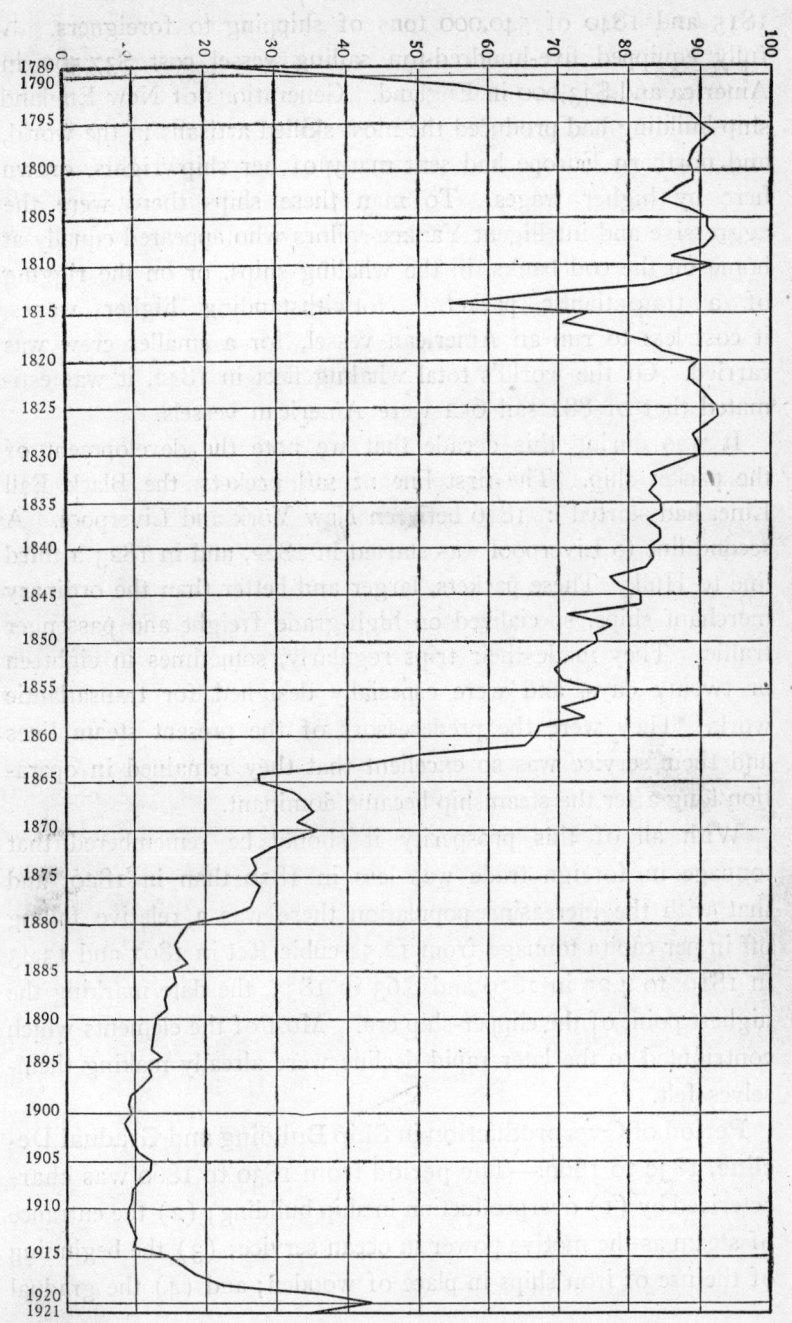

PERCENTAGE OF IMPORTS AND EXPORTS CARRIED IN AMERICAN VESSELS, 1789–1921 [1]

[1] Statistics taken from J. R. Soley in N. S. Shaler, *The United States of America*, and from the *Statistical Abstract*, 1921, p. 426.

1815 and 1840 of 540,000 tons of shipping to foreigners. A fully equipped five-hundred-ton sailing vessel cost $37,500 in America and $43,000 in England. Generations of New England ship building had produced the most skilled artisans in the world, and northern Europe had sent many of her shipwrights, drawn here by higher wages. To man these ships there were the aggressive and intelligent Yankee sailors who appeared equally at home on the cod banks, in the whaling ships, or on the rigging of a transatlantic packet. Notwithstanding higher wages, it cost less to run an American vessel, for a smaller crew was carried. Of the world's total whaling fleet in 1842, it was estimated that of 882 sail 652 were American vessels.

It was during this decade that we note the development of the packet ship. The first line of sail packets, the Black Ball Line, had started in 1816 between New York and Liverpool. A second line to Liverpool was started in 1822, and in 1823 a third line to Hull. These packets, larger and better than the ordinary merchant ships, specialized on high-grade freight and passenger traffic. They made their trips regularly, sometimes in eighteen or twenty days, and were especially designed for transatlantic work. They were the predecessors of the present steam lines and their service was so excellent that they remained in operation long after the steamship became dominant.

With all of this prosperity it should be remembered that tonnage in foreign trade was less in 1830 than in 1820, and that with the increasing population there was a relative falling off in per capita tonnage from 12.54 cubic feet in 1807 and 13.43 in 1810, to 4.25 in 1839 and 8.63 in 1855, the date marking the highest point of the clipper-ship era. Most of the elements which contributed to the later rapid decline were already making themselves felt.

Period of Overproduction in Ship Building and Gradual Decline, 1830 to 1860.—The period from 1830 to 1860 was characterized by (1) overproduction in ship building; (2) the entrance of steam as the motive power in ocean service; (3) the beginning of the use of iron ships in place of wooden; and (4) the gradual decline in the proportion of American merchandise carried in American ships.

Robert Fulton had demonstrated the practicability of steam as a motive force for propelling ships in his memorable voyage up the Hudson in 1807, and steam craft were on the Mississippi as early as 1812. Notwithstanding the fact that the *Savannah*, a three-hundred-ton vessel, built in New York and equipped with steam as well as sails, had made a twenty-five-day trip across the Atlantic in 1819, eighteen days of which she sailed under steam, it was generally considered that steamboats were adaptable only for river and coast navigation. The utilization of coal in the production of steam and the invention of the screw propeller, both of which occurred in the 'thirties, hastened an inevitable development.

Englishmen had been experimenting with steam packets on short services to Rotterdam, Hamburg, and Gibraltar for a number of years before the *Sirius* and the *Great Western* in 1838 made the trip to New York by steam alone in seventeen and a half and fifteen days, respectively. These records proved that wooden side-wheeled steamers could make the trip either way in shorter time than the fastest sailing packet. The British government was far sighted enough to realize that the motive power of the immediate future was steam, and in 1839 heavily subsidized the Cunard Company, which began its career in 1840 with four side-wheeled wooden ships. This policy of subsidization, which has been continued to the present time by Great Britain, aided materially not only in giving her maritime interests a start in the new type of ships, but in helping them win and hold supremacy on the ocean. The Peninsular Company, afterwards the Peninsular and Oriental, was established in 1837, and the Pacific Steam Navigation Company in 1840, both subsidized.

Almost as revolutionary as the gradual substitution of steam for sailing vessels was the very gradual substitution of iron and later steel ships for those of wood. With an abundance of coal and iron close to the sea, with skilled mechanics and cheap labor, Great Britain from the start forged ahead. Already by 1853 one-fourth of the total tonnage built in Great Britain were steamships, and more than one-fourth were built of iron. In the same year 22 per cent of our tonnage was constructed for steamships, but scarcely any iron ships were built here. The Yankee ship builder, overconfident in the recognized superiority of his inimitable clipper ship, was blinded to the fact that the future of the sea was for the nation which could build the cheapest and best iron steamships.

The thirty years from 1830 to 1860 were characterized by extremely rapid production in ship building. The 538,136 tons registered in foreign trade in 1831 had increased to 1,047,454 in 1847 and to 2,496,894 in 1862, a figure which (until surpassed in the recent war) represented the culmination of our ship-building tonnage. From 1848 to 1858 ship building had been maintained at an average of 400,000 tons a year. This construction was caused, in the first place, by the development after 1845 of the previously mentioned and justly famous American clipper ship, the fastest sailing ship afloat, an extraordinary product of decades of ship building and the intense rivalry between steam and canvas. Designed for speed, the clipper was built on sharp lines and carried a maximum of canvas. It was intended primarily for long voyages, and was used especially for the California and Far Eastern trade. Given a fair breeze, a clipper ship could outdistance a steamship. It was not uncommon for a clipper to sail over 300 miles a day; the *Flying Cloud* on a ninety-day run to San Francisco made 374 miles in one day. The *Comet*, on an eighty-day voyage from San Francisco to New York, averaged 210 miles a day. Records of eighty-six days from Singapore to New York, eighty-four days from Canton to New York, and ninety-six days from Manila to Salem were made between 1851 and 1853. It appeared that the American ship builder, before he relinquished his supremacy, was intent upon demonstrating to what heights of efficiency and speed a sailing ship could attain.[1]

---

[1] It is believed that the lines of the clipper ship were originally derived from a type of Chinese coasting vessel, known as the "Singapore fast boat," a half-block model of which was brought from China by Captain Robert H. Waterman, one of the most famous of the clipper-ship commanders, and now reposes in the Clark Collection at the Massachusetts Institute of Technology.

In the second place, there was an increased demand for shipping. This was occasioned chiefly by the discovery in 1848 of gold in California. The overland routes were slow and perilous, and the demand for passage and shipments to the Golden Gate swamped accommodation. The Yankee packet could make the trip around the Horn in about three months, thus giving a prolonged lease of life to the building of sailing ships. The wars between Great Britain and China in 1840-42 and 1856-60 threw a part of the China trade into American hands. The revolutionary outbreaks of 1848 interrupted European trade, with a resulting benefit to Americans, while the Crimean War, which occupied many European boats in transporting troops and supplies, gave new openings to American ships. In addition it may be said that the natural increase of commerce due to the growth in population, wealth, and production necessitated increased shipping. In no commodity was increase in production so great as in that of cotton, the production of which increased from 4,000 bales of 500 pounds in 1790 to 3,841,416 bales in 1860. Our exports had increased from $19,012,041 in 1791 to $333,576,057 in 1860, and of this last figure the exports of cotton were valued at $191,806,555. Imports increased during this same period from $48,212,041 to $353,616,119. It may also be added that increasing immigration contributed to the demand for shipping facilities.

Notwithstanding these demands, there was a decidedly unhealthy element to this remarkable activity in ship building. In the first place the demand from Europe because of the Crimean War was abnormal; between 1854 and 1859 the European nations were buying 50,000 tons of shipping as against 10,000 tons in normal years. Unfortunately, this increase in the building of wooden ships came at a time when their days were numbered, for between 1850 and 1860 the share of ocean freight carried by steamers increased from 14 to 28 per cent. When the abnormal demand for sailing ships should let up, as it did in 1858, it meant that shipyards built and equipped for the production of wooden ships and shipwrights trained for a type no longer wanted would be idle, while foreign shipyards already engaged in the building of the iron steamship would be in a decidedly superior position. The panic of 1857 precipitated the crash. In 1858 ship building, which had been maintained for the preceding ten years at an average of 400,000 tons a year, dropped to 244,000 and in 1859 to 156,000. At the same time the combined imports and exports carried in American bottoms was steadily declining, only 65.2 per cent being carried in 1861 as against 92.5 per cent in 1826. The period from 1830 to the panic of 1857 was characterized by extraordinary prosperity in ship building, while, at the same time, our carrying trade was by degrees slipping away to foreign bottoms. This was due not only to the gradual change to iron steamships in the building of which Great Britain immediately took the lead, but because of fundamental economic changes in progress throughout the United States. Capital was finding new and more profitable fields for investment. Manufacturing, which grew rapidly after the War of 1812, absorbed some of it; while considerable amounts were drawn into such internal improvements as canals and railways. Between 1820 and 1838 the states contracted debts of over $110,000,000 for the building of roads, canals, and railroads; from 1830 to 1860 over 30,000 miles of railroad were built, most of the capital coming from private investors. The minds of the venturous and ambitious turned from the sea to the unexploited west, and capital turned from ship building to the development of natural resources.

The Civil War, 1860-1865.—The elements contributing to the decline of the merchant marine were already operative before the Civil War, and the result would undoubtedly have been the same if that conflict had not come. The war, however, accentuated a tendency already existing and dealt a blow from which the merchant marine failed to recover until artificially revived during the recent war. In 1861 registered American tonnage in foreign trade amounted to 2,496,894 tons and in 1865 to 1,518,350, while the per cent of imports and exports carried in American ships dropped in the same years from 66.5 to 27.7. The decrease of tonnage in these years of some 900,000 tons was due to two causes. The first of these was the loss sustained from Confederate cruisers such as the *Alabama*, built and fitted out in England contrary to

MERCHANT MARINE AND COMMERCE 261

the laws of warfare. The second and most important was the sale during the four years 1862-65 of 751,595 tons of shipping abroad, occasioned by (1) lack of confidence, decline in profits from continual Confederate captures and high insurance rates, and (2) decline in export business due to the cessation of cotton shipments abroad.

Whaling and Fishing.—The whaling industry was distinctly retarded during the period of the Revolution and the continental wars, 1775-1815, but after Napoleon's final defeat it revived. The recorded tonnage had increased from practically nothing in 1814 to 35,000 tons in 1820, and then steadily, during the golden age of the industry, to 157,000 in 1841 and 198,000 in 1858. After that year the decline set in. Before 1791 whaling had been confined to the Atlantic, but gradually cruises were extended into other oceans, until after 1835 the industry was largely confined to the Pacific. Almost the entire whaling fleet hailed from New York and New England. Sag Harbor, New York, boasted 63 vessels in 1846, though the great center was New Bedford, Massachusetts, and the smaller towns close by, the fleet of which in 1857 numbered over 200 vessels, employing over 10,000 seamen. Nantucket ranked second and New London third as whaling ports during this period, the former relinquishing her colonial supremacy because of the shallow harbor which prevented the entrance of the large vessels. Many of the coast towns engaged in the traffic on a smaller scale and derived rich profits. Boston and New York were the chief exporting centers, while the chief foreign markets for sperm oil were the West Indies, South America, and northern Europe; and for whalebone France, England, and the Baltic region. The larger part of the oil product was absorbed in the domestic market. During the height of its prosperity the whaling products surpassed in value those of all the rest of the fishing industry combined. The average annual production from 1835 to 1860 was 118,000 barrels of sperm oil, 216,000 barrels of whale oil, and 2,324,000 pounds of whalebone, with the average annual value about $8,000,000. After 1860 the decline was rapid, owing partly to the growing scarcity of whales, but chiefly to the discovery of mineral oils.

The cod-fishing industry went through the same early setbacks that were experienced by the whalers during the two wars of independence, but eventually recovered after a series of laws had been passed remitting most of the duty on the imported salt which was used. The registered tonnage engaged in cod fishing increased from 25,000 in 1790 to 136,654 in 1860. In that year there were close to 2,500 vessels with crews numbering 18,000 fishing for cod, the annual value of which was about $3,000,000. After 1818 an increasing number of vessels went out after mackerel, the maximum tonnage before 1860 reaching 73,800 in 1849. During this period also, the commercial fishing for herring, halibut, and oysters was started. Maine and Massachusetts monopolized the cod fishing, the catch being about evenly divided between the two states. The two chief centers were Portland and Castine, Maine, and Gloucester and Marblehead, Massachusetts. The last-named state sent out the great part of the mackerel fleet.

Recapitulation of Foreign Commerce.—The development of foreign commerce went hand in hand with that of merchant shipping, but did not decline with our merchant marine. Hindered by the closing of markets during and after the Revolution, foreign commerce was stimulated mightily by the continental wars. Interrupted again by the Embargo and the War of 1812, it regained its former prosperity in the succeeding years as reciprocal trading treaties were negotiated. Though impeded by the depressions following the panics of 1819 and 1837, the increase of imports and exports as a whole was notable.

IMPORTS AND EXPORTS BY DECADES[1]

Year	Total exports	Total imports
1800	$70,972,000	$91,253,000
1810	66,758,000	85,400,000
1820	69,692,000	74,450,000
1830	71,671,000	62,721,000
1840	123,609,000	98,259,000
1850	144,376,000	172,510,000
1860	333,576,000	353,616,000

[1] Compiled from *Statistical Abstract*, 1921, table 482, p. 836.

During this entire period, about 50 per cent of the imports comprised manufactured goods ready for consumption. Textiles, metals, and earthen goods were imported from England and the Continent; wines from southern Europe; molasses, sugar, rum, and coffee from the West Indies; specie and bullion from Mexico; hides, indigo, and coffee from South America; and tea, silks, and spices from the Orient. England supplied the greater part of the imports, but as the years went on trade with continental Europe, especially France, increased, while the West Indian trade declined.

In return for imports we sent out raw materials for use in manufacturing and foodstuffs.

EXPORTED MERCHANDISE BY GROUPS, PER CENTS OF TOTALS [1]

Year	Crude materials	Manufactures for further use in manufacturing	Manufactures ready for consumption	Foodstuffs in crude condition— Food animals	Foodstuffs wholly or partially manufactured
1820	60.46	9.42	5.66	4.79	19.51
1830	62.34	7.04	9.34	4.65	16.32
1840	67.61	4.34	9.47	4.09	14.27
1850	62.26	4.49	12.72	5.59	14.84
1860	68.31	3.99	11.33	3.85	12.21

In the south, tobacco and rice gave way to cotton, which became, after the invention of the gin, the biggest single export. From the northern and middle states wheat, flour, corn, hides, wool, naval stores, and furs found their way to foreign markets. These markets were principally England, continental Europe, the West Indies, and the Orient. Great Britain during this period absorbed from 30 to 50 per cent of our exports, chiefly cotton for her mills, and foodstuffs after the abrogation of the Corn Laws in 1846. The trade with South and Central America was relatively small, Brazil being the chief purchaser. Business was carried on through New York, Philadelphia, Baltimore, Boston, and New Orleans. Preëminence in the foreign trade depended on communication with the west, and New York after the building of the Erie Canal dominated the export traffic of the Atlantic seaboard. The activities of Philadelphia and Baltimore in pushing

---

[1] Compiled from *Statistical Abstract*, 1921, table 482, pp. 848-849.

canals and railroads are attributed to their efforts to direct western products to Europe through their ports. Most of the years from 1790 to 1860 registered larger imports from Europe than exports. But this unfavorable balance of trade was compensated for by the freight profits earned by the merchant marine, by the commissions of merchants in the reëxport trade, by investments of European capital in the United States, and by specie and bullion brought in from Latin America and the Orient.

NOTES FOR FURTHER REFERENCE

The standard history of American commerce is the coöperative work of E. R. Johnson, T. W. VanMetre, G. G. Huebner, and D. S. Hanchett, *History of Domestic and Foreign Commerce of the United States* (published by the Carnegie Institution in 2 vols., 1915, reprinted in 1 vol. in 1922). At the conclusion of Vol. II there is the most complete bibliography available on the subject.

Possibly the best short account of the American merchant marine is that of J. R. Soley, one time Assistant Secretary of the Navy, in Vol. I of N. S. Shaler (ed.) *The United States of America* (1897). Good chapters are also included in A. S. Bolles, *Industrial History of the United States* (1878). The ground is covered in more detail in W. J. Abbot, *American Merchant Ships and Sailors* (1902), and *The Story of Our Merchant Marine* (1919); W. Bates, *American Marine* (1893), and *American Navigation* (1902); also A. H. Clark, *The Clipper Ship Era 1843-1869* (1911). On the whale fisheries see T. Jenkins, *A History of the Whale Fisheries* (1921); W. S. Tower, *History of the American Whale Fishery* (1907), and R. McFarland, *A History of the New England Fisheries* (1907). Delightfully written recent books are R. D. Paine, *The Old Merchant Marine* (1919), Chronicles of America Series, and S. E. Morison, *Maritime History of Massachusetts, 1783-1860* (1921), the latter one of the finest maritime studies ever made, containing also valuable bibliographies. What Morison says on whaling, however, should be amplified by reading F. R. Hart, *The New England Whale Fisheries,* to appear in the Publications of the Colonial Society of Massachusetts.

On early subsidization consult Royal Meeker, *History of Ship Subsidies* (1905), and M. M. McKee, *Ship Subsidy Question in United States Politics* (1922), in Smith College Studies VIII, No. 1. A recent book of propaganda for government-aided shipping, prefaced by a brief history of the American merchant marine, is Rear-Admiral W. S. Benson, *The Merchant Marine* (1923). R. H. Dana, *Two Years Before the Mast,* a classic of the sea, gives an interesting picture of conditions on board a clipper ship, while life on a whaler is dramatically pictured in *Moby Dick* by H. Melville. Joseph Hergesheimer, *Java Head* (1919), a story of social life in Salem in the 'forties, pictures the decline of Salem in the Oriental trade.

SELECTED READINGS

JOHNSON, E. R., et al., *History of Domestic and Foreign Commerce of the United States,* Vol. I, Chaps. XII-XIV and XIX (by T. W. VanMetre).
MARVIN, W. L., *The American Merchant Marine,* Chaps. IV-XIV.
SOLEY, J. R., in N. S. Shaler's *The United States of America,* Vol. I, pp. 518-624.
MORISON, S. E., *Maritime History of Massachusetts, 1783-1860,* Chaps. III ff.
PAINE, R. D., *The Old Merchant Marine.*

CHAPTER XII
THE INDUSTRIAL REVOLUTION IN AMERICA

Outstanding Features of the Period 1790-1860.—The years 1790 to 1860 marked a period in which the United States passed from a condition of economic dependence upon Europe to one in which the ordinary wants of manufactured goods could be supplied at home. It was also a period of industrial revolution in which household industries gave way to the factory system. These years may be roughly subdivided into three periods. The first, 1790-1815, was characterized by (1) economic dependence on Europe for manufactured goods, followed by (2) years when international disturbances brought succeeding crises in our economic life and (3) the birth of the factory system. The second period, 1815-40, showed (1) the gradual growth of manufactures and the entrance of these interests into political controversies, and (2) the dependence of these industries upon water transportation. The third period, 1840-60, profited from the discovery of the practical use of coal for smelting iron and for steam power, and was marked by (1) the rise of railroads, which allowed manufacturing to be carried on almost anywhere, (2) by the introduction of many new improvements in machinery which quickened the diversified manufacturing enterprises, both of which factors helped to bring about (3) the transition from household industry to the factory system.[1] Taking the period as a whole, agriculture still continued to occupy the chief energies of the people. During most of the period from 1790 to 1860, shipping and the industries dependent upon it absorbed a disproportionately large share of capital and labor. Milling and meat packing, industries closely allied to agriculture, rose first, followed by textiles and then the metal industries. Unoccupied public land, across which the fron-

---
[1] These divisions follow in general those made by Victor S. Clark in his *History of Manufactures in the United States 1607-1860*, chap. xi.

tier advanced, was a determining factor in the history of manufactures, especially as it affected the labor market.

Economic Dependence on Europe.—The Revolution, as we have seen, brought political independence, but not immediately economic independence. Industry which had thriven on the old "three-cornered traffic" was largely destroyed by the prohibition of trade with English possessions in the West Indies. Manufactures which had grown up during the Revolution were smothered by cheaper British goods which were thrown on the American market upon the resumption of peace. The growth of American manufacturing was held up not only by the uncertainty of foreign markets and the competition of foreign goods, but also by internal factors of equal importance. After two hundred years of settlement there remained still an apparently exhaustless supply of unoccupied land. The lure of an independent life with profits to be secured both from agriculture and from rising land values attracted the average man more than an existence as an industrial laborer. Agriculture was still the primary industry, the logical occupation from which the greatest returns were to be obtained. Unoccupied western land was the chief cause for the scarcity and consequently high cost of labor, a chief deterrent to the growth of manufacturing. Manufacturing had also to meet the competition of shipping. What loose capital was left over from agriculture was largely drawn into ship building and transportation. Marketing the agricultural products seemed the most important problem to be solved, and large amounts of liquid capital were drawn into schemes for canals, railroads, river and ocean ships. The infant manufactures had to compete with the shipping industry during the golden era of the merchant marine. The net earnings of the merchant marine between 1795 and 1801 were estimated at $32,000,000 a year, three-fourths the total value of agricultural exports during the same years; the tonnage built in American dockyards increased from 202,000 in 1789 to 1,425,000 in 1810. Nor were American carriers content with transporting their own produce. They scoured the seas in search of business. In the first decade of the century there were some years in which the value of foreign goods transported exceeded domestic products. Between 1791 and 1800 35 per cent of the goods received were reëxported, largely on American ships.

With these contending factors, it is a source of wonder that any progress whatever was made. Yet it was during the years most discouraging to industrial life that the birth of the factory system took place and American manufactures established themselves. The chief contributing causes were: first, the partial shutting off of European imports during the Revolution (1775-83), during the years of the Embargo and Non-intercourse Acts and the War of 1812-14; second, the existence of an abundance of raw materials, especially cotton, iron, and fuel, and of water power; third, the immigration in continually increasing numbers of skilled and unskilled European laborers, in many cases persons unused to agriculture and willing to engage in industry; fourth, government aid through protective tariffs; fifth, the gradual appearance of small amounts of accumulated capital; and sixth, the versatility and inventive genius of a resourceful people. These influences combined with the savings on freight tended to overcome the higher cost of labor.

Although the small amount of surplus capital in the country was largely drawn into other fields, sufficient was attracted to manufacturing to give the latter a start. Considerable amounts came from commercial firms who withdrew their capital from commerce during the uncertain years preceding and subsequent to the War of 1812. In a similar way ship owners and sea captains during the same period turned from the carrying trade to cotton manufacturing. With the decline of the East Indian trade, the capital of Salem and Providence shifted to manufacturing as did that of New Bedford with the passing of the whaling industry. Merchants with surpluses, especially those who wished to assure themselves of a supply of the finished product, invested actively in manufacturing. But undoubtedly the larger number of establishments originated from small shops and water mills whose owners reinvested their accumulations until their enterprises assumed respectable proportions. The profits from manufacturing were normally large, affording both new capital for extension and a stimulation to outsiders to invest. From the inaccurate census deductions it is probable that the capital invested in manufacturing was about $50,000,000 in 1820 and $1,000,000,000 in 1860.

The Industrial Revolution in England.—The Industrial Revolution in America was preceded and made possible by a similar transition in England. For thousands of years the economic processes were carried on in essentially the same manner. Thread was spun and cloth woven by hand. The same was true of other manufactured articles. These handicraft operations were usually carried on either in the home as a by-industry to farming or in a little shop attached to the house where the master craftsman, surrounded by his journeymen and apprentices, laboriously turned out his products. In the last half of the eighteenth century and first quarter of the nineteenth, various inventions were made which entirely revolutionized the world, changing more profoundly than had all previous ages the every-day life of mankind. England, the first of the European nations to free herself from the shackles of guild regulations, possessed of a thriving commerce and accumulated capital, free from the devastation of the Napoleonic wars, and rich in coal and iron, was a logical starting point for this great advance.

The Industrial Revolution was the result of thousands of experimenters, but certain names stand out preëminently. The beginning was in the textile industry. In 1738 John Kay invented the "fly-shuttle," a device by which a weaver, instead of reaching across to throw the shuttle back and forth, could jerk a string to accomplish the same purpose. This simple contrivance allowed the weaver to work on a much wider piece of cloth and with more rapidity. The speeding up of weaving brought increased demand for thread; but it was not until 1770 that James Hargreaves, a Lancashire weaver, patented an improvement on the old-fashioned spinning-wheel whereby eight spindles connected by a band to a horizontal wheel, and turned by a crank, could spin eight threads at a time. While Hargreaves was still working on his "spinning-jenny," Richard Arkwright, a barber of Preston, patented a water-power machine in 1769 which drew the carded cotton through a series of rollers, each set revolving at a greater velocity than the preceding set, and turned out cotton thread strong enough to be used as warp. Arkwright's "water frame" made the manufacture of cotton goods a commercial possibility and himself a wealthy man. A further improvement in spinning was made in 1779 by Samuel Crompton, who combined the work of Hargreaves and Arkwright by building a machine called a mule, which would turn out a better thread at quicker speed. Since the spinners were now far ahead of the weavers, it was quite logical that the next advance should be in weaving. The invention, strange to say, was the result of the labors of Edmund Cartwright, a Kentish clergyman, who knew absolutely nothing of machinery when he first set to work on the problem. By 1785 he had constructed a power loom propelled by water which would weave cloth successfully. The demand for cotton was now so great that the southern planters in greater numbers turned from raising tobacco to cotton, a transition made possible by Eli Whitney's cotton gin. These inventions were but the beginning of the revolution in textiles. Other inventors constructed machines for the spinning and weaving of woolens and fabrics and for the various dependent processes.

Simultaneously with the improvements in textile machinery came the practical steam engine. The properties of steam had been known to the ancient Egyptians and interesting experiments had been made toward the end of the seventeenth century by Dennis Papin, a Frenchman, and by James Savery. Thomas Newcomen patented an engine in 1705 to pump water out of the mines, an engine capable of doing the work of fifty men, but slow and expensive to operate. In 1763 James Watt, a Scotch scientist, then employed in making astronomical instruments for the University of Glasgow, was given a model of one of Newcomen's engines to repair. In 1769 he patented an improved engine which effected a great saving in fuel and time by drawing off the steam into a separate condenser, thus keeping the cylinder continually warm and making it possible by automatic controls to use the same steam in forcing the cylinder both ways. In company with Mathew Boulton, a wealthy manufacturer, Watt commenced the commercial manufacture of steam engines. To a man of Watt's genius it was not a difficult step to apply the backward and forward motion of the piston to the turning of wheels or the driving of a steam hammer. Industry was now to a great extent emancipated from water power and factories sprang up in the large cities. These engines not only facilitated the mining of coal and iron, but they provided a greater market for both when obtained.

The improvements in manufacturing which brought increased production were followed by striking advances in the methods of distribution. The years of the Industrial Revolution saw great progress in road building under the direction of such men as Telford and Macadam. James Brindley constructed the Bridgewater Canal, opened in 1761, for the purpose of carrying coal from Worsley to Manchester. The success of this canal led in the next decade to a wave of canal building similar to that seen in the United States after 1825, which connected most of the principal rivers and centers in England. Machinists early realized the possibilities of connecting the piston of the steam engine to a wheel revolving in water, but it was not until 1807 that Robert Fulton demonstrated the practicability of the steamboat in passenger and freight service. In that year a Watt engine drove the *Clermont* from New York to Albany, a distance of 150 miles, in thirty-two hours. In 1838 the *Great Western,* the first ship to cross the Atlantic by steam alone, made the distance from Bristol to New York in fifteen days. What Fulton did for steam navigation by water the English engineer, George Stephenson, building on the work of Trevithick and others, did for land transportation. Stephenson's engine, "The Rocket," in 1829 in a trial test between Liverpool and Manchester attained a speed of twenty-nine miles an hour, demonstrating beyond doubt the feasibility of steam railroads.

With the substitution of power-driven machinery for hand labor and the utilization of steam in factories and for transportation purposes the Industrial Revolution was well-nigh accomplished. Its results are the civilization in which we live. The introduction of machinery produced an immense increase in production, with a corresponding increase in wealth and a general transformation of business methods to care for this increase.

Socially it enhanced enormously the wealth, power, and numbers of the middle class. At the same time it brought into being what was virtually a new class, that of the industrial wage earner. Most of the added wealth produced by the Industrial Revolution was absorbed by the former class, while the industrial wage earners, separated from the land and concentrated in the slums of the new cities which sprang into being around the factories, were reduced to extremes of poverty and degradation. The sharp distinction in wealth and opportunity of the new era was soon felt in politics. The middle class, leaning on the *laissez-faire* philosophy of their own economists, Adam Smith, Ricardo, and others, demanded and won a commanding position in the government. To the industrial worker the displacement of the landed aristocracy by the middle class brought no advantage. Stirred to wrath by legislation which discriminated against their interests, workmen followed the example of their industrial masters and agitated for political rights. This agitation, commencing with the Chartist Movement (1838-48) and continuing to the Reform bill of 1918, has resulted in political democracy and in a long list of acts promoting social betterment. Machinery has cheapened the process of printing, a factor which, combined with the growth of urban population, has quickened intellectual life and probably been a spur to progressive thought. Upon international relations the Industrial Revolution has had an immense influence. The search for raw products and the scramble for new markets and for a place to invest some of the surplus wealth created by machines has become acute. The result is a new imperialism bringing in its wake militarism and the conquest of less developed peoples. Briefly stated, these were some of the effects of the Industrial Revolution in England and in Europe generally. As the same revolution in economic life took place in America, we are interested to see to what extent, if any, it was influenced by dissimilar conditions.

Introduction of the Factory System in America.—The Industrial Revolution took place first in England. This advantage, which was fast making her the workshop of the world, she was loath to lose and attempts were made to keep the secrets of the

AMERICAN INDUSTRIAL REVOLUTION 273

new machinery from spreading. The emigration of trained operatives was prohibited by an act of 1765. In 1774 an act, which was in force until 1845, was passed prohibiting the exportation of textile machinery, plans or models; it was supplemented in 1781 to prohibit the exportation of any utensils used in textile manufacturing. A statute of 1772 prohibited citizens engaged in printing calico, linens, or muslins or in manufacturing textile machinery, from emigrating; another of 1785 prohibited the emigration of workmen employed in steel and iron manufacture and the carrying out of tools; and another act of 1789 prohibited the emigration of coal miners.

These measures did not materially delay the introduction of machinery into this country, after interest had once been aroused in manufacturing. The years from the close of the Revolution to 1800 formed a period of experimentation. Factories in which the jenny was used were established in 1787 in Philadelphia, at Beverly, Massachusetts, and in the succeeding years at other places in New England and in New York,—undoubtedly the first cotton factories in America. Of these only the mill at Beverly survived, producing bed ticking until 1807, when it was ruined by the Embargo. The first successful Arkwright mill was built in 1789 by Samuel Slater, an English emigrant who had served an apprenticeship in one of Arkwright's factories at Belper and had been induced to come to America by bounties offered here for the improved machinery. His mill, erected at Pawtucket, Rhode Island, spun in 1790 the first machine-made cotton warp in America; the beginnings of the American factory system can truly be traced to Slater and his Pawtucket mill.

The first spinning machines set up were run either by hand or by horse power. Later, especially in New England, water power was extensively, and for a while almost exclusively, used for both spinning and weaving. Steam was undoubtedly first used in America for pumping mines in New Jersey and Rhode Island during the last decades of the eighteenth century. It is believed that the first application of steam to mill machinery was made in a sawmill in New York in 1803. In the years following, either the imported low-pressure engines of the Boulton-Watt type or the high-pressure engines of Evans were introduced in sections where water power was not to be obtained. By 1812 several of Evans's engines were in use west of the Alleghanies, and by 1817 steam engines were being manufactured not only on the coast, but at Pittsburgh, Louisville, and Cincinnati. The census of 1830 showed that 57 of the 161 plants in Pennsylvania used steam, although the factories enumerated in New York, New Jersey, and all of the New England states with the exception of Massachusetts used water power. Of 169 plants in that state 39 used steam. In general water power was used and accessibility to it determined the location of industry.

When manufacturing by machinery was once firmly established, American inventors enthusiastically took up the ideas of European engineers, adapted them to American conditions, and contributed new improvements. Labor scarcity did much to stimulate inventions, but ignorance as to what had been done in England led to duplication. Of American contributions, perhaps the most famous was that of Eli Whitney, who not only invented the cotton gin, but as early as 1807 applied the principle of standardization of parts and interchangeable mechanism in the manufacture of firearms. Another notable advance was the Goulding condenser, which greatly simplified and quickened the carding of wool. The first successful power loom for weaving in America was constructed by Francis Lowell in 1814 and put into operation at Waltham. Here for the first time the processes of spinning and weaving were brought together in the same factory. The first power loom for the weaving of woolen goods was set up in Connecticut about 1820. In the succeeding years many American inventors, of whom perhaps Samuel Batchelder and William Mason are the best known, perfected machines for knitting and lace making and for manufacturing linen and cotton in figured designs.

Not only in textiles were American machinists making progress. Oliver Evans of Philadelphia invented a high-pressure steam engine which was successfully used. Rumsay, Fitch, and Fulton made notable experiments in the application of steam to water transportation, while John Stevens of Hoboken was directing

AMERICAN INDUSTRIAL REVOLUTION

his attention to a railroad engine. Geissenhainer in 1830 successfully smelted iron ore with anthracite coal and in 1851 William Kelly of Kentucky independently discovered the principle of the Bessemer method of decarbonizing molten metal by forcing air through it. Elias Howe in 1846 invented the sewing machine, an improvement equally suitable to the home and the factory, and an invention which not only proved an immense boon to women, but revolutionized the clothing and shoe industry. The work of Morse in introducing the magnetic telegraph effected a similar revolution in methods of communication. To these men and hundreds of other inventors and *entrepreneurs* must the credit be given for the establishing of American manufacturing. The Patent Office, which reported an average of 77 inventions annually from 1790 to 1811, recorded 544 patents in 1830. In the decade 1840-50, 6,480 patents were issued, and in the next decade 28,000.

Textiles—Cotton.—Following Slater's success at Pawtucket, mills sprang up at various points; but most of them had been closed by 1800, when "the factory at Beverly, and 7 Arkwright mills, 4 within a few miles of Providence and 3 in Connecticut, represented the organized cotton industry in the United States."[1] Between 1800 and 1804 the number of mills in New England more than doubled. The Secretary of the Treasury, in an official report to Congress early in 1810, stated that at the close of the previous year there were sixty-two mills in operation, of which eighteen were located in Rhode Island and ten in Massachusetts. Either this tabulation was too conservative or else the Embargo had in the meantime greatly stimulated the industry, for the first census of manufactures taken in 1810 recorded 269 cotton mills, running 87,000 spindles. Inaccurate as this census may have been, it demonstrates the fact, fully corroborated by contemporaries, of the great activity occasioned by the Embargo; during this period cotton manufacturing was firmly established in America. The movement started by the Embargo of 1807 was further encouraged by the War of 1812.

---

[1] Clark, V. S., *History of Manufactures in the United States*, p. 535.

The mills around Providence alone, the center at that time of cotton spinning, increased from 41 to 169 during the three years. New England emigrants carried manufacturing into the Hudson and Mohawk valleys and into the central Lake region of New York, while mills sprang up in the south and west of the Alleghanies. Before the census of 1820 was taken, cylinder machines run by water power for printing cloth had been introduced, and F. C. Lowell had erected on the Merrimac his power loom for weaving.

The return of peace and the flooding of American markets by the pent-up English goods was disastrous to the infant industries. Comparatively few survived. Nevertheless, the tariff of 1816, new labor-saving devices, and a world-wide business revival, enabled new mills to rise on the ruins of the old. Although the census of 1820 showed 250,000 spindles in operation (a 213-per-cent increase since 1810), using 10,000,000 pounds of cotton (176-per-cent increase), the production in 1820 was undoubtedly less than half that of 1815. After 1820, despite the crises of 1837 and 1857, textile manufacturing continued to increase rapidly. Between 1820 and 1831, the number of spindles more than quadrupled and the factory looms increased tenfold. Cotton manufacturing increased 150 per cent between 1840 and 1860. At the same time the price of cotton yarn and cloth was reduced to less than a fourth of its cost previous to the introduction of machinery. Sixty-nine per cent of the cotton manufacturing in 1860 was concentrated in New England. The growth of the industry after 1840 as obtained from the census reports follows:

COTTON MANUFACTURING [1]

	1840	1850	1860
Mills	1,369	1,074	1,091
Spindles	2,284,631	....	5,235,727
Cotton consumed (bales)	237,000 (500 lbs.)	641,240 (425 lbs.)	1,056,726 (400 lbs.)
Capital invested	$51,102,359	$76,032,578	$98,585,269
Value of product	$46,350,453	$65,501,687	$115,681,774
Workers (average)	72,119	94,956	122,028

[1] Eighth Census, vol. iii, Manufactures, p. xix.

AMERICAN INDUSTRIAL REVOLUTION

Textiles—Woolens.—Although the census of 1860 showed 1,260 establishments manufacturing woolen goods with an invested capital of $30,862,654, consuming 83,608,468 pounds of wool, employing on an average of 41,360 operatives and producing about $61,895,217 worth of cloth,[1] we were still dependent on England for considerable quantities both of raw wool and of manufactured goods. The manufacture of woolens developed more slowly than that of cotton, notwithstanding the fact that woolen cloth was used from the early settlements and the adaptability of large parts of the country to sheep raising thoroughly demonstrated. This was due to several causes. English statesmen, acting under the policies of mercantilism, forbade in 1699 the exportation of wool or woolen cloth from the American colonies, and by other acts sought to discourage in every possible way the manufacture of woolens that the American market might be retained by England. Great Britain excelled in the colonial period as she still does in the manufacture of woolens, and it was difficult to compete even after assistance had been rendered by high tariffs in the years following the War of 1812. These factors, combined with the general influences holding back the development of manufactures, already discussed, impeded the advance of the woolen industry. Colonial sheep produced wool suitable only to coarser cloths; but after 1790 improved breeds were gradually introduced from Spain, Ireland, and England, which, mingled with the existing stock, bettered the product and made possible the manufacture of excellent woolens from domestic sheep. Simultaneously with the growing of improved wool came the gradual shift from household to factory manufacture. Little had been done outside of the home before the Revolution, and household weaving had undoubtedly increased during the troubles with the mother country. In 1788 a mill was founded in Hartford which produced 5,000 yards of cloth a year, and another mill at Stockbridge, Massachusetts. The Scholfield brothers, who had emigrated with Samuel Slater, set up factories in Massachusetts in 1793 and later in Connecticut. The census of 1810 re-

---

[1] Census of 1860, vol. iii, Manufacturing, p. xxii.

ported 24 factories and 1,682 fulling mills. Steam was first used in woolen manufacturing at Middletown, Connecticut, in 1811. The average annual factory production in these years was probably not over 10,000 yards, an exceedingly small amount in comparison with that still woven at home. The War of 1812 gave to woolen manufacturing its first real foothold, the new factories turning out military equipment, negro cloths, and even fine woolens. The passing about this time of the wearing of knee breeches also brought a wider demand for woolen goods. It was estimated at the conclusion of peace that $12,000,000 had been invested in woolen mills, the product of which was valued at $19,000,000. Connecticut, the leading state, had twenty-five factories employing 1,200 operatives. Manufacturing establishments had also been started west of the mountains at Louisville and Cincinnati.

The conclusion of the war and the subsequent heavy importation of accumulated European goods dumped on the American market ruined many of these mills and was a potent influence in bringing about the tariff of 1816, in which a 25-per-cent ad valorem rate on most woolen goods was imposed. The incomplete returns of the census of 1820 showed that over 100 factories with 700 looms had survived the post-war depression. The decade of the 'twenties showed some improvement, but as a whole it was a precarious period for the industry. After 1830 normal prosperity returned and woolen manufacturing firmly established itself, securing control in certain lines of the domestic market. The depression after the panic of 1819 had in like manner affected the farmers, thus lengthening the life of household manufacturing; as late as 1830 more woolen cloth was woven at home than in the factories and this was true even in the leading textile states of New England. It was not until the decade of the 'forties that it can truly be said that the factory-made product had the upper hand and that household industry was on a rapid decline. Although the production of worsteds remained small and the manufacture of broadcloths declined, the industry as a whole showed a healthy progress during the next twenty years,

AMERICAN INDUSTRIAL REVOLUTION

with an increasing output of cashmeres, satinets, flannels, blankets, felts, and carpets. The statistics of these years follow:

Year	Woolen factories	Capital invested	Value of product	Employees
1840	1,420	$15,765,124	$20,696,999	21,342
1850	1,817	26,071,542	43,542,288	34,895
1860	1,909	35,520,527	68,865,963	48,900

Other Textiles.—Before the introduction of the factory system flax held first place in homespun products. With the introduction of spinning and weaving machinery for cotton and wool, the latter fabrics became cheaper and displaced flax. Factories for the manufacture of cordage, sail cloth, and the finer grades of flax goods existed, but their numbers were few, their life precarious, and their product, though of good quality, not large. The duty of 1828 on imported flax practically killed the business. In the late 'forties it slowly revived, the census of 1860 showing the entire value of linens, apart from homespuns, to be $700,000.

Hemp was used extensively in the manufacture of bagging and bale cloth. The center was in the Ohio Valley, especially in Kentucky, where the manufacture of bale cloth was one of the earliest industries. Later Missouri became a competitor. The census of 1860 reported the output of bagging at 9,540,000 yards, of which Kentucky produced 5,750,000 yards, and Missouri 3,680,000.

Lack of cheap labor and indifferent success in the cultivation of mulberry trees held back the development of silk manufacturing. Spasmodic attempts were frequently made, but little of permanence was accomplished until after 1840. After this date raw silk was imported, largely from China. In the decade 1840-50, John Ryle established the silk industry in Paterson and the Cheney brothers at South Manchester, Connecticut. Silk manufacturing was still in its infancy in 1860, when the census reported forty-two silk mills in the northeast turning out thread but only one factory making woven goods. The total value of the silk products is given as $6,589,171, with the center of the industry in Connecticut and New Jersey.

Metals.—The development of modern industry is closely bound up with the story of iron, a substance which now enters into the construction of machinery, buildings, ships, and the innumerable needs of an "iron age." In the seventeenth and eighteenth centuries, when charcoal was used for smelting, the virgin forests of America provided cheap fuel at almost the same time that the supply was giving out in England. In the nineteenth century, when the possibility of smelting by anthracite coal was discovered, the existence of iron and coal in close proximity was again most advantageous to American smelters. The rapidly growing manufactures and population provided an increasing market, which was partially protected by tariffs.

Before the close of the Revolution iron was smelted from bog iron in all of the thirteen colonies except Georgia. Bog iron was gradually supplemented by iron from the rich magnetic ore belt extending from the Berkshires in Massachusetts and the Salisbury district in Connecticut across the Hudson through Orange County, New York, and into Morris County, New Jersey. Slowly the smelting industry pushed westward. By 1810 iron making had extended up the Susquehanna, into the Juniata Valley, up the Lehigh, and farther south had crossed the mountains into Tennessee and Kentucky. The census of 1810 noted 153 furnaces producing 53,908 tons of iron. The production of iron was also stimulated by the Embargo and the War of 1812, but suffered a severe setback in the years immediately following. The census of 1820 gives little information of value, but in the following decades the industry grew up in western Pennsylvania and the Ohio Valley, extending by 1860 as far as the region of Lake Superior, where smelting was carried on in northern Michigan and near Detroit. The use of mineral fuel after 1840 revived the smelting of iron east of the Alleghanies, especially in Pennsylvania and those points in New Jersey and New York reached by canal.

In the colonial period and the early decades of the century iron products were likely to be made in the same little smelting mill that produced the raw material. Thus household utensils and the few metal tools in use were generally either custom made or turned out on a very small scale. It was not until the 'twenties that specialization in metal products developed to any great extent, and even then the foundries and factories generally remained close to the source of supply. The use of the steam engine in water transportation after 1808 favored specialization in this type of ironwork and brought into being engine works on such important arteries of traffic as the Hudson, Delaware, and Ohio, although for many years engines were made only to order. With the advent of railroads, locomotives were built at many machine shops, but by 1860 the building had tended to concentrate at Philadelphia, the home of the Baldwin and Norris plants (founded in 1832 and 1834, respectively), and at Paterson, New Jersey. Rolling mills, which in the early years had been devoted chiefly to rolling and slitting nail plates, with the increased demand for new kinds of metal goods, turned their attention to other products such as iron rails and tires. In the 'fifties the iron straps used for railroads were gradually discarded for the heavy solid iron rail, a transition which brought into being a special phase of the industry centered in eastern Pennsylvania. One of the earliest of the metal industries to become specialized was the making of stoves. At first the plates were cast at the furnaces and assembled by the merchants, but eventually the whole process was carried on in the same establishment. The use of anthracite coal greatly increased the demand for the iron range, so that by 1850 the annual production was over 300,000 stoves, valued at around $6,000,000. These were manufactured chiefly in Philadelphia, New York, Albany, Cincinnati, Providence, and Pittsburgh.

In the smaller metal products, in some of which steel and iron were used, the center of manufacturing was New England. Iron axes, springs, bolts, wire, firearms, and clocks were largely made here. The factory system brought into being the manufacture of textile machinery, one-half of which in 1860 was constructed in Massachusetts. Likewise one-half of the edge tools and three-fourths of the cutlery produced in the country in that year came from New England, chiefly Connecticut. Berlin became a center for tin ware and the Naugatuck Valley for brass ware. The manufacture of sewing machines, the output of which reached 110,000 in 1860, was divided principally between the Singer factory at New York, the Wheeler & Wilson company of Bridgeport, and the Grover & Baker company of Boston.

Distribution of Industry.—The location of American industries was determined by chance in some cases, but more commonly by economic influences. The northeast asserted its leadership in manufacturing in colonial times and continued to hold first rank. New England, with poor soil for agriculture, but gifted with an abundance of water power, an active commerce, and a thrifty, energetic population settled closely together, was especially fitted for manufacturing, though it lacked raw materials. The middle Atlantic states were favored by more varied mineral resources, by direct routes to the interior, by greater supply of capital and labor, but were handicapped by competing agriculture and the constant draining off of its population to the west. Nevertheless, important centers for textiles and other manufactures arose on the water power furnished by the Mohawk and the Hudson, and in such cities as New York, Philadelphia, Pittsburgh, Rochester, and Chicago, which were located on routes of travel or favored by access to coal and iron. The early years of the century gave promise of considerable manufacturing in the Piedmont regions of Virginia and the Carolinas, but the competition of agriculture by slave labor prevented any great development until after the Civil War.[1] Far more important than the south was the region of the Ohio River, where, previous to 1860, considerable manufacturing existed. Pittsburgh specialized in many forms of iron ware; Cincinnati, the only town until 1850 west of the Alleghanies with a population of 100,000, was a great meat-packing center and manufactured machinery, clothing, and other commodities; Louisville produced cordage, bagging, and clothing, while Chicago was developing large milling and packing interests. This region produced flour, lumber, agricultural implements, meat, and the by-products of the packing industry, cordage, bagging, and distilled liquors. The products

---

[1] See chap. xxiii.

AMERICAN INDUSTRIAL REVOLUTION

of New England and the middle states in general were the type requiring detailed manufacturing with finer mechanism and involving higher labor cost. Thus textiles, boots and shoes, rubber goods, clothing, glass ware, pottery, and cutlery were centered here. The following figures from the census of 1860 give a comparison of the sections in manufacturing:

MANUFACTURING BY SECTIONS, 1860[1]

Sections	Number of establishments	Capital invested	Average number of laborers	Annual value of products
New England	20,671	$257,477,783	391,836	$468,599,287
Middle States	53,387	435,061,964	546,243	802,338,392
Western States	36,785	194,212,543	209,909	384,606,530
Southern States	20,631	95,975,185	110,721	155,531,281
Pacific States	8,777	23,380,334	50,204	71,229,989
Territories	282	3,747,906	2,333	3,556,197
Total	140,433	$1,009,855,715	1,311,246	$1,885,861,676

New England was unquestionably the center of the textile industry. The value of the product of cotton goods of various sections in 1860 follows: New England, $79,359,900; middle states, $26,534,700; southern states, $8,145,067; western states, $1,642,107, showing over two-thirds of the production in New England. The original home of cotton manufacturing, the region around Providence, had given way to Lowell and other cities on the Merrimac, with new centers developing rapidly at Fall River, New Bedford, and Holyoke. Massachusetts produced as much as the rest of the New England states combined, with New Hampshire ranking second and Rhode Island third. While textiles tended to concentrate in Massachusetts, Philadelphia continued to be the leading manufacturing city. Two hundred thousand spindles in the city and its suburbs and an equal number in near-by districts supplied yarn for hand and machine weaving. Southern cotton textile mills were to be found along the fall line of the James, Savannah, Chattahoochee, Alabama, and Tennessee rivers, at Richmond, Petersburg, Augusta, and other towns, but the total number of spindles in all of the southern

---
[1] Eighth Census, vol. iii, Manufactures, p. 725.

states in 1860 amounted to 290,000 out of 5,236,000 for the whole country. The showing of the west was even worse. Only 90,000 were to be found in the Ohio Valley, most of them at Pittsburgh and Cincinnati.

The early center of the woolen industry was Connecticut, but after the War of 1812 other states took the lead. By the middle 'thirties Massachusetts manufactured at least one-third of the woolen goods, with New York second and Connecticut third. By 1860 woolen mills were to be found in Texas, California, and Oregon, but the center still remained in New England. Massachusetts manufactured over one-fourth of the entire product of the country, followed by Pennsylvania and New York.

DISTRIBUTION OF WOOLEN MANUFACTURES IN 1860 [1]

Sections	Factories	Operatives	Value of product
New England....................	398	25,583	$40,668,498
Middle States	476	11,638	15,905,923
Western States...................	306	2,281	3,090,472
Southern States..................	78	1,768	1,995,324
Pacific States....................	2	90	235,000
Total	1,260	44,360	$61,895,217

With the decline of bog ores the center of iron smelting shifted from New England to Pennsylvania. The census of 1810 showed that state in the first place, with New York second and New Jersey third. The same ranking held until 1850, when Ohio followed Pennsylvania, with New York third. In 1860 New Jersey again ranked third.

In 1860 the ten principal iron regions [says Victor Clark] were northern New York, including Vermont, where the primitive ores of the Adirondacks were reduced by 40 bloomeries and several furnaces into high-grade iron; the highland belt from the Berkshires to Pennsylvania, including the old Litchfield, Orange and Morris County districts, where 44 charcoal and 22 anthracite furnaces and 60 bloomeries used hematite and magnetic ores; eastern Pennsylvania

[1] Eighth Census, vol. iii, Manufactures, p. xxxv.

ther to one cent a ton mile, while transportation east and west was now as practical as from north to south. Industry was no longer dependent upon seasons, for goods could be shipped in the winter months and the necessity of closed mills, idle workmen, and unproductive capital because of inability to move goods was largely eliminated. Railroads could be built more cheaply than canals and to many points impossible for the latter to reach. The railroads quickened the settlement of the frontier, the consequent increase in population creating new and greater markets; they also made it possible for different sections to specialize in the occupations favorable to them, since the products of one could now be easily transported to another. Although for some years railroad building absorbed most of the available capital and thus impeded manufacturing other than that concerned in railroad supplies, the setback was temporary, for it put in motion forces favorable to manufacturing too mighty to be halted.

Condition of Labor.—Scarcity of skilled labor is a normal condition in a new country, and as a consequence during the years under discussion the skilled workman was able to command fair wages and maintain a decent standard of living. Although urban rents were relatively high, food was exceedingly cheap and clothing reasonable. While the lot of the skilled laborer was tolerable, the unskilled laborer, although commanding greater wages than in Europe, barely made ends meet. His pay averaged about half that of skilled labor. Two shillings at the time of the Revolution was a day's pay, increasing to about ninety cents a day in 1800 and a dollar in 1825, around which point it remained for many years even during the activity of canal and railroad building. His condition at the opening of the century is gloomily pictured by McMaster:

Sand sprinkled on the floor did duty as a carpet. There was no glass on his table, there was no china in his cupboard, there were no prints on his wall. What a stove was he did not know, coal he had never seen, matches he had never heard of. . . . He rarely tasted fresh meat as often as once in a week, and paid for it a much higher price than his posterity. . . .

If the food of an artisan would now be thought coarse, his clothes would be thought abominable. A pair of yellow buckskin or leathern breeches, a checked shirt, a red flannel jacket, a rusty felt hat cocked up at the corners, shoes of neat's-skin set off with huge buckles of brass, and a leathern apron, comprised his scanty wardrobe. The leather he smeared with grease to keep it soft and flexible.[1]

Farm hands with board received from seven to fifteen dollars a month, depending upon season and locality, with the general average tending from the lower to the higher figure as time went on. Without board, the compensation of agricultural laborers rose from fifty cents a day at the opening of the century to a dollar a day in 1860. The wages of skilled labor ran from one to two dollars a day, although often going much higher.

The wages not only of skilled artisans and unskilled labor, but also of factory workers, were generally higher than in Europe. Nevertheless, in Massachusetts, where wages were highest, between 1830 and 1860 men earned five dollars a week, children between one and two dollars, and women from one dollar and seventy-five cents to two dollars a week, the latter figures including board. Wages were lower in Pennsylvania and the southern states. There was, however, a minimum of extreme poverty, for the workman could always escape from it by emigrating to the frontier. "Pauperism, that gaunt and hideous spectre, which has extended its desolating march over Asia and Europe, destroying its victims by thousands, even in the midst of luxury and wealth, has never yet carried its ravages into the United States," said an English traveler in 1839; "this is a blessing of which it is to be feared that few appreciate the magnitude, and which is, of itself, a preponderating weight in the balance of national happiness."[2]

Nor was the introduction of the factory system attended with the extreme horrors which accompanied the change in England. In America most of the cloth had either been purchased from Europe or made by women on the farm as part of their household

---

[1] McMaster, J. B., *History of the People of the United States*, vol. i, pp. 96-97.
[2] Murray, C. A., *Travels in North America* (1839), vol. ii, p. 200.

duties. The factory system, therefore, threw few men out of labor, and the factories for many years had to compete for labor with the more alluring prospects of independence on a frontier farm. The early operatives in the New England textile mills were ordinarily girls or unmarried women to whom the chance of earning money and at the same time escaping from the drudgery and dependency of farm life was looked upon as an opportunity rather than a misfortune. Living conditions, in comparison with those of similar workers in England, were not bad, being often directly supervised by the factory owners, and no loss of social standing was felt by the operatives. European observers were invariably struck by the difference of conditions in England and America. Harriet Martineau, writing of conditions as she saw them at Waltham in 1835, describes the life of the factory operatives:

> I visited the corporate factory-establishment at Waltham, within a few miles of Boston. The Waltham Mills were at work before those of Lowell were set up. The establishment is for the spinning and weaving of cotton alone, and the construction of the requisite machinery. Five hundred persons were employed at the time of my visit. The girls earn two, and some three, dollars a week, besides their board. The little children earn one dollar a week. Most of the girls live in the houses provided by the corporation, which accommodate from six to eight each. When sisters come to the mill, it is a common practice for them to bring their mother to keep house for them and some of their companions, in a dwelling built by their own earnings. In this case, they save enough out of their board to clothe themselves, and have their two or three dollars a week to spare. Some have thus cleared off mortgages from their fathers' farms; others have educated the hope of the family at college; and many are rapidly accumulating an independence. I saw a whole street of houses built with the earnings of the girls; some with piazzas, and green venetian blinds; and all neat and sufficiently spacious.
> The factory people built the church, which stands conspicuous on the green in the midst of the place. The minister's salary (eight hundred dollars last year) is raised by a tax on the pews. The corporation gave them a building for a lyceum, which they have furnished with a good library, and where they have lectures every winter,—the best that money can procure. The girls have, in many instances, private libraries of some merit and value.

The managers of the various factory establishments keep the wages as nearly equal as possible, and then let the girls freely shift about from one to another. When a girl comes to the overseer to inform him of her intention of working at the mill, he welcomes her, and asks her how long she means to stay. It may be six months, or a year, or five years, or for life. She declares what she considers herself fit for, and sets to work accordingly. If she finds that she cannot work so as to keep up with the companion appointed to her, or to please her employer or herself, she comes to the overseer, and volunteers to pick cotton, or sweep the rooms, or undertake some other service that she can perform.

The people work about seventy hours per week, on the average. The time of work varies with the length of the days, the wages continuing the same. All look like well dressed young ladies. The health is good; or rather, (as this is too much to be said about health anywhere in the United States), it is no worse than it is elsewhere.

These facts speak for themselves. There is no need to enlarge on the pleasure of an acquaintance with the operative classes of the United States.[1]

This rosy picture of Miss Martineau's did not tell the whole story. While the condition of American operatives during the Industrial Revolution was undoubtedly much better than in Europe, it was not enviable. Nominal wages gradually rose, but real wages fell. As in England, the increase of wealth due to machinery went largely into the hands of the capitalist class. The early mills were in most cases unsanitary and unhealthy places in which to work. Hours of labor were excessively long. According to Professor Ely: "The length of actual labor [1832] varied from twelve to fifteen hours. The New England mills generally ran thirteen hours, but one mill in Connecticut ran fourteen hours, while the length of actual labor in another mill in the same State, the Eagle Mill at Griswold, was fifteen hours and ten minutes. The regulations at Paterson, New Jersey, required women and children to be at work at half-past four in the morning. . . . Women and children were urged on by the use of the cowhide."[2] A committee of the Massachusetts legislature, investigating the hours of labor of children in factories as affecting

---

[1] Martineau, Harriet, *Society in America* (1837), ii, p. 57 ff.
[2] *Labor Movement in America*, p. 49.

their education, reported in 1825: "It appears, however, that the time of employment is generally twelve or thirteen hours each day, excepting the Sabbath," and naïvely added, "which leaves little opportunity for daily instruction." [1] As late as 1845 the average hours of labor in the Lowell mills varied from eleven hours, twenty-four minutes in January to thirteen hours and thirty-one minutes in April, practically from sun to sun. An observer in Lowell in 1846 describes the hours of labor:

> The operatives work thirteen hours a day in the summer time, and from daylight to darkness in the winter. At half past four in the morning the factory bell rings, and at five the girls must be in the mills. A clerk, placed as a watch, observes those who are a few minutes behind the time, and effectual means are taken to stimulate to punctuality. This is the morning commencement of the industrial discipline (should we not rather say industrial tyranny?) which is established in these Associations of this moral and Christian community. At seven the girls are allowed thirty minutes for breakfast, and at noon thirty minutes more for dinner, except during the first quarter of the year, when the time is extended to forty-five minutes.[2]

Children at first were not used to the extent in the New England mills that they were in England. The need of gathering workers together from the surrounding country, building quarters and boarding houses involved special problems with children. Nevertheless, as time went on the evil became widespread and was unchecked by law. The report of a convention of New England Mechanics and Workingmen held at Boston in 1832 estimated that the children employed in manufactories constituted about two-fifths of the whole number of persons employed. The *Mechanics' Free Press* for August 21, 1830, prints the following statement regarding children in the Philadelphia factories:

> It is a well-known fact, that the principal part of the help in cotton factories consists of boys and girls, we may safely say from six to seventeen years of age, and are confined to steady employment during the longest days of the year, from daylight until dark, allowing, at

---

[1] Commons, J. R., et al., *Documentary History of American Industrial Society*, vol. v, p. 59.
[2] *Ibid.*, vol. vii, p. 132.

the outside, one hour and a half per day [for meals] ... and that too with a small sum that is hardly sufficient to support nature, while they [the employers] on the other hand are rolling in wealth off the vitals of these poor children every day. We notice the observations of our Pawtucket friend in your number of June 19, 1830, lamenting the grievances of the children employed in those factories. We think his observations very correct, with regard to their being brought up as ignorant as Arabs of the Desert; for we are confident that not more than one-sixth of the boys and girls employed in such factories are capable of reading or writing their own name. We have known many instances where parents who are capable of giving their children a trifling education, one at a time, were deprived of that opportunity by their employers' threats that if they did take one child from their employ, (a short time, for school) such family must leave the employment—and we have even known these threats put in execution. ... [1]

The Early Labor Movement.—Local craft unions began to appear as early as 1791 with the Philadelphia carpenters, in 1794 with the Philadelphia Federal Society of Journeymen Cordwainers and the Typographical Society of New York. These societies were purely local, although organizations in the same trade sometimes arose in other cities, resulting in communication between the unions. The early local unions attempted the regulation of hours and conditions of work by collective bargaining and interested themselves in mutual insurance and in sick and funeral benefits. Strikes were rare and resulted at first in the arrest of the leaders and their trial for conspiracy under the English Common Law, a practice which ceased after 1825. By 1828 industrial life and the concentration of population in cities had progressed sufficiently to permit the associations of city craft unions and the founding of national trade unions. The movement was started in that year by the founding of the Philadelphia Mechanics' Trade Associations, which stood for free schools, abolition of imprisonment for debt, mechanics' lien laws, equal taxation, direct election of public officials and various other political and economic reforms, as well as shorter hours and better working conditions. This organization in Philadelphia inaugurated defi-

---

[1] Commons, J. R., *et al.*, *op. cit.*, vol. v, p. 61.

nitely a labor-class movement in this country and commenced a nine-year period of great labor activity. Between 1827 and 1837 the five national trade unions of cordwainers, comb makers, carpenters, hand-loom weavers and printers were founded, pursuing their purposes by means of political agitation as well as by pressure on local employers. The crisis of 1837 put a decided damper on this buoyant growth of trade unions, and the revival of the movement after 1844 took the form of humanitarian and socialistic efforts, which resulted in many experiments in communism and coöperation. Land reform, free schools, the ten-hour day, were all demands in this humanitarian wave, culminating in a new anti-slavery agitation which eventually absorbed much of the energy of the labor movement. The crisis of 1857, the rising cost of living and other factors, led in the late 'fifties to a revival of strictly trade-union activities and the reorganization of old units as well as the formation of new national unions. By 1860 the tendency of labor to unite for protection and aggression had made considerable progress, and labor unionism was prepared for a more rapid growth under the favorable conditions of the Civil War.

Summary.—Faint beginnings of factory production can be seen in the colonial period and during the Revolution, but the Industrial Revolution in America waited upon the introduction of the new inventions from England and the artificial stimulation of the years before and during the second war with England. It was at the conclusion of this conflict that factory production made such headway that the manufacturing interests were able to secure protective tariffs. Factory production, however, was still in its infancy, and the domestic system was not entirely supplanted even as late as the Civil War. After machine production had established itself, its growth was rapid, notwithstanding the strong adverse factors of lack of labor, capital, and transportation facilities, as well as easier profits in other lines. On the other hand, cheap raw materials, skillful labor, an abundance of water power, and a growing market, especially to the west, were elements in the situation which gradually enticed a certain amount of capital and placed manufacturing upon a firm basis. Although labor conditions in the early factories left much to be desired, the Industrial Revolution in America was fortunately not accompanied by the extreme evils attendant upon the transition period in Europe.

NOTES FOR FURTHER REFERENCE

The best study of manufacturing during this period is that of V. S. Clark, *History of Manufactures in the United States 1607-1860* (1916), published by the Carnegie Institution of Washington, and containing the most complete bibliography available. Of the older books the most valuable is that of J. L. Bishop, *History of American Manufactures from 1608-1860* (3 vols., 1866), containing detailed accounts of specific industries in the early stage. There is much information, not easily obtained elsewhere, crammed into A. S. Bolles, *Industrial History of the United States* (1878). See also *Eighty Years' Progress* (1869), articles under the various manufacturing industries. Recent short accounts are the chapters in C. D. Wright, *Industrial Evolution of the United States* (1897); K. Coman, *Industrial History of the United States* (rev. ed., 1910); E. L. Bogart, *Economic History of the United States* (rev. ed., 1922); and Isaac Lippincott, *Economic Development of the United States* (1921).

The census reports of the government will be found useful, as will also such special reports as those of Alexander Hamilton, *Report on Manufactures*, in F. W. Taussig, *State Papers and Speeches on the Tariff*, and of Louis McLane, *Report on Manufactures* (2 vols., 1833), House Doc. No. 308, 22d Congress, 1st Session. Much that is interesting and instructive is obtainable from the accounts of foreign travelers, extracts from which are contained in G. S. Callender, *Selections from the Economic History of the United States 1765-1860* (1909), and from the *Readings* of Bogart and Thompson. On the tariff consult F. W. Taussig, *Tariff History of the United States* (7th ed., 1923); P. Ashley, *Modern Tariff History* (3d ed., 1920); and E. Stanwood, *American Tariff Controversies in the Nineteenth Century* (1903).

On cotton see James A. B. Scherer, *Cotton as a World Power* (1916); James L. Watkins, *Production and Price of Cotton for One Hundred Years*, U. S. Department of Agriculture, Division of Statistics, Miscellaneous Series, Bulletin No. 9 (1895); Broadus Mitchell, *The Rise of Cotton Mills in the South* (1921), Johns Hopkins University Studies in Historical and Political Science, 39th Series; and M. T. Copeland, *The Cotton Manufacturing Industry in the United States* (1912).

Material for the study of labor during this period is obtainable in *Documentary History of American Industrial Society*, Vols. V and VI, by Commons and Sumner, and in Vols. VII and VIII, by Commons. Accounts of the period are incorporated in the standard labor histories, such as J. R. Commons, *History of Labour in the United States* (1898); R. T. Ely, *The Labor Movement in America* (1886); F. T. Carlton, *History and Problems of Organized Labor* (rev. ed., 1920); G. C. Groat, *Organized Labor in America* (1916); G. S. Watkins, *An Introduction to the Study of the Labor Problem* (1922);

and Mary R. Beard, *A Short History of the American Labor Movement* (1920).

SELECTED READINGS

CLARK, V. S., *History of Manufactures in the United States 1607-1860,* Chaps. XI-XXI.
TRYON, R. M., *Household Manufacture in the United States 1640-1860,* Chaps. VII, VIII.
BOGART, E. L., *Economic History of the United States,* Chaps. X, XI.
CALLENDER, G. S., *Selections from the Economic History of the United States 1765-1860,* pp. 432-478.
BOGART, E. L., and THOMPSON, C. M., *Readings in the Economic History of the United States,* Chaps. VIII, IX, XVI.

CHAPTER XIII

FINANCE AND TARIFF

Financial Laws of the New Government.—Of the numerous problems which confronted the new Republic and pressed for solution, none were more important than the financial. It was necessary first of all to determine a fiscal policy whereby old debts could be paid and money obtained to meet the expenses of the government. Duties on exports were forbidden by the Constitution, and it was natural to turn to tariffs on imports as the proper source for revenue. The act of July 4, 1789, the first tariff act under the new government, was designed primarily for revenue, but it recognized the protective feature. There were eighty-one enumerated articles, upon over thirty of which specific duties were levied, while the remainder called for ad valorem rates varying from 7½ to 15 per cent. Upon all imported articles not enumerated a 5-per-cent duty was levied. Although the rates were exceedingly low, the average ad valorem being not over 8½ per cent, some protection was given. The debates on the tariff brought out the conflicting interests of the various sections of the country. Duties were imposed to help the steel and paper mills of Pennsylvania, the brewers of New York and Philadelphia, the glass manufacturers of Maryland, the iron workers of New England, and so on. The by-products of the farmhouse were aided by duties on nails, boots and shoes and ready-made clothing; tea, coffee, sugar, wines, and other luxuries were more heavily taxed. It was soon found, however, that the tariff of 1789 did not provide enough revenue, and increases were made in 1790, 1792, and 1794. One of the last acts of the first Congress which adjourned on September 29, 1789, was to request Alexander Hamilton to make a report on the state of the finances. In compliance with this he submitted four reports; the first on January 14, 1790, which dealt with the public debt; the second, on December 13, 1790, which recommended an excise; the third, on the same date, recommending a national bank; and the fourth, December 5, 1791, his famous report on manufactures advocating protection.

In his first report Hamilton showed that the total foreign debt to France, Spain, and Holland, with arrears of interest, amounted to $11,710,378; the domestic debt with arrears at 6 per cent amounting to approximately $42,414,085, with the existing state debts about $25,000,000, totaling in all $79,124,464. Hamilton proposed that the national government take over this debt of the states which had been incurred in aid of the Revolution and that both state and national debt be refunded at par. This was necessary, he said, to place the credit of the government on a firm basis. It was sound financing and his plan was eventually adopted, but the act authorizing it went through amid violent controversy in regard to the right of the national government to assume these debts, and only by means of a political deal by which the advocates of the measure, in return for southern votes, permitted the location of the national capital on the Potomac. It was accompanied also by speculation on the part of those in a position to profit, and by bitter recrimination on the part of those who objected to the new government's assuming such large obligations under conditions which allowed great speculative profits. But the industrial and commercial interests by means of this action were firmly consolidated behind the new government.

Hamilton's advice that an excise tax be levied was in like manner followed by Congress, but only after a most strenuous opposition. It was his belief that an excise would both provide revenue for the national government and bring home to the most remote frontiersman who operated a still the power of the national government. The tax accomplished its purpose, but it rested heavily and, it was felt, unjustly upon the frontiersmen, whose bulky products could only be marketed when reduced to the more concentrated form of whisky. The opposition of the Pennsylvania frontiersmen to the tax in the "Whisky Rebellion" of 1794 demonstrated both the hatred toward it and the strength of the new government in crushing disobedience.

The third recommendation of Hamilton, that a national bank be set up modeled after that of England, was urged on the grounds that (1) it would provide a much-needed paper currency which would be safe, (2) it would furnish a safe place for keeping public funds, (3) it would benefit both the government and business by providing banking facilities for the carrying on of commercial transactions, and (4) it could act as a fiscal agency for the government in such transactions as the sale of bonds. There was a real need of such a bank, for at the time of the adoption of the Constitution there were only three banks in the United States, the Bank of North America in Philadelphia, the Bank of New York, and the Bank of Massachusetts in Boston. The Anti-Federalists, led by Jefferson, who advocated a strict construction [1] of the Constitution, opposed the bank on the grounds that it was not authorized by that document and that by the bank's superior position it might operate unfairly to the detriment of the state banks. It was, nevertheless, chartered in 1791 for twenty years with a capital of $10,000,000 of which amount the government might subscribe $2,000,000 and private investors $8,000,000, one-fourth in specie and three-fourths in governments bonds. The notes of the bank were to be limited to the amount of the capital stock and were to be receivable in taxes as long as they were redeemable at the bank in specie. Reports must be made to the Secretary of the Treasury, who was authorized to inspect the affairs of the bank at any time. The First United States Bank was a salutary influence in the financial operations of the early Republic, fulfilling amply the expectations of its advocates. Aided by the credit of the government, it was able to do business in a conservative fashion and acted as an efficient agent of the Treasury Department. By refusing to accept the notes of non-specie-paying banks it drove out fiat money and kept the paper at par.

The currency of the new bank was issued in terms of the dollar, the unit already adopted by the Congress of the Confederation.

---

[1] By "strict construction" is meant a limiting of the powers accorded to the national government by the Constitution to the exact letter of that document; by "loose construction" the broad interpretation of certain clauses of the Constitution and "the general welfare" phrase in Article I, Section 8 to include an extension to implied powers not specifically prohibited.

In 1792 Congress passed its first act placing the valuation of the new American dollar at 24.75 grains of gold, the value of the Spanish milled dollar, and establishing the decimal system. Under the belief that a grain of gold was equal to 15 grains of silver, it was provided that a silver dollar should contain 24.75 times 15 or 371.25 grains of silver, with the smaller coins of proportional weight. Free and unlimited coinage of both gold and silver was provided for in the act, and both were made full legal tender. Although a mint was established in Philadelphia which began in 1794 the coinage of silver and in 1795 that of gold, very little metal was brought in to be coined. This was due partly to the fact that the amount of precious metals mined in this country during those years was small, and partly to the fact that silver was slightly overvalued at the rate of 15 to 1, thus discouraging the coinage of gold entirely, for under the workings of Gresham's law, cheap money drives out the better currency. What gold was coined was speedily sent out of the country, and the nation was soon reduced to a silver standard. But it was difficult to keep even silver dollars in the country, because they were accepted in the West Indies in exchange for the Spanish milled silver dollars of greater value. The latter were brought in, reduced to bullion, and coined at a profit to the holders. As a consequence the coinage of dollars was stopped in 1806, and until the coinage system was changed in 1834 paper money, coins of small denominations, and foreign coins supplied the need for currency.

As a whole the recommendations of Hamilton and the laws which followed were wisely framed. The gradual return of prosperity after 1785 was accelerated in the years following the adoption of the Constitution. The measures designed to strengthen the national government, although opposed by the Anti-Federalists, were exceedingly heartening to the commercial interests. West of the Alleghanies the settlers felt the benefits and strength of the new government in the Treaty of 1795 with Spain by which the "right of deposit" at New Orleans was obtained, a right which gave to the westerners the privilege of landing their products and reshipping without the payment of duties.

The last decade of the century found the new nation happily launched.

Currency and Banking.—Hamilton's idea of a banking system, modeled after that of Great Britain, in which a great bank under the joint control of private bankers and the government, might regulate the currency and act as the fiscal agent of the government, was followed out in the First United States Bank of 1791. When its twenty-year charter expired in 1811 it went out of business. With the restraining hand of a specie-paying national bank removed, numerous state banks sprang up, the number increasing from 88 to 246 in five years, and the money in circulation from 45 to 100 million. Their various notes circulated at a discount, sometimes as great as 50 per cent. The disorganized state of the currency was accentuated by the War of 1812, which placed the government in a most difficult financial situation, and as a consequence Secretary Dallas urged the creation of another United States Bank, hoping that it might again serve the purpose of aiding the government financially and of stabilizing the currency.

With the lessons of the war still fresh, Congress chartered the Second United States Bank in 1816 for twenty years, providing that one-fifth of the $35,000,000 capital should be subscribed by the government and that five of the twenty-five directors should be appointed by the President. It was expected that the notes of the new bank, redeemable in specie on demand, would force the state banks to resume specie payment or drive them out of business. Although mismanaged for the first three years, after a reorganization in 1819 it quite efficiently handled the fiscal operations of the government and exercised a very salutary effect upon the currency. Nevertheless, the Second United States Bank was strongly disliked in many parts of the country, particularly in the south and west. Its restraining influence upon the wildcat currency of the frontier banks aroused bitter opposition, as did the feeling on the part of the west that it was a dangerous monopoly in the hands of a few eastern bankers. Certain of the states attempted to tax branch banks out of existence, but Chief-Justice Marshall in two famous decisions (McCulloch *v.* Maryland, 1819, and Osborn *v.* United States Bank, 1824) declared the acts unconstitutional, asserting that what the Constitution permitted the national government to set up no state might destroy.

The Second United States Bank, under the able direction of its president, Nicholas Biddle, had rendered a real service to the nation. Unfortunately for its future, the question of its recharter became involved in politics. Andrew Jackson, true son of the west, feared the bank as a dangerous monopoly. Furthermore, he was in favor of hard money, which he thought the bank prevented from circulating. On top of this he became convinced that Biddle and his associates were playing politics. When the supporters of Henry Clay (candidate of the National Republican party in 1832) prevailed upon Biddle to petition for the renewal of the charter, Jackson vetoed the bill. The question of the bank was made an issue in the presidential election of 1832 and the victory of the Democrats spelled the doom of the bank. Jackson, not content to wait until the charter ran out in 1836, proceeded to further weaken the institution by removing the government deposits and placing them in the so-called "pet banks." The failure to recharter the Second United States Bank was a victory for the west and it ended the attempt to control the currency by means of a central bank. In 1840 the government broke away also from the state banks by establishing an independent treasury system (made permanent in 1846) to care for its own funds. By means of sub-treasuries the government collected its revenues in specie and made all disbursements through its own officials. By using only specie it was expected that a considerable amount of coin would be kept in circulation, thus limiting the amount of bank notes. It was also hoped that the withdrawal of the government specie from the state banks might hinder the latter from issuing much paper money. This belief was well founded and the plan proved successful until the Civil War called into being the National Bank Act of 1863.

In the meantime, Jackson, in the furtherance of his hard money policy, had advocated some change in the coinage system. By laws of 1834 and 1837 the amount of gold in a dollar was changed from 24.75 grains to 23.22, but the amount of pure silver was kept at 371.25 grains. This changed the ratio from 15 to 1 as it was set up in 1792 to 15.98 to 1, or approximately 16 to 1. The new ratio overvalued gold and drove out silver from the currency, so that within a few years scarcely a silver dollar was coined. It is true gold began again to circulate, but as a whole the measure did not accomplish the end sought.

The Specie Circular and the Panics of 1837 and 1857.[1]— With the restraining hand of the United States bank removed, there ensued an orgy of speculation. Banks were started with little capital and specie, but they issued large quantities of notes and made loans freely. The number of banks increased between 1829 and 1837 from 329 to 788, their capital from $110,000,000 to $290,000,000, the note circulation from $48,000,000 to $149,000,000, and the loans from $137,000,000 to $525,000,000. The whole expansion was undoubtedly aided by the distribution of the government specie and was spurred on by the mania for internal improvements and the inordinate speculation in western land. The income from the receipt of public lands jumped from $1,880,000 in 1830 to over $20,000,000 in 1836. Jackson, who was far from an expert in finance, clearly saw the essential unsoundness of the situation when he said in a message to Congress, "It was perceived that the receipts arising from the sales of public lands were increasing to an unprecedented amount. In effect, however, the receipts amounted to nothing more than credits in bank. The banks lent out their notes to speculators. They were paid to the receivers [land agents] and immediately returned to the banks, to be lent out again and again, being mere instruments to transfer to speculators the most valuable public lands and pay the Government by a credit on the books of the banks. . . . The spirit of expansion and speculation was not confined to the deposit banks, but pervaded the whole multitude of banks throughout the Union; and was giving rise to new institutions to aggravate the evil."[2] The bubble of speculation was enlarged by the utterly reckless manner in which the states borrowed here and abroad for internal

---

[1] For a general discussion of panics, see chap. xxvi.
[2] Eighth Annual Message, December 5, 1836, in Richardson's *Messages and Papers of the Presidents*, vol. iii, p. 249.

improvements and the prodigality with which they loaned their credit to unsound institutions.

The panic of 1837 was presaged by the crop failure of 1835. This prevented the farmers from meeting their obligations to the land speculators and merchants, and the latter could not pay their loans at the banks. The crop failure eventually produced a balance of trade against the United States, a withdrawal of foreign credits, and a need of specie to pay foreign creditors. In the midst of these accumulated difficulties Jackson hastened the crisis by issuing on July 11, 1836, his "Specie Circular," an order which directed that all payments for public land must be in specie. This served as a wet blanket to dampen the ardor of speculation in western land and shook the confidence in the circulating bank notes. The situation was further complicated by the failure of important mercantile houses in England toward the end of 1836, involving many English manufacturers and cutting down the demand for American cotton.

The panic which ensued was the worst that the nation had experienced up to that date. By the end of May, 1837, every bank in the country had suspended specie payment. The bank-note circulation contracted from $149,000,000 in 1837 to $58,000,000 in 1843, and the sale of public land fell off from $20,000,000 in 1836 to $1,000,000 in 1841. In that year Congress passed a special bankruptcy law under which 39,000 persons canceled $441,000,000 worth of debt. The depression continued to be severe for five or six years, holding back the expansion of both manufactures and agriculture. Eventually recovery set in, and with the revival of business the country again experienced a period of remarkable growth. Spurred on by rising prices, due chiefly to the discovery of gold in California, railroad building was pushed on rapidly, new manufacturing establishments set up, and the westward movement accelerated. A temporary halt was called by the third great panic in our history, that of 1857. Overspeculation in the future of the country and overinvestment of fixed capital in railways and in mineral resources produced a setback, the causes of which closely resemble those of our other panics. The failure in August, 1857, of the Ohio Life Insurance and Trust Company precipitated the panic, which was primarily financial and affected chiefly the financial centers and the speculative western railroad investments. Recovery was quick and the opening of the Civil War found the nation on the upgrade to a new cycle of prosperity.

Tariff Policy to the Civil War.—Although the first American tariff was passed scarcely two months after the inauguration of Washington, until 1816 the various acts had been primarily for revenue and had afforded only incidental protection. It is true that Alexander Hamilton in his *Report on Manufactures* had given classic expression to the arguments for protection, but a strong movement for a protective tariff waited upon the War of 1812 and its effects. The collapse of prices in land and agricultural products following the deflation of 1815-18, and the fear that the dumping of European goods set free by the close of the war might snuff out the infant manufacturing, aroused an active interest in protection. Still influenced by the war, all sections of the country united in supporting the tariff of 1816. The bill was introduced by William Lowndes of South Carolina, and John C. Calhoun, later a most uncompromising opponent of protection, led the fight for its enactment. It placed duties ranging from 7½ to 30 per cent ad valorem, giving special protection to cottons, woolens, iron, and certain manufactured commodities.

From 1816 until 1833 this movement for protection grew steadily under the leadership of Henry Clay, who conceived it as a most important feature in his "American System." At the same time, diverse interests began to make themselves felt and to line up definitely the sections of the country on one side or the other. The center of the early movement for protection was in the middle and western states of that period—New York, New Jersey, Pennsylvania, Ohio, and Kentucky. They had felt keenly the disturbing effects of the war, and were anxious to develop a home market to dispose of their agricultural products. The south, on the other hand, was anxious to obtain her manufactured articles cheaply and, with her chief market in Europe, naturally opposed protection. She had unsuccessfully attempted to manufacture with slave labor and was now convinced that her future lay in agriculture. New England at first was divided. The manufacturing interests, not yet powerful, were in favor of protection, whereas the shipping group and the merchants feared that a tariff would injure their business. As a consequence, New England split her votes on the tariff until about 1830, when the manufacturers won the dominating control and lined her up on the side of protection. This change may be seen in the attitude of Daniel Webster, who opposed the tariff of 1816 but supported that of 1828.

In 1818 further protection was given to iron, and the duty of 25 per cent on cotton was extended until 1826. A general revision was undertaken after the election of 1824, at which time all of the candidates had advocated protection. Not only was additional protection given to manufacturers of woolen goods, lead, glass, and iron, but 25 per cent was now granted to hemp manufacturers, and wool growers were specifically aided. This tariff received the support of the iron interests of Pennsylvania, the wool growers of Ohio and the middle states, the hemp growers of Kentucky, and the manufacturers everywhere, but it incurred the disfavor of the northern shipper and the bitter hostility of the south, where much wool was used for negro clothing. The tariff of 1828 was the result of the agitation of the woolen interests for increased protection, aided and abetted by the Jacksonian politicians, who thought they saw a chance of promoting the interests of their candidate in the coming election. It was the intention of the latter to propose a bill so obnoxious that it could not pass, although the Jackson men of the north might vote for it and thus pose in the next election as the true friends of domestic industry. John Randolph remarked that "the bill referred to manufactures of no sort or kind, except the manufacture of a President of the United States."

To the surprise of all, this bill was passed, but it was so unpopular that it was speedily dubbed the "Tariff of Abominations." It was on the statute books only four years, but during that time aroused a storm of opposition, especially in the south. In 1832 a new bill removed many of the abominations and practically restored the tariff to the basis of 1824. Nevertheless, the bill was still essentially protective, and South Carolina in November, 1832, passed the famous Nullification Ordinance declaring the "tariff law of 1828 and the amendment of the same in 1832" to be null and void and not binding upon the people of South Carolina. President Jackson's uncompromising stand for national unity left South Carolina little hope for success, and both sides agreed to a compromise tariff in 1833, introduced by Henry Clay. The outcome was a victory for the nationalists under Jackson, but a lowering of the tariff in favor of the south. The act of 1833 provided for a decrease of all duties exceeding 20 per cent in the tariff of 1832; this reduction was to be very gradual until 1842, when a sudden lowering was to create a uniform rate of 20 per cent on all articles. This 20-per-cent level remained in force only two months, however, from July until the passage of the more strongly protective tariff of 1842 in September.

The tendency of the tariff rates from 1832 to the Civil War was downward, although the principle of protection was never relinquished. The panic of 1837 so depleted the income of the national government that the protectionists were successful in restoring the duties in 1842 almost to a level with those of 1832. Passed by the Whigs, this act was speedily repudiated when the Democrats came into power in 1845. The Walker tariff of 1846 classified imported commodities under schedules, A, B, C, D, etc. Luxuries were put into Class A and a tariff of 100 per cent imposed; semi-luxuries into Class B with a 40-per-cent tax; commercial products into the remaining classes, with duties varying from 30 down to 5 per cent. The Walker tariff changed the system from specific to ad valorem duties and introduced the warehousing system of storing goods until the duty was paid, an innovation permanently retained. The duties, while maintaining protection, were radically lowered by this tariff, and the tendency toward reduction was continued in 1857, when the free list was enlarged and a lowering of 5 per cent on the rates of the Walker tariff made. The reductions of 1857 were the result of a treasury full to overflowing from the immense expansion of business from 1846 to 1857.

Undoubtedly the tariff legislation of the first seventy years of our history aided the growth of manufacturing and industry;

it is equally true, however, that the emergence of the United States as an industrial nation was inevitable. The natural development was simply artificially stimulated by the various tariff measures, that is all. The virtual victory of the protectionists during this period in launching the nation on the road to a high-tariff policy is interesting as the first victory of the industrial interests of the north over the slaveholders of the south, and a harbinger of eventual control after the Civil War.

NOTES FOR FURTHER REFERENCE

Hamilton's reports may be found in *American State Papers, Finance*, Vol. I. The most satisfactory book for the general student is that of D. R. Dewey, *Financial History of the United States* (5th ed., 1922), containing detailed bibliography. On currency see also A. B. Hepburn, *History of Coinage and Currency in the United States* (rev. ed., 1915); J. L. Laughlin, *History of Bimetalism in the United States* (4th ed., 1897); and David K. Watson, *History of American Coinage* (1899).

On the national bank controversy there is an extended literature. In addition to the valuable accounts in many of the general histories, more detailed studies are to be found in Ralph C. H. Catterall, *The Second Bank of the United States* (1903); Charles A. Conant, *History of Modern Banks of Issue* (1896, rev. 1915); William G. Sumner, *History of Banking in the United States* (1896); William MacDonald, *Jacksonian Democracy* (1906), in The American Nation Series; and John S. Bassett, *Life of Andrew Jackson*, Vol. II (1911, rev. 1916, 1 vol.).

For the tariff consult Percy W. L. Ashley, *Modern Tariff History* (3d ed., 1920); Edward Stanwood, *American Tariff Controversies in the Nineteenth Century* (1903); and F. W. Taussig, *Tariff History of the United States* (7th ed., 1923).

SELECTED READINGS

DEWEY, D. R., *Financial History of the United States*, Chaps. III-IX.
TAUSSIG, F. W., *Tariff History of the United States*, Part I.
CALLENDER, G. S., *Selections from the Economic History of the United States 1765-1860*, Chaps. X, XI.
BOGART, E. L., and THOMPSON, C. M., *Readings in the Economic History of the United States*, Chaps. X, XV.
TAUSSIG, F. W., *Selected Readings in International Trade and Tariff Policies*, Part III, Chaps. XVII-XXI.

CHAPTER XIV

TRANSPORTATION AND COMMUNICATION TO 1860

Significance in American History.—The ever westward movement of population and the rapid conquest of a continent have made the matter of transportation the most vital problem which has faced both government and people since our history began. As the first settlers along the seacoast were dependent upon ocean transportation with Europe to market their raw materials and obtain in return manufactured products, so later waves of advancing settlers were in like manner dependent upon rivers, roads, canals, and railways to dispose of their products. Especially was the question acute in an essentially agricultural community, as that of the United States before the Civil War, where commodities to be transported were likely to be bulky and in some cases perishable.

Not only was the economic life of the new settlements dependent largely upon transportation, but their existence and location were in many cases determined by it. The first settlements were usually made upon good harbors, and the advance inland followed the numerous rivers which ran into the Atlantic and which provided the easiest facilities for transportation. As the best land was taken up along the rivers, the new settlers were forced to occupy the back country and to keep in touch with the rivers by crude roads. These roads quite often followed existing trails, for the paths of the deer and buffalo in their quest of water became the trails of the Indian hunters and the wagon roads and railroads of the white man. Where rivers met the ocean and formed good harbors, where one river ran into another or where trails connected with rivers, particularly at the head of navigation, there were the strategic points at which towns and cities sprang up. After 1850 railroad building went on so rapidly that railroads preceded the settlers and pointed, as did the rivers in earlier years, to inevitable routes. Gradual improvements in transportation weakened economic sectionalism and in turn political sectionalism. It takes no longer to travel to-day from Washington, D. C., to Portland, Oregon, than it did at the adoption of the Constitution to go between Boston and New York. Railroads bound together the east and west at the time of the Civil War, and have contributed much to the specialization of industry in certain sections and the mutual interdependence of economic groups and geographic sections.

Colonial Transportation.—The first means of colonial travel by water was the crude log dugout or the Indian bark canoe. Later for heavy transportation flat and keel boats were in use, oftentimes equipped with temporary sails. Although communication was kept up between the different colonies during the colonial period chiefly by water, the Indian trails running between the three populous centers of the north—Boston, New York, and Philadelphia—provided a route for travelers on foot or horseback, and were commonly followed by 1683. Roads developed more slowly. As late as the Revolution there were only three roads north and east of New York and only one leading west from Philadelphia. To the south two rude trails led across the mountains, one at Harper's Ferry and the other through the Cumberland Gap. In New England, where settlements were thicker, more progress had been made. The General Court of Massachusetts in 1639 ordered each town to construct a highway with the adjoining town, a command which produced several roads radiating from Boston. Road communication was established between Boston and Providence by 1654. From here the "Shore Road" connected Providence with the Connecticut settlements and New York. The overland route to Hartford led through Marlborough, Oxford, and Grafton. The roads of colonial New England followed roughly the present routes of the New York, New Haven, and Hartford Railroad. In the south the incomparable system of waterways, navigable during the entire year because of the mild climate, furnished the safest and best means of transportation and delayed the building of roads until the middle of the eighteenth century. What few roads existed at that time were of the most abominable type, constructed before the rudiments of modern road building were understood. Choked with mud in the spring, thick with dust in the summer, and heavy with snow in the winter, the overland routes were fraught with the most arduous labor and hardship, frequently by real danger. The risk was increased by the fact that until long after the Revolution there were no bridges over the principal rivers.

Delay in effective road building is explained in many sections by the fact that the colonists depended chiefly upon natural waterways for their transportation, for waterways had controlled the location of settlement and travel by water was the simplest method. Other deterrents were the extreme cost of constructing roads, due to the scarcity of labor and capital in a new country and to the necessity of building through heavily wooded and sparsely settled communities. The slow development of the use of wheeled vehicles also delayed road building. The first wheeled conveyances to appear were a few private coaches made for use in the large towns in the last years of the seventeenth century. The first horse coaches appeared in Boston in 1687, where they were frowned upon as the works of the devil. While Philadelphia had some six or eight hundred houses in 1697, it boasted of only thirty carts and other wagons, while New York had fewer. In Connecticut there were no carriages until 1750 and a chaise at the time of the Revolution created a distinct sensation; in consequence, overland transportation of goods was carried on chiefly in the winter and on sleds.

Under such circumstances it is not surprising that it was not until 1744 that the first coach and four began its trip in New England, and the first stage between New York and Philadelphia in 1756. Two days was the running time between Portsmouth, New Hampshire, and Boston, a distance of sixty miles, as it was also the fastest time made by coaches before the Revolution over the ninety miles between New York and Philadelphia. During the Revolution most of the stagecoaches professing to run with some degree of regularity ceased, but stage traveling was resumed at the end of the war. Forty miles a day in the summer and twenty-five in the winter was the average under the most auspicious circumstances. Uncomfortable as were the facilities of travel by water, they were superior to the stagecoach, although more uncertain as to time. The journey between the three great northern cities usually included both kinds of travel. The route from New York to Philadelphia was by packet to South Amboy, thence by coach to Burlington, and by packet again to Philadelphia. The journey from New York to Boston could be made by packet to Providence and on to Boston by stage, the time ranging from five days to two weeks, depending on the weather. The earliest stages made the journey between New York and Boston in seven days by traveling nineteen hours a day at a cost of two pounds and a half. By 1793 the time had been reduced to four days at a cost of six cents a mile for the ordinary traveler and an express eventually made the distance in three and a half days at eight cents a mile. The original stage-wagon of straight sides, tunnel-shaped top of linsey-woolsey, with three or four wooden benches and no springs, gave way toward the end of the century to the football-shaped coach with leather-covered seats and set on springs, a lighter, faster, and more comfortable vehicle.

The Turnpike Era.—The improvements in transportation facilities before the Civil War may be divided roughly into three periods: (1) turnpike and improved roads, (2) canals and improved rivers, and (3) railroads. The years 1790 to 1820 witnessed an intense interest and remarkable activity in road building. The effect of this movement so long delayed was felt chiefly in the well-settled states and was occasioned (1) by the demands of farmers for better transportation facilities to market their products, demands supported by the inhabitants of the coast towns desiring cheaper foodstuffs; (2) by the prospect of increasing the value of back lands; and (3) by the hope of dividends and the fascination of speculation. The lead in the movement for better roads was taken by private individuals, who organized companies and issued stock to build turnpikes, roads supported by tolls levied on all using them. The first turnpike constructed in America was the pike from Philadelphia to Lancaster, built between 1792 and 1794, a distance of sixty-six miles at a cost of $465,000. The optimism of those who had crowded to subscribe to the shares of the Philadelphia and Lancaster Turnpike Company was more than warranted. So successful was the road that a mania for turnpike building spread over the whole country. In the next thirty years 86 companies were chartered in Pennsylvania; by 1832 that state had built about 2,200 miles of roads at a cost ranging from $900 to $7,000 a mile. By 1810, 20 turnpike companies had been chartered in New Hampshire, 26 in Vermont, and upwards of 180 in all New England. New York by 1811 had chartered 137 companies with a combined capital of $7,500,000, which constructed about 1,400 miles of road. Connecticut built nearly 800 miles. Rivalries of cities for western trade spurred on such construction and led Baltimore to charter three roads leading west. The construction of toll bridges followed the building of toll roads, financed and supported in the same manner. While these turnpikes are not to be compared in construction with more modern roads, some of them were well built. Francis Baily, writing in 1797, said, "There is, at present, but one turnpike-road on the continent, which is between Lancaster and Philadelphia, a distance of sixty-six miles, and is a masterpiece of its kind; it is paved with stone the whole way, and overlaid with gravel, so that it is never obstructed during the most severe season."[1] Types of hard-surfaced roads were in operation in New England before they had been perfected in England.

The value of the new roads was clearly demonstrated, but the cost of transporting goods by land was still enormous. The freight per ton from Philadelphia to Pittsburgh by an all-land route was $125, and the average through the country for general merchandise, according to McMaster, was $10 per ton per hundred miles, making it out of the question to transport such bulky articles as grain and flour more than 150 miles. At the same time the freight on a ton from Europe was 40 shillings. These rates were kept high not only by poor roads, but by high tolls. The average toll in New England was 12½ cents per wagon for each two miles. In New Jersey the toll was 1 cent per mile for each horse, while the Pennsylvania rates varied according to

---

[1] Francis Baily, *Journal of a Tour in Unsettled Parts of North America in 1796 and 1797*, p. 107.

the width of tire and number of horses. An example of the political as well as social and economic results which arose from these high transportation costs is the Whisky Rebellion of western Pennsylvania. Here the farmers reduced their grain to whisky, a less bulky commodity, in order to market it, and the added cost of internal revenue seemed more than they could bear.

Gallatin's Plan.—The continued high cost as well as the success of the new turnpikes contributed to the demand that the government lend its aid to the transportation problem. The state governments already were contributing to road construction, and now the national government was besieged. Article I, Section VIII of the Constitution gave to the national government the power (1) to establish postoffices and post roads, (2) to raise and support armies, and (3) to regulate commerce. The loose constructionists and those who might benefit believed that ample power was given under these clauses to interfere in a practical way. It was argued that better means of communication would be of great political effect in promoting the settlement of the west and the interdependence of one section on another. Added force was given to their arguments by the surplus in the treasury in 1806 and at other times. The strict constructionists believed that this interpretation of Section VIII was too broad, and their group was augmented by those claiming it was unfair to tax the whole country for the benefit of certain sections.

The discussion brought forth on April 4, 1808, the famous report of Gallatin, made at the request of Congress, on internal improvements. While it contained little that was new, it gathered together and systematized various plans which had been long discussed. He suggested:

1. The series of peninsulas jutting out into the Atlantic should be cut by canals, thus reducing the miles to be traversed and the dangers necessary to be undergone on the outside route. A canal should be cut across Cape Cod, one across New Jersey from the Raritan to the Delaware, one from the Delaware to the Chesapeake, and another from the Chesapeake Bay to Albemarle

Sound. The last two enterprises, already undertaken by private companies, should be given government aid.

2. Further communication north and south was to be furnished by a great turnpike along the Atlantic coast from Maine to Georgia.

3. To improve communication east and west he believed that the headwaters of the four eastern rivers, the Juniata, Potomac, James, and Santee, might be joined by four great roads, respectively, with the headwaters of the western rivers, the Allegheny, Monongahela, Kanawha, and Tennessee. These rivers should be improved for navigation, and canals built around the falls of the Ohio and the Niagara rivers. He also advocated national roads from Pittsburgh to Detroit, St. Louis, and New Orleans. In addition he proposed canal connection between Boston and Lowell, between Lake Champlain and the Hudson, between the Mohawk and Lake Ontario, the Schuylkill and the Susquehanna, the Schuylkill and the Delaware, and suggested the Carondelet Canal at New Orleans.

The total cost he estimated at $20,000,000, a sum which might be realized by withdrawing from the treasury two million a year for ten years or from the sale of public lands.

Cumberland Road.—The chief result in road building of the demand for internal improvements at the expense of the national government was the Cumberland Road. Ohio was admitted to the Union under an agreement whereby federal lands sold within her borders were exempt from taxation for five years, and in return the federal government was to appropriate 5 per cent of the proceeds from the sale of such lands for the building of roads, three-fifths of which was to be expended within the state and two-fifths in building a road over the mountains to connect it with the east. Similar agreements were later made with Indiana, Illinois, and Missouri. This allowed even the strict-construction Republicans a loophole, and during Jefferson's administration, on March 29, 1806, Congress authorized the building of such a road. The first contract, however, was not signed until 1811, and the first stretch of 130 miles to Wheeling, West Virginia, was not completed until 1818. It was continued almost due west through Zanesville, Columbus, and Springfield, Ohio; Richmond, Indianapolis, and Terre Haute, Indiana, to Vandalia, Illinois, reaching Columbus in 1833 and Vandalia in 1852. The road from Wheeling was especially well built: "Its numerous and stately stone bridges, with handsome stone arches, its iron mile posts, and its old iron gates, attest the skill of the workmen engaged in its construction, and to this day remain enduring monuments of its grandeur and solidity." [1] The road was supported by a western Congressman, whose influence went far to bring about the passing of more than thirty acts for its building and maintenance between 1806 and 1838. It cost the federal government $6,821,200.

Until the coming of railroads the 834 miles of the "National Pike" provided one of the chief avenues to the west. "As many as twenty four-horse coaches have been counted in line at the time on the road, and large, broad-wheeled wagons, covered with white canvas stretched over bows laden with merchandise and drawn by six Conestoga horses, were visible all the day long at every point, and many times until late in the evening, besides innumerable caravans of horses, mules, cattle, hogs and sheep. It looked more like the leading avenue of a great city than a road through rural districts." [2]

The Cumberland Road not only furnished a great highway for emigration, but reduced transportation cost between Baltimore and the Ohio and brought great prosperity to the regions through which it ran. Communication between east and west was quickened by the Great Western Mail which followed it. So beneficial was the road that numerous local turnpikes were projected which it was hoped the national government would sponsor. Not less than 111 surveys and plans for roads, canals, railroads, and river improvements were before Congress in 1830, when Jackson called a halt by his veto of the Maysville Road bill (a proposed turnpike from Maysville to Lexington, sixty miles in length, entirely in the state of Kentucky,) in which he held it to be unconstitutional for the government to use money for such enterprises

---

[1] Searight, T. B., *The Old Pike*, p. 16.
[2] *Ibid.*, p. 16.

confined wholly to individual states. This veto contributed much to throwing future internal improvements into state hands.

The question of the constitutionality of internal improvements connected with the Cumberland and Maysville roads was not thought to concern the building of military roads. The power to raise and support an army included the power to move it from place to place and provide facilities for such purposes. As a consequence of lessons learned in the War of 1812 a number of military roads were built by soldiers receiving special pay for such work; chiefly in the territories, where the authority of the national government to do this was unquestioned.

The River Steamboat.—The steam engine had been in practical operation in textile mills since James Watt's improvements in 1769. A number of engineers, including Oliver Evans, John Fitch, James Rumsey, John Stevens, and others in the United States had been experimenting with the use of the steam engine in water and navigation, and had successfully propelled vessels, but it was left to Robert Fulton to make steam navigation commercially successful. After many discouragements, but with the backing of Chancellor Livingston, Fulton in 1807 built the *Clermont,* a 160-ton side-wheeler, and sailed it to Albany, a distance of 150 miles in thirty-two hours. "My steamboat voyage to Albany and back," said Fulton in a letter, "has turned out rather more favorably than I had calculated. The distance from New York to Albany is one hundred and fifty miles. I ran it up in thirty-two hours, and down in thirty. I had a light breeze against me the whole way, both going and coming; and the voyage has been performed wholly by the power of the steam-engine. I overtook many sloops and schooners beating to windward, and parted with them. The power of propelling boats by steam is now fully proved. The morning I left New York there were not perhaps thirty persons in the city who believed that the boat would ever move one mile an hour or be of the least utility; and, while we were putting off from the wharf, which was crowded with spectators, I heard a number of sarcastic remarks. This is the way in which ignorant men compliment what they call philosophers and projectors."

TRANSPORTATION TO 1860

Livingston and Fulton immediately obtained a monopoly of the waters of New York State for twenty years, and soon after a similar monopoly of the waters of the lower Mississippi in the territory of New Orleans. Several steamboats were in operation on the Hudson before Livingston and Fulton established a shipyard at Pittsburgh in 1811, and launched the *New Orleans,* the first steamboat on the Ohio. This boat descended the Ohio and Mississippi in the winter of 1812, but could not return against the swift current. It was three years later (May, 1815) before a steamboat succeeded in ascending the rivers from New Orleans to Louisville, a feat accomplished in twenty-five days. The development of steam navigation was held up temporarily until the monopolies in New York and Louisiana were broken by the Supreme Court decision in the case of Gibbons *v.* Ogden (1824), which held that river transportation could not be monopolized by any one state, but that it came under the powers of Congress to regulate interstate traffic. The success of the steamboat was demonstrated by the fact that the time from New Orleans to Pittsburgh was soon reduced from one hundred to thirty days, while freight and passenger rates were more than cut in half. It was some time before the old flat-bottom boats were displaced, but farmers west of the mountains could now dispose of their products. By 1825 there were 125 steamboats on the Mississippi and Ohio, and by 1860 over a thousand. The first steamboat on the Great Lakes was the *Walk-in-the-Water,* built at Black Rock, near Buffalo, in 1818, which plied between Buffalo and Detroit, but steamboat traffic on the Great Lakes did not develop rapidly until the late 'thirties. The loss of large numbers of steamboats from bars and snags led to the appropriation between 1822 and 1860 of over $3,000,000 by the national government for bettering the traffic conditions on the Mississippi, Ohio, Missouri, and Arkansas rivers. The commerce of the Mississippi Valley in 1852 was estimated at $653,976,000, a large proportion of which was made possible by the steamboat, which not only aided in the marketing of bulky foodstuffs, but thereby stimulated a more rapid settlement of the region.

The Era of Canal Building.—Although the "National Pike"

as it was gradually opened helped materially, it was far from settling the transportation problem. Costs of transportation were still too high to allow extensive handling of trans-Alleghany freight. The population in the regions tributary to the Ohio was estimated in 1800 at 400,000 and was increasing rapidly; the population of Ohio, Indiana, Illinois, Michigan, Wisconsin, and Iowa increased from 50,240 in 1800 to 792,719 in 1820 and 2,967,840 in 1840. The center of population in 1800 was eighteen miles west of Baltimore; by 1840 it was near the center of West Virginia, and by 1850 had almost reached the Ohio River in the western part of that state. As this population was entirely agricultural, its products must find a market either east of the Alleghanies or in foreign countries. The route to either market was roundabout and hazardous, being down the tributaries of the Mississippi to its mouth, where it was transshipped by sailing vessels to its destination. This route served fairly well the needs of the southwestern cotton planters, but it failed to meet the needs of the farmers north of the Ohio, although as years went by these farms found an increasing market for their products on the plantations of the south. Better communication with the east was essential.

As the limitations of the roads became evident, attention was drawn more and more to artificial waterways as a possible solution. The distinct success of James Brindley in England in constructing the Bridgewater Canal (opened 1761) and in other later projects had stimulated a great activity in that country in canal building, which was reflected here by continually increasing interest. The Erie Canal was the most notable both in extent and in results, but it was far from being the first canal built in this country; a number of local canals preceded it. A seven-mile canal between the towns of Richmond and Westham was authorized by the Virginia legislature in 1785 and constructed. The Dismal Swamp Canal authorized by Virginia and South Carolina was begun in 1787 and completed in 1794. The chief artificial waterway in the south was the Santee Canal, extending for twenty-two miles between the Santee River in South Carolina and Charleston, which was completed in 1802. Navigation of the Schuylkill in

Pennsylvania by an adjoining canal of 108 miles with 129 locks was effected in 1826 after eleven years of construction. The earliest settlers in Philadelphia had talked of joining the Susquehanna and Schuylkill rivers, but no work was performed on the enterprise until 1791. After four miles were completed in 1794 and put into use, the undertaking was abandoned until 1821, when it was resumed, completed in 1828, and eventually became part of the Pennsylvania Canal. The route for a canal between the Chesapeake Bay and the Delaware River had been surveyed in 1764 and work was commenced in 1804 with great enthusiasm, for its success would eventually mean a canal across New Jersey and thus an inland waterway from the Hudson to Hampton Roads, and by the Dismal Swamp Canal to Albemarle Sound. Work was soon abandoned, but the scheme was revived in 1822 and by 1839 thirteen miles were in use. The Middlesex Canal, thirty miles long, and including twenty locks, extending from the Merrimac to the Charles in Massachusetts, was begun in 1795 and completed in 1808. A small canal around the Patopwick Falls in Massachusetts was finished in 1797, and the Bow Canal in New Hampshire in 1812; both demonstrated the interest of New Englanders in artificial waterways, but the topography of that section prevented canals from being as useful and practicable as in the south and middle states.

The Erie Canal.—The great era of canal building may be said to have been inaugurated by the Erie Canal. No one contributed more to arousing interest in this project than Elkanah Watson, a member of the first commission appointed in 1792 to explore and lay out a possible route. In 1788 he had clearly seen the possibilities of such a canal and had written:

In contemplating the situation of Ft. Stanwix, at the head of the Bateaux Navigation on the Mohawk River, . . . I am led to think this station will, in time, become an emporium of commerce between Albany and the vast western world above. . . .
Should the Little Falls ever be locked,—the obstructions in the Mohawk River removed,—and the canal between said river and Wood Creek at this place, formed, so as to unite the waters running east, with those running west; and other canals made, and obstructions

removed at Fort Oswego,—who can reasonably doubt but that by such operations the state of New York have it within their power by a grand stroke of policy, to divert the future trade of Lake Ontario, and the Great Lakes above, from Alexandria and Quebec to Albany and New York.[1]

Followed twenty-five years of agitation, with unsuccessful efforts to obtain help from the national government and the states of Ohio and Indiana; finally in 1817, urged on by Governor Clinton, New York undertook the work alone. Fifteen miles of waterway between the towns of Utica and Rome were opened in 1819, and on October 26, 1825, cannon stationed at intervals along the canal announced from Buffalo to Albany the opening of the entire canal and the departure of the first boats. Some weeks later from the decks of the *Seneca Chief,* which had headed the procession of vessels, Governor Clinton poured the contents of a cask filled with water from Lake Erie into the New York Harbor to signify the wedding of the waters. "This solemnity at this place," he said, "on the first arrival of vessels from Lake Erie, is intended to indicate and commemorate the navigable communication which has been accomplished between our Mediterranean Seas and the Atlantic Ocean, in about eight years, to the extent of four hundred and twenty-five miles, by the public spirit and energy of the people of the State of New York, and may the God of the heavens and the earth smile propitiously on this work and render it subservient to the best interests of the human race."

As originally built, the Erie Canal was 363 miles long. It was constructed at an average cost of $20,000 a mile, the total expense amounting to approximately $7,000,000. It followed the Mohawk to Rome, and thence westward through the present cities of Syracuse, Rochester, and Lockport to Buffalo by way of the Tonawanda and Niagara rivers. Subsidiary canals connected it with Ontario, Champlain, and Seneca lakes, and eventually New York had 906 miles of artificial waterways.

The success of the Erie was immediate. Tolls exceeded the

---

[1] Watson, Elkanah, *History of the Rise, Progress and Existing Condition of the Western Canals in the State of New York,* etc. (1820), p. 15.

interest charge before it was finished, and during the first nine years amounted to $8,500,000—more than the initial cost. Nineteen thousand boats and rafts during 1826 passed West Troy in the Erie and Champlain canals. Its first and greatest effect was in providing an all-water route to the west, thus furnishing an outlet for the bulky products of the interior. Freight from Buffalo to New York dropped from one hundred to fifteen dollars per ton and the time from twenty days to eight. Farm produce of western New York doubled in value and that of the states north of the Ohio was increased, carrying in its wake a corresponding rise in land values. If the Old Northwest was mightily stimulated, the regions immediately adjoining the canal fairly boomed. Utica, Syracuse, and Rochester became thriving towns, while the terminals, Buffalo, Albany, and New York, took on new life. The last-named city doubled its population between 1820 and 1830, and took from Philadelphia its leadership as the first American seaport. The lake cities of Buffalo, Cleveland, Detroit, and Chicago entered upon a rapid growth, and commenced to rival Pittsburgh, Cincinnati, St. Louis, and New Orleans, as the produce of the western farmers was drawn through the northern route. To a large portion of the United States the Erie Canal opened a period of unprecedented prosperity. Passenger packets made the distance from Albany to Buffalo in four and a half days, and over this route passed a continually increasing stream of western immigrants. The Old Northwest seemed firmly bound to New York through the Erie, for the distance to the sea was shorter this way than through either the Mississippi or St. Lawrence valleys.

Other Eastern Canals.—The success of the Erie Canal and the prosperity resulting therefrom to New York led to similar projects in many other states. Pennsylvania, fearful lest her western trade be drawn off to New York, was first caught in the mania for canal building, and rapidly constructed (1826-34) a system of canals and portages from Philadelphia to Pittsburgh, following the Susquehanna, Juniata, Conemaugh, and Allegheny rivers. A horse railroad led from Philadelphia to Columbia on the Susquehanna, where the canal started, from which point the route was along the east bank of the Susquehanna and the west bank of the Juniata to Hollidaysburg. The mountains between Hollidaysburg and Johnstown were crossed by a portage railway thirty-three and one-half miles long upon the inclined planes of which it was possible to raise a boat 1,399 feet in less than ten miles and lower it 1,171 feet. The Pennsylvania Canal and its connecting railways was 394 miles long, and constructed at a cost of over $10,000,000. Although state owned, the government did not operate either cars or boats, simply charging toll for the use of both. To build this route to the west it had been necessary to surmount an altitude of almost 2,300 feet as against a rise of 500 feet on the Erie Canal. Not so satisfactory as the Erie, it was yet successful in its purpose and through it a share of the western trade reached Philadelphia.

Other canals were built in Pennsylvania, some as subsidiary to the Pennsylvania Canal and others designed to float down the anthracite-coal deposits of the Wyoming Valley. Work on the Union Canal, commenced in 1791 and later abandoned, was resumed, and in 1828 the eighty-two miles between Middletown on the Susquehanna and Reading on the Schuylkill were opened. Connections were made between the Wyoming Valley and the tidewater by canals along the Lackawaxen and the Delaware and along the Susquehanna, Lehigh, and Schuylkill rivers.

Two artificial waterways were cut across New Jersey and used chiefly to transport coal. The Delaware and Raritan Canal, built between 1834 and 1838 at a cost of $4,735,353, extends from Bordentown on the Delaware to New Brunswick on the Raritan. The Morris Canal, designed to connect the Hudson with the Delaware, was opened in 1836 and led from Jersey City, through Newark, Dover, Hackettstown, and Washington, to Phillipsburg on the Delaware. The old plan of a waterway from the Chesapeake to the Delaware was revived, and the Delaware and Chesapeake Canal, thirteen and one-half miles long, was completed in 1829.

Not to be outdone by their northern neighbors, the citizens of Maryland and Virginia took up with renewed vigor an old plan to join the eastern coast to the Ohio River by means of a canal running along the Potomac. The Potomac Company, with Washington as its first president, had been incorporated in 1785, but it was not until 1828 that the Chesapeake and Ohio Canal was commenced. Originally planned to extend from Georgetown, as its eastern terminal, to Cumberland, and thence by a tunnel across the range to the Youghiogheny, it was never pushed farther than Cumberland. It was completed in 1850 after many discouraging setbacks, at a cost of $11,000,000, of which $7,000,000 was contributed by the state of Maryland, $1,500,000 by the terminal cities, and $1,000,000 by the United States government. The Chesapeake and Ohio Canal was unsuccessful, due largely to the fact that it was never pushed across the mountains and to the bitter opposition and competition of the Baltimore and Ohio Railroad, which was built simultaneously over the same route.

Canals in the Middle West.—The success of the Erie Canal stimulated interest also in the middle west, and projects were immediately formulated to connect the Ohio with Lake Erie and thus provide a continuous inland waterway from New York to New Orleans. In 1825 Ohio authorized the building of two canals, one known as the Ohio and Erie, to extend from Portsmouth on the Ohio along the course of the Scioto, Muskingum, Tuscarawas, and Cuyahoga to Cleveland; the other, the Miami and Erie, to extend from Cincinnati through Middletown, Dayton, and Defiance to Toledo, following the course of the Miami and Maumee rivers. The latter was practically completed in 1829. Governor Clinton, the "father of the Erie Canal," turned the first spadeful for the Ohio and Erie, which was finished in 1833, so that Ohio at that date had over 400 miles of navigable canals. By 1850 this had extended to over 1,000 miles.

Indiana, caught with the same enthusiasm for internal improvements, commenced in 1832 (completed 1843) the Wabash and Erie, which connected Lake Erie with the Ohio River. The route led through the Miami and Erie to Defiance, entered Indiana in Allen County, and then went southwest along the Wabash to Terre Haute and south through Worthington and Petersburg to Evansville on the Ohio. Indiana also built the White Water Canal from Hagerstown, Wayne County, mostly along the White

Water River, to Lawrenceburg on the Ohio. Illinois, still sparsely settled, built the Illinois and Michigan Canal (1836-48), connecting Lake Michigan with the Mississippi by an artificial waterway from Chicago to La Salle, the head of navigation on the Illinois. Wisconsin attempted to join Green Bay, an arm of Lake Michigan, with the Mississippi by means of a canal between the Fox and Wisconsin rivers, but the project was not completed until 1856. Connection between lakes Huron and Superior was effected in 1855 by a canal around St. Mary's Falls. Constructed originally by Michigan, it was later turned over to the United States government and has become one of the most important artificial waterways in the world. Navigation on the Ohio was furthered by a short canal, around the falls at Louisville. Canal mileage in the United States was estimated at 1,270 in 1830, 3,320 in 1840, and 3,700 in 1850.

Panic of 1837 and Failure of Internal Improvements by States.—Most of these improvements had been undertaken by the several states. Constitutional objections had tied the hands of the national government, although considerable aid was rendered by donations of public lands and by the purchase of stock, as in the case of the Chesapeake and Ohio. Private capital was inadequate, but the credit of the states in the prosperous days of the late 'twenties and early 'thirties seemed inexhaustible. Land speculators and *bona fide* settlers encouraged expenditures out of all proportion to the existing wealth and population of the states, and wrote into the state constitutions of the period directions "to encourage internal improvements within the state." Rivalries between states and cities contributed to the mania. State debts which had amounted to but $12,790,728 in 1820 increased to $66,482,186 in 1835, to over $170,000,000 in 1838 and $200,000,000 in 1840, practically all of which had been incurred for banks, roads, canals, and railroads. Only seven states (Connecticut, Delaware, Georgia, New Hampshire, North Carolina, Rhode Island, and Vermont) had not contracted debts for these purposes. Much of this work had been undertaken in an era of enthusiasm without adequate knowledge of difficulties and costs. The projects were speculative and in some cases unnecessary. This too rapid investment in internal improvements, especially canals, contributed largely to the panic of 1837, and when the bubble was pricked in that year most of the states found themselves unable to pay interest or continue the work. Several of the states, including Mississippi, Louisiana, Maryland, Pennsylvania, Indiana, and Michigan, repudiated their debts. Nearly all sold out their improvements to private concerns and retired from aiding public improvements, while the people, turning to the opposite extreme, now forbade in the new state constitutions the credit of the government being used for such purposes. Consequently, private individuals and corporations shouldered the work during the era of railroad building which was just dawning.

Significance of Railroads.—The full flush of prosperity for canals had scarcely been reached before they were challenged by a new form of transportation. Simultaneously with the success of steam-driven boats, the idea had come to engineers that a steam engine might also be used to propel wheeled vehicles. Oliver Evans in 1804 had put his steamboat on wheels and driven it through the streets of Philadelphia, and John Stevens in 1820, on his estate at Hoboken, New Jersey, had built a little railroad of narrow gauge upon which he ran a locomotive and cars with himself a passenger. When George Stephenson's *Rocket* in October, 1829, pulled a train on the Liverpool and Manchester railroad weighing thirteen tons at an average speed of fifteen miles an hour, the practicability of steam railroads was clearly demonstrated. The advantages of railroads over canals were immediately realized. They were cheaper to construct and transportation over them was more rapid. Moreover, they were not confined to comparatively low districts on account of water supply and expense and delay of locks, but could be laid to reach into almost any part of the country, even to the back doors of factories. They were not affected by change of seasons, droughts, floods, or freezing; all sections could benefit at all times. The rivers ran mostly north and south, but the new railroads were able to strike directly west. The effects of railroads in opening up the west, in providing transportation for western products, in stimulating eastern manufacturing, in binding the sections together, in disseminating information and education to remote sections, provide a story intertwined with every phase of our economic, social, and political life since 1840.

Early Railroads.—"England," says Dunbar, "had been building railways for nearly two hundred years, had made iron rails since 1738 and steam locomotives since 1804." [1] Real progress came there after 1829. In America wooden rails for local purposes in moving iron and stone had been used in various places early in the century. The best known of these were the three-mile road from Quincy, Massachusetts, to Neponset, opened in 1827, and the Mauch Chunk Railway in Pennsylvania finished in the same year. The first railroads designed for passenger service were those built to supplement the canal system, such as the road from Philadelphia to Columbia and from Hollidaysburg to Johnstown.

The first stone of the railroad track of the Baltimore and Ohio was laid on July 4, 1828, by Charles Carroll of Carrollton, the last surviving signer of the Declaration of Independence. This was the first railroad in the modern sense in America, the first division of which, thirteen miles long, was opened in 1830. Even the builders of this railroad were not sufficiently convinced of the value of steam, and the motive power for the first vehicles was sails, or horse power. In 1831 Peter Cooper's engine, *Tom Thumb,* on a trial trip made the thirteen miles from Baltimore to Ellicott's Mills in an hour, and the management turned definitely to steam.

The first attempt to run a steam railroad locomotive in this country (barring the demonstration by Stevens) was made in 1829 on the Carbondale and Honesdale Railroad (now part of the Delaware and Hudson), where the nine-horse-power *Stourbridge Lion,* imported from England and set up here, was found too heavy for the rails and trestles of the road and discarded. Meanwhile a charter had been granted for a road from Charleston to Hamburg, South Carolina, and over this line in 1830 the *Best Friend of Charleston,* the first locomotive made in America

---

[1] Dunbar, Seymour, *History of Travel in America,* vol. iii, p. 906.

for regular and practical use, was put in operation, attaining a speed of thirty miles an hour when traveling alone and from sixteen to twenty-one miles with four loaded cars. In 1826 the New York legislature granted a charter to the Mohawk and Hudson Railroad Company, the earliest forerunner of the New York Central. Construction was started in 1830, and in 1831 the *De Witt Clinton,* on a trial trip, made the seventeen miles from Albany to Schenectady in an hour. The first link in the present Pennsylvania system, a strip of road connecting Philadelphia with the Susquehanna, was completed in 1834. In Massachusetts construction work on the Boston and Lowell was commenced in November, 1831, and during the next year upon the Boston and Providence and the Boston and Worcester, but it was not until April, 1834, that a locomotive was run upon any of these lines.

The practicability of the new transportation having been once established, the nation turned to railroads to settle the great problem of intercourse with the west with the same enthusiasm which a few years previously had been shown toward canals. By 1860 more than thirty thousand miles of road had been built.

RAILROAD CONSTRUCTION TO 1860[1]

Years	Miles	Years	Miles
1830	32	1838	1,913
1831	95	1840	2,818
1832	229	1845	4,633
1833	380	1850	9,021
1834	633	1855	18,374
1835	1,098	1860	30,626

The rivalry of the seaboard cities in their hope of tapping the western region commenced anew, while such cities as Boston, Charleston, Savannah, and Mobile, which had been excluded from the race during the canal-building period, joined with the rest in projecting gigantic plans for routes into the interior. As

[1] *Statistical Abstract,* 1921, table 239, p. 376.

the years went by the railroads passed out of the stage in which they were built merely as feeders for canals and connecting links between rivers and artificial waterways, and great trunk lines dependent upon themselves alone were gradually constructed. Before 1850 only one line of railroad had been completed between the tidewater and the great interior basins of the country, but a passenger upon this road, as he crossed New York, was carried by sixteen different companies. Freight was restricted by the payment of tolls, and by the frequent transfers due to different-gauge roads. The first consolidation on this road, the New York Central, was effected in 1853, after which time this route took its place as one of the two great railroad systems leading to the interior. Between 1850 and the opening of the Civil War eight other great lines were completed between the seaboard and the western system of lakes and rivers. The last link of a road from Boston to Ogdensburg was completed in 1850. The next year the New York and Erie, a rival route to the New York Central, was finished to Dunkirk on Lake Erie. By 1852 the Pennsylvania Railroad, planned to connect Philadelphia with Harrisburg, Petersburg, and Cleveland, had reached Pittsburgh. Farther south, the Baltimore and Ohio, reaching Wheeling in 1853, proved victorious over the Chesapeake and Ohio Canal and has since been the main artery by which western produce has reached Baltimore.

Similar activity was evident in the south. In 1850 the Western and Atlantic Railroad of Georgia reached the Tennessee River and, with the opening of the Nashville and Chattanooga in 1854, connected Atlanta with the river and rail system of the northwest and became the distributing agency for western grain and meat in the eastern cotton belt. By 1858 the Central Virginia, running west from Richmond, and the Southside Railroad, running west from Petersburg, had extended to a connection with the Memphis and Charleston and the Nashville and Chattanooga. The Mississippi was reached in 1859 by the Memphis and Charleston. As the traffic which had formerly flowed through New Orleans and Mobile began to be drained southeastward through the roads just named, the southwest was stirred to action, and between 1850 and

1860 about 8,000 miles were built in the region between the Ohio River and the Gulf. Of particular interest were the Mobile and Ohio, opened from Mobile to Cairo in 1859, and the New Orleans, Jackson, and Northern, opened from New Orleans to Jackson, Tennessee, in the same year by a connection with the Mississippi Central. These early roads supplemented to a certain extent the Mississippi and its tributaries in the transportation of freight between the northern and southern states west of the Alleghanies. The first locomotive west of the Alleghanies was the *Sandusky*, built in Paterson, New Jersey, to run upon the Mad River and Lake Erie. This line, begun in 1835, was the first railroad to be built in Ohio and extended eventually from Sandusky to Cincinnati. By 1840 a part of what was later the Michigan Central was in operation from Detroit to Ann Arbor. Three railroads across the state of Michigan from Port Huron, Detroit, and Munroe to Lake Michigan were planned by the legislature, and two were eventually completed after the projects were made over in 1846 to private companies. The Michigan Central and Michigan Southern after a spectacular fight across northern Indiana entered Chicago within two days of each other. By 1860 roads had been built from the Great Lakes which touched the Mississippi at ten places and the Ohio at eight. West of the Mississippi in 1860 there were a number of lines extending into Iowa and Wisconsin and a few miles of construction in Arkansas and California.

Not so vital in their economic service as the through lines to the west were the great number of roads built more or less parallel to the coast. These lines, following roughly the old post roads, had, by 1860, provided transit for mails, passengers and freight from central Maine to southern Georgia. Both north and south by 1860 were equipped with a skeleton railway system. The original tracks of what came to be the great railway systems of our day east of the Mississippi were to a great extent laid.

Problems of Early Railroad Building.—As steam locomotion was in its infancy, the early American railroad builders had to meet and conquer innumerable problems. Some ideas were obtained from England, but different conditions, such as long-distance passenger traffic and more bulky freight, soon made it apparent that development in the two countries would not be closely parallel.

Problems of track construction, gauge, friction of the wheels

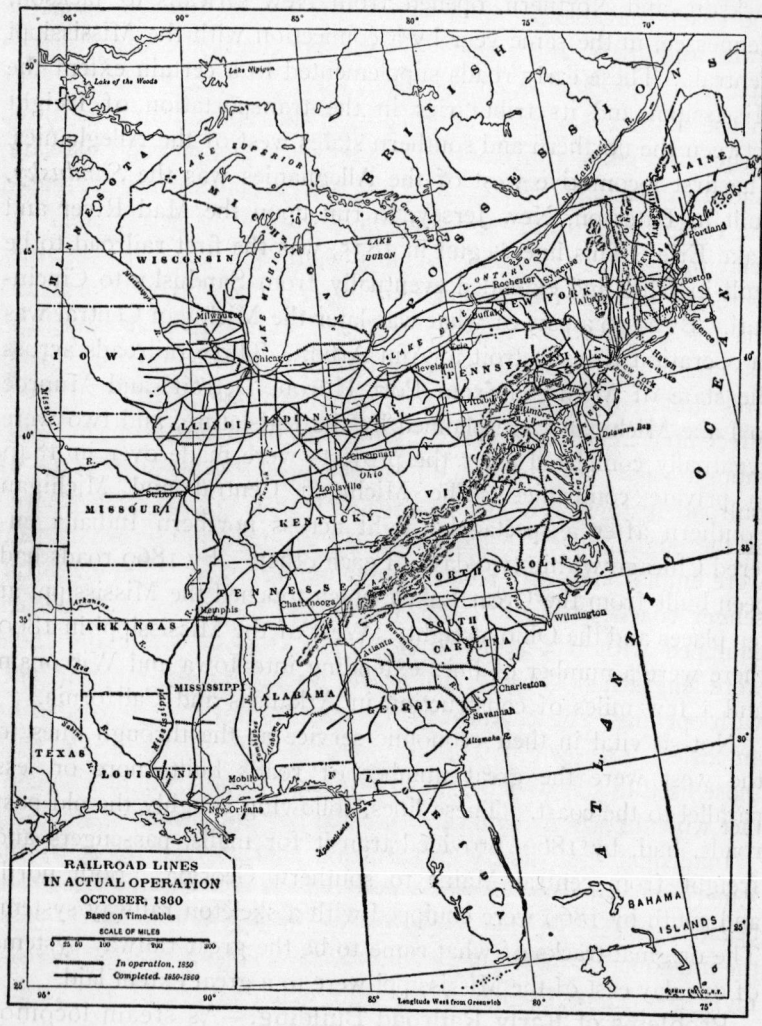

on the rails, bridges, brakes, coupling apparatus, safety devices, lighting and heating, all had to be worked out. Although decades passed before these problems were finally mastered, astonishing progress was made from the start. The first rails consisted of strips of iron laid longitudinally upon wooden beams attached to ties. These iron strips had the unfortunate faculty of coming loose and curling up, sometimes protruding through the floor of the coach and making it necessary for the engineer to stop the train and mend the track. It was not until the early 'fifties that iron rails came extensively into use. No uniform gauge was in use at first, the rails ranging in width from four feet three inches on the Delaware and Hudson to six feet on the Delaware, Lackawanna, and Western. This lack of uniformity increased the expenses in shipping freight and eventually necessitated the relaying of large portions of the roadbed. The early engines were not equipped with cabs until after 1842. Wood furnished the fuel for both engines and heating, and, until coal came into use, the destruction caused by escaping sparks in starting forest fires and igniting the wooden coaches, to say nothing of their effect upon the passengers' clothing, was a serious problem. Although passenger coaches during the first twenty-five years of railroad history were the last word in discomfort, they, as well as the engines, were usually brightly painted, highly ornate, and were emblazoned with some high-sounding name. The first passenger coaches were little more than the bodies of stagecoaches equipped with wheels adaptable to the tracks. Gradually these were lengthened to cars more nearly resembling the modern type, with openings at each end and the seats in two rows separated by an aisle. The original brakes were identical in principle with those used in stagecoaches, blocks of hard wood brought into contact with the wheels by levers operated by foot power.

Inadequate and flimsy construction characterized most of the American railroads before the Civil War. As the roadbeds began to wear out and a greater strain was put upon them by the introduction of heavier rolling stock, accidents became increasingly frequent. Complications were often caused by the failure to fence off the right of way. During the decade of the 'fifties and even later, a railroad passenger literally took his life in his hands. The day of the air brake, the automatic coupler, the block system of signals, and scientifically constructed roadbeds was still in the future. An accident was considered an "act of

God" rather than negligence upon the part of the railroad. The departure and arrival of trains were matters of pure conjecture. Even time-tables were not printed until after 1847.

Financing the Railroads.—Hardly less important than the problems of engineering were those connected with the raising of sufficient funds to construct the 30,000 miles of railroad in operation before the war. As there is proportionately little liquid capital in a sparsely settled agricultural community, railroad *entrepreneurs* met difficulties from the start, difficulties which were not lessened by the fact that the new railroads must necessarily come into competition with turnpikes, plank roads, canals, and the interests vested in these enterprises. While investments in railroads in the more thickly populated east frequently paid large dividends, the prospect of returns on the projected railways in the more sparsely settled west and south was uncertain. In any case, the investor might wait years before dividends came in. In the face of these difficulties over $1,250,000,000 were invested in railroads between 1830 and 1860. Much of this was obtained from abroad, where financiers had already forgotten their unfortunate experiences in the panic of 1837 and the repudiations incident thereto, and were willing to speculate again on the economic possibilities of the New World. Some capital was drawn from New England, where the declining whaling industry released accumulated capital, which was to be augmented in the years to come by a similar decline in shipping. Merchants and farmers in the terminal cities and along the routes subscribed, influenced oftentimes not so much by the desire for dividends as by the expectation of profit from increased business and the rise in land values.

Where private capital was insufficient, state, county, city, and national aid was freely given. Pennsylvania, Michigan, South Carolina, and Georgia undertook to finance their first railways. Other states aided by loaning money, purchasing stock, or guaranteeing in whole or in part the securities. Massachusetts loaned $4,000,000 to the Great Western, Maryland subscribed $3,000,000 to the Baltimore and Ohio stock, Illinois appropriated $8,000,000 for various internal improvements, Ohio passed a law in 1837 by which she loaned her credit in 6-per-cent stock to the amount of one-third the stock in any railroad enterprise, provided the company secured the other two-thirds, while in Indiana railroad corporations were allowed to issue paper money to pay for labor or purchase material. Continued demands that the national government lend its aid resulted in 1850 in a 2,700,000-acre grant of land to the state of Illinois to be used for the Illinois Central. Lavish gifts were also made to Illinois, Mississippi, Missouri, Michigan, Wisconsin, Iowa, Arkansas, Alabama, Florida, and Louisiana, amounting by 1861 to 31,600,842 acres. The ratio of railway mileage to population in 1860 was 1 to 861 in New England, 1 to 1,071 in the middle and northwestern states, and 1 to 1,076 in the southern. The average cost of construction per mile up to 1860 had been around $50,000 in the north and about half that in the south.

Street Railways.—The increasing concentration of population in cities developed in the fifth decade of the century a distinctively urban passenger problem. From the beginning it had been necessary in some cases to run railroads over city streets, but these were not street railways in the proper sense. It was necessary to develop the idea of the sunken rail which would not obstruct wheeled traffic before such roads could be practical. The first vehicle used for periodic transportation of city population was the omnibus, a modification of the stagecoach. In the 'fifties the sunken rail came into use, and from then until the days of the electric tramways various types of horse cars, some of which had two stories, handled as best they could the demands of city traffic.

Express.—The founder of the express business was William Francis Harnden, a former conductor and ticket agent on the Boston and Worcester. Taking his cue from the methods by which gold and silver were transferred east, he conceived in 1839 the idea that a similar business might be developed in the quick and safe transportation of small packages and valuable papers. Harnden himself carried the first express in a carpet bag between New York and Boston, but the business grew so rapidly that a partner was taken in and the firm of Harnden & Company's Express founded in 1840. Agents were hired, the business extended in 1841 to include Philadelphia and Albany, and European agencies were opened to look after the transportation of immigrants. The success of Harnden brought Alvin Adams into the field, and a rival express business was started between New York and Boston. With Ephraim Farnsworth as a partner Adams & Company was founded, whose business extended by 1843 as far west as St. Louis and New Orleans. In 1854 Harnden & Company was merged in the new Adams Express Company, leaving that organization supreme for the time being in the northeast. In the meantime, Henry Wells, Harnden's agent at Albany, had broken away and formed Wells, Fargo & Company, which by 1845 connected the east with Chicago, Cincinnati, and St. Louis. In that year Wells, Fargo sold out to the American Express and moved to the Pacific coast, where they developed business between the mining towns and the coast and between the Atlantic and Pacific.

Telegraphs.—Samuel F. B. Morse (1791-1872), trained as an artist and professor of the literature of the arts of design in New York University, evolved the idea of the electro-magnetic telegraph in 1832. Three years afterward the first crude instrument was constructed, and by 1837 his apparatus was practical. Although his invention was repeatedly demonstrated and undoubtedly contained immense future possibilities, it was impossible to interest private *entrepreneurs*. Forced to seek government aid, Morse and his friends besieged Congress for six years before that body in 1843 finally appropriated $30,000 for the construction of a line from Baltimore to Washington. This line was completed by May of the following year in time to transmit to the capital information from the Whig and Democratic conventions which met in Baltimore that spring. A private company was now formed, and with great difficulty funds were obtained for a line from Philadelphia to Newark. It was opened in 1846 and later extended to Jersey City, from which point messages were sent to New York by ferry. The practicability of the telegraph, once established, made it possible to extend its use rapidly. Ezra Cornell, a man who had been associated with Morse in the first line constructed and had demonstrated the superiority of the overhead wire, became the organizing genius of the rapid expansion which set in. The cost per mile was less than to construct railroads, and the problems to be met were fewer. In 1846-47 New York was connected with Boston, Albany, and Buffalo, and in the next year with Cleveland, Toledo, Detroit, and Chicago. Although forced to meet such new conditions as severe plain and mountain storms, as well as hostile Indians, the Western Union, spurred on by government subsidies, extended its lines to the Pacific in 1861. In that year 50,000 miles of telegraph lines were in operation.

Summary of Domestic Commerce 1790-1860.—Now that the story of the development of transportation and communication has been followed, it may be well to summarize briefly their effect upon domestic commerce. The settlement and prosperity of the Mississippi Valley and the Great Lakes region were dependent upon the marketing of their agricultural products. Before the building of canals and railroads this was possible only by way of the Mississippi and its tributaries to the Gulf of Mexico, but the route was long and, with the old flat-bottomed boats, too expensive to be practical to the farmers of the upper Mississippi. The construction of turnpikes and the National Road was of aid in stimulating domestic commerce, but the first significant expansion of western trade came with the introduction of the steamboat, which cut down the time and cost of the Mississippi-New Orleans route. The domestic exports from Louisiana increased from $1,-753,970 in 1810 to $37,698,270 in 1850 and $107,812,500 in 1860. With the building of canals and later of railroads the situation was radically changed. The region between the Ohio and the Great Lakes now had an option of two routes—east over the canals and railroads or south over the rivers. Owing to the success of the Erie Canal, the cream of the trade from west to east was drawn off by New York, whose domestic exports increased from $8,250,670 in 1820 to $80,047,970 in 1860. During the same years the domestic exports of Pennsylvania and Maryland, hampered by inferior transportation facilities with the west, barely doubled. Both the invention of the steamboat and the accelerated east-bound traffic contributed to develop the Great Lake region. Chicago, a city of 4,500 in 1840, reached 109,260 in 1860, while between 1830 and 1860 Cleveland grew from 1,070 to 43,410, and Detroit from 2,200 to 45,610. Lumber, coal, and iron comprised the chief items of traffic between the Lake ports; a government report of 1852 estimated the coastwise exports of the Great Lakes at $132,000,000 and the commerce of the Great Lakes at $312,000,000.

In addition to farm products, considerable commerce had developed west of the Mississippi in furs, gold, and other commodities. The years following the exploration of Lewis and Clark witnessed the exploitation of the fur districts of the Rockies, and St. Louis and New Orleans became the centers where the furs were collected. After the independence of Mexico had been won, commerce running in value to hundreds of thousands annually developed gradually over the Santa Fé trail between the Missouri River towns and Santa Fé, where the Mexican and American traders met. Over four hundred wagons in 1846 moved southwest over this route. After the discovery of gold in 1848 the overland trade with Colorado, Utah, and New Mexico was greatly stimulated; merchandise to the value of $10,500,000 in 1860 was shipped to settlements west of the Missouri. The years 1790 to 1860 saw not only the settlement of much of the Mississippi Valley, but the opening of that wonderfully rich region to the world's commerce.

NOTES FOR FURTHER REFERENCE

Perhaps the most valuable study of American transportation is the four-volume work of Seymour Dunbar, *History of Travel in America* (1915), somewhat diffuse but interestingly written and illustrated with many rare prints. An indispensable companion volume is that issued by the Carnegie Institution under the editorship of B. H. Meyer of the Interstate Commerce Commission and written by C. E. MacGill, *History of Transportation in the United States Before 1860* (1917), with a good bibliography. Shorter accounts may be found in E. R. Johnson and T. W. VanMetre, *Principles of Railway Transportation* (1922), and Emory R. Johnson, *American Railway Transportation* (1903). Short chapters of summary are included in E. L. Bogart, *Economic History of the United States* (rev. ed., 1922); in Isaac Lippincott, *Economic Development of the United States* (1921), and in E. E. Sparks, *The Expansion of the American People* (1900). Excellent chapters are those in J. B. McMaster, *History of the People of the United States,* Vol. IV, Chap. 33, and Vol. V, Chap. 44.

For older accounts see chapters on Travel and Transportation in *Eighty Years' Progress* (1869); H. S. Tanner, *A Description of the Canals and Railroads of the United States* (1840); A. S. Bolles, *Industrial History of the United States* (1878); and H. V. Poor, *Manual of Railroads* (1881), Introduction; and the Census of 1880, Vol. IV on Transportation. See also the *Readings* of Bogart and Thompson. Mark Twain, *Life on the Mississippi* (1883), and Mark Twain and C. D. Warner, *The Gilded Age* (1873), are history as well as humor.

On special phases the following are of value: T. B. Searight, *The Old Pike* (1894); E. L. Bogart, *Early Canal Traffic and Railroad Competition in Ohio*, Vol. XXI, Journal of Political Economy; W. F. Gebhard, *Transportation and Industrial Development in the Middle West*, Columbia University Studies, Vol. XXXIV, No. 1 (1909); R. E. Riegel, *Trans-Mississippi Railroads During the Fifties*, Mississippi Valley Historical Review, X, pp. 153-173 (1923); H. L. Haney, *Congressional History of Railroads in the United States to 1850* (1908); A. B. Hulbert, *The Old National Road* (1901); U. B. Phillips, *History of Transportation in the Eastern Cotton Belt to 1860* (1908); A. B. Hulbert, *Historic Highways* (15 vols., 1902-05), Vols. 13 and 14 on the great American canals, and his *Paths of Inland Commerce* (1920), Chronicles of America. The best résumé of commerce is that of E. R. Johnson *et al.*, *History of Domestic and Foreign Commerce of the United States* (2 vols., 1915).

Transportation maps may be found in MacGill, in McMaster, in Bogart, and in *Harper's Atlas of American History*.

SELECTED READINGS

MACGILL, C. E., *History of Transportation in the United States Before 1860*, Chaps. II-XI, XVII.

MCMASTER, J. B., *History of the People of the United States*, Vol. IV, Chap. XXXIII; Vol. V, Chap. XLIV.

HULBERT, A. B., *Paths of Inland Commerce.*

JOHNSON, E. R., *et al., History of Domestic and Foreign Commerce of the United States*, Vol. I, Chaps. XIII, XIV.

CALLENDER, G. S., *Selections from the Economic History of the United States*, Chaps. VII, VIII.

CHAPTER XV

SOCIAL BACKGROUND OF THE FORMATIVE PERIOD

Population.—A study of population statistics of the first seventy years of our history reveals three expected tendencies: first, toward rapid increase; second, toward westward migration; and third, toward concentration in cities. The total population of the nation in 1790 was probably under four million, while that given in the census of 1860 was 31,443,321. The increase by decades follows:

POPULATION OF THE UNITED STATES TO 1860 [1]

Year	White	Colored	Total
1790	3,172,006	757,208	3,929,214
1800	4,306,446	1,002,037	5,308,483
1810	5,862,073	1,377,808	7,239,881
1820	7,866,797	1,771,656	9,638,453
1830	10,537,378	2,328,642	12,866,020
1840	14,195,805	2,873,648	17,069,453
1850	19,553,068	3,638,808	23,191,876
1860	26,922,537	4,441,830	31,443,321 [2]

Most of this growth until 1820 was attributable to the natural increase in a new land where large families were economically both possible and profitable. Immigration set in after 1820 in ever larger numbers, adding over five million to the population by 1860. Not only did the normal increase of births over deaths and the additions from immigration encourage the growth of population, but the effects of the Industrial Revolution and the expansion of commerce created new possibilities of employment in trade and industry, and so provided a source of livelihood for additional population. The percentage of growth during each

[1] U. S. Census 1910, vol. i, p. 127.
[2] Including Indians, Japanese, Chinese and all others, numbering 78,954.

decade was about 34 per cent, the population almost doubling every twenty years.

Even more striking than the increase in actual numbers was the distribution. In 1790 over 94 per cent lived on the Atlantic slope of the thirteen original colonies, with less than a quarter of a million living west of the Alleghanies. By 1820 the proportion had distinctly changed. The census of that year showed about 73 per cent living on the Atlantic slope and 27 per cent west of the mountains. The southern group of states was still the most populous, but New York could boast of the greatest population of any single state. The population beyond the mountains now outnumbered that of New England. The ratio of increase had fallen in each of the Atlantic states with the exception of Connecticut and South Carolina, where it had progressed a fraction of 1 per cent. During the decade 1810 to 1820 New York had added 413,000 to her numbers, more than any other state, with Ohio coming next with 351,000. But the ratio of increase had been greatest in the new western states, whereas one eastern state, Delaware, had remained practically stationary. In the thirty years 1790 to 1820 the seaboard states had contributed almost two and one half millions to the population of the west. The census of 1850 revealed the fact that almost half of the population (45 per cent) now lived west of the Alleghanies. Professor Channing has pointed out [1] that in the thirty years 1820-1850, the inhabitants of the region west of the Appalachians more than doubled by five millions, while the population of the seaboard states, notwithstanding the immigration from Europe, failed to double by two millions. Figuring that the population should double by natural reproduction in thirty years, it would seem probable, he believes, that during these three decades the east contributed at least four million to the population of the west; with the exodus from the farms to the cities, this movement across the mountains must have placed a severe strain upon the rural population. These deductions are illuminating in any attempt to study the social and economic cross-currents of life in the seaboard states during the first half of the century. The

---

[1] Channing, Edward, *History of the United States,* vol. v, p. 49.

south furnished a large proportion of this migration; two-fifths of the inhabitants of South Carolina, one-third of those of Virginia and North Carolina, and nearly one-quarter of those of Georgia emigrated west of the mountains to form almost the entire population of the Old Southwest and the predominating element in the Old Northwest. A continual stream of New Englanders moved toward the west, sometimes pausing for a time in Vermont or western New York, but in most cases pushing on eventually to the new country. Between 1820 and 1830 population increased 32.5 per cent and the settled area 24.4 per cent; between 1830 and 1840 the figures were 32.5 per cent and 27.6 per cent, respectively.

Along with the increase and westward movement of population went its concentration in cities. The causes for this were many, most of them attributable to the Industrial Revolution. The population of the seaboard cities was largely augmented after 1820 by immigrants, many of whom were ill adapted by training for farm life and went no farther. The development of means of communication by canals and later by railroads allowed a greater distribution of agricultural produce and an expanded foreign commerce, leading to the growth of cities at collecting and transfer points. The market for agricultural products speeded up the westward movement, which in turn added to the population of important points on the routes of travel. The competition of western agriculture became so keen as to discourage eastern farmers, especially in the less fertile regions, and accentuate a movement toward the cities which the growth of manufacturing favored. In 1780 there were only five towns of over 8,000 population—Philadelphia, New York, Boston, Charleston, and Baltimore, containing 2.7 per cent of the population of the country. Of these Philadelphia alone had over 20,000. The census of 1840 showed 44 cities of over 8,000, with New York, now the largest, containing 312,710. By 1860 there were 141 towns of over 8,000, comprising 16.1 per cent of the population. In that year New York, as now constituted,[1] had a population of

---

[1] The population of New York as then constituted was 813,669 (Borough of Manhattan).

1,174,779, Philadelphia of 565,529, and Baltimore of 212,418. The following table will show the growth of city population:

GROWTH OF CITY POPULATION, 1780–1860

Year	Number of cities having a population of					Percentage of total population in cities
	8,000 or over	8,000 to 20,000	20,000 to 75,000	75,000 to 250,000	250,000 or over	
1780	5	4	1	..	..	2.7
1890	6	4	2	..	..	3.3
1800	6	1	5	..	..	4.0
1810	11	6	3	2	..	4.9
1820	13	7	4	2	..	4.9
1830	26	19	4	3	..	6.7
1840	44	28	11	4	1	8.5
1850	85	56	21	6	2	12.5
1860	141	96	35	7	3	16.1

By 1810 New York City had forged to the front and after the completion of the Erie Canal speedily became the great American metropolis, acting as a shipping center for the largest part of the western produce. Philadelphia and Baltimore, leading colonial towns, sought, by building competing systems of canals, to tap the western areas and draw to themselves a share of the produce, but with indifferent success until the advent of railroads. The commercial importance of Boston was now overshadowed by New York, but the former became the center of a rapidly growing manufacturing district whose subcenters were to be found in the thriving cities of Nashua, Lowell, Waltham, Lynn, Worcester, New Bedford, and Fall River. As New England turned to manufacturing, colonial towns like Providence, New London, Hartford, and New Haven became important cities, while hundreds of obscure villages grew into thriving towns or small cities. Pittsburgh, located at the head of navigation on the Ohio where east met west, in a position to tap vast regions and furthermore in the midst of a rich coal and iron district, became as early as 1800 a city of note. The situation of Cincinnati and Louisville on the same river insured their future. St. Louis collected the commerce of the Missouri and upper Mississippi, while Mobile and New Orleans were the shipping points at the mouth. Chicago,

Detroit, Cleveland, Buffalo, the shipping ports of the Great Lakes, had begun by 1860 to show the promise of their subsequent greatness. At other points too numerous to mention, natural advantages or fortuitous circumstances had led to urban growth.

In the south, however, many of the cities, such as Williamsburg, Charleston, and Savannah, went back absolutely or relatively. The drain of population westward, the shift of the center, and the natural outlet of cotton culture across the mountains, the drawing of western produce to Baltimore and Philadelphia rather than to the southern seaports, and the lack of manufacturing development—all contributed to the relatively small growth of urban life in the South Atlantic states.

Immigration.—The preliminary report of the eighth census estimated from a "survey of the irregular data previous to 1819" that from 1790 to 1800 about 50,000 Europeans arrived here, from 1800 to 1810 about 70,000, and from 1810 to the end of 1820 about 114,000. To determine the actual settlers, a deduction from these figures of 14.5 per cent should be made for transients. After 1819 official records were kept which are approximately correct. Immigration up to 1825 amounted to less than 10,000 a year, but gradually increased thereafter until by 1832 about 60,000 were annually coming to our shores, an increase probably due to the revolutionary disturbances of the period. This swelled to 79,000 in 1837, only to be cut in half the next year by the panic. The flow of immigration again increased to over 100,000 in 1842, to be again reduced in the next year by financial depression. The five years from 1845 to 1850 showed a tremendous gain, due to the severe winters of 1845 and 1846 on the Continent, to the subsequent spring floods which affected agriculture adversely, and to the Irish potato famine of 1845 and 1846. The influx from these causes was augmented by the revolutions of 1848 and 1849.

Economic and political influences abroad were not alone in driving hundreds of thousands to the New World; the discovery of gold in California lured many more. The immigration amounted in 1854 to 427,833, but dropped in the next year to less than half that number, and amounted in 1860 to but 153,640.

This decline has been attributed to the Crimean War and troubles in India, which absorbed some of the excess population and increased the demand for agricultural and manufactured goods. The Civil War at first adversely affected immigration, but the war prosperity, combined with the allurements of the Homestead

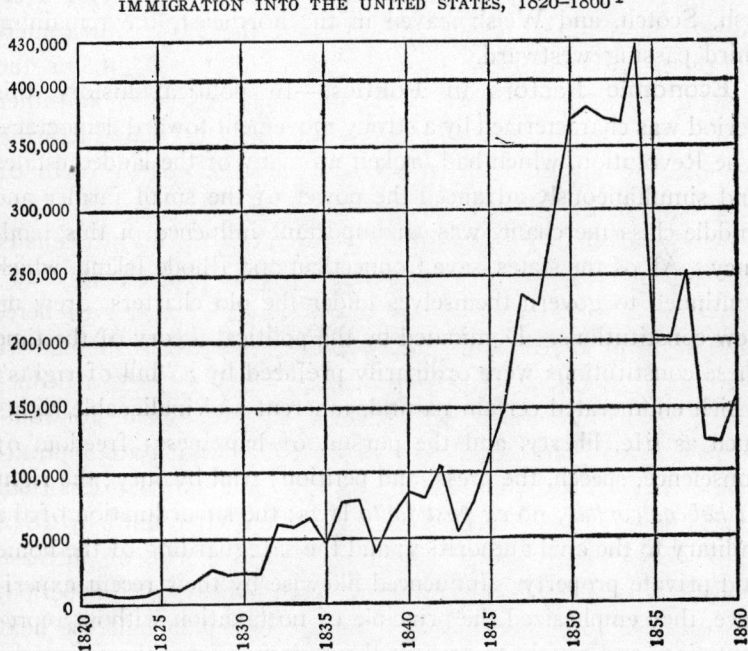

IMMIGRATION INTO THE UNITED STATES, 1820–1860 [1]

Act, renewed the flow of alien settlers. The great majority of immigrants previous to 1860 came from the British Isles, especially Ireland, and from Germany. The five leading occupations as stated by male immigrants during these years were: laborers, 872,317; farmers, 764,837; mechanics, 407,524; merchants, 231,-852; and miners, 39,967. In 1860 there were about four million foreign-born settlers in the United States.

As regards distribution it is possible to speak only in general terms. At least five-sixths of the Irish immigrants remained east of the Appalachians, most of them in the cities, where they

---

[1] *Preliminary Report of the Eighth Census*, 1860, p. 12 ff.

formed the bulk of the unskilled labor. On the other hand, the German immigrant of these years was more likely to be a farmer, and at least half of them took up lands west of the mountains, not a few on the Wisconsin and Texas frontiers. The large majority of the Scandinavians pushed west to find homes in Illinois, Wisconsin, or Minnesota. Perhaps two-thirds of the English, Scotch, and Welsh stayed in the northeast, the remaining third passing westward.

Economic Factors in Politics.—In political history this period was characterized by a strong movement toward democracy. The Revolution, which had broken up many of the landed estates and simultaneously advanced the power of the small farmer and middle-class merchant, was an important influence in this tendency. All of the states, save Connecticut and Rhode Island, which continued to govern themselves under the old charters, drew up new constitutions. Dominated by the political theory of the time these constitutions were ordinarily prefaced by a "bill of rights" which enumerated certain natural, inherent and inalienable rights such as life, liberty, and the pursuit of happiness; freedom of conscience, speech, the press, and petition; trial by jury; the right of *habeas corpus;* no *ex post facto* laws; the subordination of the military to the civil authorities; and the safeguarding of the home and private property. Influenced likewise by their recent experience, they emphasized the principle of no taxation without representation, and sought to prevent the governor from thwarting the will of the legislature. While democratic principles were thus enumerated in theory and a certain amount of mechanism created by which they might be applied, in actual practice the rule of the people was far from being accomplished. In only one state, Vermont, was there universal manhood suffrage; in the rest property qualifications, religious opinions, a tax receipt, a white skin, or a combination of these, determined the voters. In New Jersey alone, entirely without the intention of the framers, the constitution was so worded that until 1807 women and negroes voted provided they owned 50 pounds proclamation money, clear estate. The rights of franchise were narrow, but the provisions respecting officeholding were even more restricted. All of the states placed religious tests upon the office of Governor, and most required that he be a man of property. Similar requirements applied to members of each branch of the legislature. In New Hampshire it was necessary for a state senator to have a freehold estate worth £200, in Massachusetts a freehold of £300 or personal property of £600, in New Jersey and Maryland £1,000 of real or personal estate, in South Carolina 500 acres and 10 negroes, and so on.

Although little opposition was voiced at the time against this setting up of government by classes, the demand for reform came early. The shock given to economic life by the Revolution, the democratic tendencies set in motion, and the general intellectual ferment ensuing, all tended toward democracy. This change was evidenced in the Ordinance of 1787, which forbade involuntary servitude in the Northwest Territory, and contained other liberal clauses heretofore new to American constitutions; and in the Federal Constitution, which provided for representation of the lower house according to population, a plan not usually followed in the state constitutions where representation was generally based on numbers of taxpayers or arbitrarily distributed geographically. Between 1790 and 1810, however, considerable progress was made in breaking down the bars of political privilege by extending the suffrage and removing religious disabilities.

So rapid, indeed, had been the movement toward democracy from 1790 to 1800, that the conservative forces rallied and succeeded in temporarily preventing further democratizing of institutions, except in Maryland, where in 1810 the property qualification for voters was abolished and taxes for the support of religion forbidden; and in New Jersey (1807) where the property qualification was thrown aside, but the franchise restricted to white males. One new state, Ohio, drew up a constitution (1802) which illustrated both the feelings of the democratic west and the extent to which the movement had progressed. The governor was allowed neither a veto nor the right of appointment to office; the term of the tenure of judges was fixed; universal white manhood suffrage was established.

The second decade of the century was a period of constitution making in which six states entered the Union; Louisiana, 1812;

Indiana, 1816; Mississippi, 1817; Illinois, 1818; Alabama, 1819; and Maine, 1820. While democratic growth continued, it is displayed in a different way. The distrust of the governor as exemplified in the earlier constitutions was modified, and a tendency to restrict the powers of the legislature more carefully is seen. The unwise enactments of the state legislatures had proved that the welfare of the people was not always safe in their hands, and the constitution makers of the new states put in specific directions regarding charters, banks, lotteries, and school lands. Louisiana and Mississippi restricted the franchise to free white males who paid state or county taxes, while Indiana, Illinois, Missouri, and Alabama granted universal suffrage to all white men. In 1818, Connecticut after a long struggle succeeded in discarding her autocratic and antiquated form of government under the old charter, and in its stead introduced a new constitution of a democratic character with religious freedom and liberal franchise requirements. Notwithstanding the opposition of Daniel Webster, Massachusetts in amending her constitution in 1820 abolished the property qualification for voters, but retained it for senators. The struggle for democracy went on in the decade of the 'twenties, centering in the demand for representation by people rather than by political areas, for fewer appointive and more elective officials, and for universal manhood suffrage. Martin Van Buren led the fight against the conservative elements in New York in the revision of the constitution of 1821, in which the property qualification for white males was set aside and even negroes possessing a freehold of $250 [1] could vote. Maryland in 1826 opened her public offices to Jews. The new constitution of 1829 in Virginia extended the franchise, but retained property qualifications.

By 1830 the battle for political democracy had been won. The old state constitutions were being gradually revised in this direction, while those of the new western states naturally reflected the democratic ideals of the frontier. This growth of democracy was, of course, bitterly fought by the conservative element, who

---

[1] A restriction abolished in 1826.

SOCIAL BACKGROUND

looked upon the accession to power of the Jeffersonian democracy and later of the Jacksonian democracy as spelling the doom of the Republic. When it was evident that the new democracy would eventually control the executive and the legislative branches, the conservatives made every effort to continue their control of the judiciary. The decisions of the Federalist judges went far to uphold the power of the national government, to limit states' rights, to protect vested interests from the onslaughts of the new democracy, and to develop the doctrine of judicial review over legislative enactments.

Communistic Experiments.—To reformers who felt that the existing order had been built upon wrong principles, America offered a natural field in which to try out various experiments along communistic lines. The communities founded in America, in which the inhabitants lived more or less apart from the rest of the world and practiced more or less thoroughly ownership in common, were of two types: those founded primarily for religious purposes and those founded under secular auspices. Communities of the first type date back as early as the short-lived German Pietist establishment of 1694 on the Wissahickon River near Philadelphia, and the Dunker community at Ephrata, founded in 1732 by Beissel on a basis of voluntary celibacy and communism. The former disappeared quite early, but the latter continued its communistic form of life until 1814, when the settlers relinquished celibacy and communism and became incorporated with the Seventh Day Baptists. A coöperative economy was practiced for a time by the Moravians in their Pennsylvania towns of Bethlehem, Nazareth, and Lititz, founded in 1741. In these towns only members owned real estate and the church controlled both their spiritual concerns and their industrial pursuits, a policy which was totally relinquished in 1844. The Shakers, who formed the oldest sectarian community now existing in America, officially called the Millennial Church, were first established by Ann Lee and a group of English immigrants at Watervliet, New York, in 1776. Their greatest membership was 4,869 and in 1910 they still had fifteen communities. The Harmony Society, founded in 1805 by Germans under the leadership of George Rapp (1770-1847), left their first settlement at Harmony, twenty-four miles northwest of Pittsburgh, and moved west to the Wabash, where they built in 1815 the town of New Harmony, Indiana; nine years later they sold the whole establishment to Robert Owen, and, returning to Pennsylvania, established their final settlement, Economy, in Beaver County. The village was almost entirely self-sufficient, and the people were apparently contented under a system of celibacy and communal life. They throve and grew wealthy until the latter part of the nineteenth century, when their membership was so depleted by their rigid adherence to celibacy that further continuance was impracticable. The German Separatists at Zoar, Ohio, founded in 1817, practiced communal ownership, but retained the family life; they prospered for many years until the pressure of outside customs became too strong, and the society was dissolved in 1898. The revival of communities of "The Inspired" in Germany was followed by persecutions and their emigration to America. They built six villages near Buffalo, and two in Canada, and in 1859 founded the Amana Society in Iowa, which still persists under a thoroughgoing system of communism.

The Perfectionists stand out among American religious communities in that their founder, John Humphrey Noyes (1811-86), was an American, a Dartmouth graduate, and a Congregational minister. A community founded at Putney, Vermont, was forced to remove to Oneida, New York, from which other branches were formed, notably the one at Wallingford, Connecticut. Agricultural and small manufacturing industries were carried on, and the community was exceptionally well managed. Their system of "complex marriage" resulted in so much hostility that in 1881 the community organization was given up and a joint stock company formed with a paid-up capital of $600,000, which is still carrying on industries.

Less successful but more famous, perhaps, than religious coöperative commonwealths were those created under secular auspices. These may be divided into three groups; (1) those whose origin may be traced to Robert Owen and his visit to the United States in 1824; (2) those which sprang up after 1841 in an effort to practice the principles as laid down by Charles Fourier; and (3) those which were due to the communistic teachings of another Frenchman, Etienne Cabet. Robert Owen (1771-1838) believed that civilization had been constructed on the principles of individualism and selfishness, which were responsible for the evils which we suffer. To eradicate these the environment must be perfected; for decent surroundings, he thought, would make a good man. Owen had made many noble and successful experiments in the communities grouped around his mills at New Lanark, Scotland, and now wanted to try out on a much greater scale in the New World his ideas of a coöperative commonwealth. Purchasing the Rappite village of New Harmony with its mills and thirty thousand acres of land for $150,000, Owen invited "the industrious and well disposed of all nations" to join him there. Men of ability and fame came to aid in the experiment, as well as the inefficient and lazy. The personality and genius of Owen bade fair to accomplish much at New Harmony, but, unfortunately, his radical ideas on religion and marriage brought discord and criticism, and at length, financially embarrassed, he returned to England. At least eleven other communities came into existence at this time, the most famous being Nashoba, near Memphis, where Fanny Wright hoped by proper environment to prepare the negro slave for his freedom.

A second wave of interest in communistic experiments spread over the country in 1841, following the explanation by Albert Brisbane in a series of articles in the New York *Tribune* of the scheme of industrial organization advanced by the Frenchman, Charles Fourier. The hard times following the panic of 1837 had undoubtedly prepared the minds of many for further schemes of this nature. In Brisbane's book, *The Social Destiny of Man,* which had a wide circulation, he again explained Fourier's conception of organizing mankind into groups or phalanxes of from three to eighteen hundred persons who should unite to carry on industry, art, and science. A large central building was to contain the workshops, the apartments for families, and the common rooms for dining and holiday meetings, the buildings to be surrounded by farm land of 6,000 acres. Each man should engage in the occupation he most enjoyed, but the least attractive work should be best paid. Considerable freedom was to be allowed for personal taste and initiative. Fourier believed that in a large cooperative enterprise in which all worked for a common purpose at the kind of labor each enjoyed, selfishness would give place to happiness. A common workshop, granary, and dining hall, with one large farm efficiently managed, would be infinitely more economical than hundreds of little duplications and this feature would assuredly be a distinct improvement over the individual and competitive system.

At least thirty-four phalanxes were organized in various states, all of which disappeared in a few years, none having the success which attended the religious communities. Their failure may be attributed principally to three reasons: (1) inherent or developed desire on the part of humanity for personal ownership; (2) the difficulty of carrying on a social experiment surrounded by a world practicing a different economy, especially in America, where economic opportunities offered allurements outside and where the elements of a new country developed individuality; and (3) the fact that the participants in these schemes were oftentimes faddists as ready to differ with one another as they had been with society. One of the experiments made at this time, that at Brook Farm, has a literary as well as an economic interest. On a little farm about nine miles southwest of Boston gathered some of the most famous of the intellectuals of the time, including George Ripley, Nathaniel Hawthorne, Charles A. Dana, and George William Curtis. Other well-known persons, Ralph Waldo Emerson, A. Bronson Alcott, Horace Greeley, William H. Channing, Theodore Parker, and Margaret Fuller, though not residents, were intensely interested in the scheme. At Brook Farm the morning was devoted to work and the afternoon to culture and amusement. Expenses were partially paid by conducting a school which for some years was crowded with students. Eventually (1847) popular prejudice against the community life, fanned by lies and slanders, broke up the Brook Farm association.

Unlike the schemes of Fourier, because they were not based

upon phalanxes, were the Icarian communities, the name of which was derived from Etienne Cabet's *A Voyage to Icaria* (1840). Having obtained a grant of 40,000 acres on the Red River, Cabet set out for Texas with several hundred followers in 1848. This first settlement did not prosper, and in the next year, with a greatly depleted following, he moved to the deserted Mormon village at Nauvoo, Illinois. Here the Icarians met with indifferent success until internal dissensions caused part of the group to break away and found other communities at Chelthenham near St. Louis, in Iowa, and in California. The last of these communities dissolved in 1895. With the failure of these various schemes the ideal of social betterment found vent in other ways, partially in the anti-slavery agitation and in the organization of labor.

Economic Influence on Education and Religion.—During the colonial period education was quite largely a class affair. Only the wealthy could afford it, and it was hardly considered necessary for others than magistrates and clergy. Illiteracy was widespread. However, many of the early settlers were men of education and some efforts were made to extend the benefits of knowledge. The last half of the eighteenth and the early nineteenth century in many respects was a period of educational decline. This was caused by (1) the Revolution and the demoralizing influences of war, (2) the breaking up of old communities and the dispersal of the people as the frontier line moved across the Alleghanies, and (3) the lowering of intellectual standards resulting from frontier life where the struggle with Indians and nature left little time or energy for education, even if the difficulties and expense of maintaining schools had permitted it. Notwithstanding these disintegrating tendencies, there were influences at work which led during the second quarter of the century to a remarkable advance and laid firmly the foundations of our free public school system of to-day. Among these influences may be mentioned (1) the growth of city population after the Industrial Revolution had made itself felt in America; (2) the efforts to offset the increase in crime and pauperism caused partly by the rapid growth of these cities; (3) the growth of democracy which by enfranchising the common man put into his hands the means to provide for education; (4) the possibility in the new states of endowing a system of education from public lands without drawing on the taxpayer; and (5) the awakening feeling that in a democracy education must be widespread. As the triumph of democracy seemed assured and the control of government gradually passed from the handful of educated aristocrats, where it had rested during the colonial and Federalist periods, the demand for more schools on the part of far-seeing men became insistent.

Notwithstanding the obvious need of better educational facilities, the movement for public schools open to all and supported by taxation was bitterly fought by many groups—proprietors of private schools, sectarians, large taxpayers, and the conservatives in general. In consequence, the battle was long drawn out and painfully won, even with such noted leaders as DeWitt Clinton, Thomas Jefferson, and Horace Mann to direct the fight. While in New England such direct local taxes as the rate bills (a per-capita tax levied on parents of children attending school) were well known, the people of the nation as a whole were yet unaccustomed to the idea of taxation for school support. As a consequence, indirect taxation was resorted to at first—the income from the sale of certain lands, the proceeds from bank taxes, from liquor licenses and from lotteries. In the case of new states one section, and usually two, out of the thirty-six in a township, was set aside for schools. Connecticut received $1,200,000 for the sale of her "Western Reserve" in Ohio.[1] By 1830 it was evident that the income from state funds would not solve the problem of education and that direct taxation must be resorted to. Eventually this was done.

The battle which raged around the sectarian question was fought even more bitterly than that of free schools. Early schools had usually been closely allied with the churches, and the first American colleges were chiefly to train clergymen. Furthermore, it seemed sacrilegious to provide schooling without religious instruction. With so many sects demanding recognition or a di-

---

[1] See pp. 193, 194, 196.

vision of the school funds, most of the states (beginning with New Jersey in 1844) introduced constitutional amendments forbidding a diversion of school funds. The principle of free, public, non-sectarian schools supported by taxation was generally accepted by the time of the Civil War.

Many who could appreciate the need of elementary schools supported by state taxation did not believe that state education should be carried farther. Although various cities had previously established high schools, the "real beginning of the American high school as a distinct institution dates from the Massachusetts law of 1827, enacted under the influence of James G. Carter,"[1] requiring a high school in every town of 500 families or over where algebra, geometry, bookkeeping and United States history should be taught; and one in every town of 4,000 inhabitants or over where, in addition, Greek, Latin, history, rhetoric and logic should be given. By 1860 there were probably between three and four hundred high schools in the country, practically all north of the Ohio and east of the Mississippi. It was during these years before high schools were numerous and in good standing that many seminaries were established, of which there were over 6,000 in 1850—valuable institutions in this period of transition until the high school should become universal.

In addition to a general revival of interest in education there was great activity with respect to institutions of higher learning, especially after 1820. To the nine colleges of colonial days there were added fifteen more by 1800; by 1860 at least 246 colleges had been founded, 17 of which were state institutions, the remainder largely denominational. While many of these colleges were small and of doubtful scholastic standing, their mere numbers testify to the appreciation which church leaders had of the need of an educated clergy and laity. Some of the older colleges, Harvard, William and Mary, and Yale, which had originally been affiliated with the state, eventually became largely independent. The real beginnings of the state universities, which capped the system of state education, are to be found in the University of

---

[1] Cubberley, E. P., *Education in the United States*, p. 193.

Virginia founded by Thomas Jefferson (1819), of North Carolina (established in 1789, taken over by the state in 1821), of Vermont (chartered 1791, rechartered as a state institution in 1838), of Indiana (1820), and of Alabama (1831).

The religious life of the formative period was both influenced by economic factors and in turn had an effect upon our economic history. The American Revolution, largely an economic phenomenon, decidedly disturbed the religious life of the colonies both by lowering the effectiveness of the churches and by disrupting organizations closely connected with European bodies, tendencies augmented by the wide dissemination of atheistic doctrine during the era of political revolution in France and America. Furthermore, the disintegrating influences of frontier life contributed to weaken religious activity.

Nevertheless, there was, as the years went on, an increasing interest in religion which took many forms. On the one hand, there was the questioning of the old faith as seen in the Unitarian movement in New England and in the formation of numerous sects, some of which were offshoots of other religions and some with tenets quite new. Of the latter the most famous were the Mormons, who were persecuted and driven out of Ohio, Missouri, and Illinois, only to lay the political and economic foundations of the state of Utah. On the other hand, there was the missionary zeal of the evangelical churches—the Presbyterian, Baptist, and Methodist—whose home missionaries or circuit riders attempted to minister to the religious needs of the frontiersmen. Especially were the organization and methods of the last-named church suitable to a region where population was too scattered, restless, and poor to support a settled minister. With a closely knit, almost military, system of government, it successfully divided the country into circuits, upon which was placed a rider who preached in the various communities on his circuit as often as he could make the rounds. In this way the maximum religious stimulation was provided at the minimum cost, and by groups of noble men whose accomplishments and experiences in their ordinary round of duty read like fiction. Naturally these evangelical churches drew to themselves the great majority of the frontiersmen. The most rapidly growing church on the eastern seaboard was the Roman Catholic, which profited from the heavy Irish immigration in the late 'forties and the 'fifties.

Conditions of Living.—The second quarter of the century witnessed the beginning of revolutionary changes in the everyday life of the people. By that time the Industrial Revolution had sufficiently progressed, so that much which had formerly been made at home could now be more cheaply bought. In the more thickly settled regions of the northeast domestic manufacture gave way to factory-made goods. This tendency brought more specialization of work on the part of the men and more leisure for the women. Eventually it took many men and women away from agriculture and into factory work. Improved and factory-made machinery simplified and to some extent made easier agricultural labor, but most of the changes in the actual living conditions came of necessity first to the city dwellers. After the 'twenties, in the larger cities candles and oil as a means of lighting slowly gave way to gas. Street lighting by gas was adopted in Boston in 1822, in New York in 1823, and Philadelphia in 1837. The existence of anthracite coal in the Wyoming Valley had been known since 1762, and in the Lehigh Valley since 1791, and shipments had been made to Philadelphia as early as 1805; but its use for home heating began after 1815. Its extensive use was delayed by the difficulties of transportation and the expense of installing grates and stoves. Canals and railroads solved the first problem and the increasing cost of wood in the large cities made the second inevitable. By 1825 wood had been replaced by coal in a large number of homes of New York and Philadelphia.

The problem of water supply was naturally a pressing one in the growing cities. Until well into the century most of the water for the city dwellers was obtained from cisterns, house pumps, or various community pumps scattered throughout the city. After 1799 water from the Schuylkill River was raised by steam pumps to a reservoir and distributed through log pipes to a small part of Philadelphia. This system was improved and extended in 1822 by the opening of the Fairmont Waterworks, which conveyed the water by iron pipes through the entire city. Wooden pipes had in a similar way carried water to lower New York up to 1842, when the aqueducts were finished and Croton water brought to the city.

The old and inefficient night watch which lit the lamps, cried out the hours of the night, and gave the alarm for fires was superseded in 1845 by an organization of day and night watchmen more nearly approximating our police force of to-day. As a matter of fact, the modern police system extends back hardly more than seventy years. The problems of urban transportation were attacked first in New York City, where an omnibus line was established before 1828, running between Wall Street and Greenwich Village. In Philadelphia the first omnibus appeared in 1831.

Very influential in stimulating the higher life of the people was the act of 1845 which introduced cheap postage. The charge on letters of not over a half ounce in weight going less than three hundred miles was now five cents, over that limit ten cents, with additional charge for extra weight. Further reductions were made in 1851, when a half-ounce letter prepaid would be carried 3,000 miles for three cents, or, if not prepaid, for five cents; for 3,000 miles and over the rate was six and twelve cents. By 1840 the penny newspaper had made its appearance upon the streets, competing newspapers had commenced their keen rivalry for news, and the mass of Americans, from the unskilled laborer to the powerful capitalist, became the slave of the daily papers—an institution purporting to exist to carry news, but usually spreading propaganda for some interest political or economic. The news-spreading function of the papers and their ability to keep the citizens cognizant of what was happening in the world at large were made possible by the invention of the telegraph and its introduction after 1844. The effect of the invention of the steamboat and the steam railway was also considerable in promoting travel, in breaking up the intellectual isolation, and in eliminating to a slight extent intellectual provincialism.

As a whole the American people during these early decades were not given to spending much for amusements. De Tocqueville remarked that people who spent every weekday making money and every Sunday in going to church "have nothing to invite the muse of comedy." [1] Other European visitors spoke of the haste and intense application of Americans to business, to the neglect of amusements. Possibly the Puritan antecedents of many may have accounted for the neglect in some cases, but the conquest of a continent in less than a century left little time for else than work. Quilting parties, husking bees, house raisings, and church affairs continued to afford opportunities in the rural districts for social life, while occasionally in the east a traveling group of actors gave a performance. On the frontier the social and economic life of colonial days was duplicated again and again as the line advanced. In the south the social life had changed little since colonial days. Visits back and forth on the plantations and occasionally a winter season in the city provided a change for the women, while hunting parties and horse racing gave amusement to the men. In the cities the period from the 'thirties to the 'fifties was the golden age of the lyceum. Public lectures on all kinds of subjects were very popular; it was a time when curiosity in anything unusual was highly developed, and exhibitions in phrenology, mesmerism, and the like, drew great crowds. The theater, however, in America was in its infancy, and an American school of actors had not yet risen. Foreign artists appeared and were welcomed in New York and Philadelphia, the centers of what drama there was; but the crowds were apparently drawn by curiosity rather than by understanding of the dramatic art.

NOTES FOR FURTHER REFERENCE

L. B. Schmidt, *Topical Studies and References on the Economic History of American Agriculture* (rev. ed., 1923), contains a bibliography on "Pioneer Life and Ideals, 1830-1860." Source material with special reference to labor is available in the *Documentary History of American Industrial Society* (10 vols., 1910-11), edited by J. R. Commons and associates. Some of the most interesting material on American life during this period is furnished by foreign travelers. See S. J. Buck, *Travels and Descriptions (Illinois), 1765-1865* (1914). Selections from the work of several of these are given in the *Readings* of Bogart and Thompson. The work of J. B. McMaster, *History of the People of the United States,* is a mine of information in which the social and

---

[1] De Tocqueville, Alexis, *Democracy in America,* vol. ii, chap. xix.

economic are kept continually in mind. Edward Channing in the fifth volume (1921) of his *History of the United States* gets away to a considerable extent from the political and devotes valuable chapters to the social and economic as well as helpful footnotes and bibliographies. In Chapter II of this volume there is a fine study of the westward movement of population. A brilliant contribution to the study of the development of metropolitan economy in America is the last chapter of N. S. B. Gras, *An Introduction to Economic History* (1922). H. C. Emery in Vol. VII of the *Cambridge Modern History* gives a short survey of American economic and social history.

On immigration consult the *Preliminary Report on the Eighth Census* (1862), partially reproduced in Bogart and Thompson; and such books as R. Mayo-Smith, *Emigration and Immigration* (1912); J. R. Commons, *Races and Immigrants in America* (1907), and I. A. Hourwich, *Immigration and Labor* (1922). On the growth of democracy, Vol. II, Chap. 17 and Vol. V, Chap. 50 of McMaster are helpful, as is the excellent summary of C. E. Merriam, *A History of American Political Theories* (1903). Short chapters on the communistic experiments are included in E. E. Sparks, *Expansion of the American People* (1900), and in Morris Hillquit, *History of Socialism in the United States* (1910). More detailed studies are those of J. H. Noyes, *History of American Socialisms* (1870), and of Albert Shaw, *Coöperation in a Western City,* American Economic Association Publications, Vol. 1, No. 4 (1886); *Icaria: A Chapter in the History of Communism* (1884), *Coöperation in the Northwest,* Johns Hopkins Studies (1888). See also Charles Nordhoff, *The Communistic Societies of the United States from Personal Visit and Observation* (1875). A picture of Brook Farm is given in O. B. Frothingham, *George Ripley* (1882). On the most famous of the American communists see G. W. Noyes (ed.), *Religious Experiences of John Humphrey Noyes, Founder of the Oneida Community* (1923).

Brief résumés of education during this period are contained in C. F. Thwing, *A History of Higher Education in America* (1906); E. G. Dexter, *History of Education in the United States* (1911); S. C. Parker, *History of Modern Elementary Education* (1912), and E. P. Cubberley, *Public Education in the United States* (1919).

A picture of frontier religion and the life of the greatest of the circuit riders is in E. S. Tipple, *Francis Asbury, the Prophet of the Long Road* (1916), and in Edward Eggleston, *The Circuit Rider* (1874), remarkably true to life.

SELECTED READINGS

Channing, E., *A History of the United States,* Vol. V, Chaps. II, VI-VIII.
McMaster, J. B., *History of the People of the United States,* Vol. II, Chap. XVII; Vol. V, Chaps. XLIII, XLIX, L; Vol. VI, Chap. LVI; Vol. VII, Chaps. LXXIII, LXXIV.
Merriam, C. E., *American Political Theories,* Chaps. III-V.
Cubberley, E. P., *Public Education in the United States,* Chaps. IV-VII.
Bogart, E. L., and Thompson, C. M., *Readings in the Economic History of the United States,* pp. 537-558.

CHAPTER XVI

ECONOMIC CAUSES OF THE CIVIL WAR

The Development of Slavery.—From whatever angle the study of the Civil War is approached, the causes relate themselves in varying degrees to the system of slavery. This is distinctly evident when we approach it from the standpoint of economic history. One of our great historians has said, "of the American Civil War it may safely be asserted that there was a single cause, slavery."[1] The economic order of the south became so intertwined with the system that eventual withdrawal from the Union seemed to southern leaders the only means of preserving an institution believed to be necessary to the prosperity of that section. Slavery, which had flourished in the colonial south, was on the defensive at the close of the Revolution. Losses incurred by the planters in the War of Independence, the exhaustion of the soil in the coast states, the influx of white settlers from the north into the mountainous regions, all tended to make the system less profitable. These influences, augmented by the Revolutionary theories unfavorable to slavery, had led many southerners to question its economic and moral basis, but it was still firmly intrenched in 1781 in the rice and indigo fields of the Carolinas and Georgia, although its hold on the tobacco plantations was weakened.

The factors which contributed beyond all others to revive an apparently dying institution were the introduction of sea-island cotton and the invention of the cotton gin (1793). The first gave the planters of the coast regions an opportunity to recoup their waning fortunes; the latter made it possible to raise profitably the inland short-fibered variety, and both led to the rapid extension of cotton culture into the uplands and westward.

In cotton the south found a crop that apparently paid with slave labor, for the requisite conditions necessary to make it prof-

---
[1] Rhodes, James Ford, *Lectures on the American Civil War*, p. 2.

itable seemed ideally combined. The first of these was simplicity of operation; slavery thrives under a one-crop system of agriculture, the methods of which may be learned and mechanically repeated year after year. To cotton, a comparatively easy plant to raise, the labor of the negro could be adapted. Few tools and little equipment were needed, so that small loss was sustained even from inefficient labor. At the same time, the owner could invest almost his entire capital in slaves, and thus invested, it returned a higher profit when the slaves were employed on cotton plantations. Cotton culture extends over three-fourths of the year, and in its production, more than that of many other staples, it was possible to give employment to women and children, thus obtaining the maximum return from the whole family.

Still another advantage was that the slaves could be more compactly massed in the raising of cotton than in that of many other products. A single laborer could handle only three acres of rice and only five to ten acres of cotton, while he might cultivate thirty or forty acres of corn, a significant fact when it is appreciated that the labor of slaves is given only under fear of punishment and that constant supervision is necessary. The supervision, moreover, was expensive. "To diminish the inducement for overdriving," says Professor Phillips, "the method of paying the overseers by crop shares, which commonly prevailed in the colonial period, was generally replaced in the nineteenth century by that of fixed salaries." [1] An overseer's salary in 1863 was about $1,300, a considerable amount in cash. The comparatively high cost of white overseers contributed as much as anything in shifting slave labor largely to cotton plantations.

A further condition necessary to the profitable employment of slave labor is cheapness and ease of subsistence. The expenditure for shelter, fuel, and clothing was naturally not great in the warm climate of the cotton belt. The chief food of the slave was bacon and corn; consequently some corn was usually raised on the plantation to provide food for slaves and hogs, although in later

---

[1] Ulrich B. Phillips, *American Negro Slavery* (1918), p. 281.

years considerable corn and pork were obtained from the states north of the Ohio. The cost of keeping a slave ranged from fifteen dollars a year under the most favorable conditions to from thirty to forty dollars a year on the border states. The average was about twenty dollars a year.

Land values in the older states were, according to Professor Dew, dependent upon the institution of slavery. "It is, in truth, the slave labor in Virginia which gives value to her soil and her habitations;—eject from the State the whole slave population, and we risk nothing in the prediction, that on the day in which it shall be accomplished, the worn soils of Virginia would not bear the paltry price of the government lands in the West, and the Old Dominion will be a waste howling wilderness." [1]

A final factor as important as any in the development of slavery was the abundance of unoccupied land. Slave labor, incompetent and ignorant as it was, condemned the cotton planter to a one-crop system. Although cotton was less destructive to the soil than other staples, especially tobacco, its uninterrupted growth without the use of fertilizer meant the wearing out of the soil. As land was cheaper than slaves, the tendency was to "butcher the land" by continued crops of cotton until it was exhausted, and then to push on to new and fertile fields to repeat the process. This procedure was aided by the fact that most of the land was suitable for cotton. Thus slavery was dependent on a one-crop economy, which in turn, under the agricultural methods pursued, depended on the opening up of new lands. These facts explain the rapid westward advance of the southern cotton planter and the apparently insatiable hunger of the slaveholder for new land. Until fresh cheap land was exhausted, slavery seemed able to hold its own against the competition of free labor.

The southern states during the first half century produced about seven-eighths of the world's cotton supply. The demand for this staple was steady and increasing. Whoever had slaves or could buy them turned more and more to the culture of cotton. DeBow in 1850 estimated that 2,500,000 of the 3,204,313 slaves

---

[1] Dew, Thomas R., in *The Pro-Slavery Argument* (1852), p. 358.

were engaged in agriculture, and of these 72.6 per cent (1,815,-000) were employed on cotton.[1] As the cotton planter pushed westward into the fertile lowlands of Alabama and Mississippi and a continually greater area was put under cultivation, the demand for more slaves increased.[2] The wealthy planters were able to obtain the choicest lands, thus driving the poor whites to the small and less fertile farms, where they raised food crops; the latter would not or could not compete with slave labor, carried on as it was upon the big plantations. Not only were the efforts of the poor native whites largely withdrawn from cotton raising, but the labor situation in the south repelled the immigrant. The slaveholding states contained 378,205 foreign born in 1850, the non-slave states 1,866,397, constituting 3.91 per cent and 13.89 per cent, respectively, of the aggregate population of the section.

The demand for slaves was met by natural increase and by importation from Africa, though the latter was illegal after 1808. There was also considerable internal slave trade. The surplus negroes of the border states and the eastern Carolinas were shipped south to be sold to the cotton planters of the new states. The increase of slave population by natural means on the cotton plantations was very slight while on the sugar plantations of Louisiana the waste of life was greater than the annual increase. On the other hand, the border states developed a hardy type of negro, longer lived and more prolific, so that the increase on the Virginia farms amounted frequently to twenty per cent. With the growing demand for negroes there was a corresponding rise in the price. The average value of slaves in 1798 has been estimated at about $200; in 1815, at $250; in 1840, at $500; and

---

[1] J. D. B. DeBow, *Statistical View of the United States . . . Being a compendium of the Seventh Census* (1854), p. 94, note.

[2] APPROXIMATE NEGRO POPULATION, 1740-1860.

(The following estimate of negro population is taken from *The South in the Building of the Nation*, vol. v, p. 111, note.)

Year	Population	Year	Population
1740	140,000	1820	1,777,000
1776	300,000	1830	2,328,000
1790	750,000	1840	2,873,000
1800	1,002,000	1850	3,638,000
1810	1,380,000	1860	4,441,000

in 1860, at $700. "Prime field hands," however, that sold for $200 in 1780 brought from $350 to $500 in 1800, $700 to $1,000 in 1818, $1,200 to $1,800 and $2,000 in 1860.[1] The increase was not steady, but varied with periods of prosperity and depression, which in turn were dependent on the price of cotton.

Not only was it now possible for the border states of Virginia, Maryland, and Kentucky to get rid of their surplus laborers, but it became more and more profitable for the slaveholders there to raise negroes to "sell south." Professor Dew of William and Mary College asserted in 1832 that Virginia was "a *negro* raising State for other States; she produces enough for her own supply, and six thousand for sale,"[2] while Olmsted estimated that in the ten years preceding 1860 the annual export of negroes from the slave-breeding states was about 25,000. Negroes formed 50 per cent of the population of Virginia in 1782 and only 37 per cent in 1860. The following figures presented by J. E. Cairnes illustrate the constant draining off of slaves from the border states by the internal trade as well as the decline of the system in the northern states.

PERCENTAGE INCREASE OF POPULATION IN THE DECADE ENDING 1850 [3]

State	Whites	Slaves
Virginia	20.77	5.21
Maryland	31.34	0.70
Kentucky	28.99	15.75
Arkansas	110.16	136.26
Mississippi	65.13	58.74
Louisiana	61.23	45.32

The decade before the war saw both the maximum expansion of slavery and the early indications of its decline. Of the 12,-000,000 people in round numbers in the fifteen slave states, 4,000,000 were slaves. The production of cotton, the great staple of the south, was closely tied up with the institution, while the

---
[1] *South in the Building of the Nation*, v, p. 127.
[2] Dew, Thomas R., in *The Pro-Slavery Argument* (1852), p. 359.
[3] Cairnes, John E., *The Slave Power* (2d ed., enlarged, 1863), p. 130.

border states, with land exhausted or not adaptable to cotton, were joined to the system as breeding grounds for slaves. At the same time, it was evident that the progress of slave labor was practically at a standstill in most of the south and actually declining on the Atlantic seaboard and in the border states. Throughout the south white farmers and free laborers were increasing faster than slaves. The more rapid growth of the white population presaged even in the south an approaching struggle between white and slave labor, a contest in which slavery would be at an increasing disadvantage as the better land was absorbed.

Slavery in the South.—The census of 1860 gave the white population of the slave states at 8,099,760 and the slaves at 3,953,580. The slaves were owned by only 384,000 whites, of whom 107,957 owned more than ten slaves, 10,781 owned fifty or more, and 1,733 owned a hundred or more. At least 6,000,000 southern whites were not interested directly in slave ownership. Nevertheless, the fact that the great staple upon which the wealth and prosperity of the south depended was raised largely by slaves on a plantation system gave to the institution an importance which the number of slaveholders would not seem to warrant. The slaveholding aristocracy produced able politicians who so molded the opinion of the south that when the break came in 1860 the great majority of whites were behind the secession movement. There is no better example in modern history of a handful of the ruling class so shaping public opinion as to bring on a war to preserve an institution which benefited themselves alone.

As three-fourths of the slaves were engaged directly in agriculture and most of these in the production of cotton, the typical life of the slave was that on the plantation. The most intelligent and trusted negroes, often those with a mixture of white blood, were employed as household servants. On the plantation there were often to be found negro carpenters, blacksmiths, and drivers, but the great mass of slaves—men, women and children—were occupied in the fields. On the smaller plantations the owner was usually his own overseer, but on the larger he was forced to turn over the direct management to hired white overseers.

These were in turn assisted by drivers, trusty negroes who set the pace or supervised small groups. Work on a plantation was carried on in one of three ways: by the task system, in which a definite amount of work for the day was assigned to each slave according to his ability; by the gang system, in which a good driver set the pace and the rest followed; or simply by setting the slave to work with no incentive but the fear of the lash. It was not uncommon for owners to hire out slaves at from $100 to $200 a year, depending on the work and skill of the negro. Absentee ownership was not widely prevalent in the south, but where it existed was a great curse, involving as it did the use of a hired overseer. Good overseers were scarce. They were ordinarily of the poor-white class, not recognized as the social equal of their employer, and often working temporarily as overseers to obtain money to become slaveowners themselves. As he was judged chiefly by his ability to produce a large crop, the typical overseer was likely to drive the slaves to the limit and to abuse the land more than an owner directly supervising his plantation. He was inclined to be free with the whip and was intent upon immediate results rather than eventual gain. He was but one step removed from the professional slave dealer, who socially was decidedly under the ban in the south. Where the plantation was supervised by the owner, and especially on small establishments, the condition of slaves was likely to be better. Speaking broadly, the economic condition of a slave was not far above that of a well-kept domestic animal of the present day. House servants inherited the cast-off clothing of the master's family, had enough to eat, but few comforts. The field hands worked usually from sunrise to sunset, cooked their own meals, and were lodged in cabins to the rear of the plantation, where furniture and cleanliness were noted by their absence. The ordinary food was corn bread and bacon. More humane masters sometimes allowed the slaves a small garden patch to cultivate and a few chickens or a pig to keep, varied their diet at times, and gave them holidays and presents. Under the more intelligent owners and upon the most scientifically conducted plantations great care was exerted with respect to the sanitation and health of the slaves, and some provision made for their amusement and spiritual welfare. While the teaching of the negro to read and write was forbidden in five states, under the belief that it made him discontented, religious instruction was often given. The negro undoubtedly delighted in religious expression, but an attempt to inculcate habits of honesty and chastity through religious teaching, to men and women who did not own their own bodies or the fruit of their labors, was bound to be futile. A marriage ceremony was often performed, although the law in none of the states recognized slave marriage.

In the abolition literature of the time much attention was given to the cruelties practiced upon the slaves. In a system in which the absolute ownership of some human beings rests in others, and in which labor is rendered chiefly through fear of the lash, very naturally flogging existed and wanton cruelty was often perpetrated. It was not, however, the general practice, certainly not with the house servants. The worst conditions existed upon the sugar plantations of Louisiana, the rice fields of Georgia and South Carolina, and the large cotton plantations of the lower south, where gangs of slaves worked under the direction of white overseers, described by Patrick Henry as "the most abject, degraded, unprincipled race." Certain other forms of cruelty which went with the system, in particular the separation of families and their disposal on the auction block, were more often seen in the border states, where the breeding of slaves for the southern market was carried on. On the other hand, there were many estates upon which the slaves were comparatively happy and contented, where kindly relations existed between master and slave, and where the economic condition of the negro was undoubtedly better than after emancipation. Manumission was restricted as dangerous, but many masters set free their slaves. The incomplete figures of the census of 1860 gives the number of slaves manumitted in that year at 3,018, a ratio of one to every 1,309 slaves. In that year there were about 262,000 free negroes in the slave states.

Economic Advantages and Disadvantages of Slavery.— The advantages of slave labor, which had been questioned before the introduction of cotton culture, were more and more reaffirmed as the century progressed. It was maintained that the absolute ownership of the workmen by the employer was advantageous to the latter because it allowed him to enjoy the entire fruit of the product of labor, to organize his labor force as he thought best, and to control his workmen through his single will to a definite end. The control of the full time of men, women, and children seemed to be the last word in the elimination of waste power. After the initial purchase, in case that was necessary, the only expense was to keep the slave in health and strength. It was furthermore sincerely believed that negro labor under slavery was the only kind that could be employed in the unhealthy work of the rice fields. While some enlightened southerners believed that it was possible to raise cotton with free labor, the great majority were convinced that only by means of slaves could large-scale cotton production be carried on; in fact, cotton culture seemed providentially designed to enhance the advantages of slave labor. Southern writers argued, not without grounds in some localities, that the slave in the south was better housed, better fed, and happier than the free unskilled laborer either in Europe or in the northern states. Not only was the slave better off, it was contended, but the master and his family, freed from the necessity of manual work, could devote their abilities to the amenities of life and to intellectual development, a contention certainly borne out in the case of the eight or ten thousand families who were able to live in luxury on the labor of negroes. As attacks upon slavery grew stronger the institution was defended upon the authority of Scripture; upon the theory of the inferiority of the colored race, which doomed it to economic dependence; and finally, as necessary to the safety of the whites, for southerners professed to see in the disturbances in Haiti the future civilization of the south if the slaves were freed.

The advantages of slave labor were more apparent than real. In the first place, the labor was given reluctantly and without interest. There was small incentive to increase production, for that meant a greater expectation on the part of the owner, and, therefore, the tendency of the slave was to hide his true ability and to render labor less than capacity. The continued eye of the overseer and the fear of the lash were necessary to produce results. The hired overseer was expensive, as regards both salary and methods; for his business was to turn out a big crop, and this he was tempted to do (especially when cotton profits were large) to the detriment of both land and slaves. The work of the negro slave was not only given reluctantly, but it was essentially unskillful. Shortly removed from African barbarism, with little incentive for intellectual growth, and his character brutalized by the system under which he lived, it is little wonder that his labor was ignorant and wasteful. Only the simplest tools could be used, and it was ordinarily difficult to train him to use machinery. The slave lacked versatility and had to be kept at the simplest tasks and upon operations constantly repeated. Any cheapness which slave labor might have was more than offset by its wasteful character and the reluctant way in which it was given. Furthermore, although the cost of clothing and feeding a slave was small, ranging from fifteen to forty dollars a year, when to this is added the interest on capital, depreciation, taxation, and insurance against sickness, flight and death, the yearly expense of an able-bodied slave was not far from $135. Slaves were hired out in Georgia before the war at from $140 to $150 a year, but afterward, under a system of free labor, negroes could be obtained for $120 a year with board.

With these handicaps, it was found that slave labor could compete with free only where there was an unlimited extent of highly fertile soil. Consequently, as has been explained, the cotton planter, fortified with the wealth of the south, appropriated the richest land, used it up, and passed on to take more. Whatever the immediate gains might be, the eventual effect was bound to be disastrous and out of all proportion to the gain. The percentage of unused and exhausted land was very high in the seaboard states, almost resembling the havoc wrought by an invading army. The following table will give the relationship between developed and undeveloped land. The proportion of the latter was great in the south, for only the best land could be made to pay with slave labor:

AGRICULTURAL DEVELOPMENT IN FREE AND SLAVE STATES[1]

	Free states and territories	Border states (Ill., Md., Ky. and Mo.)	Slave states
Improved land, acres	88,730,678	17,547,885	56,832,157
Unimproved land, acres	72,983,311	27,474,315	143,644,192
Total quantity, acres	161,713,989	45,022,200	200,476,349
Cash value	$4,091,818,132	$702,518,382	$1,850,708,493
Average value per acre	$25.30	$15.60	$9.28
Agricultural implements, value	$142,077,802	$21,068,903	$82,971,436
Livestock, value	$574,067,208	$133,484,109	$381,778,598
Agricultural capital	$4,807,963,142	$857,071,394	$2,315,458,529

Social Disadvantages of Slavery.—Slavery was not only undermining the economic foundations of southern agriculture, but it was ruinous to the white population. If it elevated a few thousand rich and cultured families, it drove hundreds of thousands of whites into a lower economic status. Most of the southern wealth was invested either in land or in slaves. Land was cheap, but slaves were costly; only the rich could buy them, a fact which made it difficult for a poor white to push himself up in the social scale. The wealth in slaves was largely concentrated in the hands of a few, who were thus in a better position to absorb also the best land. The poor whites were consequently driven to the less desirable locations, especially in the mountain regions, where they raised on small farms foodstuffs and sometimes a little cotton. Unable to compete in raising cotton with the slaves on the richer lowlands and unwilling to work beside the colored man, they had little left but an inferior farm and pride in their white skin. Upon the minority of slaveholders the system worked evil as well as good. The unquestioned obedience rendered by slaves made the master impatient of having his will thwarted, quick with the gun, and too ready to take the law in his own hands. Morally the effect of slavery was as deleterious to the white as to the black. Pride in race as well as laws governing sex relationships tend in a free society to keep the races apart, but under a system of ownership these influences

---

[1] Seaman, Ezra C., *Essays on the Progress of Nations* (2d series, 1868), p. 572.

were apparently not felt. The census figures of 1860 put the proportion of mulattoes at 12 per cent (518,360) of the colored people, but even that figure probably does not record accurately the extent of the mixture of blood.

Anti-Slavery Agitation.—The first negroes were brought to Virginia in 1619 by a Dutch privateer. Slave labor proved itself adaptable to the rice and tobacco plantations, and seemed to be the best solution to the problem of labor scarcity which invariably confronts a new country. Imported first in Dutch and then in English and New England ships, the number of negroes increased until at the opening of the Revolution there were 300,000 slaves in the colonies, constituting one-tenth of the population. This was a larger proportion than at any subsequent period, so large, in fact, that the wisest men were fully alive to the danger. Maryland, Virginia, and the northern colonies were in favor of ending the foreign slave trade, but efforts directed toward this end by Virginia were vetoed by Great Britain, although slavery was not tolerated at that time in England. With the wearing out of the tobacco lands slaves became less profitable in Maryland and Virginia, but were still believed to be indispensable in the rice fields farther south. Undoubtedly a majority felt that slavery had outlived its usefulness and looked toward its eventual disappearance. Franklin asserted in Congress that "slaves rather weaken than strengthen the state," while Jefferson hoped to see "an entire stop forever put to such a wicked, cruel and unnatural trade." Washington was likewise opposed to it and provided for emancipation in his will. Vermont, which separated from New York in 1777, forbade slavery forever, and in 1780 Pennsylvania voted a scheme for gradual emancipation. By 1787 seven states had either abolished slavery or were preparing to do so (Vermont, Pennsylvania, New Hampshire, Massachusetts, Connecticut, Rhode Island, and New Jersey).

The Ordinance of 1787[1] which prohibited slavery in the Northwest Territory (out of which were later formed the states of Ohio, Indiana, Illinois, Michigan, Wisconsin, and part of Minnesota), clearly reflected the anti-slavery feeling of

---

[1] See chapter on "The Westward Movement from the Revolution to the Civil War."

the time. In the federal Constitution the word "slave" does not occur, but the subject of slavery is touched on in three places. On the question of who should be enumerated in apportioning representation and direct taxes, "the whole Number of free Persons" were to be counted and "three-fifths of all other Persons." (Art. I, Sec. 2). While Congress was given power over interstate commerce, it was prevented from abolishing the external slave trade until 1808 (Art. I, Sec. 9). It was further provided (Art. IV, Sec. 2) that persons "held to Service or Labour in one State" who escaped to another should be "delivered up on Claim of the Party to whom such Service or Labour may be due." Upon this authority the first Fugitive Slave Act of 1793 was passed. Nevertheless, at the opening of the Republic slavery was a decaying institution.

It was Eli Whitney's cotton gin that gave slavery its new lease of life, but there was still enough vitality in the first anti-slavery movement to bring about under Jefferson's leadership in 1807 the federal act which forbade the foreign slave trade after 1808. But with this act the anti-slavery societies felt that their mission had been fulfilled, and for the time being their activity ceased. As the south turned to the cultivation of cotton, slavery seemed more necessary, and gradually the new southern leaders came out strongly in its defense. It was not, however, until Missouri petitioned for admission to the Union as a slave state that the nation awoke to the fact that slavery had become a growing rather than a decadent institution. The entrance of Missouri made it necessary to decide the question as to whether slavery was to be held back east of the Mississippi or allowed to overflow into the vast territory obtained through the Louisiana Purchase of 1803. In the treaty by which we acquired Louisiana, slavery was to be permitted in the district of New Orleans, but its status in the remaining territory was left to Congress to decide. Many even in the south were opposed to further extension of slavery. The result was the "Missouri Compromise" of 1820, by which Missouri was admitted as a slave state with Maine as a free state to preserve the status in Congress; but slavery was prohibited north of 36° 30′ north latitude in the rest of the Louisiana Purchase. The Missouri Compromise appeared at the time to be entirely one-sided in that it favored the south. It brought again to the fore the matter of slavery. The Missouri question, wrote Thomas Jefferson, "is the most portentious one which ever yet threatened our Union. In the gloomiest moment of the revolutionary war I never had any apprehensions equal to what I feel from this source."[1]

The spirit of sectionalism aroused by the Missouri Compromise, increasing in the next few years, culminated in the attempted nullification of 1832-1833. As the south increasingly devoted itself to agriculture, it became more hostile to the protective tariff, which its leaders in earlier years had supported. The tariff of 1828, the so-called "tariff of abominations," aroused the bitterest indignation in South Carolina. Following the doctrine of John C. Calhoun, who held that a state might interpose a veto upon an act of Congress, South Carolina declared null and void the tariff acts of 1828 and 1832 as affecting that state. Jackson's stand was firm. South Carolina was not supported by the rest of the south, and this act of rebellion subsided in another compromise which produced both the "Force bill," giving the President the power to use the army and navy to enforce acts of Congress, and Clay's Tariff bill, an act reducing the tariff gradually for ten years until in 1842 it should be 20 per cent.

The anti-slavery movement was revived in the 'thirties. In 1831 William Lloyd Garrison published the first issue of *The Liberator* with the famous announcement: "I shall strenuously contend for the immediate enfranchisement of our slave population. . . . I am in earnest—I will not equivocate—I will not excuse—I will not retreat a single inch—and I will be heard." In the same year, but with probably no connection with Garrison's movement, occurred the Nat Turner negro insurrection in Virginia, an event of small importance in itself, but a suggestion of terrifying possibilities. Two events happened in the year 1833

---

[1] Letter to Hugh Nelson, February 7, 1820. See also letter to John Holmes, April 22, 1820, in *Writings of Thomas Jefferson*, edited by Paul Leicester Ford, vol. x, pp. 156-157.

to give strength to the abolition movement—the emancipation by the British Parliament of the West Indian slaves and the organization in Philadelphia of the American Anti-Slavery Society. Despite bitter persecutions, the abolition movement continued to grow, and eventually attracted to itself the talents of Channing, Whittier, and Wendell Phillips. Petitions by the hundred were presented to Congress through John Quincy Adams in the House and Webster in the Senate, though opposed by protests and the use of a "gag" rule. This second wave of anti-slavery agitation reached its climax about the year 1838, when there were at least 2,000 societies with 200,000 members. Again in the 'forties the movement subsided to some extent, only to blaze forth again in the decade preceding the war.

The anti-slavery agitation of the 'thirties had been scorned by both political parties, but events soon tended to tie up the question with politics. Texas, freed from slavery under Mexican rule, had been overrun by Americans—first by ranchers and then by cotton planters with their slaves. Having won their independence from Mexico, the Texans opened negotiations looking toward the annexation of the Republic of Texas to the United States. This annexation was advocated by a majority of southern Democrats to extend slave power, and opposed to a considerable extent by northern Whigs. The election of 1844 turned on the annexation of Texas, and James K. Polk, the Democratic candidate in favor of immediate annexation, won in an exceedingly close contest. Following the annexation of Texas in 1845, Polk and his imperialistic advisers, intent upon still greater territorial accessions, forced the war with Mexico. The opposition to the annexation of Texas and the war with Mexico was but the opening of a new slavery controversy which led straight to the Civil War. While the Mexican War was still in progress, an attempt had been unsuccessfully made in the Wilmot Proviso to close to slavery any land that might be taken from Mexico at the end of the war. The discovery of gold in California in 1848 and the petition of that territory to enter the Union free brought up immediately the status of slavery in the newly acquired territory. In a last attempt to settle the question Henry Clay came forth in 1850 with the compromise scheme which called for the admission of California as a free state, the organization of the territories of New Mexico and Utah with or without slavery as the settlers there might determine, the passage of a strong fugitive-slave law, and the abolition of the slave trade in the District of Columbia.

These Acts, known as the "Compromise of 1850," were supported by many who hoped that it would fix the status of slavery for the whole country and that the excitement might thus subside. This was not to be. The new fugitive-slave law brought into the limelight some of the worst features of slavery and aroused both north and south. Harriet Beecher Stowe's *Uncle Tom's Cabin* (1852), a book which pictured both the good and bad sides of slavery, was read by millions and undoubtedly did much to prepare the way for emancipation. The four-year truce ended in 1854 with the passing of the Kansas-Nebraska bill. This bill, introduced by Stephen A. Douglas and founded on the conception of squatter sovereignty, declared that the Missouri Compromise was superseded by the principle of the legislation of 1850, and divided the rest of the Louisiana Purchase into two territories, Kansas and Nebraska, where squatter sovereignty should work itself out. The hope which many entertained that the Compromise of 1850 had laid to rest the slavery controversy was utterly wrecked by the Kansas-Nebraska bill, for the matter of slavery could not be dismissed while pro-slavery men and abolitionists struggled to gain control of "bleeding Kansas." On top of this came the Dred Scott decision of 1857 asserting that neither Congress nor the territorial legislatures could legally forbid slavery in a territory, and that, as the Constitution recognized property in slaves, such property must be protected in any part of the common national domain. This decision virtually declared unconstitutional the Missouri Compromise. Two years later John Brown's raid disturbed still further an already overcharged atmosphere.

The Kansas-Nebraska bill threw the slavery question definitely into politics. It smashed the declining Whig party and split asunder the Democratic organization. The abolitionists had attempted to form a political party around the slavery question, but the Liberty party in 1840 and 1844 had polled a mere handful of votes. In 1854 the Republican party was formed and presented its candidate, Frémont, on a platform demanding the exclusion of slavery from the territories. Already in 1856 prominent southerners threatened that the election of Frémont would mean the dissolution of the Union. The victory of Buchanan put off the crisis temporarily, but the mere appearance of a major party pledged to circumscribe slavery was fast bringing the issue to a climax. This came four years later when the Republican party, now grown strong, nominated as its candidate Lincoln, a man who in 1858 had said in accepting the nomination for senator, "A house divided against itself cannot stand," and "I believe this government cannot endure half slave and half free." Events beginning with the Kansas-Nebraska bill had divided north and south beyond reconciliation. The election of Lincoln, who polled fewer votes than the divided Democratic party, brought the long-threatened secession.

Economy of the South.—A most important effect of the investment of a large portion of southern wealth in land and slaves was the one-sided economic life which it produced. Slave labor was too ignorant for industrial enterprise, while skilled white mechanics in general avoided the south. Although water power and cotton were both at hand, there was little liquid capital to invest in manufacturing, and the planter preferred to send his products to the mills of New England or Europe. The great industrial progress which encompassed the north in the two decades before the war largely passed the south by.[1] Except in cotton manufacturing, development had been trivial.

COTTON MANUFACTURING IN NEW ENGLAND AND THE SOUTH, 1840–1850

	Census	Plants	Capital	Operatives
Southern States	1840	248	$4,331,078	6,642
	1850	166	7,256,056	10,043
New England	1840	674	34,931,399	46,834
	1850	564	53,832,430	61,893

[1] See Woodrow Wilson, *Division and Reunion*, pp. 104-108.

Even the slight beginnings of manufacturing evidenced in cotton were not duplicated elsewhere. The southern states, as was only too well illustrated after the war broke out, were hopelessly dependent upon the outside for the simplest of manufactured products. The planters of the southwest were even importing their corn and bacon from north of the Ohio in order to devote their whole plantation to cotton. The one-sidedness of southern economic life is thus pessimistically described by a citizen of North Carolina:

In one way or another we are more or less subservient to the North every day of our lives. In infancy we are swaddled in Northern muslin; in childhood we are humored with Northern gewgaws; in youth we are instructed out of Northern books; at the age of maturity we sow our "wild oats" on Northern soil; in middle life we exhaust our wealth, energies and talents in the dishonorable vocation of entailing our dependence on our children and on our children's children, and, to the neglect of our own interests and the interests of those around us, in giving aid and succor to every department of Northern power; in the decline of life we remedy our eye-sight with Northern spectacles, and support our infirmities with Northern canes; in old age we are drugged with Northern physic; and, finally, when we die, our inanimate bodies, shrouded in Northern cambric, are stretched upon the bier, borne to the grave in a Northern carriage, entombed with a Northern spade, and memorized with a Northern slab.[1]

In transportation alone had the south awakened to her needs and opportunities, but even in railroad mileage she was surpassed in 1850 by New England. In 1860 there were 9,517 miles of railway in the southern states compared with 11,114 in the north central states, and 30,000 for the country as a whole. In 1859 the real and personal property in the country amounted to $16,159,000,000, of which $10,957,000,000 was credited to the northern states. Northern farms and factories contributed $2,818,000,000 of the $3,736,000,000 wealth produced in 1859. The population of the north in 1860 was 19,083,927 as against 12,315,374 in the south, although the numbers had been about even in 1800. Slavery and the plantation system were preventing

---

[1] Helper, H. R., *The Impending Crisis of the South*, pp. 22-23.

the growth of population, making the civilization one-sided and causing the region to fall behind the north in material welfare.

In a discussion of the plantation economy of the south the fact should always be kept in mind that the cotton planters constituted a small minority of the white southerners. Fully nine-tenths of the landowners were small farmers, cultivating without slave labor from fifty to a hundred acres, and raising livestock and grain for their own use and for the near-by plantations. Although in a great majority, the small farmer was at a disadvantage without slaves. Forced to flee before the slave system, he relinquished the rich lowlands and retreated to less desirable ground. He rarely took part in politics, accepted his ideas from his wealthy neighbors, came to believe that the prosperity of the south depended upon slave labor and fought hard in the Civil War to preserve it, though it had brought him nothing but misfortune.

The Growth of Sectionalism.—Although slavery was the great underlying cause of the Civil War, it is possible to restate the economic causes in the growth of sectionalism. The economic differences between the north and south were accentuated by innumerable differences in the political and intellectual point of view of the two sections. The agricultural life of colonial times, founded on the plantation system and perpetuated by the introduction of cotton, tended to make the south an agricultural exporting section with scarcely any manufacturing. On the other hand, the north was continually developing a commercial and a manufacturing life. The opposition of interests first made itself felt on the question of tariff. Before the south became wedded to cotton, southern leaders like Madison, Jefferson, and Calhoun had supported up to 1816 a protective tariff; succeeding protective tariffs they steadily opposed. The western states, desiring to build up a home market, had supported the tariffs of 1824, 1828, and 1832, but with the growth of the southern market for their products they had been inclined to shift in their attitude. The doctrine of nullification, or state sovereignty, which had been earlier set forth by Kentucky in protest against the Alien and Sedition Acts and by New England in opposition to the

War of 1812, was now reaffirmed by South Carolina against the "tariff of abominations" of 1828 and the tariff of 1832, and the doctrine was never lost sight of until it resulted in actual secession. The compromise tariff of 1833 registered a victory for the south, whose representatives with the aid of the west were able to prevent the adoption of the protective principle on a large scale until the Morrill bill of 1861.

Not only was the south victorious over the north in regard to the tariff, but also with respect to the public-land policy and western expansion. Southern agriculture, tied up as it was with cotton and slavery, needed room for rapid expansion. Northern manufactures, on the other hand, desiring a more concentrated population, opposed measures which might encourage migration to the west. The south favored rapid sale of western lands in large tracts at cheap prices, while New England insisted on smaller and more restricted sales at higher prices. These conflicting views resulted in a compromise, a preëmption bill in 1841, which provided for restricted sales to actual settlers at a very low price. In actual practice, however, the southern planter found this law liberal enough for his needs.

The same differences were to be seen in the matter of expansion. Southern activity and leadership had been responsible for the acquisition of Louisiana, Florida, and Texas, the lands won from Mexico, and the conspiracies to annex Cuba. The southern system of agriculture needed room for free expansion. Furthermore, as the anti-slavery movement grew stronger and as the north began to surpass the south in population and wealth, it became more essential to the south that they maintain an equal number of senators. It was from these acquisitions that they hoped to carve out new slave states. The opposition to the Mexican War came, not so much from opposition to extension of territory as opposition to the expansion of the slave power. Agriculture under slavery depended for its very existence upon a continued supply of fresh land, and the demand for new slave territory was literally pushed on by its own weight. As long as the states of the northwest found their market in the south they usually supported that section. The building of canals and rail-

ECONOMIC CAUSES OF CIVIL WAR

roads eventually provided the trans-Appalachian states north of the Ohio with an eastern and a European market. By the decade of the 'fifties, the Old Northwest was securely linked with the east rather than the south. The loss of western backing in the fight to extend slavery brought home the fact that the south was losing her political supremacy and that in the future the fight would be a losing one.

NOTES FOR FURTHER REFERENCE

Source material, extracts from contemporary sources, etc., have been collected by G. S. Callender, *Selections from the Economic History of the United States 1765-1860* (1909); by E. L. Bogart and C. M. Thompson, *Readings in the Economic History of the United States* (1916), and by Ulrich B. Phillips, editor of the first two volumes of the *Documentary History of American Industrial Society* (1910), the whole work prepared under the direction of J. R. Commons and associates. The introduction to the first two volumes on slavery by Professor Phillips is, in the opinion of Channing, "the best brief survey of the system that has been written."

Other short studies of slavery are included in the standard histories—*e.g.*, J. F. Rhodes, *History of the United States*, Vol. I, Chap. 4, and Vol. III, Chap. 1; J. B. McMaster, *History of the People of the United States*, Vol. VII, Chap. 76; and Edward Channing, *History of the United States*, Vol. V, Chap. 5. There is also an excellent chapter in E. L. Bogart, *Economic History of the United States* (rev. ed., 1922). Thumbnail sketches of various phases of southern economic history are included in Vol. V of *The South in the Building of the Nation*, and Alfred H. Stone's article, *The Negro in the South*, Vol. X. An interesting and well-balanced résumé with an excellent bibliography is A. B. Hart, *Slavery and Abolition*, in the American Nation Series. Recent histories of the American negro are those of Ulrich B. Phillips, *American Negro Slavery* (1918), an expansion of many earlier studies; Benjamin Brawley, *A Short History of the American Negro* (1919), and *A Social History of the American Negro* (1921).

The contemporary material on slavery is large. Quite valuable are the works of J. E. Cairnes, *The Slave Power* (2d ed., enlarged, London and Cambridge, 1863), an impersonal study by a famous English economist; Hinton R. Helper, *The Impending Crisis of the South* (1857), the best known denunciation of the system by a southerner; and J. S. Buckingham, *The Slave States of America* (1842), by an English traveler. A typical defense of slavery is that of Professor C. F. McCoy of Columbia, South Carolina, in *Eighty Years' Progress* (1869), and *The Pro-Slavery Argument* (1852), by several writers, including Thomas R. Dew. A mine of information on life and conditions in the pre-war south is J. D. B. DeBow (ed.), *The Industrial Resources of the Southern and Western States* (3 vols., 1852), and the various works of Frederick L. Olmsted, the "best-known writer on conditions in the south prior to the outbreak of the Civil War," including *Journey in the Seaboard Slave States*

(1859), *A Journey Through Texas* (1857), *A Journey Through the Back Country* (1860), and *The Cotton Kingdom* (1861). Interesting also are the observations of the famous actress, Frances Kemble, *Journal of a Residence on a Georgia Plantation in 1838-1839* (1863); of Edward Ingle, *Southern Sidelights* (1896), and of J. B. Angell, *Reminiscences of James Burrill Angell* (1912), who in Chapter II tells of a horseback journey and winter spent in the south in 1850 and 1851. Mrs. Harriet Beecher Stowe's *Uncle Tom's Cabin* (first published in 1852) should be read for its tremendous historical significance as an influence upon the generation which fought the Civil War.

The best study of cotton is that of M. B. Hammond, *The Cotton Industry*, publications of the American Economic Association, New Series, No. 1 (1897). A more recent book, previously cited, is James A. B. Scherer, *Cotton as a World Power* (1916). The whole story has been excellently summed up in W. E. Dodd, *The Cotton Kingdom* (1919), in the Chronicles of America.

SELECTED READINGS

PHILLIPS, U. B., *American Negro Slavery*, Chaps. XI-XX.
RHODES, J. F., *History of the United States*, Vol. I, Chap. IV.
McMASTER, J. B., *History of the People of the United States*, Vol. VII, Chap. 76.
CHANNING, E., *History of the United States*, Vol. V, Chap. 5.
HART, A. B., *Slavery and Abolition*, Chaps. IV-XI.
CALLENDER, G. S., *Selections from the Economic History of the United States 1765-1860*, Chap. XV.

CHAPTER XVII
THE CIVIL WAR

Depression of 1861.—The first economic effect of the Civil War was to throw the north and west into a severe panic. The agricultural south owed northern merchants at the outbreak of the war close to $300,000,000, practically all of which was a total loss. Uncertainty as to the future, and the forebodings incident to the commencement of the war brought about a wave of retrenchment and economy; the banks were caught with cash reserves far too small to meet such an emergency; all of these factors united in bringing on the depression of 1861. The Dun reports listed in 1861 nearly 6,000 failures of northern firms for sums of $5,000 or more (a larger number than in the panic year of 1857); and probably 6,990 more failures for sums under that amount.[1] The northern banks in general were able to maintain specie payment until the latter part of December, 1861, when they were forced to suspend, followed almost immediately by the federal government. In the south, outside of New Orleans, suspension occurred immediately after the opening of the war and continued until the end. The wild-cat banks in the west were especially hard hit, not only because of their methods of banking, but also because of their more intimate relations with the south. In Illinois, out of 110 banks, 89 failed, while in Wisconsin 39 and in Indiana 27 went under.

Revival of Prosperity—Agriculture.—The depression of 1861 gave way in the following spring to a revival of prosperity in the north and west. Although thousands of farmers were drawn into the Union army and thousands more deserted agriculture for the mines of the far west, the effect was offset by the work of women in the fields, by the influx of immigrants from Europe, and by the use of labor-saving machinery. The eastern

---
[1] Fite, Emerson D., *Social and Industrial Conditions in the North During the Civil War*, pp. 105-106.

farmer, finding that he could not compete in certain crops with the westerner, emigrated in large numbers to take up lands in the west, a movement stimulated after 1862 by the Homestead Act,[1] which granted free 160 acres to almost anyone who would cultivate them for five years. Immigrants from abroad during the five war years flocked to take advantage of this act, 45,000 declaring their intention, upon arrival in New York, of continuing to Illinois and 23,000 to Wisconsin.[2] There was also a considerable exodus to the western farmlands from the harassed border states. Never was there such interest displayed in labor-saving farm machinery, now the center of attention at the county fairs. The number of mowers manufactured increased from 20,000 in 1861 to 70,000 in 1865, and a similar story could be told of horse rakes, grain drills, threshers, and other improved machinery.

Not only were the crops saved, but there was considerable increase in agricultural production. The wheat crop was greater during the war than at any time previous. Corn production, although it fell off slightly for the country as a whole, showed a remarkable increase in the western states. More hogs were put on the market than before the war; the wool production rose from 40,000,000 to 140,000,000 pounds and the number of sheep from 16,000,000 to 32,000,000.

The principal factors which brought about this agricultural prosperity were: first, the necessity of feeding an army which numbered at the close of the war a million men; second, the increasing population of the country as a whole; third, the prosperity of the north, enabling a greater expenditure for farm products; fourth, the rapidly growing manufactures, which employed increasing numbers in non-agricultural pursuits; fifth, the stimulation of high prices, which came from an expanding paper currency; and, finally, the heavy demand from foreign countries, especially England, where the harvests of 1860, 1861, and 1862 had been below normal. As regards this last point, it was freely asserted during the war, and with much foundation in

---

[1] See chap. xviii.
[2] Fite, *op. cit.*, p. 11.

fact, that the food supplies of the north and west were influential in preserving the neutrality of Great Britain.

Manufacturing in the North.—Manufacturing, stimulated by war needs and liberally protected by tariff acts passed at every session of Congress, prospered enormously during the last three years of the war. Although the record is stained with the usual story of the sale of inferior goods to the government at high prices, of speculation, of lobbying, and of the amassing of fortunes by the so-called "shoddy aristocracy" out of the needs of the soldiers and the exigencies of war, it is still a significant fact that the period marked a notable advance in American manufacturing. In only one important industry, that of cotton manufacture, was production decreased. Southern statesmen expected that the cutting off of the cotton supply would bring Europe to their aid and the north to its senses, but they were mistaken. Surplus stock, already in the north at the beginning of the conflict, and a certain amount which found its way there through illicit traffic or from captured parts of the Confederacy, actually enabled many cotton manufacturers to run their mills on part time, and the enhanced prices of cotton goods occasioned considerable profits.

Perhaps no industry received a greater impetus from the war than that of manufacturing woolen cloth. Anything that looked like wool was purchased by the government for uniforms.[1] At the height of the war over 200,000,000 pounds of wool per year were being woven as against 85,000,000 in times of peace, and yearly dividends of from 25 to 40 per cent were frequent. Simultaneously came the rapid development of the ready-made clothing industry, made possible by the sewing machine first put on the market by Elias Howe in 1849. The leather industry was also stimulated by war needs and in turn was aided by the application of the sewing machine to leather through the patents of L. R. Blake and Gordon McKay. It was during the war that

---

[1] The origin of the fortune of the late John Wanamaker may be traced to his purchase in the early part of the war of large quantities of blue flannel, sold to the government for soldiers' uniforms.

Chicago took the lead in pork packing, and Pittsburgh increased enormously the manufacture of iron.

The production of machinery of all kinds must have been extensive, for the increase in manufacturing and transportation facilities during these years was unprecedented. Philadelphia, the largest manufacturing center in the country, boasted of 58 new factories in 1862, 57 in 1863, and 65 in 1864, while other large cities showed similar progress. Even the government went into ship building and the manufacture of implements of war. The fact that a government rifle manufactured at Springfield in 1860 cost nine dollars, while a similar product made by private contractors cost twenty, throws light both on the success with which government embarked in industry and on the profits of the munition makers. The war apparently contributed to develop the inventive genius of the nation, for the number of patents issued yearly more than doubled between 1860 and 1866.

The two basic mineral products, coal and iron, more than held their own in production during the war years. Michigan continued to turn out copper at the rate of about 6,000 tons per year, and at the same time the copper industry in California was developing rapidly. The striking of oil on the Drake farm at Titusville, Venango County, Pennsylvania, in 1859, was the beginning of a great industry which went through the first stages of its development during the war. Thousands of wells were soon bored along Oil Creek in Pennsylvania, as well as near Wheeling, West Virginia, and in Ohio. By 1862 the production amounted to 128,000,000 gallons. Coincident with the excitement in the oil fields was the speculation in the cities, where 1,100 oil companies with a capital stock of $600,000,000 sold $90,000,000 worth of securities.

In 1859 the famous Comstock Lode of gold and silver was discovered in Nevada, and the Gregory Lode of gold in Colorado; these mines and others must have produced during the war at least $8,000,000 worth of the precious metals. The rush for the mining towns, which commenced in 1859 with the announcement of the discoveries of new deposits, continued during the war years. The population of Colorado jumped from 32,227 in

1860 to about 100,000 in 1864. The year 1863 alone brought over 30,000 to Idaho. Virginia City, Nevada, grew from nothing to 18,000 in a short time, and the population of the state from 6,857 in 1860 to 42,491 in 1870. The overland routes during the summer months were marked by a continuous stream of prairie schooners. One traveler on the Kansas route in 1863 testified to meeting on a sixteen-day journey on an average of 500 wagons a day. Omaha, the great point of exodus, in 1864, saw 75,000 emigrants pass through toward the golden west.

Labor and the Cost of Living.—Leaving until a later chapter the story of how the war was financed, let us here note the fact that the successive issuing of legal-tender notes (greenbacks) and short-term treasury notes, which filled the country with a paper money fluctuating in value, tended to drive prices up. According to the "Aldrich Report," [1] the relative course of prices and money wages was as follows:

Year	Prices	Money wages
1860	100	100
1861	100.6	100.8
1862	117.8	102.9
1863	148.6	110.5
1864	190.5	125.6
1865	216.8	143.1

Labor, which had enjoyed prosperity in the years immediately preceding the war, was hard put to it to make ends meet in the face of the rapidly rising cost of living. It has been estimated that the cost of sixty articles of prime necessity, aggregated according to quantities of such articles consumed, increased 125 per cent during the four years of the war. Although specie payment was suspended by the government, prices measured in terms of gold are not without interest.

---

[1] Senate Reports, 2d Session, 52d Congress, 1892-93 (special session, March 4, 1893), vol. 3, part i, pp. 9 and 13.

PRICES MEASURED IN GOLD DURING THE CIVIL WAR [1]

Year	Food	Clothing	Fuel and lighting	Metals and implements	Building materials	Drugs, chemicals	Home furnishings	Miscellaneous	All articles
1860	100	100	100	100	100	100	100	100	100
1861	95.8	94.9	103.5	102.5	108.9	101.3	96.8	100.7	100.6
1862	107.7	121.1	94.8	114.3	145.6	113.6	87.3	101.2	114.9
1863	91.7	132.0	73.8	96.5	122.1	101.0	84.8	89.0	102.4
1864	106.1	167.7	115.9	115.6	142.3	109.5	105.9	99.3	122.5
1865	100.1	138.4	110.0	88.5	84.2	125.6	83.8	93.8	100.3

While commodities measured in paper money doubled in price, wages lagged behind. The big jump in prices came in the year 1863, and the winter of 1863 and 1864 saw both the beginning of trade unionism on a large scale and numerous desperately fought strikes. Any rise in wages up to this time had been staved off by capital by means of labor-saving machinery, by the employment of women and children, by the systematic importation of cheap European labor, and by negro strike breakers; but capital was now forced to give in, and wages generally went up. During the war 800,000 immigrants entered the United States, a number sufficiently large to fill a big part of the gap which the war had made in the labor force of the country. While it would be an exaggeration to say that the Union army was an army of boys, it is true that a remarkably large number of the recruits were too young to have entered industry; this fact and the large immigration explain the slight disturbance in the labor market and the failure of wages materially to rise until 1863. Professor Fite believes that the average advance in wages during the war amounted to 60 per cent, a point which left the laborer at its conclusion worse off than in 1860.[2] Those most severely affected were, as usual, the professional classes, particularly clergymen and teachers, government employees, and women. The pay of soldiers remained at $13 a month until June 20, 1864, when it was advanced to $16. Many of the most efficient of the government officials in civil occupations were forced to sever their connections with the government because of low pay.

---

[1] Aldrich Report, part i, p. 13.
[2] Fite, *op. cit.*, p. 185, note.

Women, especially seamstresses, were not able, apparently, to hold the mailed fist over the employer as did the men, and in 1865 the average wage of women working for contractors on army clothing was $1.54 a week. Low wages in the industrial world were eked out considerably by military wages, of which a part was often sent home, by military bounties, and by charitable aid given to soldiers' families. The increased buying capacity of the nation during the last years of the war, the sale of luxuries, and the prosperity of popular amusement enterprises demonstrate, on the other hand, that labor shared to some extent in the flush times. Some light is thrown both on the rapid advance in manufacturing and on the demand for labor by the census figures of 1870, in which the increase in the number of industrial establishments for the decade is placed at 252,148, an advance of 79.6 per cent, the greatest in our history. The number of wage earners in industrial establishments advanced from 1,311,-246 to 2,053,996, or an increase of 56.6 per cent, a rise not equaled even in the decade of the Great War.

Capital in the North.—Although labor held its own with difficulty, capital found itself in a most flourishing position. Before the war the millionaires of the country might have been counted on the fingers of both hands; at its conclusion there were hundreds. War taxation favored the larger industry, and the process of consolidation, so marked in subsequent years, had its beginning during the war in the union of various telegraph lines and transportation companies. The American Telegraph Company and the United States Telegraph Company, the last important rivals of the Western Union, were acquired by the latter in 1866, giving the Western Union control of 75,000 miles of telegraph lines. In like manner numerous smaller railroad companies in various parts of the country were brought under a single direction. The tendency toward centralized control in railroads was emphasized, no doubt, by the increased need of transportation facilities occasioned by the war, and by the rivalry of different cities to be termini or collecting points for produce. The conduct of the war was hampered by the frequent transference of freight, necessitated by the many little independent systems with different-gauge tracks and different types of rolling stock. The Pennsylvania, Lehigh, Erie, and most of the important roads absorbed minor companies just before, during, or immediately after the war, and laid the foundations of the great railroad systems of to-day.

In addition to the consolidation of various units the war witnessed considerable activity in railroad building. The longest railroad constructed during the war was the Atlantic and Great Western, now part of the Erie system, which ran from Salamanca, New York, to Cincinnati, Ohio. By means of the Erie in the east and the Ohio and Mississippi in the west, it now linked up New York with St. Louis in a one-gauge railroad. This road, constructed at the rate of a mile a day by European capital and imported labor, demonstrated both the tireless business energy of the nation and the faith of northern and European financiers in the future, whatever might be the fate of the conflict. Chiefly through state and county aid a railroad was built from St. Louis to Kansas City. The Philadelphia and Erie, which connected the new oil fields with Philadelphia, was completed in 1864. Massachusetts, hoping to save part of the western trade, took up in 1863 the unfinished work on the Hoosac tunnel of the Troy and Greenfield and pushed to completion the connection between Albany and Boston. Other pieces of railroad construction were completed or planned, including the laying of the first rails for the Union Pacific, the first of the transcontinental lines to be built. Never were the railroads so prosperous. The Erie, the Hudson River, the Cleveland and Pittsburgh, and the Illinois Central, none of which had ever paid dividends, were paying by the end of the war 8 per cent or over. Erie stock rose from 17 to 126½, Hudson River from 31½ to 164, Cleveland and Pittsburgh from 52½ in 1861 to 138 in 1864, and the Illinois Central from 6½ to 132 during the same period. Many railroads sought to camouflage their real earnings by stock dividends. Capital was fully alive to its power and possibilities. The value of monopoly was understood and the railroads began to tighten their grip upon the coal producers of the anthracite region. In the cities capitalists were indefatigable in planning street-railway monopolies and pressing for long-term franchises in the legislatures. In twenty-seven cities street cars started for the first time.

Although the cost of building was almost doubled, the decline in construction for the whole country was very slight, a striking difference from the conditions existing during the World War. In such cities as Philadelphia, Chicago, and San Francisco, where population was rapidly mounting, or in cities like Lynn and Springfield, where industry was immensely stimulated by war needs, there was extensive building. The Capitol at Washington was completed during the war, as were many state and municipal buildings. "On every street and avenue," said the Chicago *Tribune* on October 8, 1863, "one sees new buildings going up, immense stone, brick, and iron business blocks, marble palaces and new residences, everywhere the grading of streets, the building of sewers, the laying of water and gas pipes, are all in progress at the same time."

Social Background.—The unprecedented war prosperity plunged most of those who benefited into a riot of extravagance and pursuit of pleasure. More charitable observers might attribute this to an attempt to forget the horrors of war and to keep up a brave front, but the true explanation was the accumulation of sudden wealth in the hands of those unused to it. The race tracks were crowded and stakes offered on a scale never before seen. Athletics were enthusiastically patronized, and the leading actors played to packed houses. The most expensive jewelry, clothing, and furniture found the readiest sale.

To picture the life of those who remained at home in the north as a mad scramble for wealth to be spent in extravagant living while the "boys in blue" were fighting and dying for the Union would be far from correct; high living was largely confined to certain classes and to the cities. Furthermore, if the northerner was whole-hearted in his spending, he was also whole-hearted in his giving. At least fifteen colleges were founded during the 'sixties, including Vassar, the first institution of collegiate rank exclusively for women, the Massachusetts Institute of Technology, Cornell, Lehigh, Swarthmore, Bates, and the state universities of Kansas and Minnesota. Private benefactors contributed heavily to endowment funds and financed new buildings in most of the already existing institutions. Voluntary contributions of at least $5,000,000 to education are recorded. That the minds of the national legislators were turned to educational needs, even during the terrible strain of war, is demonstrated by the Morrill Act of July 2, 1862, by which the national government gave to each state 30,000 acres of public land for each of its senators and congressmen, the income from which was to be devoted to mechanical and agricultural schools, with provision for military training in their curricula.

Sanitary and welfare work among the soldiers, taken care of in subsequent wars by the Red Cross and other organizations, was consolidated chiefly under two bodies—the United States Sanitary Commission and the United States Christian Commission. The former, approved by the government to supplement the medical department of the army, affiliated the local societies engaged in soldier welfare work. Through their instrumentality clothing, bandages, medicines, food, and tobacco to the value of $25,000,000 were distributed to the soldiers. The most important service of the Commission was that rendered on the battlefield and in the actual campaign, but the other phases of its work were most valuable. These included twenty-five soldiers' homes maintained in the leading cities, where passing soldiers might find meals and lodging; agencies to advise and help soldiers in regard to back pay, bounties, and pensions; convalescent hospitals; the publishing of a hospital directory; and, in fact, innumerable means to make the lot of the soldier easier. Newspapers collected funds for the Commission; huge sanitary fairs were held in the large cities; theaters contributed from their receipts, as did public-service utilities; school children gave entertainments and contributed their pennies. The spiritual welfare of the soldiers (in addition to the work of the regular chaplains) was taken care of by the United States Christian Commission, which sent clergymen and bibles, as well as hospital supplies and food.

Larger amounts than those given for soldier welfare

work were the donations and bounties to the dependent families of the soldiers, contributed by the state, county, and local authorities and by individuals. The Provost Marshal General of the United States estimated that $600,000,000 was distributed by the national, state, and local authorities in bounties, and $100,000,000 more by individuals. Probably half of this found its way back to the dependent families. Municipalities contributed generously to the support of needy relatives, Philadelphia having at one time 9,000 on its list at an annual expenditure of $600,000. Considerable sums were also raised in the north for the support of negroes and southern refugees, and for the relief of the starving cotton operatives in Lancashire. The story of this wholehearted generosity and sacrifice goes far to offset the more sordid details of war speculation, profiteering, and extravagance.

The Rôle of Cotton in the Struggle.—One of the most interesting phenomena of the Civil War in its economic phase is the rôle of cotton. Southern statesmen made their greatest mistakes on that subject. Scherer, in his very illuminating book, *Cotton as a World Power,* says on the authority of the expert M. B. Hammond, "that had it not been for the reliance which the architects of the 'Great Rebellion' placed on cotton as a means of obtaining revenue, it is doubtful if the war would ever have been undertaken." Their confidence in the power of cotton was unbounded, as the peroration of a speech by Senator Hammond on March 4, 1858, clearly demonstrates: "Without firing a gun, without drawing a sword, should they make war on us, we could bring the whole world to our feet. . . . What would happen if no cotton was furnished for three years? I will not stop to depict what every one can imagine, but this is certain: England would topple headlong and carry the whole civilized world with her save the south. No, you dare not to make war on cotton. No Power on the earth dares to make war upon it. Cotton *is* King."[1] This was the slogan that affected so powerfully southern economic thinking.

That events did not bear out their sanguine hopes was due to

---

[1] Quoted by Scherer, James A. B., *Cotton as a World Power,* p. 239.

two chief causes: first, to a surplus of cotton on hand, due to overproduction; and second, to the European need of northern wheat. The decade 1840 to 1850 had been one of overproduction, but during the decade 1850-60 both planters and manufacturers "enjoyed a period of unexampled prosperity."[1] This raised their hopes. In 1860 a record crop of cotton was produced, "which the southern planter marketed with unusual haste on account of the threats of trouble, England taking 1,650,000 bales before the war broke out. The British market was glutted to such an extent that many of the mills actually shut down in 1861 and prices remained practically stationary. The blockade, when it came, was laughed at as a paper blockade; and indeed it seemed to be so, for it is estimated that 3,127,568 bales were exported during the year ending August 31, 1861. Mill owners even longed for an effective blockade to relieve the glut of the market."[2] The blockade saved English middlemen from actual bankruptcy. The north had stock also, and "in the first year the American mills ran on two-thirds time, the next year on from one-quarter to one-half time, in these two years consuming cotton that remained over from the heavy purchases of 1860."[3] Thus to cotton manufacturers the cutting off of the supply of cotton was at first a benefit, for it enabled them to dispose of their surplus stock, and to keep up prices. When the mills in the north lowered production, operatives readily found employment in war industries. In England there was unemployment and great distress among factory operatives, but they realized that in the terrific struggle in America was being fought their battle as free laborers, and they stood stanchly by the side of the north, opposing any recognition by the British government of the Confederacy. The north realized the value of their good will, and showed appreciation by fitting out three ships with relief supplies for the Lancashire sufferers.

The second deterrent to English aid for the Confederacy

---

[1] Watkins, James L., *Production and Price of Cotton for One Hundred Years*, U. S. Dept. of Agriculture Bulletin No. 9, Miscellaneous Series (1895).
[2] Scherer, *op cit.*, p. 265.
[3] Fite, *op. cit.*, p. 86.

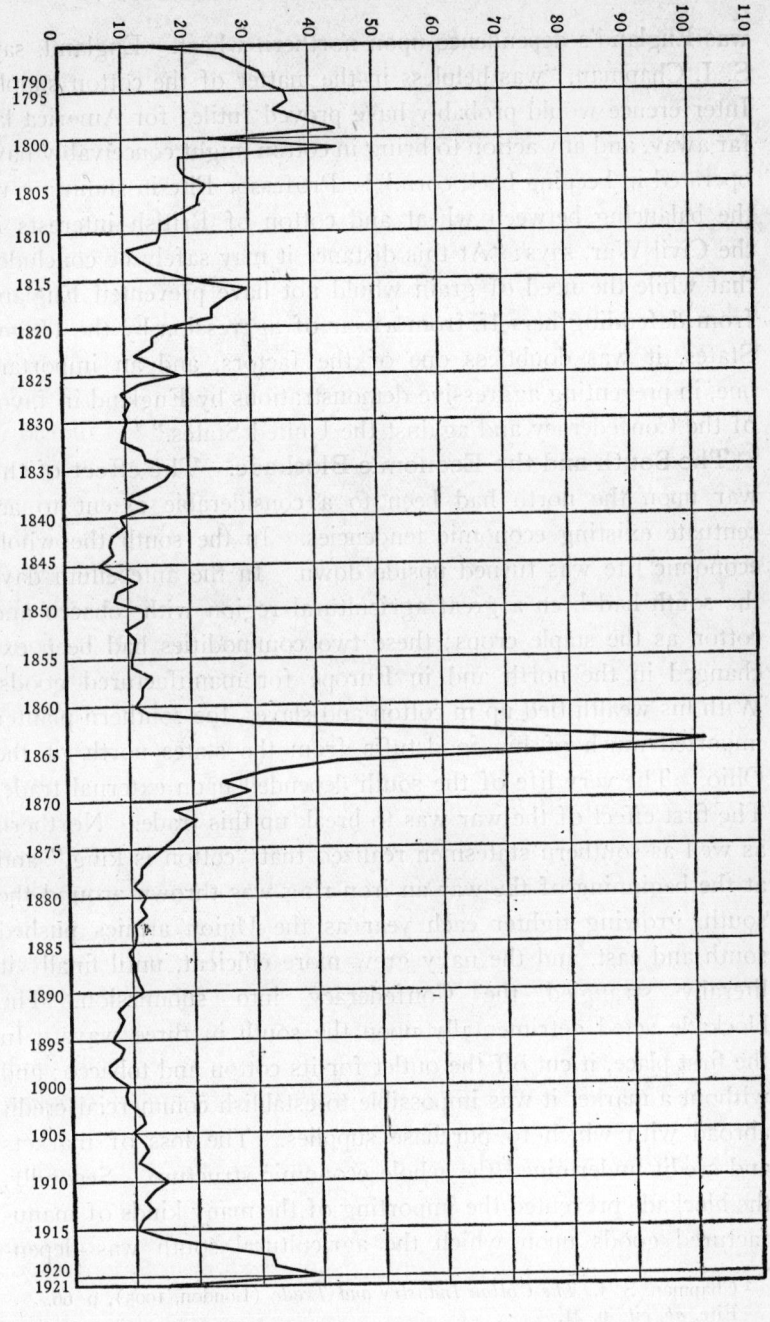

AVERAGE ANNUAL NEW YORK COTTON PRICES FOR MIDDLING UPLANDS, 1790–1921 [1]

[1] Hammond, M. B., *The Cotton Industry*, Appendix I, and *Statistical Abstract*, 1921, p. 878.

was England's dependence upon northern wheat. England, says S. J. Chapman, "was helpless in the matter of the cotton supply. Interference would probably have proved futile; for America lay far away, and any action to bring in cotton might conceivably have operated in keeping back corn." [1] Professor Fite in summing up the balancing between wheat and cotton of British interests in the Civil War, says: "At this distance it may safely be concluded that while the need of grain would not have prevented England from defending herself from a war of aggression by the United States, it was doubtless one of the factors, and an important one, in preventing aggressive demonstrations by England in favor of the Confederacy and against the United States." [2]

The South and the Economic Blockade.—The effect of the war upon the north had been to a considerable extent to accentuate existing economic tendencies. In the south the whole economic life was turned upside down. In the antebellum days the south had been a great agricultural region with tobacco and cotton as the staple crops; these two commodities had been exchanged in the north and in Europe for manufactured goods. With his wealth tied up in cotton and slaves, the southern planter imported much of his foodstuffs from the states north of the Ohio. The very life of the south depended upon external trade. The first effect of the war was to break up this trade. Northern as well as southern statesmen realized that "cotton is king," and at the beginning of the war an iron ring was thrown around the south, growing tighter each year as the Union armies pushed south and east, and the navy grew more efficient, until finally it literally strangled the Confederacy into submission. The blockade acted detrimentally upon the south in three ways. In the first place, it cut off the outlet for its cotton and tobacco; and without a market it was impossible to establish commercial credit abroad with which to purchase supplies. The loss of markets and credit undermined the whole economic structure. Secondly, the blockade prevented the importing of the many kinds of manufactured goods upon which the agricultural south was depen-

---

[1] Chapman, S. J., *The Cotton Industry and Trade* (London, 1905), p. 66.
[2] Fite, *op. cit.*, p. 21.

dent. Thirdly, it forced slave labor into unaccustomed occupations, resulting in a tremendous decline in the value of both land and slaves.

The south strove heroically to counteract the effects of the blockade. Arms, ammunition, shoes, blankets, medicines, and various luxuries were shipped from Europe to various ports in the West Indies, and from there low-built, lead-colored, side-wheeled steamers crept into the ports of Wilmington, Charleston, Savannah, Mobile, or Galveston. The profits were enormous and attracted a disproportionate amount of capital; 30,000 pounds each way was a not uncommon profit on a voyage. Although captures were frequent and the chances of success grew smaller as the war drew to a conclusion, the fact that two successful voyages compensated the owners for total loss on a third made blockade running still a good gamble. Even the state governments invested in companies operating blockade runners. The North Atlantic blockading squadron reported the names of fifty runners captured between August 1, 1863, and September 30, 1864, but a large volume of business must have been carried on if southern sources are to be believed. According to the Secretary of the Treasury of the Confederate States, forty-three runners entered the ports of Wilmington and Charleston in May and June of 1864.

The temptation for excessive profits from certain types of imports was more than the patriotism of some could stand, and the Confederate government finally passed an act (February 6, 1864) forbidding the importation of such luxuries as wines, spirits, laces, carpets, toys, furniture, and jewelry. By later laws the government sought to control for its own use a certain amount of space in incoming and outgoing boats. It is impossible to determine accurately the amount of goods exported and imported by blockade runners. In the years 1862, 1863, and 1864 the exportation of cotton to Europe was probably less than one-tenth that of the pre-war years. The comparatively small amount exported reached its destination either through the medium of blockade runners or by way of Matamoras, Mexico, to which point considerable cotton was taken overland through Texas.

Southern cotton found its way to the markets of the north by other routes than those of the blockade runner. The first policy of the Confederate government was to prevent any cotton whatsoever reaching the north, on the theory that its lack would bring the war to a speedy conclusion, and in the furtherance of this policy it passed an act on May 21, 1861, prohibiting exportation of cotton except through seaports. This plan broke down under the exigencies of the situation and contraband trade in the staple was carried on all through the war. There were times when the Confederate armies were actually provided with food obtained in exchange for cotton. This widespread illicit trade was a source of demoralization to civilians and officers of both north and south, and brought numerous protests from commanders in the field to their respective governments. As before the war, cotton was raised by negro slaves and "one of the strange things in this eventful history," says Mr. Rhodes, "is the peaceful labor of three and one-half million slaves whose presence in the South was the cause of the war and whose freedom was fought for after September, 1862, by the Northern soldiers." [1] The overland cotton trade to the north during the years 1862 to 1865 [2] amounted to about 1,108,000 bales, an amount larger than that obtained by Great Britain from blockade runners. This trade was more essential to the south than to the north and helped prolong the war. The following table shows the cotton movement at New Orleans, the principal port of the Confederacy:

RECEIPTS AND SHIPMENTS OF COTTON FROM NEW ORLEANS DURING THE WAR [3]
(in bales)

Year	Receipts	Total exports	Export to Liverpool	Export to Havre	Export to New York	Export to Boston
1859–60	2,235,448	2,214,296	1,348,163	303,157	62,936	131,648
1860–61	1,849,312	1,015,852	1,074,131	384,938	29,539	94,307
1861–62	38,880	27,678	1,312	472	4,116	109
1862–63	22,078	23,750	2,070	1,849	17,859	1,418
1863–64	131,044	128,130	1,155	4,023	109,149	12,793
1864–65	271,015	192,315	31,326	5,952	144,190	15,993

[1] Rhodes, J. F., *History of the Civil War, 1861–1865*, p. 381.
[2] Freedom of trade was restored in May, 1865.
[3] Hammond, M. B., *The Cotton Industry*, p. 263.

Manufacturing.—But the amount of cotton run through the blockade or smuggled through the lines, and the manufactured goods and gold received in return, softened but slightly the grip of the iron ring. The one-sided civilization of the south put the Confederate states at a decided disadvantage. Manufacturing concerns were comparatively few, and many of those existing were destroyed by the Union armies. Up to the opening of the war practically all of the machinery had been imported, and the rich coal and iron deposits of the south had scarcely been tapped. It is true that cotton to the value of $7,000,000 was spun in the south in 1860 and that cotton manufacturers, although handicapped by depreciating machinery, continued production during the war and made extensive profits. Nevertheless, the only large-scale manufacturing was carried on by the Confederate government itself, which took over and worked establishments producing whisky, salt, guns, small arms, gun-powder and other munitions of war, and for these, after the first two years of the war when the factories were in practical operation, the Confederate armies did not suffer. Except for immediate war needs, manufacturing production declined.

In general, there was a reversion to hand industry. The hand looms and spinning-wheels were brought out and much of the clothing and shoes for civilian and soldier were plantation made. The sacrifice and painstaking labor of the women of the south, who worked night and day at these unaccustomed tasks, was as heroic as the labor of the slaves was faithful, and both were in bright contrast to the speculation and extravagance all too evident in such cities as Charleston, Mobile, and Richmond. This rapid return to household production was like turning the hands of civilization back a hundred years, an undoing of the Industrial Revolution. The development of war manufacturing was severely handicapped by the lack of surplus capital. Previously all of the excess wealth had been invested in land and slaves, and much of the liquid capital at hand during the war years was attracted almost exclusively to blockade running, where profits were enormous. "Fifty or sixty millions of dollars," complained the president of a manufacturers' convention in Augusta in 1864, "have gone into blockade running, while scarcely a new dollar has gone into manufacturing."

Production of Food and the Cost of Living.—There was a decided reluctance on the part of many planters to substitute grain crops and meat production for cotton, but it became more and more necessary as the war progressed. In this respect an agricultural revolution was to a certain extent temporarily brought about by the war. The cotton crop of 1862 was but a little over a quarter that of 1861 and that of 1864 only about one-eighth, while the production of cereals, especially corn, increased yearly. Nevertheless, the price of bread and meat, measured in gold, mounted steadily, and in many parts of the south there was a distressing lack of food at various times, a lack which was especially acute in 1864. The pinch was greatest in the cities, for the farmer could somehow worry along. "Meal is the only food now obtainable except by the rich," commented an observer in Richmond on February 23, 1864, and added: "We look for a healthy year, everything being so cleanly consumed that no garbage or filth can accumulate."[1] In the early months of 1864, when a dollar of gold was worth $22 in Confederate money, flour sold in Richmond at $300 a barrel, Confederate money, and shoes at $150 a pair. There were bread riots in Atlanta, Mobile, and other places.

The food scarcity was caused not so much by absolute lack as by poor distribution, for there was no failure of crops during the war. There were over 6,000 miles of railroad in the southern states, sufficient at that time to distribute the food under normal conditions. But the equipment of rolling stock and rails wore out rapidly under the severe strain to which they were subjected by war needs, and it was impossible to replace them. The government used the iron mills for other work, and commandeered much of the rolling stock for military purposes. Necessarily, all but the main lines were abandoned. The result was that while corn sold for a dollar a bushel in Georgia in 1864, it was $15 in Virginia. The Union forces eventually obtained control

---

[1] Jones, J. B., *A Rebel War Clerk's Diary at the Confederate Capitol*, vol. ii, p. 186.

of a large portion of the southern railways, thus further reducing the food supply. Sherman's sixty-mile swath of destruction in Georgia and the desolation left by Sheridan in the Shenandoah Valley added to the woe of the south and the scarcity of provisions.

Financing the Confederacy.—When the war broke out the southern states were in debt both to the north and to Europe, and it was thus almost impossible to obtain further credit from the outside. Of surplus capital there was little. Some metallic currency existed, of course, but it was difficult for the government to get possession of it, and specie payment was suspended almost immediately after the outbreak of the struggle. In addition to taxation the Confederate government up to 1863 endeavored to support itself by bond issues and fiat money in the form of treasury notes. Of the latter, close to one billion dollars were issued. With only the credit of a rebel government behind them, they speedily depreciated in value until on February 17, 1864, the Confederate Congress passed an act providing for either the compulsory funding of these notes into 4-per-cent bonds, or the exchange of all notes under $100 for new notes at the rate of three dollars' worth of the old for two of the new.[1] This amounted to virtual repudiation and drove the people to primitive methods of barter for the final months of the war. The situation was made more hopeless by the state, municipal, and corporation issue of paper money.

A $15,000,000 loan, floated in Europe early in 1863 by the French firm of Erlanger & Company, and secured by cotton purchased by the Confederate government, was designed to sup-

---

[1] G. C. Eggleston in *A Rebel's Recollections* (p. 84) quotes a friend as saying: "Before the war I went to market with the money in my pocket, and brought back my purchases in a basket; now I take the money in the basket, and bring the things home in my pocket." By the end of 1864 there was probably over one billion dollars in treasury notes in circulation, "but the issues grew so enormously," says Rhodes (vol. v, p. 344), "that apparently no exact account of them was made public; it is even possible that the treasury department itself did not know the amount afloat." One gold dollar, according to a table given by Schwab (p. 167), would purchase 61 Confederate paper dollars. These figures, of course, do not seem impressive when compared to the post-war inflation in central and eastern Europe.

ply the specie with which to purchase naval and military supplies in Europe. Although the bonds maintained a surprisingly high value, the actual return to the south was slight; $5,000,000 was expended on vessels that were never delivered, $6,000,000 in attempting to keep the market price of the bonds higher, another million in the three semiannual drawings for redemption of bonds. Considering these expenditures, says Professor Schwab, and the fact that "the government had to go heavily into debt at home and wrecked the currency in order to gain the necessary cotton on which to base the foreign loan, the net gain from the loan sinks to an insignificant sum."[1] On their return from Europe, the Confederate agents reported to the Richmond government that "the loan was unsuccessful as a source of revenue, but very successful as a political demonstration." Professor Schwab says that "as late as September 7, 1864, the London *Times* considered the holders of the cotton bonds better off than those of Federal securities."[2]

In 1863 the Confederate Congress, realizing that the financial system was weakened beyond repair, passed a law providing for a tax of one-tenth on agricultural products and authorized any officer of the army within certain limits to seize such property. Without specie, without credit abroad, with their foreign trade cut off, and with a worthless paper currency, the south was forced to levy on agricultural products. It was the agricultural resources freely given or forcibly taken that provided the chief strength of the rebellion.

Conclusion.—In contrast to the north, the Civil War did not bring in its wake for those in the south financial prosperity or a period of booming business. Instead of inaugurating a new epoch of unprecedented expansion, it marked a period of destruction in which the old economic life was torn up by the roots. It is true that the fiat money made it a heyday for the speculator, and

---

[1] Schwab, John C., *The Confederate Foreign Loan: An Episode in the Financial History of the Civil War*. Yale Review, vol. i, 1893, p. 185.

[2] *Ibid.*, p. 183. He goes on to give the reasons for this optimism, largely due to the idea of investors that the cotton for redeeming the bonds was certain to be forthcoming, and that there would be no repudiation of the debt.

THE CIVIL WAR

that some fortunes were made by blockade runners, by merchants and others in the cities, and for those who had gold there was plenty of the best to be obtained. To the average southerner the war meant the sacrifice of luxuries and many necessities, in addition to the distress and bitterness entailed by the disastrous outcome; to the economy of the south it meant destruction and chaos, and eventually a fresh start. In summarizing the war Woodrow Wilson said truly: "On the part of the North it was a wonderful display of spirit and power, a splendid revelation of national strength and coherency, a capital proof of quick, organic vitality throughout the great democratic body politic. ... But its material resources for the stupendous task never lacked or were doubted; they even increased while it spent them. On the part of the South, on the other hand, the great struggle was maintained by sheer spirit and devotion, in spite of constantly diminishing resources and constantly waning hope. Her whole strength was put forth, her resources spent, exhausted, annihilated; and yet with such concentration of energy that for more than three years she seemed as fully equal to the contest as the North itself. And all for a belated principle of government, an outgrown economy, an impossible purpose." [1]

NOTES FOR FURTHER REFERENCE

The digested material on the economic history during the Civil War is still quite meager. The most satisfactory study of the south is that of John C. Schwab, *The Confederate States of America* (1901), a financial and industrial history of the south during the conflict, with special attention to the financial phase and containing a good bibliography. What Schwab has done for the south, Emerson D. Fite in his *Social and Industrial Conditions in the North During the Civil War* (1910) has done for the north. The latter book lays more emphasis upon social and industrial and less upon financial history. Further summaries of conditions in the south are in J. F. Rhodes, *History of the United States,* Vol. V (1905), Chaps. 27 and 28, and in his *History of the Civil War, 1861-1865* (1917), Chaps. 11 and 12. In Vol. V of *The South in the Building of the Nation,* there are a number of contributions on the economic background, notably those of J. C. Reed, *Economic Conditions in the South During the Civil War, The Finances of the Southern Confederacy,* and *The Labor Force and Labor Conditions.* James A. B. Scherer, *Cotton as a World Power, a Study in the Economic Interpretation*

---

[1] Wilson, Woodrow, *Division and Reunion* (rev. ed., 1898), p. 239.

of *History* (1916), contains interesting material on the rôle of cotton in the struggle.

The financial history of the period is of special interest. That of the north is summarized in Davis R. Dewey, *Financial History of the United States* (rev. ed., 1922); and in more detail in A. S. Bolles, *Financial History of the United States*, Vol. III, 1861-85 (1886); also in W. C. Mitchell, *History of the Greenbacks* (1903). On the finances of the Confederacy, the book by John C. Schwab previously mentioned, and the following articles: *Finances of the Confederacy*, Political Science Quarterly, Vol. VII, pp. 38-56 (1892); *The Confederate Foreign Loan*, Yale Review, Vol. I, pp. 175-186 (1893); and *The Financier of the Confederate States*, Yale Review, Vol. II, pp. 288-301 (1894), in part a review of Henry D. Capers, *The Life and Times of C. G. Memminger* (1894). See also his Chap. XIX in *Cambridge Modern History*, Vol. VII. For prices during the war consult the Report by Mr. Aldrich for the Committee on Finances and Prices, Wages, and Transportation, *Senate Reports*, 2d Session, 52 Congress, 1892-93, Special Session, March 4, 1893, Vol. III, four parts. See also E. P. Oberholtzer, *Jay Cooke* (2 vols., 1907).

The influence of railroads on the course of the Civil War may be traced in E. A. Pratt, *The Rise of Rail Power* (1916), Chaps. II, III, and IV.

SELECTED READINGS

SCHWAB, J. C., *The Confederate States of America*, Chaps. IX-XII.
FITE, E. D., *Social and Industrial Conditions in the North During the Civil War*.
RHODES, J. F., *History of the United States*, Vol. V, Chaps. XXVII, XXVIII.
DEWEY, D. R., *Financial History of the United States*, Chaps. XII, XIII.
SCHERER, J. A. B., *Cotton as a World Power*, pp. 228-297.
HART, A. B., *American History Told by Contemporaries*, Vol. IV, Part V.
SEMPLE, E. C., *American History and Its Geographic Conditions*, Chap. XIV.

CHAPTER XVIII

THE LAST FRONTIER

The Mining Frontier.—Missionary activity and ranching had brought the first permanent white occupation of California. But the handful of Spaniards were quickly submerged in the flood of gold seekers who overran the country in 1849 and created from it an American state in 1850. Only a small percentage of the "'forty-niners" actually acquired wealth from the golden metal, but a large number remained to exploit the possibilities in agriculture and lumber. With thousands of white settlers in California and a scattered population of farmers and trappers in the Oregon territory, the American possession of the Pacific coast was beyond dispute. The frontier line, which in 1850 had run from eastern Minnesota down through eastern Nebraska and Kansas and then cut through Arkansas to Texas, had suddenly leaped a thousand miles to the Pacific coast. The story of the "last frontier" is, then, the history of how miners, ranchers, and farmers, advancing from both the east and west, gradually filled in the intervening space.

Some beginnings of actual settlement had already been made in this vast region by Mormons around the Salt Lake basin, and by the Spanish ranchers pushing north from Mexico. The discovery of gold in California was of course followed by the activities of numerous prospecting parties in many parts of the Rockies. Rumors that the precious metal was to be found in the Pike's Peak region were confirmed in 1859, and that year saw a great exodus to what later became the state of Colorado. Professional prospectors and miners of the west hastened to the scene, their ranks augmented by thousands from the east who had felt the pinch of the depression following the panic of 1857. "Pike's Peak or bust," had been the motto of the "'fifty-niners," and not far from 100,000 reached Colorado during the rush of the first year. Although half returned "Busted! By gosh!" those who remained laid the foundations of a new state. Pay dirt was discovered at El Paso (the present site of Colorado Springs), and around the confluence of the South Platte and Cherry Creek; mining camps appeared at Boulder, along the Clear Creek near Denver, and along the Arkansas from Pueblo to Leadville. The great Colorado deposits, discovered a few months after the first rush, were embedded in quartz lodes requiring heavy machinery and large capital for working, and were not developed immediately. Nevertheless, enough people remained to organize the Territory of Jefferson in 1859, which was re-formed into the Territory of Colorado in 1861 and admitted as a state in 1876. Mining, which brought the first rush to Colorado, has continued to be its chief industry with coal, gold, silver, zinc, uranium, and vanadium ores important in the order named (1919).

The rush which laid the foundations of Colorado was but one of a series of booms which planted mining camps on many a lonely creek and forbidding hillside of the Rockies. Small deposits of gold had brought about the founding of Carson City in 1858 at the extreme western part of the Territory of Utah, close to the California border line and near the old overland route from Salt Lake City to San Francisco. The discovery of the famous Comstock Lode in the next year on the eastern slope of Mt. Davidson, not far from Lake Tahoe and about twenty miles east of the California state line, brought an influx of immigrants who transformed the region into the Territory of Nevada (formerly part of the Territory of Utah) in 1861 and into a state three years later. The great silver deposits were easily accessible, located as they were on the main route to California, and Carson Valley was speedily entered from both the west and the east. The first spurt brought the population of Nevada to 42,491 in 1870, but this number had less than doubled in 1910, when it reached 81,875. The Comstock Lode from 1860 to 1890 yielded $340,000,000 worth of silver, and was the economic backbone of the region for those years. The yield declined rapidly after 1890, seriously affecting the prosperity of the state, until new discoveries of gold, silver, and copper in 1906 and the years following at Tonopah, Goldfield, and other places renewed industry and brought a fresh influx of population. Copper, the most valuable product of the state, was not mined in great quantities until after 1908, but Nevada in 1919 ranked fifth among the states in its production. In point of value (1919) silver ranked second and gold third among her products; among the states she rated third in her output of silver, and fourth in that of gold.

Handicapped by climate and inaccessibility, the western part of the Territory of New Mexico (now Arizona) was prospected slowly and individual miners found it difficult to operate successfully. Mining companies, however, with laborers recruited from the dregs of California, opened up a few deserted shafts near the old Spanish town of Tucson soon after the Gadsden Purchase was consummated. The Civil War closed the Tucson mines for the time being, but further discoveries in 1862 and 1863 along the left bank of the Colorado near Bill Williams' Creek brought a new burst of gold enthusiasm, which was encouraged by Col. James Henry Carleton, who had occupied the region for the Union cause. Arizona was created a territory in 1863 (a state in 1912), but for over ten years after territorial government was set up troubles with the Apaches made mining a dangerous occupation. Gold and silver, which drew the first prospectors to Arizona, have since become a small part of the mineral production of the state. Copper, the valuation of which in 1919 was over $100,000,000, amounted to almost ten times that of gold and silver combined. Arizona now leads the Union in copper production.

Just as the influx of gold diggers had brought sufficient population to create the new Territories of Colorado, Nevada, and Arizona, so discoveries of the precious metal brought the organization of Idaho Territory in 1863 from the Territories of Washington, Dakota, and Nebraska, the new territory including the present area of Idaho, Montana, and Wyoming. The Territory of Washington had already been divided from the Oregon Territory in 1853, owing to the inconvenience of administering the distant settlements on Puget Sound, but Washington was not admitted as a state until 1889. In 1860 gold was discovered on the reservation of the Nez Percé Indians near the juncture of the Clearwater and Snake Rivers. The next year thousands of miners poured into these river valleys and the town of Lewiston sprang up as a center. The discoveries here were followed by others on the Salmon River, at Boisé, and in the Owyhee district, south of the great bend of the Snake. The trappers of the Hudson's Bay and the American Fur Company first roamed over this country, and the farmers who followed McLoughlin and Whitman first seriously occupied it; but it was the gold seekers of '61 and '62 who gave Washington a new impetus and founded Idaho. As gold had brought in 1861 a development of the region which is now western Idaho, so new discoveries in 1863 gave birth to Alder Gulch, to Virginia City, and to a new group of mines in eastern Idaho between Beaver Head River and Madison River. Ten thousand came to Virginia City in 1864 and the same year saw the founding of Helena, "the last of the boom towns of the period." Such an ingress of miners about these points caused the cutting away in 1864 of northeastern Idaho into the Territory of Montana, and in 1868 the organization of Wyoming Territory from land previously included in the Territories of Dakota, Idaho, and Utah. Idaho with 400 mines still claims the extraction of minerals as her chief industry, ranking second among the states in the production of her chief metal, lead, and fifth in that of silver. The discovery of gold in 1882 at Coeur d'Alene brought a new rush to Idaho; but the production of gold, in relation to that of other metals, has not been large. Her next-door neighbor, Montana, produces copper in the Butte region, the amount of which is surpassed only by Arizona, while silver, chiefly a by-product of the copper industry, is mined to an extent which gives her second place among the states in the production of this metal.

The decade of the 'sixties saw the Rockies at least partially occupied from Mexico to the Canadian border, with most of the population scattered on the mining claims which dotted the hillsides and valleys or gathered into the raw towns which had sprung up near the more valuable deposits. Professor Paxson

THE LAST FRONTIER

has caught the spirit of the picturesque but demoralizing life of the mining frontier:

> The shifting population which inhabited the new territories invites and at the same time defies description. It was made up chiefly of young men. Respectable women were not unknown, but were so few in number as to have little measurable influence upon social life. In many towns they were in the minority, even among their set, since the easily won wealth of the camps attracted dissolute women who cannot be numbered but who must be imagined. The social tone of the various camps was determined by the preponderance of men, the absence of regular labor, and the speculative fever which was the justification of their existence. The political tone was determined by the nature of the population, the character of the industry, and the remoteness from a seat of government. Combined, these factors produced a type of life the like of which America had never known, and whose picturesque qualities have blinded the thoughtless into believing that it was romantic. It was at best a hard, bitter struggle, with the dark places only accentuated by the tinsel of gambling and adventure.
>
> A single street meandering along a valley, with one story huts flanking it in irregular rows, was the typical mining camp. The saloon and the general store, sometimes combined, were its representative institutions. Deep ruts along the streets bore witness to the heavy wheels of the freighters, while horses loosely tied to all available posts at once revealed the regular means of locomotion, and by the careless way they were left about showed that this sort of property was not likely to be stolen. The mining population centering here lived a life of contrasts. The desolation and loneliness of prospecting and working claims alternated with the excitement of coming to town. Few decent beings habitually lived in the towns. The resident population expected to live off the miners, either in way of trade or worse. The bar, the gambling house, the dance hall have been made too common in description to need further account. In the reaction against loneliness, the extremes of drunkenness, debauchery, and murder were only too frequent in these places of amusement.[1]

Yet upon such unpromising foundations were laid the beginnings of many of our far western states.

The Ranchers' Frontier.—Between the eastern frontier line and the mining settlements of the west there stretched from

---
[1] Paxson, F. L., *The Last American Frontier* (1910), pp. 170-172.

Texas to Manitoba a vast territory of rolling land. Grass-covered but lacking in rainfall, this country was generally believed to be unfit for cultivation and unlikely ever to be occupied. It was accidentally discovered about 1866 that cattle not only could withstand the severe winters of northern Nebraska, but would thrive on the pasturage of wild grass afforded there. This discovery opened the country almost immediately to cattlemen and ranchers, who occupied it for the next two decades, until driven aside by the advancing frontier of farmers.

Since the days of the Spaniards cattle had been bred on the Texas plains, where, exposed to the weather and running free on the wide ranges, a sturdy stock had developed. Heretofore there had been little incentive for ranchers to market their cattle, for the Mississippi Valley and the Atlantic coast farmer had been easily able to supply the local need. But the rapidly growing population of the east and the railroads advancing to the very doors of the cattle ranch now offered both a market and the means of transportation, opportunities quickly taken advantage of by the Texas cattlemen.

In the spring the cattle were rounded up and divided among the owners, according to the existing customs and laws. The yearling steers were separated from the rest, branded, and then started on the long trail north to Nebraska, Wyoming, or Dakota, to be fattened for the market, while the remainder were turned back upon the range to multiply. At Dodge City, Kansas, on the newly built Atchison, Topeka, and Santa Fé, there grew up during the 'seventies the greatest of the cattle towns. Here southern owners often sold out rather than continue the drive farther north, either to purchasers for immediate slaughter, to shippers who sent the cattle to the stockyards of Kansas City, or to buyers whose cowboys would continue the herd upon its northward journey. At Ogallala, Nebraska, a second great cattle center grew up, and 400 miles northwest of that town, at Miles City, Montana, still another great center developed for the northwest. Upon their arrival at these points the cattle were fattened for the market, often upon inclosed ranches. As time went on, more and more thought was given to experimental breeding and to the production of the most profitable type for this method of disposal. Stock growers' associations appeared for mutual protection against lawbreakers and thieves and to guard as best they might against the contamination of their herds from Texas fever, the hoof and mouth disease, and other ailments.

The ranchers' frontier lasted about two decades, from the late 'sixties to the late 'eighties. By 1890 the great drives were a thing of the past. Transient as this episode was, the hard but romantic life of the cowboy has become immortalized as part of our history through the association of Roosevelt, the novels of Owen Wister and others, and the pictures of Frederic Remington.

Three agencies were chiefly responsible in the ending of the long drives and the breaking up of the last frontier of the ranchers: illegal inclosures, the westward advance of the farming frontier, and the quarantining of one section against the diseases of another. As the permanent establishments for the slaughter of cattle grew up near the centers of transportation, large stretches of public land were inclosed. This was helped by the competition among wire manufacturers, which reduced by 1874 the price of barbed wire to twenty cents a pound. Over 8,000,000 acres were known to be illegally inclosed by 1888. As the westward advance pushed farther into the range, new settlers fenced in their own land and clamored for that illegally inclosed by the ranchers. Upon the appearance of cattle diseases the more northern states passed quarantine laws against Texas cattle. In the face of these obstacles the open range was bound to disappear.

Although it has been quite usual in our history for the farming frontier to be preceded by that of the ranchers, never before had cattle raising been conducted on such a large scale or with such far-reaching results. The extent to which illegal inclosures had gone eventually forced the national government to take action and also to open Oklahoma to white settlement. Furthermore, the demands of the western cattlemen were chiefly instrumental in the establishment in 1884 of the Bureau of Animal Industry, under the Department of Agriculture. The struggle between the cattlemen, on the one hand, and the packers and railroads on the other, for the profits of the industry was no small factor in the discontent against railroads and middlemen and in the rising tide of Populism. It was during these years that the packing industry shifted westward and centered in Chicago, St. Louis, Kansas City, and Omaha, where it speedily became monopolized in the powerful hands of Armour, Hammond, Morris, and Swift. The transportation of meat products necessitated the solution of the problems of refrigeration and packing, while the competition for the carrying trade from the packing cities to the seacoast tended to accelerate the movement both toward railroad consolidation and toward government regulation.

The Farmers' Frontier.—For years representatives of the west had clamored for a more liberal policy as regards the public lands. Their demands were upheld by northern abolitionists, who believed that free land would encourage anti-slavery men to take up claims, and by humanitarians who looked upon the unoccupied west as a haven for the oppressed of the eastern states and foreign nations. The preëmption law of 1841 was a victory for the west, and the subsequent bill introduced by Andrew Johnson in 1845 for the free granting of public land was a logical development. From then until 1860 bills to that effect were passed several times by the House, only to be thrown out by the Senate, where the south held the balance of power. Finally, an act passed both houses in 1860 giving to each head of a family the right to take up a quarter section of unappropriated public land at twenty-five cents an acre, the title to be granted after a proved residence of five years. President Buchanan immediately vetoed the bill, recapitulating in his message the stock arguments against free distribution. It would be unjust, he said, to the old settlers who had paid $1.25 per acre to grant land near them for twenty-five cents; by stimulating emigration it would decrease the value of land in the older states; it would lessen the revenue of the national government; and it would be of little use to the artisans, for their habits of life made them unfit for agriculture. Two years later the bill came up again, and on May 20, 1862, received the signature of Abraham Lincoln. This famous

Homestead Act granted a quarter section (160 acres) free to a head of a family or a person over twenty-one who was a citizen of the United States or to anyone who had filed his intention of becoming one. Residence of five years was required, good faith to be evidenced by cultivation. After fourteen months, however, the entry might be commuted by the payment of $1.25 an acre. Later amendments have further liberalized the act by permitting veterans of the Civil and succeeding wars to count the time served in the army against the five year required residence period. The Timber and Stone Act of 1878 opened to citizens at the appraised value, but in no case less than $2.50 an acre, 160 acres of public lands valuable chiefly for timber and stone and unfit for cultivation at date of sale. The Dawes Act of 1887 provided for individual ownership on the part of the Indians of small amounts of land instead of tribal ownership, and has thus opened up large areas of reservations to settlers. In 1909 an act was passed providing for enlarged homesteads of 320 acres of non-irrigable land where dry farming was necessary, one-fourth of which must be cultivated in two years.

It has been easy to evade and misuse these liberal laws. At one time it was possible for a settler to secure 1,120 acres of arable land; 160 acres under the Homestead Act, 160 acres under the old Preëmption Act, 160 acres under the Stone and Timber Act, and 640 acres of desert land. By collusion with individuals it was a simple matter for mining and lumber companies to secure immense holdings by violating the intent, if not the letter, of the law. The situation respecting the public lands was so notorious that in 1879 Congress appointed a commission to examine and report on the land system. Their suggested reforms were ignored, and until the rise of the conservation movement about 1901 only the courageous stand of Arthur and Cleveland can be pointed to as real evidences of a national desire to enforce the existing laws. A new Public Land Commission, appointed by Roosevelt, submitted elaborate reports and suggested salutary reforms along the same line as the previous commission. Little was done and the policy of the government as regards public land has remained essentially as before, and far from subserving (according to the National Conservation Commission of 1909) the best interests of the nation.

Not alone is our land policy open to the criticism that the laws have been criminally evaded and negligently enforced, but the wisdom of the whole system itself has been questioned. Has it been to the best interests of the nation to give away so rapidly its heritage of land in so free a manner that it has encouraged wasteful methods of farming and decreased the value of the agricultural lands of the east? On the other hand, farms have been provided for the excess population of the east and for multitudes drawn from Europe by the prospect of free land. It has enormously stimulated the rapid occupation of the trans-Mississippi country and the founding of new commonwealths. This, in fact, was the main purpose of the Homestead Act, and the intention was that the settlement should proceed under a democratic system. The Public Land Commission said of the act, "It protects the government, it fills the state with homes, it builds up communities and lessens the chances of social and civil disorder by giving ownership of the soil, in small tracts, to the occupants thereof." These hopes were but partly realized. Ownership in small tracts, as originally contemplated, was frustrated by the fraudulent evasion of the acts, resulting in large holdings and landless workers—for example, the lumberjacks. The regions chiefly influenced by the Homestead Act have during the past fifty years been the seat of the most acute economic and political discontent.

Almost as potent as the Homestead Act in promoting the advance of the farming frontier was the construction of the transcontinental railways. After exhaustive government surveys, twenty years of agitation, and arduous labor, the last spike was driven May 10, 1869, at Ogden, Utah, in the roadbed connecting the Union Pacific and the Central Pacific, and the first railway across the continent was completed. Other transcontinental lines were authorized in the 'sixties, and the years from the close of the war until the panic of 1873 were characterized by feverish railroad building which was renewed with the revival of business after 1878. Under the direction of Henry Villard the Northern

THE LAST FRONTIER

Pacific was completed in 1883, and the same year saw the linking of the Atchison, Topeka, and Santa Fé to the Southern Pacific in a southern route to California. In 1882 the Texas Pacific and the Southern Pacific had met at El Paso and connections were thus made between the Pacific and New Orleans or St. Louis. By the middle 'eighties there were at least four main routes to the Pacific, while such roads as the Chicago, Burlington, and Quincy, completed to Denver in 1882, opened up much additional territory to immigrants. The Union Pacific had been looked upon as a national project and had been aided lavishly by the government, a policy which was generally followed until 1871 with other prospective western roads. Approximately 122,000,000 acres have been granted by the national government to states or to private corporations for internal improvements, of which the larger part has gone to the transcontinental roads. These grants included a right of way and alternate sections on each side of the track, ranging from five sections per mile to as high as forty in the case of the Northern Pacific. The sections retained by the United States government were withdrawn from entry until the railway allotment had been made, the expectation being that the rise in land values resulting from the building of the railroad and the occupation of settlers would more than recompense the government for its original gifts.

In any case it was to the advantage of the railroads to populate the country, from the point of view of both transportation revenues and sale of land grants. Their literature and advertisements were spread over the eastern states and Europe, and their agents competed in directing the newly arrived immigrant to the respective routes. Although heavily subsidized in their early years by lavish gifts and operated wholly for private gain, the railroads have contributed notably to the building of the west. They carried the immigrant to his new home and later transported his products to the market. Much of the settlement west of the Mississippi was preceded and directed by railroads. Where the railways penetrated, permanent settlements followed.

A phase of the occupation of the last frontier, upon which it is unnecessary to enter in detail, is that of the dispossession of the

Indians. Until 1861 the trans-Mississippi Indians were generally on friendly terms with the United States, notwithstanding the fact that their lands were continually traversed by surveyors and miners. At the slightest mention of gold, white men by the thousands entered their reservations. Driven to desperation by the obvious fact that the end was near, the Indians made their last stand against encroaching civilization. The Sioux uprising in 1862 was followed by that of the Cheyenne and other tribes in the 'sixties, and the Indian struggle culminated in the Sioux war of 1876 and the defeat of their chieftain, Sitting Bull. In the elimination of the western Indian, it is only fair to say that the outstanding feature is not the treachery of the red men, but rather the ruthless greed of the white invader backed by the ever-ready rifles of the regulars. A subsequent chapter of this long-drawn-out tragedy was written in 1887 in the Dawes Act, when the government wisely moved to break the Indian to civilization by abolishing the tribal ownership of land and allotting to each head of a family a quarter section of 160 acres, an eighth section to single adults and orphans, and a sixteenth to each dependent child. To protect the new owner the right to mortgage or dispose of the land for twenty-five years was withheld, and the land was to be tax free for that period. The Dawes Act carried with it the right of citizenship for those Indians who voluntarily left their tribes and took up homesteads under its provisions.

The frontier (technically, a region of more than two and less than six people per square mile) stretched in 1860 not far from the ninety-fifth meridian. By 1880 the line ran along the westward boundary of Minnesota, jutting out in Nebraska and Kansas to beyond the hundredth meridian, back to the western boundary of Arkansas and west again through Texas beyond the ninety-seventh meridian. The next two decades marked the rapid occupation of Minnesota, the Dakotas, Montana, and western Nebraska, giving warrant in 1890 for the declaration of the Superintendent of Census that the frontier existed no longer. By that time all of our good arable land had been taken up. The farmers' frontier had met and expropriated the cow country of the ranchers and now reached to the miners of the mountains. To the west of the Sierras on the Pacific coast, fertile farming land was now yielding more than the precious metal ever had. In 1904 the government still owned 700,000,000 acres of land, but most of it was valueless except for dry farming, irrigation or drainage projects. Economically this great region between the Mississippi and the mountains is primarily agricultural, the newly populated territories separating roughly into the wheat country of Montana and the Dakotas; the corn belt of Kansas, Iowa, and Nebraska; and the cotton fields and grazing lands of Texas. Occupation and economic development was followed by admission to statehood—North Dakota, South Dakota, Montana, and Washington in 1889, and Idaho and Wyoming in 1890. The abandonment by the Mormon church of polygamy in 1890 paved the way for the admission of Utah in 1896; Oklahoma became a state in 1907, and New Mexico and Arizona in 1912. With the passing of the frontier a new era in American history began.

Western Problems in American Politics Since 1860.—The economic and political problems since 1860 resulting from the advancing frontier line are strikingly similar to those of previous periods. The radicalism of the youthful west has displayed itself principally in continued clashes with eastern capital. The cost of westward migration was largely paid for with borrowed money, and the west through all its stages has been in debt to the east. The conflict of interests between the debtor west and the creditor east, strongly in evidence throughout the colonial period and clearly seen in Jackson's spectacular war on the Second United States Bank, was renewed in the years following the Civil War. It seemed unjust to the pioneer, who had braved Indian dangers and suffered the hardships of a new country, to relinquish to eastern money lenders much of the results of his arduous labors.

The Civil War had hardly closed, and the question of paying for it taken up, before this clash of interests arose. During the war, Congress had provided for the issue of $450,000,000 of United States legal-tender notes, commonly known as greenbacks. Although in November, 1864, greenback dollars were worth only 43 cents, measured in terms of gold, they were legal tender for all debts, public and private, except customs duties and interest on the public debt. The depreciation of the currency showed a certain lack of confidence in the successful prosecution of the war, but to the debtor class it proved a godsend. It was easy to pay former debts and interest in the depreciated paper money, and the boom times of the war period encouraged the contraction of new debts. The successful termination of the struggle and the evident intent of the Republican administration both to redeem bonds in specie at par and to call in gradually the legal tenders, aroused a storm of protest in the west. As the greenbacks rose to par the prices of farm products fell and it became increasingly difficult to pay interest on borrowed money. Caught in this period of deflation, the farmer believed that the hard times were due to a lack of sufficient circulating currency, and against the wishes of eastern Democrats, he forced into the Democratic platform of 1868 the "Ohio Idea," a plan to pay for the redemption of all bonds in paper money, except where the agreement was to the contrary. Their failure to accomplish this served only to stiffen the fight against the further contraction of the greenbacks and to inaugurate a campaign for an increase in the amount of circulation. Upon this issue the Greenback party was formed, whose candidate, Peter Cooper, polled 81,000 votes in 1876, and the congressional candidates over a million in the campaign of 1878. This strong showing had its effect in a law of 1878, which forbade further contraction of greenbacks, leaving the amount permanently in circulation at $346,681,016.[1]

Although the Greenbackers ran candidates in the next two presidential elections,[2] the agitation for a more inflated currency during the next twenty years was drawn off in a demand for the free and unlimited coinage of silver. The law of 1834 had fixed

---

[1] The Greenbackers, however, were not powerful enough to prevent the act of 1875, calling for resumption of specie payments on January 1, 1879.

[2] James B. Weaver polled 307,306 votes in 1880, and Benjamin F. Butler 133,825 in 1884, but these figures by no means represent the true strength of the party.

the ratio of gold and silver at 16 to 1, a ratio which had so overvalued gold as to drive silver out of circulation. This fact was recognized in 1873, when the coinage of the silver dollar was discontinued. For various reasons, chiefly the demonetization in certain foreign nations and the discovery of new sources of supply, the price of silver dropped steadily during the 'seventies. If free and unlimited coinage could be established at the old ratio of 16 to 1, money would be cheapened and currency inflated, to the obvious advantage of the west.[1]

Western debtors and representatives of the silver states combined to advocate the possibilities of silver. Senator Jones of Nevada, speaking in the Senate on May 12, 1890, after outlining the development of time contracts as a phase of the revolution in industry and commerce of the nineteenth century, said:

> The natural concomitant of such a system of industry is the elaborate system of debt and credit that has grown up with it and is indispensable to it. Any serious enhancement in the value of the unit of money between the time of making a contract or incurring a debt and the date of fulfillment or maturity always works hardship and frequently ruin to the contractor or debtor.
> Three-fourths of the business enterprises of this country are conducted on borrowed capital. Three-fourths of the homes and farms that stand in the name of the actual occupants have been bought on time, and a very large proportion of them are mortgaged for the payment of some part of the purchase money.
> Under the operation of a shrinkage in the volume of money, this enormous mass of borrowers, at the maturity of their respective debts, though nominally paying no more than the amount borrowed, with interest, are, in reality, in the amount of the principal alone, returning a percentage of value greater than they received, more than in equity they contracted to pay and oftentimes more, in substance, than they profited by the loan. To the man of business this percentage in many cases constitutes the difference between success and failure. Thus a shrinkage in the volume of money is a prolific source of bankruptcy and ruin. . . .
> It is a remarkable circumstance, Mr. President [he continued], that throughout the entire range of economic discussion in gold-standard circles it seems to be taken for granted that a change in the value of the money unit is a matter of no significance, and imports no mischief

---

[1] For a further discussion of currency history, see chap. xxi.

to society so long as the change is in one direction. Who ever heard from an Eastern journal any complaint against a contraction of our money volume, any admonition that in a shrinking volume of money lurk evils of the utmost magnitude?[1]

While Senator Jones was voicing the interests of his own state, he undoubtedly expressed the views of the farmer group.

Before the question of silver had been fought to a finish two compromises were effected—the first known as the Bland-Allison Act of 1878, and the second as the Sherman Silver Purchase Act of 1890. The first act ordered the Secretary of the Treasury to buy not less than $2,000,000 and not more than $4,000,000 worth of silver a month to be coined into silver dollars, while the Sherman Act, which superseded it, called for the purchase of 4,500,000 ounces of silver for which treasury notes of the United States were to be issued, legal tender for all debts.

The monetary problem was not a clear-cut party question. Southerners, struggling through the burdens of the reconstruction period, combined with western and mid-western Republicans to form the Greenback party and later the People's party. The People's party was an outgrowth of the farmers' alliances of the 'eighties, which between 1889 and 1892 attempted union with the labor groups. Denouncing demonetization of silver as "the vast conspiracy against mankind" and demanding both the free and unlimited coinage of silver at the ratio of 16 to 1 and the issue of legal-tender currency until the circulation should reach fifty dollars per capita, the Populists polled over a million votes for their presidential candidate in 1892 and elected two senators and eleven representatives. Four years later the Populists allied themselves with the Democratic party on the silver issue, after the radical wing had captured the convention and nominated William Jennings Bryan, following his classic utterance of the discontent of the south and west. Denying that business would be disturbed by the free and unlimited coinage of silver, and maintaining that the coming campaign was a struggle between the holders of idle capital and the great masses who produced the wealth, Bryan continued:

---

[1] Fifty-first Congress, 1st Session, vol. 21, part ii, Appendix, pp. 239, 247.

If they come to meet us on that issue we can present the history of our nation. More than that, we can tell them that they will search the pages of history in vain to find a single instance where the common people of any land have ever declared themselves in favor of the gold standard. They can find where the holders of fixed investments have declared for a gold standard, but not where the masses have. . . .

You come to us and tell us that the great cities are in favor of the gold standard; we reply that the great cities rest upon our broad and fertile prairies. Burn down your cities and leave our farms, and your cities will spring up again as if by magic; but destroy our farms and the grass will grow in the streets of every city in the country. . . .

Having behind us the producing masses of this nation and the world, supported by the commercial interests, the laboring interests, and the toilers everywhere, we will answer their demand for a gold standard by saying to them: You shall not press down upon the brow of labor this crown of thorns, you shall not crucify mankind upon a cross of gold.

The campaign of 1896 was the culmination of the economic discontent which had been seething since the Civil War. As such it is the key campaign of the whole period. Byran's defeat settled the question of a double standard of currency and in 1900 the gold standard was definitely adopted. But the disposal of the silver issue did not settle the broader questions involved. The old cry of the west for a more elastic currency and for fairer treatment at the hands of eastern capital was a most potent influence, leading to the Aldrich Report of 1913, and the subsequent modification of the old National Bank system of 1863 by the Federal Reserve system of 1913, and by the establishment of the Federal Farm Loan banks in 1916. The unfortunate position in which the agricultural interests generally found themselves as a result of the deflation at the end of the World War has again revived the old question of farm credits and was a strong factor in the congressional elections of 1922.

Not alone in respect to monetary theories did the farmer of the south and west clash with the interests of eastern capital. The "granger movement," emanating from the "Patrons of Husbandry," an agricultural organization, denounced the railroads as the chief cause of their troubles. Dependent as they were upon the railways, the farmers felt keenly such abuses as stock watering, pooling and rate discrimination, and strenuously advocated government regulation of rates and supervision over the handling and warehousing of grain. In the states where the Grangers gained control of the legislature, laws were passed to bring public carriers under state control; but when this was made ineffective later by court decisions, the demand for government regulation was shifted from the states to the national government. The Interstate Commerce Act of 1887 and its subsequent amendments was a direct result of this agitation. The Greenback party demanded the regulation of interstate commerce, and the Populist platform of 1892 called for the government ownership of railroads, telegraphs, and telephones.

While it was against the railroads that the farmer hurled his bitterest denunciations, his antipathy was not limited to one type of monopoly. The feeling that he was exploited unjustly and unreasonably by "big business" has been ever present. "Never in our history," said the Greenbackers in prefacing their platform of 1884, "have the banks, the land-grant railroads, and other monopolies been more insolent in their demands for further privileges—still more class legislation. In this emergency the dominant parties are arrayed against the people and are the abject tools of corporate monopolies."[1] The Populist platform of 1892 declared that "The fruits of the toil of millions are boldly stolen to build up the colossal fortunes of a few, unprecedented in the history of mankind; and the possessors of these in turn despise the republic and liberty. From the same prolific womb of governmental injustice are bred two classes of tramps and millionaires."[2] In his "cross-of-gold speech" at the convention of 1896 Bryan expressed the same feeling. "Mr. Carlisle said, in 1878, that this was a struggle between the idle holders of idle capital and the struggling masses who produce wealth and pay the taxes of the country; and, my friends, it is simply a question that we shall decide upon which side shall the Democratic party fight? Upon the side of the idle holders of idle capital, or upon

---

[1] McPherson's *Handbook of Politics* for 1884, p. 216.
[2] *Ibid.* for 1892, p. 269.

the side of the struggling masses?" It was but a step from the hostile feeling toward the railroads to the anti-trust planks of the Democratic platforms of 1896 and 1900, and to similar economic causes may be traced the present-day hostility of the Non-Partisan Leaguers to the bankers, middlemen, and grain brokers, who, they believe, absorb an unfair percentage of the profits which should accrue to the original producer.

In a similar way the roots of many of our recently adopted political and economic innovations may be found in the demands of these early and so-called "radical" groups. The proposal for a graduated income tax and for labor legislation is to be seen in the Greenback platform of 1888 and in that of the Populist of 1892, while the latter includes also the demand for postal savings banks. The more aggressive and hopeful democracy of the west has sponsored the recent advances along this line. The Greenbackers advocated the reduction in the terms of United States senators, while the Populists demanded the Australian ballot, a single term for President and Vice-president, direct election of United States senators, and commended the initiative and referendum. The Progressive movement of 1912 drew from much the same groups that had voted the Populist ticket, but went even farther than that party in its advocacy of democratic innovations when it supported direct primaries, including preferential presidential primaries, the short ballot, the initiative, referendum, and recall (the latter not alone of elected officials, but of judicial decisions), and woman suffrage. The advocacy of government ownership of public utilities and the hostility to the trusts of the earlier radical parties was toned down by the Progressives, however, to merely strict governmental supervision.

These demands are progressive or radical according to the point of view, but enough has undoubtedly been said to point out clearly that the desire for economic and political changes has its home chiefly in the west. The consciousness of economic bondage to the east has been a cause of continual unrest and has combined with the more democratic influences of frontier life to promote changes toward democracy.

NOTES FOR FURTHER REFERENCE

The history and significance of the westward movement since the Civil War has never been adequately summarized, although much has been done since F. J. Turner in 1893 pointed out the *Significance of the Frontier in American History* in the American Historical Association Annual Report. Other essays of Professor Turner, especially those in *The Frontier in American History* (1921), touch on this recent phase. The most complete bibliography of the period is in F. J. Turner and F. Merk, *List of References on the History of the West* (rev. ed., 1922). The earliest phases of the trans-Mississippi west are developed in Katherine Coman, *Economic Beginnings of the Far West* (2 vols., 1912), and the more recent in F. L. Paxson, *The Last American Frontier* (1910). See also F. L. Paxson, *The Pacific Railroads and the Disappearance of the Frontier*, Annual Report of the American Historical Association (1907, Vol. I, pp. 105-118); *The Cow Country*, Am. Hist. Rev., Vol. XXII, pp. 65-82 (October, 1916); Edward Everett Dale, *The Ranchman's Last Frontier*, Mississippi Historical Review, Vol. X, No. 1, pp. 34-46 (June, 1923). On the public-land policy consult McLoughlin and Hart, Cyclopædia of American Government (1914), articles on Public Land, Land Grants, Homestead Act, etc.; also G. M. Stephenson, *The Political History of the Public Lands from 1840 to 1862* (1917); T. Donaldson, *Public Domain* (1881), inaccurate, but the only available detailed account; the *Public Land Report* (1880) and the *Report with Appendix* (1905). Two recent popular accounts are to be found in the Chronicles of America: S. E. White, *The Forty-Niners* (1920), and Emerson Hough, *The Passing of the Frontier* (1918).

The farmers' protest is interpreted in F. L. McVey, *The Populist Movement* (1896); in S. J. Buck, *The Granger Movement* (1913), Harvard Historical Studies XIX, and *The Agrarian Crusade* (1920), Chronicles of America; opposed in A. A. Bruce, *Nonpartisan League* (1921) and championed in H. E. Gaston, *The Nonpartisan League* (1920), in C. E. Russell, *The Story of the Nonpartisan League; a Chapter in American Evolution* (1920) and in Arthur Capper, *The Agricultural Bloc* (1922).

The Indian wars are recounted in N. A. Miles, *Serving the Republic* (1911). Material on the contribution of the railroads to the opening of the west may be found in H. K. White, *Union Pacific Railway* (1898); J. P. Davis, *Union Pacific Railway* (1894); E. V. Smalley, *The Northern Pacific Railway* (1883); in the *Memoirs of Henry Villard* (2 vols., 1904); in E. P. Oberholtzer, *Jay Cooke* (2 vols., 1907), and in J. G. Pyle, *Life of James J. Hill* (2 vols., 1917).

SELECTED READINGS

PAXSON, F. L., *The Last Frontier*, Chaps. II, IX, X, XI, XIX, XXII.
PAXSON, F. L., *The Cow Country* in the American Historical Review, Vol. XXII, pp. 65-82 (October, 1916).
BUCK, S. J., *The Agrarian Crusade*.
HOUGH, E., *The Passing of the Frontier*.
WHITE, S. E., *The Forty-Niners*.

CHAPTER XIX

THE AGRARIAN REVOLUTION

Outstanding Factors Since 1860.—So momentous have been the developments of American agriculture since the Civil War that nothing less than an agrarian revolution has taken place. The last half century has witnessed both the introduction of agricultural machinery on a large scale and the increased adoption of scientific farming. It has seen the rapid growth of government interest and aid to agriculture and a widespread movement toward agricultural education. Spurred on by the Homestead Act and by migration from Europe, the frontier has been pushed westward until most of the arable land has been preempted. But with the end of the frontier, interest has been stimulated in conservation and in adding new lands to cultivation through dry farming and irrigation projects. This westward expansion, however, has been accompanied by circumstances which have brought hardship and discontent to the farmers and which have been reflected in political and economic unrest. With rising prices of products and land values after 1896, and with better roads, electric trolleys, automobiles, and farm machinery, the economic condition of the farmer has improved and rural life become more satisfying. Nevertheless, these years have seen a constant increase in urban population and manufacturing, and a relative decline in agriculture.

Agrarian Discontent.—Although the period since the Civil War has been one of great agricultural expansion, it has not been marked by uninterrupted prosperity; on the contrary, the years from 1867 to 1897 were years of uncertainty and discontent. During the flush period of the war when prices soared, owing to greater demand for foodstuffs and an inflated currency, many farmers extended their operations by increasing their holdings and equipment. Ex-soldiers, tradesmen, and mechanics, encouraged by the Homestead Act, hastened to take up land; but in all of these cases capital was usually lacking and the land was mortgaged to provide for the necessary equipment. All went well until the inflated war prices collapsed. The government's policy of calling in some of the greenbacks and ultimately raising the paper currency to a parity with gold put the farmers at a disadvantage. Unable to meet his interest payments, which continued at the old rate, while prices fell and the value of money increased, the farmer was often forced to see his mortgage foreclosed and the results of years of labor wiped out, with the option of going into industry, entering the ranks of the tenant farmer or agricultural laborer, or moving to the frontier. He felt strongly that eastern capital was benefiting from his misfortune. Even the fortunate farmer who was able to hang on during the lean years of falling prices had serious troubles. The railroads upon which he was dependent for the marketing of his products were often careless and inefficient, discriminating in their favors to industries at the expense of the agricultural sections. Where there was no discrimination, freight rates still seemed needlessly high to pay dividends on heavily watered stock; furthermore, the farmer felt that an undue share of the profits was taken by the middlemen and by the speculators on the grain and cotton exchanges. While he bore the hardships of a lonely and arduous frontier life, eastern capitalists took from him the profits of his toil. The feeling was especially bitter in those sections where the pinch was greatest, notably upon the wheat farms of the northwest. Minor elements in the prevailing unrest were the conflict between the land-hungry pioneer farmers and the cattle raisers; and the fraudulent methods employed by individuals and companies in obtaining large blocks of land.

In the south the situation was equally discouraging. Here the whole economic structure had crashed with the Civil War and was painfully being reconstructed on the ruins. Bankrupt planters, ignorant colored labor, and declining cotton prices were the elements out of which a new system must be erected. In the northeast, deflation and western competition severely affected the agricultural interests, accentuated the movement to the cities, and increased the area of the deserted farms. Throughout the country the general decline in land values was a factor in the agrarian

THE AGRARIAN REVOLUTION 425

discontent; for the American farmer, it must be remembered, is a land speculator as well as an agriculturist. Added to all this was the high tariff of the Civil War, continued during the years of peace, which aided the manufacturing interests, at the same time increasing the cost of living and jeopardizing the foreign market for foodstuffs.

Relief was sought (1) by pressure upon the government and (2) by coöperative attempts at buying and selling. The first notable legislative efforts were made under the leadership of the Patrons of Husbandry, an organization founded in 1867 by Oliver H. Kelley. The local chapters, or "granges," included in their membership both men and women and were intended to be principally social and educational in their purpose. Their activities soon extended into politics, and in the early 'seventies, the period of their greatest influence, the membership of the organization numbered 758,767 (1875) with over 19,000 granges. It was during these years that efforts were made by the states to control the railroads, and, upon their failure, by the federal government. These state railroad acts, usually spoken of as the "granger laws," were not necessarily the work of the granges, for the "granger movement" itself was a result of the agrarian discontent. Later the political interests of the farmers were directed by the Greenback party, the Farmers' Alliances, the Populist party, the Democratic party of 1896, and the Farmer's Non-Partisan Political League, founded in 1915. The most important results of the agrarian agitation, as reflected in national legislation, may be briefly summarized as follows:

I. Railroad control:
    State granger legislation;
    Interstate Commerce Act of 1887.
II. Agricultural education and research:
    Land Grant Act of 1862, its additions and amendments;
    Hatch Act of 1887;
    The creation of the Department of Agriculture with its subsidiary bureaus.

III. Financial legislation:
  Retention of the greenbacks;
  Bland-Allison Act of 1878 and the Sherman Silver Purchase Act of 1890;
  Federal Reserve Act of 1913;
  Federal Farm Loan Act of 1916.[1]

In addition to political pressure the farmers have sought relief by entering the field of business for themselves. Examples of these attempts are organizations for coöperative buying and selling, and farmers' insurance companies. The movement has been notably strong in the control of grain elevators, 4,000 of which, it is estimated, are owned by 400,000 farmers.

After the depression of 1893 agricultural conditions improved, due to the fact that the demand for agricultural products had again caught up with the supply. The first decade of the new century was a period of expansion and prosperity. The total value of the crops in 1899 was $2,998,704,412, and in 1909 was $5,487,161,223. The value of all farm property, including land, increased 100.5 per cent from 1900 to 1910. The value of the land alone increased from an average of $15.57 per acre in 1900 to $32.40 in 1910, or 108.1 per cent, a greater increase than has taken place in all the previous years since the discovery of America. The World War with its inflated prices brought even greater prosperity, but no group felt more keenly the subsequent depression than the farmer.

The New Machinery.—Virgin soil and scarcity of labor, those two forces which heretofore had directed the development of farm machinery, continued to be operative after the Civil War. The first great improvements in the plow, reaper, and thresher had already demonstrated their practicability before 1860, but it was not until the war period that the last two inventions came into wide use. It is correct to say, therefore, that the agricultural

---

[1] While agrarian agitation was an important, perhaps the most important, single influence, it was, of course, by no means the only one, especially in producing the financial legislation listed under III.

revolution in America, as far as machinery is concerned, has come in the last half century.

The climate in the wheat regions of the middle west necessitated rapid harvesting when the crop was ripe, and the amount planted was dependent upon the farmer's ability to harvest before the grain spoiled. Consequently the attention of inventors was directed most of all toward methods to speed up harvesting. Already in 1858 C. W. and W. W. Marsh had patented the "Marsh harvester," a reaping machine designed by means of an endless apron to deliver the grain upon a table, where two men could bind it. This reaper almost doubled the amount of grain that could be harvested in a given time. Even more important was the invention in 1878 by John F. Appleby of a "twine binder," a machine which took the place of the crude and unsatisfactory wire binders already in use, and increased eightfold the speed in harvesting. "The invention of the twine binder," says Professor Carver, "therefore, by increasing the amount which a farmer could harvest, increased by that precise amount the quantity which he could profitably grow. In other words, it was the twine binder more than any other single machine or implement that enabled the country to increase its production of grain, especially wheat, during this period. The per capita production of the country as a whole increased from about 5.6 bushels in 1860 to 9.2 bushels in 1880."[1] Further improvements have been made by the addition of a bundle carrier and in dry climates by a header. On the great wheat farms of the west are now to be found monster machines drawn by a score or more horses or propelled by gasoline tractors which cut, thresh, clean, sack, and weigh the grain without the touch of human hands.

Improvements in machinery for planting and cultivating appeared simultaneously with those for harvesting. During this period there came into use the straddle-row cultivator; the sulky plow; spring-tooth sulky harrows of various types; seeders that plant, cover, and fertilize at the same time. The lister, which

---

[1] Carver, T. N., *Principles of Rural Economics*, p. 99.

plows and plants the seed at the same time, was introduced in 1880. The mowing machine has been perfected, and improvements in haying have included the spring-tooth sulky rake and machines for loading, stacking, and baling. Hand shelling of corn gave way after 1850 to machine shelling. The failure of the hay crop several times in the 'eighties, when the dairying industry was being rapidly developed, directed attention to corn raising, and the combined work of many inventors resulted in a machine by which one man can cut and bind from six to ten acres a day. This enables the farmer to cut his fodder corn green with the juice still in the stock, and store it in the silo for winter food, whereas before he was often forced to leave it standing in the fields to dry.

As man power in industry had given way to machinery, so now in agriculture animal power was superseded by steam, electricity, and gasoline. On the large prairie farms steam tractors were used soon after the invention of steel plows, and at present it is possible to see steam tractors drawing a tandem of plows, harrows, and seeders which can prepare and plant strips twenty feet wide as they move along. More practical than steam, because less heavy and bulky, are the new gasoline tractors, which are hardly more than the union of the gasoline engine to the old horse-drawn machinery. Almost as revolutionary in its effects has been the use of the gasoline truck and pleasure car, which have brought the farmer into closer touch with urban life and thus facilitated both marketing and purchasing. Not only the automobile, but gasoline pumping and lighting outfits have helped to bring the advantages of the city to the farmer and to decrease the household drudgery. Where the farm is close to an electric supply, much of the smaller indoor machinery, such as milk separators, churns, and washing machines, can be operated by that power.

It is here possible merely to indicate the tremendous effects of machinery upon agriculture. It has been estimated that the efficiency of the average farm worker increased over four-fifths, and the average increase in the nine most important crops during the period 1830 to 1895 was nearly 500 per cent. In 1870, of those engaged in gainful occupations, the percentage in agriculture was 47.36, and in 1900 it was 35.7; yet agricultural production far outran the growth in population. The new machinery made possible larger farms. Between 1880 and 1900 in the seven leading cereal states the average farm acreage increased from 64.4 acres to 102.5, an increase of 59.2 per cent; while the average acreage of crops cultivated by one person in the same states increased from 40.6 acres to 62.4, an increase of 53.7 per cent.

The advent of farm machinery has raised the farmer's standard of living. His work has become less arduous and his economic condition, with the possible exception of the south, has improved. This economic betterment has included the agricultural laborer, whose real wages have steadily increased, although they have not kept pace with the income of the farm owner. The gradual concentration of wealth in the hands of the proprietor class has been accompanied in the seven leading cereal states by an increase of agricultural laborers greater relatively than the increase in proprietors.

SEVEN LEADING CEREAL STATES
(Illinois, Iowa, Kansas, Nebraska, Minnesota, North and South Dakota)

	1880	1900	Percentage of increase
Proprietors (owners or tenants)	836,967	1,073,911	28
Agricultural laborers	363,233	631,740	74

In the country as a whole, the percentage of increase for the two classes was about the same, due to the growth of tenant farming in the south. The man with capital was obviously at an advantage in the cereal states, where expensive machinery was becoming the order of the day. As a result, the poorer farmer was reduced to the status of tenancy or that of an agricultural laborer. The rise of a constantly growing landless agricultural proletariat has been the most unfortunate concomitant of the agricultural revolution, and is becoming characteristic of America as well as Europe. Radical groups like the Industrial Workers of the World have found ready adherents among this class.

Summarizing, it may be said that machinery on the farm has (1) released men for other work; (2) increased the production of agricultural products and the output per capita; (3) eliminated much drudgery from farm life; (4) allowed the cultivation for other purposes of many acres which had hitherto been used to produce fodder for horses; (5) enlarged the real income of proprietors and the real wages of laborers. On the other hand, the new machinery has undoubtedly (6) increased relatively the landless agricultural laborer and made it more difficult for the man without capital to engage in agriculture.

Agricultural Education.—The agrarian revolution has been hastened and to some extent directed by education. The decade of the 'fifties saw a rapidly growing interest in agricultural education which found vent in the establishment of several state agricultural schools. An impetus to the movement was given by the passage in 1862 of the Morrill Act. Introduced by Justin S. Morrill in 1857 and vetoed by President Buchanan, it was brought up again during the war and passed. The act provided that 30,000 acres of public land be given to each state for each senator and representative in Congress, the funds from the sale of these lands to be accumulated and the interest used to support, endow and maintain "at least one college where the leading object shall be, without excluding other scientific and classical studies, and including military tactics, to teach such branches of learning as are related to agriculture and the mechanic arts, in such manner as the legislatures of the states may, respectively, prescribe, in order to promote the liberal and practical education of the industrial classes in the several pursuits and professions of life." This first "land grant" act constituted the greatest single piece of legislation ever passed in the interest of agricultural education, and under its provisions there were gradually established institutions in each of the states and in Hawaii and Porto Rico. In some states the agricultural or mechanical schools are attached to the state universities or other colleges. In Massachusetts, the income was divided to help found two schools, the Massachusetts College of Agriculture and the Massachusetts Institute of Technology. There were sixty-eight land-grant colleges in

THE AGRARIAN REVOLUTION

1916 teaching agriculture. The Morrill Act has been extended by subsequent legislation, notably in 1890 and in 1907, when additional appropriations were voted to increase to $50,000 the annual income of each school subsidized by the government. The Hatch Act of 1887, which provided funds for experiment stations in the various state colleges, has turned the attention of these schools to investigation as well as to teaching.

Almost as valuable as the actual instruction given in the colleges is the diffusion of information among those not regularly attending. The scope of the agricultural colleges has been extended to include special short-term winter courses and extension work. The latter is carried on by correspondence, by publications, by lectures, by itinerant schools sometimes conducted in special trains, by farmers' institutes, and by coöperation wherever possible with farmers' organizations. The value of this extension work was recognized by Congress in the Lever Extension Act of 1914, under the provisions of which $480,000 was appropriated to be divided equally among the states, and in addition $600,000 was granted to be increased annually by $500,000 until 1923, when the annual appropriation of the national government for this purpose would amount to $4,500,000.

Agricultural education is carried on by the United States Department of Agriculture, which is engaged in the twofold task of experimentation and the dissemination of information. The latter is accomplished by means of more than a dozen publications, among which may be mentioned the *Year Book of Agriculture,* the *Farmer's Bulletin,* the *Journal of Agricultural Research,* the *Monthly Crop Reporter,* and the *Weekly News Letter.* The state agricultural departments, which exist in most of the states, function in a somewhat similar way. The educational influence of the county and state fairs is still potent. Hundreds of agricultural societies have grown up to promote knowledge and spread information on almost every conceivable phase of plant and animal culture. One or more of their organs or of the general farm journals, of which there are nearly 500 published, reach almost every farmer. Of these several boast of a circulation of over 500,000. Gradually agricultural instruction is being introduced into high schools, and in several states is required in the rural schools. While the channels for diffusion of scientific agricultural information are many, it is a discouraging truth that as yet but a small percentage of farmers are seriously influenced in their daily farming by the information thus available.

The Washington Bureaus.—The American farmer has not lacked aid from either the state or the national government. This is attributable to three causes. In the first place, the fundamental importance of agriculture has always been recognized. Although the census of 1920 places the value of agricultural products at only $21,000,000,000 and those of manufacturing at $62,000,000,000, the fact remains that more than half of the important manufacturing industries—as, for example, slaughtering and meat packing, milling, the production of cotton and woolen cloth, boots and shoes, and many others—are dependent upon agriculture. Farm products are also an important, in some sections the most important, item of railroad freight. Agriculture still remains the foundation of our economic life. In the second place, the farmer has exerted during most of our history a very potent influence upon the legislative branch of the government. As late as 1870 47 per cent of the gainfully employed population were farmers; although this proportion had fallen off in 1910 to 33, and in 1920 to 26 per cent, the fact that the industrial population is largely centralized has given the farmer an especial weight in the upper house, where southern and western senators are extremely susceptible to the demands of agriculture. In the House the so-called "agricultural bloc" is quick to coalesce when the farmers' interests are at stake. In the third place, the policy of *laissez-faire,* so strong during the first decades of the Industrial Revolution, has been gradually breaking down, and nowhere has this change of attitude been more apparent than in the relations of the government to agriculture. This has been due not alone to the political strength of the farmer, but to a realization of the economic importance of agriculture and to the fact that the farmer has been at a disadvantage in his dealings with other economic groups, and so needed special protection. Consequently, government aid has taken three forms: first, educational and ad-

THE AGRARIAN REVOLUTION 433

visory; second, protection by legislation against other groups; and third, help in reclamation and irrigation.

Some mention has already been made of government aid to education and the scientific study of agricultural problems. The work of the schools and experiment stations is augmented and to a certain extent directed by the activities of the Department of Agriculture. George Washington as President had recommended a governmental board, but it was not until 1839 that Congress voted $1,000 to the Commissioner of Patents for the "collection of agricultural statistics and other agricultural purposes." In 1862 these activities were removed from the Patent Office and a Commissioner of Agriculture was created to direct a bureau whose duty it was "to acquire and diffuse among the people of the United States useful information on subjects connected with agriculture in the most general and comprehensive sense of the word, and to procure, propagate, and distribute among the people new and valuable seeds and plants." In the year 1889 this bureau was elevated to the rank of the other departments, and its head made secretary with a cabinet position.

The activities of the Department of Agriculture have extended into many fields and gradually divided into separate bureaus. The Weather Bureau, whose duties in 1891 were transferred from the Department of War, carries on investigations in meteorology, climatology, and seismology, and from numerous observation points collects data by which weather is forecast. The Bureau of Animal Industry has charge of meat inspection and animal quarantine. It has done remarkable work in studying and checking such animal diseases as cattle fever, pleuro-pneumonia, and hoof and mouth disease. The Bureau of Plant Industry is engaged in combating plant disease, in studying better agricultural methods and plant acclimatization, in seed distribution, and in similar lines. More than 30,000 new plants have been brought into the country through its agency, notably Kaffir corn, durum wheat, and drought-proof alfalfa. Closely allied to the last named bureau is the Bureau of Plant Entomology. Its work is to study insects and thus direct the work of combating pests and of introducing the beneficial insects. Campaigns have been waged against the Hessian fly, the gypsy and brown-tail mouth, and the boll weevil. The federal bird and game preserves are administrated by the Bureau of Biological Survey, which also studies and protects wild animals and experiments in the elimination of animal pests. The Bureau of Soils investigates the chemical and physical properties of soils, studies fertilizers, and makes soil surveys. Other chemical investigations, particularly in foodstuffs, have been carried out by the Bureau of Chemistry, which is also engaged in enforcing the Food and Drug Act. Statistics of crops and livestock are gathered and published monthly by the Bureau of Crop Estimates. Valuable work has been done by the Bureau of Markets in studying the problems of transportation, distribution, warehousing, grading, coöperative marketing and purchasing, rural credits, and other factors involved in the handling of farm products as they move from the field to the consumer. In a similar manner such farm problems as labor, finance, and housing are investigated by the Office of Farm Management.[1] This work is supplemented by the Office of Home Economics, which is concerned with questions of food, household management, and general family welfare. The protection and development of the national forests are handled by the Forest Survey, which studies tree life and carries on educational work in the scientific treatment and preservation of trees. A Bureau of Public Roads and Road Engineering has been created to carry on investigations in road construction and maintenance, farm irrigation and drainage, and other engineering problems. The Smith-Lever Act of 1914 authorized a States' Relation Service under the direction of the Department of Agriculture, to supervise the funds voted to the state colleges.

On a much smaller scale most of the states, through their departments of agriculture, financial appropriations, and protective legislation, have sought to aid agriculture after the manner of the federal government. Some states have gone to the extent of offering subsidies to encourage the production of certain agricultural products, as, for example, the efforts of Kansas to promote

---

[1] The Bureaus of Markets and Crop Estimates and the Offices of Farm Management and Farm Economics were combined on July 1, 1922 into the Bureau of Agricultural Economics.

the growing of beet sugar. Perhaps the crowning example of state resources applied to the interests of the farmer was the sanction by the North Dakota legislature in 1919 of the complete program of the Farmer's Non-Partisan Political League, which called among other things for state-owned flour mills and terminal elevators, a state owned and operated bank, and state loans to home builders and land purchasers. These were to be financed by state credit and supervised by the Governor, the Attorney General, and the Commissioner of Labor and Agriculture.

Government Aid—Protection Against Other Economic Groups.—Protection of the farmer against other powerful interests has been afforded: (1) by laws to regulate and control public carriers; and (2) by legislation designed to aid the farmer in his financial operations. Both types of legislation are a distinct outgrowth of the agrarian discontent of the decades after the Civil War. The farmer is peculiarly dependent upon the railroads. The railroad may have opened up the country in which he lives, it may have sold him his farm, and later the railroad transports his products to market. The efficiency of the roads and the freight rates largely determine his profits. The farmer was naturally sensitive to unjust rate discriminations and resented his utter dependence upon a group of non-resident railroad owners. This hostility of the farmers to the railroads led between 1870 and 1880 to the "granger laws," which created railroad commissions to study and, in some cases, to regulate rates. The courts were at first inclined to support the validity of these laws, but eventually extracted their teeth (1) by upholding the right of judicial review as to the reasonableness of a rate, and (2) by limiting the activities of the commissions to purely intrastate commerce. Indignant at what they considered the subserviency of the courts to the railroads, the farmers pressed for federal supervision and obtained it in the Interstate Commerce Act of 1887 and its various amendments and additions.[1] Government regulation has safeguarded the interests of the farmer as regards rates, but has naturally not solved his difficulties with inefficient freight handling.

The voice of the farmer during the whole course of our history

---
[1] See chap. xx.

has been raised against the financial system, which he has looked upon as advantageous to the capitalist and oppressive to the debtor. The agricultural group, normally a debtor class, opposed the First and Second United States Banks, brought about the downfall of the latter, opposed greenback contraction, favored paying the Civil War debt where possible in paper, advocated the free and unlimited coinage of silver at the ratio of 16 to 1, and favored generally an expanded currency and more liberal banking laws. The farmers criticized the national-bank system on two grounds: first, that it did not adequately serve small communities, because the minimum capital required was too great; and second, it encouraged the flow of accumulated capital from the country to the city, where it was used for industry or speculation. The Federal Reserve system, inaugurated by a Democratic administration in 1913, was expected to be of particular aid to the agricultural class in the greater facilities afforded for the expansion of the currency when needed, and the more rapid movement of funds from one section of the country to another. For the first time under this act national banks were permitted to loan money on farm mortgages. Furthermore, agricultural paper running six months could be rediscounted at the Federal Reserve bank, while commercial paper, to be eligible for rediscount, must mature within three months.

More direct in its application to the farmers was the Federal Farm Loan Bank system, originated by an act of July 17, 1916, and designed to eliminate some of the chief disadvantages under which the farmer labored. The commercial banks loaning for thirty, sixty, or ninety days were inadequate to the needs of the farmer, who must borrow from seed time to harvest, a period of six months or more. To supply this need, private institutions had grown up to loan money on farm mortgages, but the interest rates were high. If the loan was negotiated through the local merchant, as was quite common in the south, the borrower was too much under the latter's control. Some efforts had already been made by the states to extricate the farmer from his dependence upon these forms of borrowing, but it was left to the federal government to make the first important move.

The Federal Farm Loan Act of 1916 provides (1) for a Federal Farm Loan Board to administer the system and report to Congress concerning its operation, to be composed of the Secretary of the Treasury and four other members appointed by the President; (2) the establishment of Federal Land Banks in twelve sections of the country to be located in Springfield, Massachusetts; Baltimore, Maryland; Columbia, South Carolina; Louisville, Kentucky; New Orleans, Louisiana; St. Louis, Missouri; St. Paul, Minnesota; Omaha, Nebraska; Wichita, Kansas; Houston, Texas; Berkeley, California; Spokane, Washington; and Washington, D. C. These banks are chartered under the terms of the act and are managed by a board of directors, six elected by the Farm Loan Associations and three appointed by the Federal Farm Loan Boards. The capital for each of these banks, which was to be not less than $750,000, might be subscribed by the public with the provision that after thirty days from the opening of the subscription books the balance should be taken up by the national government. Of the $9,000,000 required by the twelve banks, the United States subscribed $8,891,-270. These banks are to loan money to the farmers, the loans not to exceed 50 per cent of the value of the land and 20 per cent of the value of the permanent improvements, for the usual agricultural purposes—purchase of lands, equipment, livestock, etc.

In addition to the Federal Farm Loan Board and the Federal Farm Loan Banks, a third organization was called for by the act, namely the National Farm Loan Associations. Ten or more persons, owners or prospective owners of farm lands upon which mortgages may be placed, form an Association managed by a board of five directors. Shares in the Association have a par value of $5, to be deposited in the Farm Loan Bank; the member may borrow from $100 to $10,000, taking out in the Farm Loan Bank one share for each $100 which he desires to borrow. These shares are retained to form a fund to cover bad debts. The borrower offers his farm as security and the loan is negotiated through the secretary-treasurer of the Association from the Federal Loan Bank of the district. The Federal Farm Loan Bank itself secures its money chiefly from the sale of bonds secured by mortgages. Proceeds from the sale of bonds may be used to buy more mortgages from the Farm Loan Associations, a process which may be repeated until the amount of the outstanding bonds is equal to twenty times the capital of the Farm Loan Bank. The borrower arranges to liquidate his debt and pay for interest and administration costs by annual or semiannual payments. Joint Stock Land Banks, similar to the Federal Farm Loan Banks, were authorized by the same act, to be incorporated by private individuals; they obtain funds by selling on the market first-mortgage, tax-exempt farm-loan bonds. These banks may deal directly with the farmer rather than through associations, may loan for any purpose and in excess of $10,000.

Even the credit facilities provided by the legislation already described did not prove entirely satisfactory and they were supplemented by the Agricultural Credits Act of March 4, 1923. This act was designed to aid the agricultural and livestock industries of the country by establishing twelve Federal Intermediate Credit Banks (as adjuncts to the existing Federal Farm Loan Banks), which do not deal with individual borrowers or lend directly on land security, but are banks of rediscount of agricultural and livestock paper for periods of six months to three years. Five millions of capital for these banks was furnished by the United States Treasury. The act also established National Agricultural Credit Associations, without government capital or limitation in numbers, to do business directly with the public.

Under this system the farmer may obtain his money at close to the market rate for the period that he wants it. It promises escape from the "loan shark" and seems likely to increase the morale of the borrower by the method of gradual amortization which it provides. Furthermore, the Federal Farm Loan Associations in their financial dealings are not without their educational advantages to the farmer.

Irrigation.—Not alone has the national policy been liberal with regard to arable land, but it has aided in the reclamation of desert areas. Under the Desert Land Act of 1877, permission was granted to take up 640 acres (reduced in 1891 to 320 acres) at 25 cents an acre, with the additional payment of $1 per acre within three years, the theory being that successful irrigation necessitated larger grants and that the postponement of final payment would encourage reclamation schemes. Little was accomplished by this act but the stimulation of land frauds, and it was supplemented in 1894 by the Carey Act. Under the Carey Act states in the arid region might appropriate 1,000,000 acres of public land and authorize irrigation through private enterprises; but must reserve authority to pass on the plans submitted and the charges for water rates. The land was sold at 50 cents an acre, and the water rights of these irrigation projects have averaged from $30 to $40 an acre, paid for usually in ten annual installments; the irrigation companies retained the control of reservoirs, dams, and other equipment until full payment had been made, when control was turned over to the landowners.

Artificial irrigation, however, proceeded slowly, due chiefly to the large amount of capital required. Partly because of this and partly because of the increasing interest in conservation, the Reclamation Act of 1902 was passed. It provided for the setting aside of proceeds from the sale of public lands in sixteen designated states to be used as a fund for irrigation projects. When money is available, the Secretary of the Interior may award contracts for such works. The farmers, who have taken up the land either by purchase or through the Homestead Act, defray the cost of the work, usually by annual payments, thus perpetuating the fund. Under the Reclamation Act of 1902 approximately $120,000,000 had been spent up to 1919 in the examination, construction, and operation of works for the reclamation by irrigation of arid regions, and a million acres were under actual cultivation. At least twenty-five separate projects were then in operation, the most notable being the Roosevelt dam in Arizona, distributing water to the valley lands in the vicinity of Phœnix; the Arrowrock dam in Idaho; the Elephant Butte dam in New Mexico; and the Gunnison tunnel in Colorado. The total acreage under irrigation in the United States in 1880 was considerably under a million acres, by 1890 it had increased to approximately 4,000,000, by 1900 to 8,000,000, by 1910 to 15,000,000, and by 1920 to 19,000,000. Notwithstanding the continued government aid, nearly four-fifths of the irrigation has been accomplished by private initiative.

Scientific Farming.—Nothing has been more significant of the agrarian revolution in America than the advent of scientific agriculture. The beginning of scientific farming in England took place in the eighteenth century, but the abundance of unoccupied land and of rich virgin soil, as well as the scarcity of labor, held back its advent here. Although individual farmers as early as Washington had sought to improve methods, the typical American farmer continued to "butcher" the land and neglect his livestock in the careless methods of earlier years. The era of scientific farming came in after the Civil War—in fact, largely during the past thirty years. Among the chief influences which have promoted it are: (1) state aid in research and experimentation; (2) agricultural education; (3) the gradual disappearance of unoccupied arable land; (4) new machinery; and (5) greater markets.

The contributions of state and national governments through the departments of agriculture, the experiment stations, the colleges of agriculture and the educational and extension work carried on by all these agencies, have been most important and have been discussed elsewhere. Studies by private individuals, of whom Luther Burbank is perhaps the most famous, have supplemented their work. These studies have been carried on especially in the fields of animal husbandry, botany, bacteriology, and chemistry. Studies in animal husbandry have aimed not only at general improvement, but also at the production of stock for special purposes; distinct strains of cattle, for instance, have been developed for cream and butter, for quantity of milk, and for beef. Studies in plant breeding have resulted in improved species of fruit and vegetables as well as in new varieties. Scientific bacteriology and plant pathology have discovered the nature of many animal diseases and plant blights and provided the cure; beneficent bacteria have also been discovered which have been used either to fight destructive bacteria or to renew the soil. Entomologists in the same way have discovered insects which are the enemies of various pests. Chemists in their studies of soils were able to determine what ingredients were lacking and to prescribe the proper fertilizer for the specific crop. Until well into the period under discussion, fertilizer was not used extensively by the American farmer, for it was easier to cultivate new land than to renew the old. The value of artificial fertilizer sold in 1859 was $891,344; the output of the factories in 1909 amounted to over $100,000,000. It is estimated that the manure supply is worth many times that amount annually, but from one-fourth to one-half of it is wasted. The use of this type of fertilizer has been encouraged by the invention of manure carriers and manure spreaders.

The first agricultural experiment station in the United States was established under the direction of Professor W. O. Atwater in 1875 at Wesleyan University, Middletown, Connecticut, through appropriations of the state and donations made by Orange Judd, proprietor of the *American Agriculturist*. The notable work accomplished here encouraged Congress to pass the Hatch Act of 1887, under the terms of which experiment stations have been established in each of the states, where scientists quite often specialize on some problem connected with their particular section; as, for example, diseases and improvement of the cotton plant in Alabama, the pineapple in Florida, the proper feeding of cattle in Texas, new varieties of sugar cane in Louisiana, rust-resisting wheat in Minnesota, diseases of potatoes in Vermont. Agricultural research has increased by hundreds of millions the annual value of the crops, but there is still much to be done in this direction. The average yield of wheat per acre in America is but half that of England, Germany, or Holland, although the fields in those countries have been cultivated for centuries.

Especially interesting have been the effects on scientific agriculture of the pressure of population on land. As the arable land was occupied settlers pressed westward into the semi-arid country between the region of adequate rainfall and the Rockies. This settlement was stimulated by a series of wet years in the early 'eighties. In later years, when normal weather returned, a partial solution for the lack of rainfall was realized in dry farming.

The principles of dry farming call for the deep plowing of the land after harvest, the deep disking after each rainfall, the pulverizing of the top soil, and the keeping of it free from weeds, and in alternate years tilling through the summer without raising a crop—all of these expedients to lessen evaporation. There are about 300,000 acres still available for dry farming, but already population has pressed beyond the semi-arid lands and into the regions where artificial irrigation is necessary.

Recent Agricultural Tendencies.—Speaking broadly, the history of American agriculture until about 1890 was chiefly the story of the westward movement and the continuous opening of new land for speculation and production. In detail this story has been modified by the invention of new machinery, by the necessary adaptation of crops to new soil, and by the gradual shifting of production to new regions whose superiority severely handicapped older communities. The placing of large areas of new land under cultivation has in turn been attended by a decline in agricultural prices and by the production of a surplus for exportation.

With the opening of the twentieth century, American agriculture enters a new period in its history radically different from the old. As most of the land immediately available for farming had been preëmpted by 1900, further additions to arable land must come from irrigation or drainage or from putting into use woodland or other unimproved land on the farm. While the farm area increased by 15,000,000 acres a year for the thirty years previous to 1900, the increase from 1900 to 1910 was but 4,000,000 acres a year, an addition of 4.8 per cent. The decade 1910 to 1920, however, shows an increase in farm acreage of 8.8 per cent and of improved land of 5.1 per cent, figures which must be accounted for by the unusual stimulus of the war. The percentage of the land area of the United States included in farms has increased from 44.1 in 1900 to 46.2 in 1910 and 50.2 in 1920. On the other hand, the percentage of improved land on farms has slightly decreased from 54.4 in 1910 to 52.6 in 1920, indicating a tendency which is probably temporary.

American agriculture to 1910 was extensive in its nature and there was no hesitancy to sacrifice land to labor. The farmer resembled the miner who took out riches from the soil without giving anything in return. The increasing value of farm products and of land has turned the farmer to more intensive agriculture. This movement seemed well under way during the first decade of the century, but has not been so pronounced during the past ten years. Nevertheless, large areas of corn, barley, and buckwheat have been turned over to the growing of more intensive crops, and the increasing use of fertilizer demonstrates more careful husbandry. While the production percentages in the 1920 census do not keep pace with the increase in acreage devoted to many of the products, this may not by any means mean less intensive farming. Considering the poorer land brought under cultivation, the case may be just the opposite.

An outstanding factor in agricultural life has been the decided movement toward the cities; neither rural population nor farm products have advanced as rapidly as the increase in urban population. As late as 1910 the majority of the population was rural, but in 1920 urban population (that is, those living in towns of 2,500 or over) amounted to 51.4 per cent of the total. In 1870 about 47.36 per cent of those over ten years of age gainfully employed were in agriculture; in 1910, 32.9; and in 1920, only 26.3. During the past decade the rural population increased only 3.2 per cent, and the urban 28.8, and this during years in which immigration was greatly restricted. To put it differently, population in the United States increased 13,000,000 between 1910 and 1920, and of that increase the cities got 11,000,000 and the country 2,000,000. At the same time the agricultural wealth of the country, which was 56 per cent of the total in 1850, declined to 25 per cent in 1890 and to 21.8 in 1912.

In production the farms have lagged behind the needs of the urban population. The per capita acreage of improved land in 1880 was 5.7 and in 1920 was 4.8. Until about 1895 the production of food gained upon population; since that time population has overtaken food supply. Formerly we exported foodstuffs heavily, but now we produce largely for our own needs and import about as much food as we export.

The natural result of many of these factors enumerated has been a sudden and enormous rise in land values and prices. Between 1900 and 1910 the price of agricultural products increased 46 per cent and the average value of the crops 66.8. At the same time farm lands jumped in value 100.5 per cent, more than during our entire previous history, with another increase of 90.1 per cent in the decade 1910 to 1920.

Since 1900 there has been an increase in tenancy, but not so marked as during the final decades of the last century. In the raising of those products that require expensive machinery, and on those lands whose value has risen rapidly, tenancy has increased because of the inability of the poor man to meet such expenses. On the other hand, it should be pointed out, the growth of tenancy may be an encouraging sign, for to many a man tenancy is an intermediate step between being a farm laborer and an owner.[1]

INCREASE OF TENANCY IN THE UNITED STATES, 1880–1920 (PER CENT)

	1880	1890	1900	1910	1920
North Atlantic States	16.0	18.4	20.8	18.2	17.1
South Atlantic States	36.1	38.5	44.2	45.9	46.7
North Central States	20.5	33.4	27.9	28.9	31.2
South Central States	36.2	38.5	48.6	51.7	51.2
Western States	14.0	12.1	16.6	14.1	17.7
United States	25.5	28.4	35.3	37.0	38.1

Recent Agricultural Tendencies—The Northeast.—Influences which were at work even before the Civil War to modify agricultural conditions in the northeast have since been operative in an increasingly greater degree. A century ago New England and the middle Atlantic states were agriculturally self-sufficient.

---

[1] There are a few who think that farming in the future is to be like industry, a business to be carried on by large corporations with hired labor working at the height of the season in night and day shifts as in factories. This counsel of despair comes from those who do not see how agriculture as a whole can carry the shiftless marginal farmer who does not and seemingly will not treat farming as a business enterprise.

THE AGRARIAN REVOLUTION

To-day New England imports 80 per cent of her food. Competition of western products, made possible through the development of transportation facilities, and the growth of industrial life, have radically altered the nature of the agricultural products. Livestock raising for wool and meat has given place to dairying and to vegetable and fruit growing. Where this type of farming is impossible the competition of the west has quite often forced out of cultivation the poorer land. In New England between 1860 and 1910 farm land under cultivation decreased by over 5,000,000 acres, or 42 per cent, resulting in the thousands of deserted farms to be seen in this section. During these years cattle decreased from 56 to 20 per 100 of the population, and sheep from 60 to 4. While rural population declined relatively, the whole population increased by 110 per cent. The richer soil in New York, Pennsylvania, and New Jersey has kept a greater amount of land under cultivation than in New England, but the transition to vegetables, dairy products, and fruit has been the same.

Recent Agricultural Tendencies—The South.—Before the Civil War the south was almost entirely agricultural, engaged in raising some foodstuffs, but dependent chiefly upon the great staple cotton, which was produced upon large plantations by the labor of negro slaves. The war and the freeing of the slaves changed the whole system. The products remained the same, but the manner of production was altered. Ruined by the war, the great planter had neither resources nor equipment to continue the old plantations under the wage system. On the other hand, some manner of livelihood had to be found for the new freeman. The result was the gradual breaking up of the large holdings into small farms ranging from twenty to fifty acres, which are operated by negroes, usually as tenant farmers. While in some cases (about one-fourth) cash rentals are paid, in most cases the farms are let out under a system by which the owner furnishes the tools and sometimes the seeds and a mule, and in return takes one-half of the corn and cotton raised.

The effect of this tenant farming on the part of the more or less shiftless blacks has been almost as destructive to the soil as that of the old plantation system. There was actually less improved land in the south Atlantic division in 1900 than in 1860. Not only has this tenant farming been ruinous to the soil, but it has tended to perpetuate the one-crop system. The owner or local merchant who provides tools and supplies and takes in return a lien upon the future crop insists that it be cotton, as that is the product most salable. It is in fact the crop which the negro has been trained for generations to raise, and the simplicity of its production, as in days of slavery, fits in with the stage of his agricultural advancement. On the other hand, it must be admitted that the system of tenant farming has not worked wholly in favor of the negro. While it might be an overstatement to say that slavery in the south was followed by a period of serfdom in which the negro by being constantly in debt to the landowner or cotton factor was held in bondage, it would not be far from the truth. This has been a normal evolution in society—a development made necessary after the war by the economic situation and the intellectual and moral status of the negro at that time. At the height of the tenant farm system, conditions were thus described:

The agricultural land of the cotton states has little sale. Merchants will ordinarily not accept it as security for debt unless they are compelled to do so when crops, mules, cattle, and other personal property are insufficient. This is one reason why mortgages on Southern farm land are so few. The blacks prefer a tenancy to selling their labor for wages; and in some regions, at least, the white owners who cultivate their farms find that only the inferior laborers can be hired, because the superior ones prefer tenancies. As the planters become independent of the merchants, they are unfriendly to these tenancies, but, in some instances, have to grant very small ones in order to hold the services of the blacks, who, under such circumstances, work for wages during a part of the year on the plantation cultivated by their landlord. If the white landlords arrive at independence from debt before the black tenants do—as it may be assumed they will—if either class is to improve, it seems likely that the blacks will see a service for wages encroaching on the tenant system. . . .

The plantation owners, most of whom are landlords, often live in towns, having abandoned their plantations to irresponsible tenants, who care to work only indifferently and for a bare subsistence of the poorest sort. A tenant whose crop by chance more than suffices to meet his obligations, will pick enough cotton to discharge his debts to the landlord and merchant, and abandon the remainder, knowing that he can live on the next crop until it is harvested. The merchant who has a lien on his share of the crop pays his taxes, buries his wife or child, buys him a mule if he needs one, and feeds and clothes him and his family.[1]

Though this system seems destined to live for a long time to come, the increasing number of negro owners bespeaks eventually a different future. It is estimated that over 200,000 negroes own their own farms aggregating 20,000,000 acres and valued at over $500,000,000. But the tenant system for both blacks and whites is still predominant, with the yield per acre discouragingly small and the average annual income of the tenant farmer only about $150. The methods are still crude and wasteful, and so little cereal and meat is raised that this agricultural region still imports foodstuff. The salvation of the southern farmer has been the usually steady demand and the high price paid for cotton. The cotton crop has increased from 4,490,000 bales (500 pounds) in 1861, to 13,440,000 in 1920, with the south still producing from 60 to 65 per cent of the world's yield. This supremacy has recently been challenged by the northward advance of the boll weevil; but this has not been an unmixed curse, for it has forced certain sections of the south to turn from one staple to diversified farming.

The center of cotton production continues to be west of the Alleghanies, with Texas as the largest producer; the same is true of tobacco also, with chief centers Kentucky, Tennessee, Mississippi, and Alabama. Sweet potatoes, peanuts, and semi-tropical fruits are grown in this section in considerable quantities. The potential agricultural possibilities of the south are enormous and it is the belief of many students of agriculture that the greatest progress in the future along this line will be made south of the Ohio and the Mason and Dixon line.

---

[1] Holmes, G. K., *Annals of the American Academy of Political and Social Science*, September, 1893. See also article by same author on *Tenancy in the South*, reprinted in Carver's *Selected Readings in Rural Economics*, pp. 494, 495.

PRODUCTION OF COTTON IN THE UNITED STATES BY 500-POUND BALES, 1865-1921 [1]

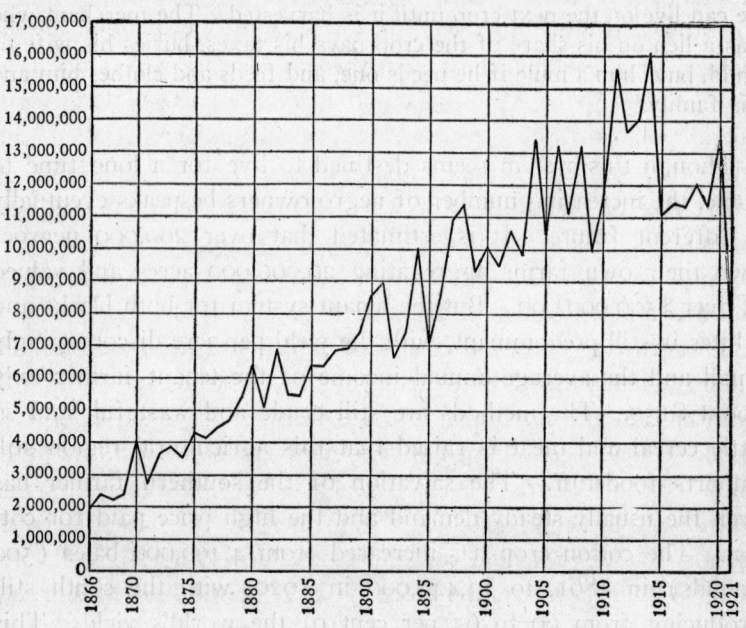

Recent Agricultural Tendencies—The North Central States.—By 1890 the frontier line of farms had pushed west until practically all of the arable land had been occupied. In this great region between the Alleghanies and the Rockies has arisen the most productive agricultural area in the world. Its chief products are corn, wheat, and livestock. The central states from Ohio to Iowa were found well adapted to corn, and its production has laid the foundation for the raising of hogs, cattle, and poultry. While the center of corn production since 1860 has to a certain extent remained fixed, the center of wheat has moved steadily west. In 1860 most of the wheat was grown east of the Mississippi, but by 1890 there were great centers in Missouri, Nebraska, Kansas, Oklahoma, Minnesota, the Dakotas,

---

[1] *Statistical Abstract*, 1921, p. 864.

Washington, and California. The westward advance was accelerated by certain inventions in flour manufacture, notably the "middlings purifier" of La Croix and the substitution of rollers for stones in crushing the grain. These inventions allowed the making of fine white flour from the hard spring wheat, the type best grown on the Minnesota and Dakota prairies. Under the liberal policy of the government large claims were staked out, and often through actual fraud immense farms were built up where the soil was mined for wheat alone by the wonderful new machinery. These "bonanza farms" of a thousand acres or more not only wore out the soil, but they glutted the market with wheat and drove down prices.

The growing value of farm land which has necessitated more intensive cultivation, the wearing out of the land, and the immigration into the northwest of new farmers have all helped to break up the large holdings and in many cases to substitute other kinds of agriculture. In Wisconsin, where twenty years ago wheat was the chief product, the farmers have turned to other cereals, to livestock, and to dairying. Though not so well situated, Wisconsin has taken from New York the lead in dairying and now produces one-fourth of the butter, cheese, and condensed milk. At the same time the competition of Canadian wheat is turning the attention of the farmers of Minnesota and the Dakotas to the same products. As the reaper and binder have made possible the great wheat farms of the west, no less have the Babcock tester, the power churn and mixer, made possible the rapid expansion in butter and cheese making.

Recent Agricultural Tendencies—The Far West.—It was the gold stampede of 1849 that first brought large numbers to the Pacific coast, but the great wealth of the region was destined to be agricultural. The mild and even climate of California has made it ideal for fruit and meat products, as well as for wheat. In Oregon and Washington, where there is a variety of climate, agriculture is not specialized, but fruit, wheat, dairy products, and wool are all produced in considerable quantities. Farther east the valleys and foothills of the Cordilleras, which hitherto had been supposed unarable, have been brought under cultivation by dry farming or irrigation. The introduction of durum wheat, Kaffir corn and a different type of alfalfa, all suitable to dry soil or cold climate, have helped to vegetate these regions.

AGRICULTURE BY SECTIONS, 1920 [1]

	Rural population	Number of farms	Value of farm property	Value of farm crops
New England........	1,535,836	156,564	$ 1,173,019,594	$ 275,175,536
Middle Atlantic.......	5,588,549	425,147	3,949,684,183	914,499,927
East North Central...	8,426,271	1,084,744	17,245,362,593	2,818,367,792
West North Central...	7,816,877	1,096,951	27,991,434,545	3,676,902,149
South Atlantic........	9,651,480	1,158,976	6,132,917,760	2,083,808,429
East South Central...	6,899,100	1,051,600	4,419,466,237	1,306,179,989
West South Central..	7,271,395	996,088	7,622,066,027	2,168,622,649
Mountain	2,121,121	244,109	4,083,137,939	562,954,399
Pacific	2,095,388	234,164	5,307,011,460	948,854,024
Total...........	51,406,017	6,448,343	$77,924,100,338	$14,755,364,894

NOTES FOR FURTHER REFERENCE

See notes at end of preceding chapter. A full bibliography on agriculture during this period is available in L. B. Schmidt, *Topical Studies and References on the Economic History of American Agriculture* (rev. ed., 1923). The Annual Reports of the United States Department of Agriculture contain a mine of information. Consult also the Introduction to the volume on *Agriculture* in the Eighth Census, the special reports on *The Cereals*, on *Flour Milling*, on *Meat Production;* and in the Tenth Census, Vol. III, the report on *Tobacco*. In Vol. V, pp. xvi-xxvii of the same census, there is a brief review of the *Agricultural Progress of Fifty Years, 1850-1900.* See also volumes on *Agriculture* in the Thirteenth and Fourteenth Census. The Institute for Government Research has a series of *Service Monographs of the United States Government* (1922-23), 27 volumes, each giving the history, development, functions, organization, etc., of the particular service treated.

Valuable contributions on this period are to be found in Bailey's *Cyclopedia of American Agriculture*, Vol. IV, especially those of T. N. Carver, *Historical Sketch of American Agriculture;* F. H. Fowler, *Abandoned Farms;* and David Kinley, *The Center of Agricultural Production*. A popular and well-written account is that of A. H. Sanford, *The Story of Agriculture in the United States* (1916). Excellent chapters of summary are included in the volumes of both Bogart and Lippincott. Another brief résumé is that in T. N. Carver, *Principles of Rural Economics* (1911), and much necessary information is in J. E. Boyle, *Agricultural Economics* (1921). See also N. S. B. Gras, *History of Agriculture* (1925).

---

[1] Compiled from the Fourteenth Census, vol. vi, pp. 30, 31, 32, 69.

THE AGRARIAN REVOLUTION 451

Two worth-while books of readings are those of E. G. Nourse, *Agricultural Economics* (1916), and T. N. Carver, *Selected Readings in Rural Economics* (1916). See also the selections in Bogart and Thompson.

On the public lands consult T. Donaldson, *The Public Domain* (1884), and L. H. Haney, *A Congressional History of Railways in the United States, 1850-1887*. On the farmers' uprising, see S. J. Buck, *The Granger Movement* (1913), and F. L. McVey, *The Populist Movement* (1896), in Economic Studies, Vol. I, No. 3, Publications of American Economic Association. Two phases of southern agriculture are presented in M. B. Hammond, *The Cotton Industry* (1897), in the Publications of the American Economic Association, and by M. Jacobstein, *The Tobacco Industry* (1907), Columbia University Studies, Vol. XXVI, No. 3, and a phase of western agriculture in R. A. Clemen, *The American Livestock and Meat Industry* (1923).

On farm machinery see H. N. Casson, *The Romance of the Reaper* (1908), and H. W. Quintance, *The Influence of Farm Machinery on Production and Labor* (1904), Publications of the American Economic Association, 3rd Series, Vol. V, No. 4. Interesting as a foreigner's viewpoint is P. Leroy-Beaulieu, *The United States in the Twentieth Century* (1906). On special aspects see C. R. Van Hise, *Conservation of Natural Resources in the United States* (1912); C. E. Russell, *The Story of the Nonpartisan League* (1920), and K. L. Butterfield, *The Farmer and the New Day* (1919).

The recent effort of the federal government to extend agricultural credit is discussed in Herbert Myrick, *The Federal Farm Loan System* (1916), which includes the full text of the Federal Farm Loan Act; and A. C. Wiprud, *The Federal Farm Loan System in Operation* (1921). See also L. C. Gray and H. A. Turner, *Buying Farms With Land-Bank Loans* (1921), Bulletin 968 of the United States Department of Agriculture, as well as other bulletins of the Department.

On irrigation see George Thomas, *The Development of Institutions Under Irrigation* (1920).

SELECTED READINGS

SANFORD, A. H., *The Story of Agriculture in the United States*, Chaps. XVIII-XXIX.

BAILEY, L. H., *Cyclopedia of American Agriculture*, Vol. IV, pp. 64 ff., 102 ff., 113 ff., 174 ff., 386 ff., 422 ff., etc.

CARVER, T. N., *Selected Readings in Rural Economics*, pp. 254-337; 487-536; 645-699.

HAMMOND, M. B., *The Cotton Industry*, Chaps. IV-VI.

BUCK, S. J., *The Granger Movement*, Chaps. I-III, VIII-IX.

CHAPTER XX

INTERNAL TRANSPORTATION AND COMMUNICATION SINCE 1860

Rapid Development After 1860.—The history of the United States for decades after the Civil War might almost be written in terms of railroads. Our industrial and agricultural development was dependent upon internal transportation, of which the major part was furnished by the railroad. The very settlement of large parts of the west was promoted by the railroads, built in many instances through unoccupied regions with the settlers following in their wake. In 1860 the railroad mileage amounted to 30,625, most of which had been constructed in the prosperous years preceding the panic of 1857. The effect of the Civil War upon the railroads was both disastrous and stimulating. While the rolling stock and other equipment in the south either depreciated or was destroyed, the war spurred the north on to fresh construction. It was during the midst of the conflict and partially as a war measure that the first transcontinental railroad was commenced. Most of the construction of the decade, however, came after 1865, the mileage amounting in 1870 to 52,922.

In the succeeding decades the increase was rapid and, except in periods of acute depression, continuous. Thirty-three thousand miles were built between the years 1867 and 1873, before the first great spurt had played itself out and the panic had halted further construction. The panic of 1873 was itself chiefly attributable to overbuilding and overcapitalization of railroads. The period from 1860 to 1875 witnessed the extension not only of the first transcontinental lines, but also of five great railroads from the Atlantic seaboard to Chicago—the New York Central, the Pennsylvania, the Erie, the Baltimore and Ohio, and the Grand Trunk. After the recovery from the crisis, the country entered into another phenomenal period of railroad growth.

The mileage of 1880, amounting to 93,261, grew to 167,191 in 1890—an increase of over 70,000 miles in a decade. The panic of 1893 again hampered construction, but it picked up in the prosperous years after 1898, and for twelve years thereafter new mileage averaged five thousand a year (2 per cent annual increase). Railroad miles in operation amounted in 1900 to 198,964; in 1910 to 249,992; and in 1920 to 253,152. This growth had far exceeded that of population; since the Civil War the population has trebled, while railroad mileage has grown eightfold. In 1860 there was one mile of road to every 1,087 people; in 1880 one mile to every 571, and in 1920 one mile to every 417. The United States boasted in 1914 of more mileage than all of Europe, and more than one-third of the entire world.

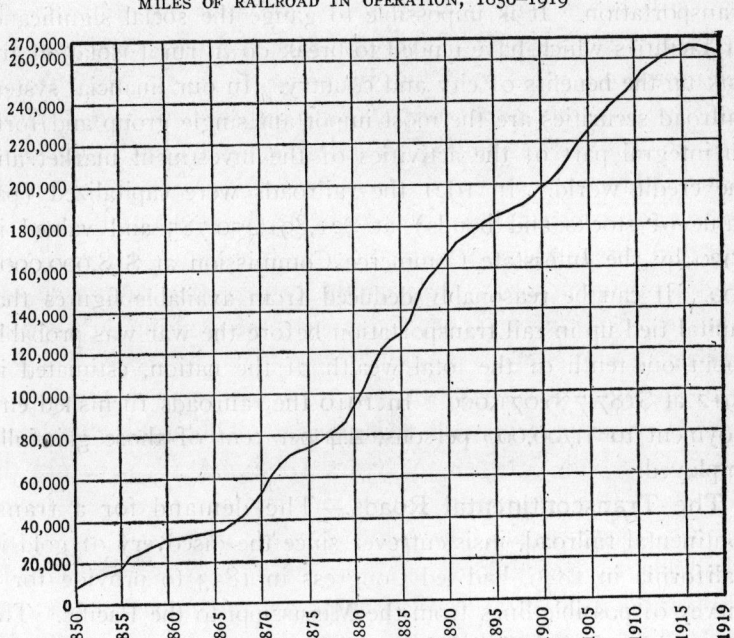

MILES OF RAILROAD IN OPERATION, 1850–1919 [1]

On the other hand, by 1914 the country seemed to have approached a point of saturation in railroad building. While the

---

[1] *Statistical Abstract*, 1878, p. 151, and *Statistical Abstract*, 1921, p. 875.

average of construction was over 3,000 miles a year between 1910 and 1913, it declined rapidly thereafter until in 1920 only 314 miles were built. Between 1916 and 1920 more mileage was abandoned than built, leaving the total at the end of the period less than at the beginning. Many causes have contributed: (1) the competition of gasoline motor traffic, (2) economies of war time, (3) low railroad profits and the precarious financial condition of many of the roads, which have suffered in the adjustments to higher costs, (4) the approach to the saturation point. During the past half dozen years the weak roads have exerted all their efforts to keep alive while the prosperous companies have improved their equipment rather than lengthened their lines.

The economic significance of railroads is far wider than merely transportation. It is impossible to gauge the social significance of facilities which have tended to break down rural isolation and link up the benefits of city and country. In our financial system railroad securities are the most important single group and form an integral part of the activities of the investment market and the credit world. In 1921 the railroads were capitalized (par value of stocks and bonds) at $21,891,450,785 and valued in 1920 by the Interstate Commerce Commission at $18,900,000,-000. It can be reasonably deduced from available figures that capital tied up in rail transportation before the war was probably about one-tenth of the total wealth of the nation, estimated in 1912 at $187,739,071,090. In 1910 the railroads furnished employment to 1,700,000 persons, 4.4 per cent of those gainfully employed.

The Transcontinental Roads.—The demand for a transcontinental railroad, insistent ever since the discovery of gold in California in 1849, had led Congress in 1853 to provide for a survey of possible lines from the Mississippi to the Pacific. The exigencies of the Civil War, political and military as well as economic, led eventually to the construction of the first line. The Union Pacific Railroad Company was created by Congress in 1862 for the purpose of building a road from Nebraska west to California; and the Central Pacific, under the leadership of

Leland Stanford, Collis P. Huntington, and other famous railroad men, was organized to build from the Pacific coast eastward to meet the Union Pacific. Both railroads were granted subsidies of $16,000 a mile for construction on the level country, $48,000 a mile through the mountain ranges and $32,000 for the sections between the ranges, the government taking a second lien on the property. Land grants of alternate sections contiguous to the railroads were offered in addition. Stimulated by these substantial aids and urged on by the great popular interest, both roads built frantically toward one another in the hope of obtaining as much of the subsidy as possible. Neither expense, hostile Indians, nor the severity of the mountain winters was permitted to hold up the work. Twenty thousand men were laying two miles of track a day in the concluding weeks of an effort which brought the two roads together at Promontory Point, Utah, on May 10, 1869, three years after the work had commenced and much sooner than the most sanguine had expected.

Hardly less romantic in its conception, but less successful in its immediate realization, was the effort to construct the Northern Pacific. Chartered by Congress in 1864 and subsidized with land grants as large as a European country, it was finally commenced in 1867 through the financing of Jay Cooke & Company. Five hundred miles had been built when the road was thrown into a receivership by the failure of Jay Cooke. Eventually aggressive construction was resumed and, largely through the genius of Henry Villard, backed by German capital, was completed in 1883. The Atchison, Topeka, and Santa Fé obtained from the national government in 1863 a grant of 6,400 acres for every mile built; but construction did not begin until 1869, and had proceeded no farther than the eastern boundary of Colorado when building was stopped by the panic of 1873. Construction was resumed in 1880, following in general the old Santa Fé Trail, and in 1884 the tracks reached the Pacific coast. In 1878 James J. Hill, a man who later developed into one of the greatest of railway executives, but at that time an unknown storekeeper in St. Paul, interested influential Canadians in a bankrupt little two-hundred-mile railroad supposedly without a future, known as the St. Paul and Pacific. No sooner had the new group taken over the property than the road prospered, due chiefly to a succession of good harvests. The St. Paul and Pacific was now lengthened to the Pacific under the oversight of Hill and developed into the Great Northern System. In the meantime the Southern Pacific (1883) had opened up a line east to New Orleans.

Financing the Railroad and Government Aid.—Not the least interesting phase in the history of American railroads is the story of early financing, which can be merely touched upon here. The urgent need of internal transportation, the remarkable success of the Erie Canal, and the sectional rivalries were chief among the causes leading the states to finance many of the early projects. State aid was overdone and was brought to a disastrous close by the panic of 1837. Thereafter, although government aid was rendered, particularly in the case of the transcontinental roads, the financing and management of internal transportation has been largely in the hands of private capital.

An appreciable share of the yearly surplus of wealth has been invested in railroads. The fundamental fact that transportation was essential to the building up of the nation was apparent, and nowhere is the American willingness to speculate on the progress of the country better seen. But before private capital went into railroads, the government was expected to do much to make the way easy. Railroads were built under charters voted by the state legislatures, all of which contained valuable rights and concessions and many of which were obtained by corrupt means. The right of eminent domain—that is, the power to lay out a road and condemn the land needed if impossible to obtain it otherwise—was invariably granted. In certain cases a monopoly or protection against competition was conferred, as were special banking privileges to aid in raising money. Tax exemption was permitted in some charters forever, in others for a stated period, and in still others until the dividends should reach a certain per cent. On the other hand, even in these free-and-easy early railroad charters, sections were sometimes inserted providing for a reduction in rates when the dividends exceeded a normal yield. The attitude on the part of the legislatures and public was that transportation should be encouraged by every possible means, and the charters reflected this attitude. The charter having been secured, the railroad was built by money or credit obtained through national, state, county, municipal, or private subscriptions. National aid was rendered by (1) tariff remission on rails, (2) land grants, and (3) direct financial aid. Almost 200,000,000 acres were originally granted, but this was reduced to 155,000,000 by forfeitures resulting from inability to meet the requirements of the law. These land grants included one-fourth of the states of Minnesota and Washington; one-fifth of Wisconsin, Iowa, Kansas, North Dakota, and Montana; one-seventh of Nebraska, one-eighth of California, and one-ninth of Louisiana. In all 242,000 square miles, a region larger than Germany or France, was given to the railroads. Of the individual donations, the Northern Pacific received 44,000,000 acres, the Southern Pacific 24,000,000, the Union Pacific 20,000,000, the Santa Fé system 17,000,000. National bonds also to the amount of $64,623,512 were issued on the security of second mortgages to help certain of the transcontinental lines—loans eventually almost entirely repaid.

Aid on the part of the various states was rendered chiefly by: (1) subscriptions to the capital stock, a method resorted to by a number of states; (2) the loan of state credit by such methods as direct purchase of railroad bonds or indorsement of construction bonds; (3) by state land grants; (4) by bearing the expense of survey. Close to 55,000,000 acres have been turned over by the states to transportation companies. The counties and municipalities followed the state in encouraging construction by subscribing for stock, by exchanging municipal or county bonds for railroad securities, or by actually donating money and land. In New York State 294 cities, towns, and villages contributed $29,978,206 to railroads and 51 counties gave subsidies or amounts varying from $5,000 to $3,000,000. In Massachusetts,

171 towns and cities had issued bonds to aid railroads up to 1871.

The sum total of these various aids, amounting in most cases to outright subsidies, was very large; nevertheless, private backing has been even greater. Two classes of speculators have purchased railroad securities—those buying for dividends and a market rise, and those buying to promote building through their own region. This latter group in their enthusiasm were glad to exchange labor, land, and money for stock which often proved worthless. A considerable amount of financing was done in Europe, where American transportation securities have been exceedingly popular. In 1907 over $6,000,000,000 of railroad stocks and bonds were held abroad, representing over one-fourth of the entire value of the roads at that time. Of this amount Great Britain owned $4,000,000,000, and Germany $1,000,000,000. In 1914 the *Wall Street Journal* estimated $3,400,000,000 worth of bonds, one-third of the outstanding railroad mortgage indebtedness, as held abroad; but the war has brought many of these securities back to America, rendering our railroads for the first time virtually independent of foreign capital.

Chaotic Conditions and Early Abuses.—In a period of such rapid extension a chaotic situation was likely to develop, and abuses to creep in. Especially was this probable in an age in which business morality was at a low ebb. For some years the demand for transportation was so insistent that little attention was given to anything else; but by the early 'seventies railroad abuses had transformed hearty coöperation on the part of the people to a feeling of distrust and to a demand that the transportation companies be curbed. The abuses were indeed many. In the first place, there was complaint that money was wasted in unneeded and purely speculative enterprises. In their enthusiasm for railroad building, the American people apparently believed that the country could support an unlimited mileage. Roads were driven parallel in direct competition and into uninhabited country where the future was doubtful.

While some excuse might be made for overconstruction, none whatsoever can be discovered for the reckless graft practiced by the promoters through the medium of construction companies. These construction companies in themselves were not necessarily bad. In fact, there was considerable justification for them, since the rapid extension of railroads in the south and west, where the promise of adequate returns were not sufficient to attract capital, made it necessary to build railroads through organizations willing to take land and railroad securities in payment. The construction companies were generally composed of the same group who controlled the projected railroad. The railroad directors voted themselves, as members of the construction company, contracts to build the roads, thus reaping profits both as builders and later as stockholders. The cost of building the Central Pacific was $58,000,000, but a construction company was paid $120,000,000 for the work. The most famous of these companies was the Crédit Mobilier, formed to build the Union Pacific. To prevent any interference which might arise because of the aid received from the government, Oakes Ames, a representative from Massachusetts and prominent in the Crédit Mobilier, was given 343 shares of the stock, to be placed among Congressmen where they would "do the most good." Ames's activities resulted in an investigation which showed that he had sold stock below face value to a number of Congressmen to influence their votes, and that the trail of bribery reached as high as the Vice-President. The exorbitant and reckless expenditures as exemplified by the construction companies was typical of methods in many channels of railroad finance, but it is only fair to say that some railroads condemned these practices.

Even more exasperating than the unbusiness-like and inefficient methods of construction were the reckless irresponsible manipulations of the finances of the roads, once constructed. Railroad magnates of the early decades looked upon the whole matter as a private business for personal gain. Apparently no feeling of public responsibility swayed them, and their conception of common honesty was exceedingly flexible. The attitude is illustrated by the famous story about Cornelius Vanderbilt, perhaps the greatest of the early railroad builders, who is alleged to have replied to a remonstrance in regard to the feelings of the public over an arbitrary act, "The public be damned." Men like Jay Gould and Daniel Drew controlled railroads not to serve the public, improve the property, and make legitimate profits, but to manipulate the stock to build their fortunes. No industry has suffered more from stock-watering than the railroads; for time and again an expanded capitalization has been placed upon a road without equivalent addition in actual capital. This has been done chiefly to pay expenses which the roads did not want to carry under regular expenditures, and to camouflage earnings. In many cases the increasing valuation of the properties has soaked up the water, but in more cases watered stock lies like an inert weight upon the real earnings; and stockholders clamor for dividends that can be paid only by unreasonably high charges. It was estimated that of the $7,500,000,000 indebtedness of the railroads in 1883 as much as $2,000,000,000 represented water. In four years, 1868 to 1872, Erie stock was watered from $17,000,000 to $78,000,000 in market speculation. In 1897 only 29.9 per cent of the railroad stock of the country paid dividends; in the prosperous year 1890 less than 50 per cent and even in the war year 1918 only 58.09 per cent. In the words of Charles Francis Adams, Jr., a railroad president, "The system was, indeed, fairly honeycombed with jobbery and corruption."[1]

An abuse which struck at the very foundation of our democracy was the continued assault upon the integrity of government by the railroad interests. The Crédit Mobilier scandal was notorious and famous because it implicated the highest legislators in the land, but similar activities on a smaller scale were quite common. The pressure upon legislators to grant favorable charters, and upon courts to interpret them broadly, was exerted by every means. The pass system in the height of its glory took care of most influential persons. The most powerful of the legislators were frequently employed as counsel at large salaries. Where this was not sufficient many roads followed the example of the Erie, which in one year expended $700,000 as a corruption fund and for legal expenses, the amount carried on the books as the "india-rubber account." The general attitude was much like

---
[1] Adams, C. F., Jr., *The Railroad Problem*, p. 126.

that of the railroad magnate who was reported as saying that in Republican counties he was a Republican, and in Democratic counties he was a Democrat, but everywhere he was for the railroad. The low political as well as low business morality of the period made corrupt practices possible, and the blame rests not alone with the roads.

More closely concerned with the general prosperity was the straits to which competition had reduced the railroads. Between competitive points rate wars had lowered transportation costs to ruinous figures. Passenger fares between Cleveland and Boston in August, 1876, were down to $6.50, cattle in the same year were carried from Chicago to New York for a dollar a carload. While the shippers at competitive points (especially where there was an option of water transportation) profited at the expense of the roads, the farmer suffered from the general practice followed by the roads of raising their tariffs at non-competitive points to recoup their losses. The same causes led to another form of discrimination equally galling, the custom of charging more for a short haul than for a long. Still another unfair practice which was exceedingly prevalent was that of granting rebates on freight charges. Where competition was bitter or the shipping concern strong, substantial refunds were obtained. Thus the Standard Oil and other large shippers procured rebates giving them an advantage the smaller organizations failed to obtain and, like the farmer, the smaller organizations had to pay higher freights to make up for the losses of the railroads. Cut-throat competition led, eventually, to the system of "pooling," by which the available business was allotted proportionately at agreed rates. Although the railroads broke the pooling agreements almost as soon as they were made, the mere attempt to make them was looked upon as unfair, monopolistic, and contrary to the common law. A mere enumeration of the grievances against the railroads in the decade of the 'seventies enables one to understand the strong reaction against them which culminated in the "granger movement." While many of these abuses have been eliminated and others softened, the old distrust on the part of the public has never entirely died out.

The Granger Movement and the Railroads.—The first strong agitation against the railroads occurred in the early 'seventies among the farmers, especially those of the middle-western states of Illinois, Minnesota, Iowa, and Wisconsin. This activity is known as the "granger movement," a name originating from the "granges," or local lodges, of the Patrons of Husbandry. In the grange the isolation of the individual farmer broke down, and he was able to voice his grievances. The attack on the railroads, which was waged fiercely from 1869 to 1875, was but a phase of a mighty movement of agrarian unrest which was inaugurated by the granger movement and surged through the west until 1896.

Inasmuch as early legal decisions such as that in the Dartmouth College case had interpreted a charter as a contract, the western states had been careful to insert in the state constitutions provisions declaring that laws creating corporations might be altered or repealed; or to specify in the charters that the railroad rates must be equal and reasonable. Backed by these specific rights and by the common-law conception that a business which was a public calling came under the regulatory power of the state, the representatives of the farmers passed laws in an attempt to control the roads. The first act passed was that of Illinois in 1869, which limited the roads to "just, reasonable, and uniform rates." In the new Illinois state constitution of 1870 the legislature was ordered to "pass laws to correct abuses and to prevent unjust discrimination and extortion in the rates of freight and passenger tariffs." Laws of 1871 attempted to do this by providing maximum fares and freight rates, by regulating warehouses and the transportation of grain, by establishing a board of railway and warehouse commissioners, and by the enactment of a general railway incorporation act. Minnesota followed in the same year with laws fixing freight and passenger schedules and providing a railroad commission. Iowa and Wisconsin passed similar acts in 1874, the latter state enacting the Potter law, the most radical of the granger acts. During this decade a demand appeared for the regulation of railroads, and there seems to be no doubt that the aggressive activity of the "granger states" of the upper Mississippi Valley gave an impetus to the whole movement for control which in some states was not consummated until the next decade. Most of the states passed some kind of railroad legislation; while in practically all of the new or rewritten state constitutions of the decade, provisions are inserted making it the duty of the legislatures to regulate rates and prevent discriminations, and declaring the railroads public highways and the companies common carriers.

As a whole the granger acts sought (1) to establish, either by direct legislation or through a commission, schedules of maximum rates; (2) to prohibit a greater charge for a short haul than for a long one; (3) to preserve competition by forbidding the consolidation of parallel lines; and (4) to eliminate the evil of granting free passes to public officials. "Several of the principal features of American railroad legislation," says Professor Buck, "can be looked upon as primarily Granger in their origin." [1] Where railroad legislation was passed it was customary to set up commissions of experts. These were of two kinds: the strong commission, as in Illinois, with power to regulate rates and enforce the law; and the weaker commission, as in Massachusetts, with powers merely advisory and the duty to make reports to the legislature. It may be said that the latter type in the long run often proved to be the most successful.

These first attempts to regulate the railroads were of course vigorously opposed by the companies and immediately fought in the courts. In general the railroad laws were attacked from two angles. It was maintained first that the exclusive power to regulate interstate commerce rested with Congress, and that, as the bulk of the commerce was interstate, the national government should legislate if it was necessary. Secondly, the effort to regulate rates was maintained to be contrary to that portion of Section I of the Fourteenth Amendment which declares, "No state shall make or enforce any law which shall abridge the privileges or immunities of citizens of the United States; nor shall any state deprive any person of life, liberty, or property, without due process of law; nor deny to any person within its jurisdiction the equal protection of the laws."

---

[1] Buck, S. J., *The Granger Movement*, p. 205.

The first of the so-called "granger cases" was that of Munn v. Illinois, decided by the Supreme Court in 1876 and involving the Illinois law of 1871, which had declared grain elevators to be public warehouses and had established maximum charges. The plaintiffs sued on the ground that (1) warehousing was not a public calling and the business was therefore not within the regulatory power of the state; (2) the fixing of rates deprived the owners of the power to establish higher rates and thus deprived them of their property without due process of law. They further maintained that if the courts did decide that their business was a public calling, it was the work of the judiciary and not the legislature to determine a fair charge. All of these contentions were thrown out by the decision of Chief-Justice Waite, who held that this section of the Constitution did not invalidate the old English common law, generally accepted when the amendment was passed. "Property," said he, "does become clothed with a public interest when used in a manner to make it of public consequence, and affect the community at large. When, therefore, one devotes his property to a use in which the public has an interest, he, in effect, grants to the public an interest in that use, and must submit to be controlled by the public for the common good." He further held that the fixing of rates was a legislative and not a judicial matter, asserting that "it has been customary from time immemorial for the legislature to declare what shall be reasonable compensation under such circumstances." [1]

The firm attitude of the court in the Munn case was maintained in the case of Peik v. the Chicago and Northwestern Railway Company, handed down in the same year. The contention of the railroads that state regulation was an infringement of interstate commerce (and so unconstitutional), as most of the railroad traffic crosses state boundaries, was thrown out. Ignoring the effect of such laws upon those outside the state, the court declared that "until Congress acts in reference to the relations of this company to interstate commerce, it is certainly within the power of Wisconsin to regulate its fares, etc., so far as they are of domestic

---

[1] Munn v. Illinois, 94 U. S., 113.

concern."[1] The decisions of the Munn and Peik cases were supposed to have settled the main points of constitutional law involved in railway regulation, and it was a matter of surprise when, ten years later (1886), in the case of the Wabash, St. Louis, and Pacific Railway v. Illinois,[2] the Supreme Court reversed itself. The case arose over a violation of a law forbidding a greater charge for a short haul than for a long one, when it was discovered that the railroad rates were higher on freight from Gilman to New York than from Peoria to New York, although the latter point was eighty-six miles farther away. The decision now held that no state could exercise any control over commerce beyond its limits. The new interpretation remained as the legal one.

The Inauguration of National Control—The Interstate Commerce Act and Amendments.—The granger attack upon the railroads brought with it the demand for federal as well as state regulation. At the recommendation of President Grant in 1872 a committee under the chairmanship of William Windom of Minnesota was appointed, which made a report[3] in 1874 advising government construction and extension of transportation facilities in order to reduce rates by making the government a competitor to private roads. The "Regan bill," aiming to do away with some of the worst abuses of the railroads, had passed the House in 1878, but was not acted on in the Senate, and the question of railroad legislation lay dormant until 1885. In that year a Senate committee was appointed, headed by Shelby M. Cullom of Illinois, an indefatigable worker for railroad regulation. In its report the following year[4] it reviewed carefully the various methods by which a remedy for the situation might be found, and indorsed some form of federal regulation and control to obviate what seemed to them the greatest evil, namely "unjust discrimination between persons, places, commodities, or particular descriptions of traffic."

---

[1] Peik v. Chicago and Northwestern R. R., 94 U. S., 164.
[2] 118 U. S., 557.
[3] Senate Report No. 307, Forty-third Congress, 1st Session.
[4] Senate Report No. 46, Forty-ninth Congress, 1st Session, vol. 2.

In 1887 federal regulation of railroads was inaugurated as a necessary compromise between government ownership and unrestrained private operation. The Interstate Commerce Act of that year (1) provided that all charges should be just and reasonable; (2) forbade personal discriminations in the form of special rates, rebates, or otherwise; (3) forbade discriminations between localities, classes of freight, and connecting lines; (4) forbade a greater charge for a short haul than for a long; (5) prohibited pooling; and (6) ordered that all rates and fares should be printed and publicly posted, and no advance be made except after ten days' notice. The administration of the law was placed in the hands of an Interstate Commerce Commission of five members, which was given power to collect data from the carriers, call witnesses, hear complaints, and render decisions. If the commissioners believed the law was violated and the roads refused to abide by their decisions, it was their duty to institute proceedings in the circuit courts. They were required to submit annual reports to Congress.

The passage of the Interstate Commerce Act was strongly opposed by the railroad officials, who predicted dire results; but their continued opposition and evasion, aided by various judicial decisions, effectively pulled the teeth from the act. The railroads successfully avoided giving full testimony until 1896, when the Commission [1] eventually obtained compulsory power of investigation. In 1897 the Supreme Court [2] held "that the power to prescribe rates or fix any tariff is not among the powers granted to the Commission," thus limiting the Commission's power over rates to deciding what was unfair, without the right to prescribe fair rates. In a similar manner the heart was cut out of other sections, and the courts further hampered the work of the Commission by going beyond the reports submitted, allowing the introduction of new testimony, and then rendering a decision. The greatest weakness in the position of the Commission, however, was the fact that its decisions were not compulsory and that

---

[1] Brown v. Walker, 161 U. S., 591.
[2] Interstate Commerce Commission v. Cincinnati, New Orleans, and Texas Pacific Railway Company, 167 U. S., 479.

upon it rested the burden of initiating action in the courts. The attitude of the railroads and courts prevented the Interstate Commerce Act of 1887 from receiving a fair trial. Abuses continued and the Commission became little more than a bureau of statistics. Although some good work was done in securing rate publicity and in reducing the number of freight classifications, the act was important chiefly in its educational value and in introducing federal legislation, the system under which we have worked until the present time.

That the Interstate Commerce Act of 1887 had achieved no solution of the problem was apparent to all, and during the administration of Roosevelt efforts were again vigorously pushed to strengthen the hands of the Commission. The Elkins Act of 1903, aimed at the practice of rebates, declared the deviation from published rates to be discrimination, and held both giver and receiver guilty. The Expediting Act of the same year gave preference in the circuit courts to cases arising under the Interstate Commerce Act of 1887 and the Sherman Anti-trust Act of 1890, on the theory that such cases were "of general public importance." An important amendment to the legislation of 1887 was the Hepburn Act of 1906, which enlarged the scope of the Interstate Commerce Act to include express and sleeping car companies, pipe lines, switches, spurs, tracks, and terminal facilities. The Commission, now increased to seven, of which only four could be of the same political party, was empowered to determine just and reasonable rates and to order the carrier to adhere to them, leaving to the latter the burden of initiating court action. The act also instructed the Commission to prescribe methods of bookkeeping for the railroads and made their adoption compulsory. To obviate a certain type of discrimination, railroads were forbidden to carry commodities which they had themselves produced, except timber and goods needed in the conduct of their business. Free passes were forbidden, and rates must be published thirty days before change. Although that part of the act respecting the right of railroads to carry commodities in which they had an interest has been largely nullified by court action, in general the court has limited itself to determining the legality of the orders of the Commission rather than their wisdom or expediency. The Hepburn Act went far to obviate the faults of the act of 1887, and since 1906 the Commission has been a responsible and powerful body.

In 1910 the Mann-Elkins Act was passed. It clarified the short and long haul clause of the Interstate Commerce Act and enlarged the powers of the Commission by granting to it the right to suspend for six months the operation of a new scale of rates to give time for an investigation. It set up a special Commerce Court to hear railroad cases arising from the Commission's activities, considered a much-needed innovation, for the necessity was obvious of having railroad cases tried by judges who were experts in such questions.[1] Of the subsequent minor legislation, mention might be made of the amendment of 1913 to the Interstate Commerce Act, requiring the Commission to report the value of all property owned or used by all the common carriers; and of the Newlands Act of 1913 (amending the Erdman Act of 1898), providing for voluntary settlement of railroad disputes. In 1916 the government in the Adamson Eight-hour Act entered into a new sphere of activity in regulating the hours of labor in interstate traffic.[2]

The opening of the World War found the principle of government regulation firmly established and the railroad industry stabilized as never before. At the same time inefficient management and the inevitable results of early excesses, coupled with dislike of government regulation, had helped to instil in many units of the industry a feeling of disquiet and uncertainty as to the future. Railroad labor was, moreover, becoming more restless and the public more exacting. All factors pointed to further developments when war broke out.

Railroad Consolidation.—Parallel to the combination of capital in other lines of industry there has developed a consolidation of railroads. This has taken place primarily to insure greater

---

[1] This court was allowed to lapse in 1912, by the failure of Congress to appropriate funds for its continuance.
[2] Dodd, W. E., *Woodrow Wilson and His Work* (1920), pp. 164, 189, 190.

efficiency, to eliminate competition, and to secure larger profits. It has taken two courses—that of uniting railroads to form a continuous line of travel, and that of consolidating roads in a given geographical division. The first type of consolidation began before the Civil War and continued for many years. It is best exemplified by Vanderbilt's work in combining (1853) eleven little roads, hitherto most inefficiently handling the traffic between Albany and Buffalo, into the New York Central, and in adding from 1855 to 1858 five more roads to the system.

As already noted, the late 'sixties and the decade of the 'seventies was a period of disastrous and unbridled railroad competition. By this time the through lines had taken form and were bitterly fighting for traffic. As this competition bade fair to ruin the roads, repeated efforts to eliminate it were made by means of pools or traffic associations, which sought to apportion arbitrarily among the roads the available business at rates mutually agreed upon. Such methods proved inadequate because the railroads did not keep their own agreements, and on account of the interference of the government. The Interstate Commerce Act of 1887 forbade "any contract, agreement, or combination . . . for the pooling of freights of different railroads," and later the Supreme Court, in the case against the Trans-Missouri Freight Association (1898), declared that they violated the Sherman Anti-trust Act of 1890 as agreements "in restraint of trade and commerce."

With pooling forbidden by law, the railroads, like other industries, turned again to consolidation to save themselves from the evils of too great competition. During the 'eighties, and again from 1898 to 1904, consolidation went on rapidly. By purchase, by lease, by the ownership of a majority of the stock, the larger railroads absorbed many of the smaller competing lines, thereafter often operated as separate but subsidiary companies. Thus the Pennsylvania owns many subsidiary lines, such as the Long Island, the Philadelphia, Baltimore, and Washington, the Pittsburgh, Fort Wayne, and Chicago; controls through stock ownership the Pittsburgh, Cincinnati, Chicago, and St. Louis, the Grand Rapids and Indiana, and others, and owns large blocks of stock in such important roads as the Baltimore and Ohio, the Chesapeake and Ohio, and the Norfolk and Western.

The period of rapid consolidation was brought to halt in 1904 by the Supreme Court decision in the Northern Securities case. A fight between the Hill-Morgan group and the Harriman interests to gain control of the Northern Pacific had resulted in the victory of the former, who now dominated transportation in the northwest by majority holdings in the Great Northern, the Northern Pacific, and the Burlington system. The Hill-Morgan stock was turned over to a holding corporation, known as the Northern Securities company, which was adjudged by the Supreme Court a violation of the Sherman Act and required to dissolve. "If Congress has not," said the Court, "by the words of this Act, described this and like cases, it would, we apprehend, be impossible to find words that would describe them."[1]

Since 1904 consolidation has gone on by the development of "community of interest," furthered by interlocking directorates, stock ownership, and similar means, until to-day easily recognized regional combinations have developed. Of the 228,000 miles of railroad in 1906 about 176,000 were divided among seventeen systems, of which the six most important are the Vanderbilt system (21,333 miles), handling much of the traffic in New York State and along the Great Lakes; the Pennsylvania system (20,370), serving the Middle Atlantic and west to the Mississippi; the Morgan group (17,810), controlling the Erie, the New England roads, and a large proportion of the southern mileage; the Hill group (21,303), dominating the northwest routes to the Pacific; and the Gould roads (16,902), serving the southern transcontinental routes. For example, both the New York Central (Vanderbilt group) and the New Haven (Morgan road) were from 1911 until 1914 jointly interested in the Boston and Albany,[2] and similarly the Philadelphia and Reading is jointly controlled by the Lake Shore and Michigan Southern (a subsidiary of the New York Central), and the Baltimore and Ohio (affiliated with the Pennsylvania system). It was charged by

---

[1] Northern Securities Company v. U. S., 193 U. S., 197.
[2] Now leased and controlled by the New York Central.

Senator La Follette [1] in 1921 that twenty-five directors linked together 99 Class 1 roads, operating 211,280 miles, or 82 per cent of the country's transportation system, and that these were likewise closely allied with the leading equipment companies. The truth of these charges would support the belief that consolidation through community of interest has proceeded far.

In the face of this tendency the government has been able to do little. The Federal courts broke up the Northern Securities company (1904), ordered the Union Pacific to dispose of its Southern Pacific stock (1912), and dissolved the New Haven monopoly in 1914. Congress in the Panama Canal Act of 1912 attempted to prevent the control of domestic water transportation by competing railroads, and in the Clayton Anti-trust Act of 1914 forbade a corporation's acquiring, "directly or indirectly, the whole or any part of the stock or other share capital of another corporation engaged also in commerce, where the effect of such acquisition may be to substantially lessen competition between the corporation whose stock is acquired, and the corporation making the acquisition." [2] The apparent futility of preventing an accomplished fact, however, was recognized in the Transportation Act of 1920, which gave to the Interstate Commerce Commission power to permit the carrier to acquire control by lease or purchase of another carrier in any manner which does not involve their consolidation into a single system of ownership and management. The act went further when it empowered the Commission to prepare plans for the consolidation of the roads into a number of systems in which competition might "be preserved as fully as possible." This may be the next development.

Improvements in Service.—A number of factors has prevented a more rapid advance in improvements to railroad rolling stock. The financial hardships, the elimination of competition through consolidation, and the fact that express companies and the government have taken over the responsibilities of handling the more exacting traffic have all contributed to hold back this phase of development. Nevertheless, great progress has been

---

[1] *Congressional Record*, March 14, 1921, accompanied by diagrams.
[2] 38 Stat., 730, Sec. 7.

made. Probably no traveler rides more luxuriously than the American, and increasingly large orders are being placed here by Asiatic countries for equipment. Before great advance could be made steel rails had to be substituted for iron, just as the solid iron had been substituted for the strips. The first steel rails were imported in 1863, and their manufacture begun in 1865, impelled by the discovery of the Bessemer process. After that date much of the mileage was so equipped. With better rails came more scientific roadbeds, and with heavier rolling stock better bridges and other structures.

Freight earnings are over three times as great as passenger earnings on American roads, and there are forty freight cars to one passenger car. A majority of the freight consists of heavy, bulky products such as coal, grain, timber, and petroleum, a factor of great influence on the development of rolling stock. In contrast to European equipment—which is light and small, designed to carry compact freight for short hauls—in America monster engines and large cars, built to carry heavy freight for long distances, have been the rule. A present-day engine turned out by the Baldwin Locomotive Works weighs with tender over 426 tons, and draws trains of 3,000 tons made up of eighty-ton cars. Passenger coaches have followed this tendency in size, and our more democratic customs have prevented the adoption of the European compartment train. The tremendous weight of engines and cars has caused a continually increasing strain on roadbeds, but the abundance of raw materials has held down the expense of keeping them in repair.

The comfort of passengers is far beyond the dream of earlier travelers. Better springs, heating and lighting facilities, Pullman day and sleeping coaches and dining cars have all contributed. Safety devices have considerably eliminated the old dangers. Air brakes, block signals, and steel cars have furthered the safety of passengers, while the automatic coupler has lessened some of the risks for employees. Nevertheless, the loss of life on American railroads is very great, and railroading continues among the most hazardous occupations. In 1920 the roads reported 6,958 killed and 168,309 injured in accidents; of employees, 2,578 were killed and 149,414 injured; of trainmen one in every 391 was killed and one in every eleven was injured.

Rates and Fares.—Equitable and fair freight rates are essential in a country whose economic life depends upon the transportation of large amounts of freight. Railroad transportation is a business that ought to yield increasing returns, for the first cost is highest, and with the growth of population business should automatically increase without proportionate new expenditure. As population has grown and new industries started, transportation has been stimulated, resulting in more efficient service, the application of improved machinery and the construction of better roadbeds. These factors, together with competition, had kept freight rates on the decline until 1899, but since then government regulation has been necessary to keep rates as low as possible. The average receipts per ton mile measured in gold were 1.92 cents in 1867; 1.24 in 1883; .941 in 1890, and .724 in 1899. Since then they have increased—.766 in 1905, .753 in 1910, and 1.052 in 1920.

Passenger fares have not decreased to the same extent, but travelers have benefited by improved service. The average rate per passenger per mile was 2.63 cents in 1871, 2.42 in 1883, 1.99 in 1898, and 1.94 in 1910. Increases in fares were made during the World War by the United States Railroad Commission, and in 1920 by the Interstate Commerce Commission. The average receipts per passenger per mile on Class 1 carriers was 2.75 cents in 1920.

The old theory of charging "all that the traffic would bear" has been modified first by the desire not further to antagonize public opinion, and later by government interference. The present rate-making theory as exemplified by the Transportation Act of 1920 is to allow rates high enough to give a fair return on the actual value of the property. Farmers, miners, and other producers of raw materials have been the great gainers from decreased freight rates, although the people as a whole have profited. The increase of both rates and fares since the war has had a depressing economic effect.

Railroads and the War.—Government regulation has been

the accepted policy since 1887, but it was only during the war that the experiment of government operation was tried. The inability of the railroads to cope successfully with the exigencies of war needs, and the absolute necessity of subordinating transportation facilities to the one purpose of winning the war, brought about this step. The position of the roads at the opening of the European conflict was far from strong. For fifteen years, since about 1897, the costs of railroad operation in maintenance, materials, and labor had been increasing. On the other hand, attempts to gain higher rates had not been successful. In 1910 certain of the roads petitioned for a 10-per-cent advance in freight rates; but this was refused by the Interstate Commerce Commission, which based its decisions on the prosperous year of 1910. Further efforts in 1913 and 1914 were partially successful, when the Commission eventually permitted substantial increases, but these came too late to be of much immediate value. The year 1914 had been disastrous to the railroads, and 1915 found one-sixth (42,000 miles) of the railroad system of the United States in the hands of receivers. The war brought temporary prosperity with the enormous stimulation of the freight business in 1915 and 1916. But the rising cost of materials and the greater expenditures for wages, necessitated partially by the Adamson Eight-hour Act of 1916, absorbed much of the profits. Eventually German submarines and the necessity for using the European merchant marine for war needs decreased the tonnage available to American shipping to such an extent that first the terminals, and later inland cities, were congested with freight that could not be moved. The railroads for some time had been buying little new equipment and were without sufficient rolling stock to meet the pressure of war needs. It was in such a predicament that the roads found themselves when the United States entered the war.

As a whole, the railroads did their best to rise to the occasion. Daniel Willard, president of the Baltimore and Ohio, was appointed as the transportation expert on the Advisory Commission of the Council of National Defense. Under his direction the railroad executives organized a Special Committee of National

Defense, which elected an executive committee of five known as the Railroads' War Board. This board opened offices in Washington and made every effort to coöperate with the government, but as the months went on there were increasing complications due to special army needs, to the orders of the Priorities Board, to lack of equipment, to the inability to force all of the roads to follow the orders of the committee, and to labor difficulties brought about by demands for bigger wages and the departure of men to join the army—all of which pointed to the breaking down of private control and the need of government operation during the war. The Esch bill of May, 1917, had given the Interstate Commerce Commission power to regulate freight cars, and finally on December 26, 1917, upon the advice of the Interstate Commerce Commission after a thorough investigation of the situation, President Wilson issued a proclamation providing for the government operation two days later.

The President's proclamation was followed on March 21, 1918, by a Railroad Control Act which provided (1) that each road taken over should receive an annual payment not to exceed its average net operating income for the three years ending June 30, 1917; (2) that a revolving fund of $500,000,000 should be created to finance the operation; (3) that the roads should be returned to their owners within one year and nine months following the ratification of the treaty of peace; (4) that each road should be returned "in substantially as good repair and in substantially as complete equipment as it was at the beginning of government control"; and (5) that the Interstate Commerce Commission be deprived of its power to suspend rates, but that it retain most of its other powers. The Pullman and express companies were taken over, but not all of the short lines. William G. McAdoo, Secretary of the Treasury, was immediately made Director General of the Railroads. Eventually the whole transportation system was handled through eight administrative divisions at Washington: division of public service and accounting, division of law, division of finance and purchase, division of capital expenditures, division of operation, division of traffic, division of labor, and division of inland waterways. For operating purposes the country was divided into seven regional organizations: Eastern, Alleghany, Pocahontas, Southern, Northwestern, Central Western, and Southwestern. Over each district was a regional director and over each road a federal manager, the latter in some cases a former railroad president. Below the federal manager the regional organization was in general kept intact.

In dealing with labor the government found it necessary to take cognizance of the fact that railroad wages were not so high as other lines of work. Without legislation to require railroad men to stay on the jobs, the only other course was to increase wages, which was done upon the recommendation of a non-partisan board of adjustment. Increased rates were necessary, and on May 25, 1918, all classes of freight were raised 25 per cent, and passenger fares to three cents a mile.

The gains to an efficient prosecution of the war by government operation were many. Joint use of terminals, equipment, repair shops, etc., relieving of congestion by arbitrary routing; better control of traffic at the source; a more efficient handling of troop and war supplies; the elimination of duplicate passenger service and the standardization of equipment in purchasing; all these aided in carrying on the war. On the other hand, it was discovered, when the roads were returned on February 28, 1920, that the excess of operating costs over revenues for the twenty-six months of government control was about $900,000,000. As to how much of this was due, as alleged, to "wasteful government administration," how much to the exigencies of the situation in a war period of mounting costs, and how much to a previous run-down condition of the railroad system, will never be known. Further claims of the roads for alleged under-maintenance and deterioration during government control will probably be largely written off by counter-claims of the government for loans and additions to equipment. Without entering upon a discussion of the efficiency of government operation, it must be said that it accomplished the purpose aimed at and was a necessary war measure. That the government erred in not increasing rates and fares proportionately to the advancing cost of materials and labor is generally conceded.

The Transportation Act of 1920.—One of the most important problems of reconstruction had to do with railroads. Director-General McAdoo advised continued government operation for five years after January 1, 1919. The labor group strongly advocated government ownership and worked hard to popularize the "Plumb plan," which called for the government purchase of the roads and their operation by a board of directors upon which labor would have a one-third representation. The country at large was not prepared for such an innovation and eventually Congress was spurred to action by the President's threat to return the roads on March 1, 1920, whether Congress legislated or not.

The Transportation Act of February 28, 1920 (otherwise known as the Esch-Cummins Act), made up largely of amendments to the Interstate Commerce Act, is important as providing the law under which the roads operated during the post-war period. Among other provisions, it guaranteed the roads for a period of six months after March 1, 1920, a net return equal to one-half the rental paid during government operation. The Interstate Commerce Commission was authorized to divide the country into rate districts, and in each of the districts to prescribe rates which "under honest, efficient, and economical management" would give a "fair return upon the aggregate value of the railroad property." The duty of appraising the property and determining a "fair return" was left to the Commission, but was temporarily fixed for two years at $5\frac{1}{2}$ per cent, with the addition of $\frac{1}{2}$ of 1 per cent to provide for improvements and additions if the Commission thought best. In order that the weak roads might be preserved without permitting too great profits to be reaped by the strong, it was stipulated that any carrier receiving in any year a net income in excess of 6 per cent should turn over one-half of the excess to the Interstate Commerce Commission. This was to be held as a revolving fund to be lent to the weak roads. The half retained by the carriers must be placed in a reserve fund until the sum accumulated amounted to 5 per cent of the value of its property, after which the annual excess income might be used at will. In any year in which the income failed to reach 6 per cent, the reserve might be drawn on for dividends.

The Commission was authorized to work out plans for the consolidation of the roads into not less than twenty nor more than thirty-five systems, among which some competition might be attained. The Commission was given new power in regulating the capitalization of the roads and was now given the right to prescribe minimum as well as maximum rates. Additional powers were conferred upon it in the control of rolling stock and the use of terminal facilities and in respect to the control of new road construction. It will be seen that the authority of the Commission, now increased to eleven members, was considerably augmented by this act.

The railroad strikes of 1919 had brought home to the public the belief that transportation was so essential to the economic life of the nation that repetition of such a catastrophe should be made impossible. On the other hand, the unions insisted on the inalienable right of a laborer to quit his job if he thought it necessary. Congress straddled the issue by authorizing in the act the optional creation of railroad boards of adjustment between one road or group of roads and employees, and in addition created a Railroad Labor Board of nine members, three representing the railroad employers, three representing labor, and three the public. The last three were to be appointed by the President, the representatives of the first two groups choosing their own members. The Railroad Labor Board was authorized to take up questions involving wages, rules, and working conditions which had not been settled by the adjustment boards. Its decisions were not binding, as no penalty for refusal to obey its findings was provided.

Present and Future.—The policy of control, not ownership, was reaffirmed in the Transportation Act of 1920; but in addition a new principle was recognized, namely, the responsibility of the government, after assuming control, of providing the carriers with a fair return upon their investment. This was a notable gain for the holders of railway securities, and it was hoped the future might be brighter. Hardly had the act been passed before the Railroad Labor Board was called upon to consider demands for higher wages made necessary by the mounting cost of living. On July 30, increases were granted which were estimated to add over $700,000,000 to the annual operating expenses. The day previous, however, the Interstate Commerce Commission had allowed an advance in freight rates of from 25 per cent in the mountain states and southern regions to 40 per cent in the eastern districts. This additional income left the roads in an improved position, but their prospects were dashed by the post-war business depression which set in almost immediately. Since 1920 the roads have labored unceasingly to have the Labor Board set aside the working agreements made during the war, which they claimed were wasteful and disadvantageous, and to permit a decrease in wages consonant with the fall in the cost of living. The first of these demands was eventually granted under rules protecting labor. After considerable study and deliberation the Labor Board commenced, in the spring of 1922, to order wage reductions, beginning with the less organized and weaker unions. The statistics, as presented by the railroad executives, showed the scale of living of the railroad worker as improved by the augmented wages of the war and post-war period; but these gains were somewhat dubious, especially when contrasted with those of labor in other industries; and at all odds the laborer was loath to lose them. The dispute was brought to a crisis in the strike of 400,000 shopmen in July, 1922—a strike that dragged through the summer and ended unsatisfactorily without any general settlement. In the height of the strike the President addressed Congress, strongly urging further legislation after the passion of the moment had subsided, to prevent, if possible, the future obstruction of transportation. Such legislation, taking perhaps the form of compulsory arbitration or that of increasing the powers of the Labor Board, may possibly result.

Internal Water Transportation.—With the advent of railways the importance for transportation of internal waterways, both artificial and natural, has constantly diminished; notwithstanding the yearly grants by the government of large sums for the improvement of river channels. Only upon the Great Lakes has the tendency been otherwise. While the tonnage carried by the railroads has increased enormously, that of the rivers has shown an absolute as well as a relative falling off. This is true even of the Mississippi traffic. The high-water mark of river transportation for the lower Mississippi was in 1880, when over 1,000,000 tons were received and shipped at St. Louis from

NAVIGABLE STREAMS OF THE UNITED STATES [1]

	Number	Navigable miles
Tributary to the Atlantic	148	5,365
Tributary to the Gulf (Exclusive of Mississippi)	53	5,213
Mississippi and Tributaries	54	13,912
Flowing into Canada	2	315
Tributary to Pacific	3	1,606
Total	260	26,410

and for the lower Mississippi, a figure which fell to 141,000 in 1905. Receipts and shipments at St. Louis to and from the upper Mississippi declined from 340,000 tons in 1870 to less than 70,000 in 1905. Cotton receipts by river at New Orleans were reduced from 1,087,000 bales in 1880 to 231,000 in 1906. On the Ohio, however, there has been an absolute increase, owing to coal shipments, but a relative decline.

What is true of rivers is even more so of canals. Of the 4,633 miles of canals built before 1909 in the United States, 2,444, or over half, have been abandoned. Of the 2,189 miles in operation, 194 are owned by the national government, 135 by state governments, and 63 are private canals. Of all artificial waterways, the Erie has been the most important, and the statistics of this canal are indicative of the general tendency. The annual tonnage carried on the Erie Canal increased to a high point of over 4,500,000 tons in 1880, only to decline to 2,000,000 in 1905 and to 1,159,000 tons in 1918; this, too, in spite of the fact that tolls were abolished in 1882. At the same time the tonnage of the Erie Railroad in the same state had increased by 1905 to over 30,000,000, and that of the New York Central to over 40,000,000. In 1853 the New York State canal system carried 81 per

[1] Report of Commissioner of Corporations on Transportation in the United States, 1909, part i, p. 28.

cent of the total traffic; in 1873 it carried 35 per cent, and in both 1907 and 1908 only 4 per cent. The traffic which waterways still carry is largely bulky and low-class freight—iron, coal, lumber, grain, and building materials. The Sault Ste. Marie for the year 1908 reported 59.6 per cent of its traffic iron ore, 23.9 per cent coal, and 7.7 wheat. Over half the Mississippi River traffic and over three-fourths of that of the Ohio in recent years has been coal. The tonnage on the Monongahela, Allegheny, and Kanawha is similar. Nearly 90 per cent of the entire tonnage of the Chesapeake and Ohio Canal is made up of coal.

The causes for the decline of water transportation in the United States are many. While ordinarily it is cheaper than by land; this advantage has been lost through other factors. In the first place, American railroads also have been designed to handle large, bulky traffic. As the railroads have improved and enlarged, they have been able in many instances to lower their rates to a point approximately as low as those charged for water transportation. Where this has been impossible the roads have obtained possession of the steamship lines and canals, and either operated or discontinued them. The speed of the railroads and their superiority in handling high-class freight has been influential in diverting traffic to them, for America as a nation likes speed. The many branch lines of the railroads touch an infinite number of points inaccessible to canals. Many of the early canals were unwisely located, others are too short to be used for through freight, still others are built in regions where the original products, such as lumber, have become exhausted. The cost of trans-shipment often eats up what advantages might accrue from cheaper rates. The difficulty of river navigation owing to shifting sand, snags, and other impediments, to say nothing of winter interference, has discouraged development. The large rivers, furthermore, flow in a southerly direction, whereas the bulk of the transportation moves east and west.

Notwithstanding the decline in water traffic, interest has revived in recent years. Temporary inability of the railroads to handle freight, or an increase in rates, invariably turns the attention of shippers to water facilities. Considerable stimulus was given to the interest in artificial waterways by the building of the Panama Canal. In 1903 the people of New York State authorized the expenditure of $101,000,000 to widen and deepen the Erie, Oswego, and Champlain canals, a figure which has been approximately doubled by subsequent appropriations. In 1909 a private company started a canal across Cape Cod to shorten the water route south from Boston, a project completed in 1914. Houston, Texas, in the same year was given more adequate access to the sea by the Houston Ship Canal.

Although the results of the large expenditures for the New York Barge Canal have been disappointing, they have not deterred the enthusiastic advocacy of other similar projects. The most practical of these is the Lakes-to-the-Gulf Deep Waterway, which it is planned to carry along the line of the old Illinois and Michigan Canal and thence to the Mississippi. A more recent project and one of absorbing interest to the middle west is that of developing a route from the Great Lakes to the Atlantic by way of the St. Lawrence.

While river and canal traffic has diminished, that on the Great Lakes has increased. Here the ordinary advantages of water transportation are evident, and in addition long-distance conveyance is provided during a larger part of the year than is possible on canals; the cost of maintenance is smaller, too, than that of canals and rivers. This is seen in the freight rates, which in 1900 were 4.42 cents per bushel of wheat from Chicago to New York by lake and canal and 9.98 by railroad, and in 1920 were 14.60 and 16.68 cents, respectively. Furthermore, the traffic is especially suitable to water transportation, being composed almost entirely of bulky raw material. Anthracite and bituminous coal are transported north and west, while the return shipments are composed of flour, grain, iron ore from the Lake Superior mines, copper, and lumber. The tonnage engaged in the Great Lakes trade increased from 467,700 tons in 1860 to 2,595,062 in 1920. At the same time the tonnage of vessels passing through the Sault Ste. Marie Canal had risen from 403,659 tons in 1860 to 58,194,083 in 1920.

Electric Railways.—The development of electric railways

has taken place almost entirely in the last thirty years. The first practical overhead trolley line was built in Kansas City in 1884, and by 1888 there were thirteen electric railways with forty-eight miles of track in the United States. After 1890 the development was extremely rapid, most of it being in street and interurban service, the single-track mileage of street and electric railways in 1920 amounting to 47,705. There were numerous street railways before 1890, but the motive power was chiefly steam or animal; to-day, however, practically all street railways are operated by electricity.

The electric railroads have supplied an important economic need made imperative by the growth of urban life. So far their chief business has been the transportation of passengers in thickly populated sections, but there are freight possibilities not yet exploited, discouraged hitherto by the competition and hostility of the railroads. The type of traffic thus handled has had an important significance, social as well as economic. The trolley has helped to break down the isolation of country life, decrease its disadvantages, and improve the economic and cultural opportunities. By stimulation of travel and intercourse, it has done more than anything else except the more recent automobile to unite the suburban with the urban communities.

Electric railways have been able to compete quite successfully for passenger traffic with steam roads. Greater cheapness in construction and operation has made possible lower fares, while the ability to send off cars singly has allowed frequent service and thus greater facilities and convenience. Built ordinarily in densely populated regions, they have an advantage over a railroad which is forced to carry the burden of long stretches of thinly peopled country. Nevertheless, the history of electric railways has been a checkered one. The rising costs of fuel and equipment combined with the constant introduction of improvements, rendered necessary in many cases by municipal ordinances, has made operating more expensive. The habit of the five-cent fare, which had become grounded in custom and law, was difficult to break. High financing, overoptimism in construction, and the growth of gasoline motor traffic all added to the discomfiture of electric roads.

Between 1900 and 1913 a considerable portion of the street railways passed through either financial reorganization or actual receivership. The recent war years were particularly disastrous, and only radical fare increases and the most strenuous efforts in economizing have kept some of the companies in operation. The attitude of the people in regard to street railways has been a repetition of their attitude toward the steam carriers—first a period of encouragement and aid through liberal franchises and stock subscriptions; then a period of dissatisfaction and criticism, in many cases justly deserved; and finally a realization that the roads are an essential that must be preserved and regulated for the benefit of the whole community.

An interesting recent development is the introduction of electricity upon some of the steam railroads. This has been tried on parts of the New York, New Haven, and Hartford and the New York Central which serve densely populated communities, and on the Chicago, Milwaukee, and Puget Sound, which has 400 miles of electrified track in the Rockies. The electrification eliminates smoke and noise in the terminals, increases power on grades, and adds to the safety and comfort of the passengers. Resultant savings have so far been small and have been canceled by the heavy costs of installation. The latter factor more than any other has prevented the railroads, in their precarious financial situation, from extended electrification; but since 1904 about 2,700 miles have been so equipped.

Motor Traffic and Road Building.—More pregnant of future transportation possibilities than even the electric and street railroads is the gasoline-driven motor vehicle. After a century of experimentation, mostly with steam, practical cars were produced in 1893; but until 1903 the industry was in an experimental and unstable position. Since that time it has grown so fast that it ranks third among American manufactures and first in finished products. Nothing so well illustrates the wealth of America, the purchasing power and optimism of her people, as the development of the automobile industry. At the end of 1923 there were approximately 14,000,000 automobiles registered in the United States, of which about nine-tenths were passenger cars and one-tenth trucks. The proportion of cars to people is about one to nine, as against one to two hundred in England and France. There are probably seven times as many machines in this country as in the rest of the world.

To the city man the passenger car is a vehicle of convenience, and to the farmer an economic necessity. More important in its social and economic implication than even the trolley has been the automobile in joining the city and country and enlarging the life of the rural communities. The Bureau of Census reports almost a third of the farmers as owning automobiles. Electric and steam railways have been forced in many cases to abandon their service under the competition of hundreds of public motor passenger buses and express trucks with their obvious advantages over cars running on fixed tracks. Motor taxi-cabs have practically eliminated the horse cab, while the motor trucks care for a large part of city and suburban short-haul business. In fact, the motor truck on the short hauls has become a dangerous competitor to the railroad.

The coming of the automobile has brought in its wake a renascence of road building which is reminiscent of the turnpike era of the closing years of the eighteenth century and the early nineteenth. Beginning in 1891 with New Jersey, one state after another has taken up the matter of state aid and supervision in road building, until to-day practically every state government has contributed. In 1916 the federal government reverted to the old policy which built the national pike, and passed a Federal Aid Act, supplemented in 1919 by an act appropriating $200,000,000 to be available in three years. The new act of 1921 appropriated $75,000,000 to be distributed among the states, dollar for dollar contributed by the state, on the basis of population, area, and mileage. Up to November, 1921, there were 28,135 miles of highways completed or in process of construction under the Federal Aid Act, involving a cost of close to a half billion dollars of which the federal government contributed about 40 per cent. In the year 1919 cash expenditures for roads and bridges by federal, state, and local authorities were close to $390,000,000. Paved roads now extend for miles through the open country, while excellent highways traverse the old trails from the Atlantic to the Pacific.

The Post Office.—One of the most important social and economic functions of the government in modern times is the proper collection and distribution of mails. So important is this felt to be that any interference with the mails is a criminal offense. The great development of the postal system has come since 1860. The postal law of 1816, which lasted until 1845, charged six cents for one piece of paper going not over thirty miles, prepayment being optional. In 1847 the rates were lowered and postage stamps of five- and ten-cent denominations introduced, requiring prepayment. The rates were again reduced in 1851 to three cents per half ounce for distances under 3,000 miles and in 1883 to two cents an ounce for all first-class mail. The registration of letters was commenced in 1854, and during Lincoln's administration the free delivery of mail was instituted (1863), the railroad post office (1862), and the money-order system (1864). Under McKinley in 1897 the rural free delivery was commenced with 87 routes, the number of which by 1921 had grown to 43,752, with a mileage of 1,163,896. In 1838 the country declared every railroad a mail route, but the post office pushed ahead and beyond the railroads, its operations in the sparsely settled west of the pioneer days forming a romantic and inspiring chapter.

In recent years the activities of the government have been extended into both the express and the banking businesses, although its entrance in both cases was bitterly fought by the interests affected. A postal-savings system was inaugurated in 1910 to provide absolute safety at low interest for the comparatively poor man. The interest of 2 per cent upon a full year's deposit was in fact so low that the amount placed was relatively small. The postal-saving certificates, issued as evidence of deposit, may be exchanged for postal-saving bonds in denominations of $20, $100, and $500, payable in twenty years and bearing interest at 2½ per cent, exempt from all taxes.[1]

---

[1] During the war the Post Office joined in the campaign for thrift by selling Thrift Stamps in twenty-five-cent denominations, exchangeable for five-dollar

Of more general use, the domestic parcel-post system, introduced in 1913 after successful operation in Europe, has been an immense blessing to the people, transporting as it does small packages more rapidly, cheaply, and safely than by any other means. Some industries, the most notable of which are the great mail-order houses of Chicago, rest a large bulk of their business upon its facilities. Again the farmer has been the great gainer. The following figures show the growth of the postal service:

GROWTH OF POST OFFICE

Year	Gross revenue	Gross expenses
1860	$ 8,518,067	$ 19,170,610
1880	33,315,479	36,542,804
1900	102,354,579	107,740,267
1920	437,150,212	454,322,609

Among the recent innovations which give the most promise for expansion are the pneumatic-tube service in the cities and airplane mail service over long distances.

Telephone.—The closing years of the nineteenth century saw the rapid expansion in the use of the telephone, an invention of epoch-making significance in the history of communication. Although a number of experiments in transmitting the human voice by electricity had been made, it was not until 1879 that Alexander Graham Bell successfully carried on a reciprocal conversation over a line which he had erected between Boston and Cambridgeport, Massachusetts. Bell's first patent was taken out in that year and marked the beginning of the telephone industry, almost wholly American in its origin and in its improvements;

War Savings Stamps, the latter bearing interest at 4 per cent compounded quarterly, and maturing in five years. In 1921 these stamps were discontinued, but a new issue of stamps in denominations of ten cents and one dollar was put out which might be exchanged for Treasury Saving Certificates of $25, $100, or $1,000. The last named were sold at 20 per cent less than their face value and matured in five years. Approximately 75 per cent of depositors in postal savings are of foreign birth and a determined effort has been made to acquaint new immigrants with the advantages of the system by means of leaflets, printed in twenty-four foreign languages, and distributed before the immigrants are released from the ports of debarkation. This information is of real service to the immigrant and lessens the tendency to hoard, due to distrust of banks.

this nation to-day has approximately two-thirds of the telephones in the world. In 1880 there were 34,305 miles of telephone wires in the United States; in 1921 the American Telephone and Telegraph Company, the huge holding company controlling most of the important American systems, reported over 28,000,000 miles under its direction. At the same time it claims 13,380,000 stations and assets of over a billion, making it one of the largest corporations in the world.

Spurred on by the Civil War, a telegraph line was strung across the continent in 1862, and since that time the growth of telegraphy has been rapid. In 1919 there were in the United States 245,560 miles of telegraph lines and 1,433,978 miles of wire, a substantial percentage of the world's telegraph service, while in that year about 167,000,000 telegrams were sent. Not only has the industry grown enormously, but numerous inventions have enabled it to expand in various ways. Beginning with the invention of J. B. Stearns, in 1872, of the duplex method by which two messages might be sent simultaneously in the same direction, the wires were further utilized by the discovery of the quadruplex system of transmitting four messages, two each way, simultaneously, and later by multiplex telegraphy. Submarine telegraphic communication after many unsuccessful efforts was achieved in 1858 and has been in extensive use since 1865. Printing telegraphy has been developed, as exemplified in the stock ticker, and various forms of writing telegraphy have been put in operation. Practically all of the telegraph lines in the United States are privately owned and 98 per cent are under the control of two affiliated companies.

The last twenty years have been marked by the rapid developments of wireless telegraphy, which for over a decade has assumed a practical and commercial rôle in the transportation of messages over both short and long distances. More astonishing than wireless telegraphy and more spectacular in its mushroom growth is that form of wireless telephony known as radio. Comparatively inexpensive and simply constructed bits of apparatus have allowed literally hundreds of thousands of amateur and professional radio enthusiasts to sit in their own homes and catch out of the air news, lectures, music, and other amusement programs to their hearts' content. The radio was hardly two years old before it was estimated that 2,000,000 were making use of the broadcasting services and that $150,000,000 was invested in apparatus.

The development of heavier-than-air flying machines was immensely stimulated by the World War, which can almost be said to have lifted aëronautics out of its earlier experimental stage and airplanes into a practical military and commercial instrument. Although aëronautical progress has not been spectacular since the war, advancement has been made, especially in the direction of mail and passenger transportation and in mechanical tests and experiments made by the army and navy fliers. Notwithstanding the economic chaos of post-war Europe, greater progress has been made there than here in the practical application of various types of airplanes.

NOTES FOR FURTHER REFERENCE

In addition to the various railroad magazines, Hunt's *Merchant's Magazine*, 1835-70, and the *Commercial and Financial Chronicle* since 1870 will be found useful; also the yearly reports of the Interstate Commerce Commission. Seymour Dunbar, *History of Travel in America* (4 vols., 1915), well illustrated and interestingly written, is not so full on the period since 1860. Short résumés are E. R. Johnson, *American Railway Transportation* (1903); E. R. Johnson and T. W. Van Metre, *Principles of Railway Transportation* (1922); C. F. Adams, Jr., *Railroads: Their Origin and Problems* (1878, rev. ed., 1893); A. T. Hadley, *Railroad Transportation* (1886), F. L. McVey, *Railway Transportation* (1921); I. L. Sharpman, *The American Railroad Problem* (1921); George R. Chatburn, *Highways and Highway Transportation* (1923). Exceedingly valuable are the standard studies of W. Z. Ripley, *Railroads, Rates and Regulations* (1912) and *Railroad Problems* (rev. ed., 1912). Summarizing chapters are to be found in E. L. Bogart, *Economic History of the United States* (rev. ed., 1922), and Isaac Lippincott, *Economic History of the United States* (1921). For the reconstruction period in the south, see Carl Russell Fish, *The Restoration of the Southern Railroads* (1919), in University of Wisconsin Studies in the Social Sciences and History, No. 2. For the present situation, see A. T. Hadley, *Factors in the Railroad Situation,* in the *Yale Review,* April, 1923.

On special phases consult L. H. Haney, *A Congressional History of Railroads in the United States 1850-1887* (1910); S. J. Buck, *The Granger Movement* (1913), and the *Agrarian Crusade* (1920), in Chronicles of America; F. Cleveland and F. W. Powell, *Railroad Promotion and Capitalization* (1909); W. F. Gephart, *Transportation and Industrial Development in the Middle West,* Columbia University Studies (1909); M. B. Hammond, *Railway Rate Theories of the Interstate Commerce Commission* (1911); E. R. Johnson and G. G. Huebner, *Railway Traffic and Rates* (1911); W. C. Noyes, *American Railroad Rates* (1905); Leonor F. Loree, *Railroad Freight Transportation* (1922); Rogers MacVeagh, *The Transportation Act, 1920* (1922); Frank H. Dixon, *Railroads and Government; Their Relations in the United States 1910-1921* (1922); W. J. Cunningham, *American Railroads: Government Control and Reconstruction* (1922).

The history of certain of the specific railroads has been written—*e. g.*, H. S. Mott, *Story of the Erie* (1900); C. F. Adams, *Chapters of Erie* (1886); J. P. Davis, *The Union Pacific* (1894); E. V. Smalley, *History of the Northern Pacific* (1883); Stuart Daggett, *Chapters in the History of the Southern Pacific* (1922). Biographies which throw much light on early railroad building are: H. G. Pearson, *An American Railroad Builder* (1911), relating the career of J. M. Forbes; E. P. Oberholtzer, *Jay Cooke, Financier of the Civil War* (2 vols., 1907), J. G. Pyle, *Life of J. J. Hill* (2 vols., 1917), and the *Memoirs of Henry Villard* (2 vols., 1904).

Two legislative reports giving insight into early abuses are those of the "Hepburn Committee," New York State Assembly Document, No. 38 (1880), and of the "Cullom Committee," Senate Reports, Forty-ninth Congress, First Session, Serial Number 2356 (2 vols.).

The testimony of Hines, McAdoo, and others on government operation during the war is in *Extracts from Hearings before the Committee on Interstate Commerce,* United States Senate, Sixty-seventh Congress, Second Session, Senate Resolution 23, Washington, 1922.

On waterways consult C. L. Jones, *Economic History of the Anthracite-Tidewater Canals,* University of Pennsylvania Series in Politics, Economics, and Public Law, No. 22 (1908); H. G. Moulton, *Waterways vs. Railways* (1912); and H. B. Hepburn, *Artificial Waterways of the World* (1914). For the history and organization of the postal system, see D. C. Roper, *The United States Post Office* (1917), by a one-time First Assistant Postmaster-General.

The general histories of the period and the biographies of the Presidents will be found helpful—*e. g.*, C. R. Lingley, *Since the Civil War* (1920); F. L. Paxson, *Recent American History* (1921); C. S. Olcott, *William McKinley* (2 vols., 1916); R. M. McElroy, *Grover Cleveland; the Man and the Statesman* (2 vols., 1924); Theodore Roosevelt, *An Autobiography* (1913); Lord Charnwood, *Theodore Roosevelt* (1923); W. E. Dodd, *Woodrow Wilson and His Work* (1920); and C. Seymour, *Woodrow Wilson and the World War* (1921), in Chronicles of America.

SELECTED READINGS

JOHNSON, E. R., and VAN METRE, T. W., *Principles of Railway Transportation,* Chaps. II-IX; XIV-XXXI.
MOODY, JOHN, *The Railway Builders,* Chaps. VI-XII.
MOULTON, H. G., *Waterways Versus Railways,* Chaps. I-V; XV-XIX.
RIPLEY, W. Z., *Railroads: Rates and Regulations,* Chap. I.
RIPLEY, W. Z. (ed.), *Railway Problems,* Chaps. I-IV.

CHAPTER XXI

FINANCIAL HISTORY SINCE 1860

Financing the Civil War.—The effects of the Civil War upon the financial structure of our nation were revolutionary in their scope. The needs of war financing enormously increased the rates of the protective tariff of 1861, created a new banking system, injected into the currency fiat paper money, and inaugurated a bitter half-century conflict between the currency inflationists and contractionists. The cost of the war was stupendous for that period, and much greater than was believed possible at the opening of the struggle. David A. Wells, Special Commissioner of Revenue, estimated in 1869 that the total war expenditure of the national government in the eight and a quarter years of the war and post-war period was $4,171,914,498.33. In addition should be reckoned the payment for pensions amounting to $2,000,000,000, and the further direct and indirect losses which he estimated as follows:

Increase of state debts, mainly on war account.....	$ 123,000,000
County, city, and town indebtedness increased on account of the war (estimated)..............	200,000,000
Expenditures of states, counties, cities, and towns, on account of the war, not represented by funded debt (estimated)	600,000,000
Estimated loss to the loyal states from the diversion and suspension of industry, and the reduction of the American marine and carrying trade....	1,200,000,000
Estimated direct expenditures and the loss of property by the Confederate States by reason of the war	2,700,000,000

"These estimates, which are believed to be moderate and reasonable," said Wells, "show an aggregate destruction of wealth, or diversion of industry, which would have produced wealth, in the United States since 1861, approximating *nine* thousand *mil-*

lions of dollars. . . . What does it measure? It is substantially a thousand millions a year for nine years; or, at the wages of five hundred dollars a year, the labor of two millions of men exerted continuously during the whole of that period. It is three times as much as the slave property of the country was ever worth. It is a sum which at interest would yield to the end of time twice as much as the annual slave product of the South in its best estate." [1]

In four years the government expenditures had been greater than during the whole previous history of the nation. In a frantic effort to meet these expenses, Congress used every known device for obtaining revenue. Since the tariff had supplied most of the federal income up to that time, it was regarded as the most important source. A new tariff, framed largely by J. S. Morrill, had been passed by the House the previous year, and ratified by the Senate on March 2, 1861, just before Lincoln's inauguration; it aimed to supplant the low rates of 1857 and restore the general level of the Walker tariff of 1846, the average rates of which were about 25 per cent. But the Morrill tariff did not produce the income anticipated, and legislation during the succeeding years gradually raised the average of duties, until in 1864 it reached 47 per cent.

The continued tariff increases had been made to balance in some degree the high internal taxes, imposed upon a wide variety of manufactured articles and business transactions. The method of distributing the excises was likened by Wells to that of the Irishman at Donnybrook Fair: "Whenever you see a head, hit it; whenever you see a commodity, tax it." [2] By the end of the war the internal revenue was yielding twice as much as the tariff. Between 1861 and 1865 over $1,280,000,000 was pro-

---

[1] Report of the Special Commissioner of the Revenue, 1869, p. vi, Executive Document No. 27, House of Representatives, Forty-first Congress, Second Session.

The direct and indirect cost of the war to both north and south, including the interest on the national debt and pensions up to 1909, is placed at $15,500,000,000 by the Encyclopædia Britannica, article on United States History 1783-1865, by Alexander Johnston and C. C. Whinery.

[2] Quoted by Taussig, F. W., *Tariff History of the United States*, 5th ed., p. 164.

duced from internal taxes, and over $910,000,000 from customs duties. Another form of taxation, that upon incomes, was levied for the first time in our history in 1861, when 3 per cent was imposed on incomes above $800. Increased in 1862 and 1865 until incomes between $600 and $5,000 were taxed at 5 per cent and those above $5,000 at 10 per cent, this tax yielded about $347,000,000 before it was abolished in 1872.

While taxes were levied with a free hand, a large proportion of the income derived therefrom did not come in until after the war was over. In the meantime the conflict was prosecuted upon borrowed money obtained through the flotation first of short-time loans and later, when it became evident that the war would not be over in a few months, by long-term borrowing. On September 1, 1865, the public debt reached its highest point, when it amounted to $2,846,000,000, made up of many types of notes. In June, 1866, the "interest-bearing debt consisted of loans bearing five different rates of interest and maturing at nineteen different periods of time."[1] Eight-ninths of these were short-time notes.

One other method of financing the war expenses of the national government brought results which were to plague the country for decades. This was the issuance of paper money "on the credit of the United States." Treasury notes had been issued at times of stress previously in our history, but they had been interest-bearing, had not been legal tender, had been issued for the most part in large denominations, and hence had had a very small circulation as currency. Other than these, no paper money had ever before been issued by the national government. Bank notes, supplied by some 1,600 state banking institutions, and the metal coined by the United States mint, together with the "old demand notes," issued in 1861, made up the currency at the beginning of the war. The drain upon the metal in the Treasury had occasioned the suspension of specie payments in December, 1861, first by the banks and soon afterward by the government. Metal currency was hoarded (except on the Pacific coast) and the

---

[1] Dewey, D. R., *Financial History of the United States*, 8th ed., p. 332.

issues of state banks were inadequate to meet the increased needs for a circulating medium. Since an issue of currency backed by metal was impossible, inconvertible paper seemed to Congress the only solution. On February 25, 1862, an act was passed authorizing an issue of $150,000,000 in notes on the credit of the United States, and making them legal tender. Supplementary acts followed in July, 1862, March, 1863, and June, 1864—altogether authorizing greenbacks to the extent of $450,000,000; though no more than $400,000,000 was used for actual currency purposes at any one time, inasmuch as $50,000,000 of the first issue was for redemption of the Treasury demand notes of 1861, and the same amount by the law of June, 1864, was for redemption of a temporary loan. The colloquial term greenbacks originated from the green ink used in printing the backs of these notes. The greenbacks were supplemented by an authorization of $50,000,000 for fractional currency in denominations as low as three cents, to replace the subsidiary coins that, as the war went on, had been hoarded and withdrawn from circulation. On September 1, 1865, $433,160,000 in United States notes was outstanding, besides $26,344,000 in fractional currency.[1] Other government obligations were also employed as currency, but were interest-bearing. The war left the country with the problem of fiat money still to settle, and the greenbacks became a burning issue, both legally and financially.

National Bank Acts of 1863 and 1865.—One of the beneficial results of the war was the elimination of the chaotic paper currency of the state banks, and the substitution for them of the national bank notes. In 1862 there were about 1,600 banks established under the laws of the various states and circulating at a discount varying with the distance from the bank of issue. "It

---

[1] Earlier efforts to replace small coins by those who had to make change had resulted in the use of stamps, old Spanish quarter dollars, bank bills cut in halves or quarters, and the issuance of tickets, due bills, and other forms of obligation by individuals, firms, banks, and even municipalities. These were called "shinplasters." Congress interfered with this private issuance of money and authorized first the use of postage stamps; and later, to prevent the inconvenience of using gummed stamps as currency, authorized the Post Office Department to issue postal currency, in denominations of five to fifty cents. By May 27, 1863, over $20,000,000 had been placed in circulation.

was estimated," said A. Barton Hepburn, "that there were 7,000 kinds and denominations of notes, and fully 4,000 spurious or altered varieties were reported." [1] It was difficult to estimate the value of the various notes, and the annoyances and losses attendant upon doing business with them was great. Those advocating the creation of national banks believed they would (1) create a market for United States bonds, (2) drive out the numerous notes of the banks operating under the state laws, (3) create a powerful financial interest to back the government, and (4) provide the country with a standardized paper currency. In the long run, these hopes were realized. The National Bank Act of 1863 granted charters to groups of not less than five stockholders (the amount of the capital stock being graduated for cities of different sizes), who might buy government bonds, deposit them with the Treasurer of the United States, and receive in return bank notes up to 90 per cent of the current market value of the bonds. These notes were receivable for all government dues except duties on imports. Numerous provisions, including reserve requirements, liability of stockholders, and strict national supervision, protected the depositor. Banks were slow in taking out charters, and in 1865 state bank issues were driven out by a 10-per-cent tax. Supplementary legislation in 1900 liberalized the National Bank Act by permitting bank notes to be issued up to the full par value of the bonds; by reducing the capitalization of banks for cities of 3,000 or under from $50,000 to $25,000; by refunding the existing national debt in thirty-year 2-per-cent bonds; and by reducing the tax from 1 to ½ per cent per annum on all bonds yielding not over 2 per cent. The bank-note circulation, which reached $339,000,000 in 1873, declined to $168,000,000 in 1891 as the bonds fell due and were retired. Eventually further loans, issued to cover the Spanish-American War and other expenses, increased the note circulation by 1913 to $715,754,236 issued from 7,473 national banks.

---

[1] Hepburn, A. Barton, *History of Coinage and Currency in the United States*, p. 177. An interesting old book compiled to aid in detecting spurious notes, and published in New York, 1863, is *Hodge's Bank Note Safeguard;* giving facsimile descriptions of upwards of ten thousand bank notes, embracing every genuine note issued in the United States and Canada.

Financial and Economic Results of War Financing.—It would be incorrect to attribute wholly to the legal-tender greenbacks the rise in prices which took place during the war; contributing factors were the numerous short-term Treasury notes which passed almost as money and the enlarged issues of local bank notes. The increased demand for commodities of all sorts likewise made prices go up. But issuing nearly $450,000,000 worth of irredeemable paper money at a time when prices were already on the rise greatly emphasized the movement. Measured in terms of gold, as will be seen from the accompanying graph, the greenbacks at no time during the war reached par, dropping in the summer of 1864 as low as 39. The fluctuations in the value of the greenbacks was partly caused by speculation and by varying confidence in the credit of the government, and do not necessarily approximate the rise in living costs.

The chaos in the currency of post-war Europe has illustrated very clearly how the value of money fluctuates with the credit of the government and the metallic reserve maintained. Without some tangible backing (ordinarily gold) the value of money is dependent largely on confidence in the government. Inasmuch as the most stable government keeps on hand gold sufficient to cover only a fraction of the paper currency, the value of any paper currency rests in the last analysis on the belief of the public that the financial system of the government is secure. American paper money circulates at par because there is at the present time no doubt as to the ability of the United States to redeem it; paper currency of some European states is practically worthless, because it is clear that the governments cannot possibly redeem their issues.

While the government during the Civil War gained as an employer of labor from the issuing of the greenbacks, it lost in the long run from the rise in prices. It has been estimated that the cost of the war was increased by over $500,000,000 by the decline in value of legal-tender paper. The rise in prices during the war has been more fully treated elsewhere.[1] Between 1860

---

[1] See chap. xvii.

and 1865 textiles quadrupled in price; groceries and flour doubled; meat, fuel, and rents increased over 50 per cent. At the same time real wages, especially for salaried men, lagged far behind.

The Greenback Movement.—After the suspension of specie payments at the end of 1861 metallic money became scarce. It went entirely out of circulation with the appearance of the legal-tender notes, a demonstration of Gresham's law—that bad money drives out good.[1] Just as gold had driven out silver after 1834, so now paper drove out both silver and gold.

Three main problems in connection with the greenbacks pressed for settlement at the close of the war: (1) Had Congress the power under the Constitution to issue legal tender? (2) Should the existing issues be enlarged or contracted? and (3) Should specie payment be resumed? The first question was settled in 1871 by the Supreme Court in the case of Knox v. Lee[2] (overruling a previous decision in Hepburn v. Griswold, 1870),[3] by affirming that greenbacks could be presented to satisfy any debt contracted before the legal-tender acts were passed. The other questions reopened the conflict between inflationists and contractionists which has, to a certain extent, been ever present in our history. The inflationists were generally the debtors, chiefly westerners who had pioneered during the war period of rising prices and were now finding it increasingly difficult to swing their mortgages and pay the debts incurred in buying stock and machinery. Declining prices for foodstuffs and a contracting currency were bringing real hardship to the debtor farmer and cheaper money seemed to be the logical way out. The bondholding creditor looked upon currency inflation as an attempt to rob him of his just interest and undermine the credit of the government.

The first move of the inflationists was to urge the payment in

---

[1] When two currencies of the same face value but of different intrinsic worth circulate, the more valuable will be hoarded and the less valuable will remain in daily use and determine the worth of the circulating medium.

[2] Knox v. Lee, 12 Wallace, 457.

[3] Hepburn v. Griswold, 8 Wallace, 603. This decision is especially interesting inasmuch as it was delivered by Chief-Justice Chase, who had been Secretary of the Treasury when greenbacks were first authorized, and who here passed adversely on the constitutionality of his own acts.

paper of both interest and principal upon all Civil War bonds unless the bond definitely specified gold. This plan, known as the "Ohio idea," was incorporated in the Democratic platform of 1868, but the election of Grant insured the payment of the debt in gold. Having lost this battle, the inflationists attempted not only to prevent the further contraction of the greenbacks, but even to increase the amount. In 1864 $400,000,000 had been fixed as a maximum, with a further reserve of $50,000,000 to redeem a temporary loan. In 1866 the Treasury was authorized to retire the greenbacks, and Secretary McCulloch did withdraw about $77,000,000, reducing the volume to about $356,000,000 when in February, 1868, further contraction was suspended. This amount remained practically constant for nearly five years, except for reissues of retired notes under Secretaries Boutwell and Richardson, who held that the $400,000,000 maximum passed in 1864 still obtained, and issued from what they called the greenback reserve of $44,000,000 such amounts as they thought advisable. In 1874 Congress fixed the maximum at $382,000,000. The Resumption Act (1875) provided for reducing the volume of United States notes to $300,000,000, but the greenbackers succeeded again in 1878 in suspending further contraction. The amount then outstanding, $346,681,016, is the number of greenbacks still circulating to-day.

The fight now turned to the resumption of specie payments. The exchange by the government of gold for paper would bring the greenback to par, stabilize the currency, and raise the credit of the government. On the other hand, it was believed it would further depress prices and it was doubtful if gold in sufficient quantities could be obtained for the purpose. The defeat of the Republicans in the congressional election of 1874 caused that party to hasten through in the following year a Resumption bill calling for the restoration of specie payments on January 1, 1879. This controversy over paper money, especially in its last phase, gave birth to a new political organization, the Independent National, or Greenback, party. Formed in 1876, it presented national tickets in three presidential campaigns, calling for a number of reforms radical for that day, particularly the redemption of war bonds in paper, and the non-resumption of specie payments. Its greatest strength was exhibited in the congressional elections of 1878, when it polled over a million votes.

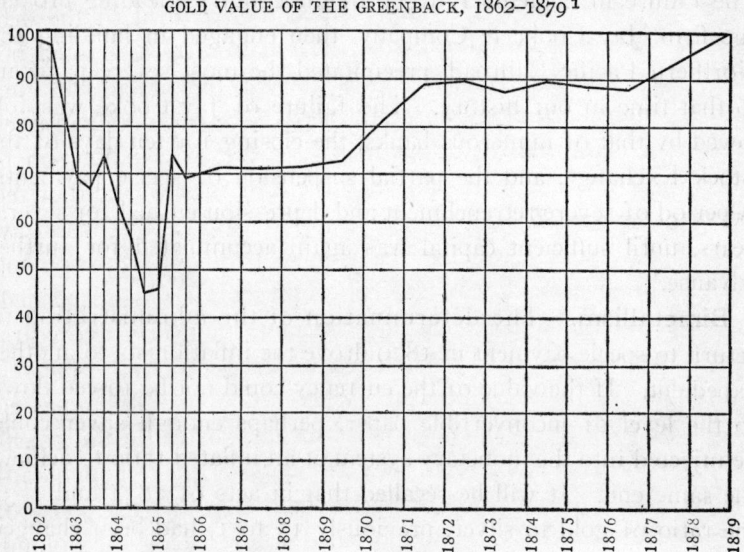

GOLD VALUE OF THE GREENBACK, 1862–1879 [1]

Panic of 1873.—In the meantime the distress of the debtors had been accentuated and the greenback movement stimulated by the panic of 1873. The feverish industrial and agricultural activity in the north during the Civil War, aided by the rising prices, had inaugurated a period of unprecedented prosperity. Immense regions in the west had been opened up to agriculture, while the easy profits of war prosperity had been invested freely in fixed forms of capital, notably transportation facilities. The prosperity had been too rapid, the expenditures too lavish, to be healthy; and the decade of the 'seventies opened with underlying conditions far from encouraging. Enormous amounts of capital had been sunk in railroads to finance the 30,000 miles built between 1867 and 1873, from which small immediate returns could be expected. The opening of western lands had thrown the older areas out of cultivation and decreased their value.

---

[1] Figures from D. R. Dewey, *Financial History of the United States*, 1st ed., pp. 293, 376.

Speculation and extravagance were rampant, and the business morality of politicians and capitalists, as witnessed by the Crédit Mobilier and the Black Friday scandals, left much to be desired. The failure in September, 1873, of the country's leading brokerage firm, Jay Cooke & Company, then engaged in building the Northern Pacific Railroad, precipitated the most severe panic up to that time in our history. The failure of Jay Cooke was followed by that of numerous banks, the closing for ten days of the Stock Exchange, and the partial suspension of specie payments. A period of severe retrenchment and depression ensued for several years, until sufficient capital was again accumulated for further advance.[1]

Bimetallism.—The determination of the administration to return to specie payment in 1879 drove the inflationists to another expedient. If the value of the currency could not be forced down to the level of inconvertible paper, perhaps enough silver could be injected into the monetary system at an inflated ratio to achieve the same end. It will be recalled that in acts of 1834 and 1837 the ratio of gold to silver, previously 15 to 1, had been changed to 16 to 1, the gold dollar thereafter containing 25.8 grains nine-tenths fine, and the silver dollar remaining 412.5 grains. As this slightly overvalued gold, under the workings of Gresham's Law gold rapidly came into circulation and silver disappeared. In 1873 the silver dollar was worth $1.02 in gold and it was no longer profitable to coin it. So scarce was silver and so long had it been since any had been presented to the mints for coinage that Congress dropped in 1873 the further minting of the silver dollar. Far from being a deep-dyed plot to demonetize silver, this act, afterward denounced as the "Crime of '73," was merely a legislative recognition of the fact that silver dollars were not being coined.

The situation in regard to silver, however, changed rapidly after 1873. Germany in 1871, Holland and the Scandinavian peninsula in 1875, had adopted the gold standard, while the Latin Monetary Union (France, Switzerland, Belgium, Italy, and

---

[1] For a further discussion of panics and business cycles in our history, see chap. xxvi.

Greece) had limited in 1873 the coinage of silver. This threw a large supply of bullion on the market, which was augmented by the discovery of large deposits in Nevada. The price of silver dropped so sharply that in 1876 it was worth ninety cents, with the prospect of further decline.

As silver grew cheaper, it was evident that if enough could be coined at the old ratio of 16 to 1 the working of Gresham's Law would drive out the gold and reduce the currency to the value of silver. The demonetization of silver was now called the "Crime of '73," and the debtor west, backed by the silver states, demanded that the government "do something for silver."

The Bland-Allison Act of 1878 and the Resumption of Specie Payment.—The silver sentiment had grown so strong by 1876 that a commission had been appointed to study the currency problem, but before it had presented its report Richard Bland of Missouri offered in 1877 a bill for the free and unlimited coinage of silver at the old ratio of 16 to 1. In the more conservative Senate the bill was toned down to limit the purchase of bullion to not less than $2,000,000 and not more than $4,000,000 a month, to be coined into silver dollars of 412.5 grains. Vetoed by the President, it was passed over his veto, and during the twelve years of its operation 378,166,000 silver dollars were coined. The act of 1878 provided for the issuing of silver certificates in amounts of ten dollars and upward upon the deposit of silver dollars; but the metal money proved unpopular in business centers and in 1886 the denomination of the certificates was reduced to include one, two, and five-dollar bills. As the banknote circulation decreased by $126,000,000 between 1886 and 1890, while at the same time the normal needs of business increased, the certificates of the Bland-Allison Act were absorbed into the currency without disturbing the financial system or halting the downward trend of prices.

On the first of the year following the passing of this act the Treasury went back to specie payment. Secretary Sherman had a slender supply of $140,000,000 in gold on January 2, 1879, accumulated with great difficulty, to meet the expected rush of holders of paper, but the credit of the government was demonstrated by the fact that only $125,000 was presented for gold, while $400,000 in gold was turned in for paper.

The Sherman Act of 1890.—Although the Bland-Allison Act was in force for twelve years, it was unsatisfactory to both the inflationists and their opponents. The former looked upon it as simply an opening wedge to be pushed farther if the purposes for which it had been passed failed to materialize. Prices of agricultural products continued to fall, and the distress of the debtor farmer who had borrowed on a fifty-cent dollar and must pay his debts with an eighty- or ninety-cent one, became keener. The silver in the dollar by 1889 had declined to seventy-two cents and hope was still alive that if more silver was forced into the currency, inflation would take place.

This was exactly what the gold advocates feared and both President Arthur and President Cleveland urged the repeal of the act, the latter pointing out to Congress that the continued coining of silver dollars would eventually increase the currency beyond the needs of business, after which the unnecessary portion would be hoarded and thus the gold gradually eliminated.

Notwithstanding the opposition of the Chief Executive and the Treasury Department, the pressure for more silver became so great that the Republican party, as a matter of political expediency, and as a means of insuring the passage of the McKinley tariff, sponsored and passed the Sherman Silver Purchase Act in 1890. This bill required the Secretary of the Treasury to purchase 4,500,000 ounces of silver bullion a month and to issue in payment for it Treasury notes having full legal tender. These notes were to be redeemed in gold or silver at the discretion of the Secretary, "it being the established policy of the United States to maintain the two metals on a parity with each other," a provision later interpreted by the Executive as a promise to redeem all notes in gold. The amount of silver purchased under this act was practically the entire output of the American mines and was almost double that required by the Bland-Allison Act, amounting to $155,931,002 in the three years of its operation. This proved to be more than the currency could stand without endangering the gold standard, and the only saving feature of the bill was the provision to purchase by ounces rather than by dollars, which meant that the amount of silver purchased by the government would be kept at a uniform level. If the law had provided for the purchase by dollars (as in the Bland-Allison Act) rather than by ounces, a decline in the value of silver would automatically have increased the coinage of that metal. Since the value of silver fell steadily during the life of the act until it reached sixty cents in 1893, the significance of this provision is apparent.

The Panic of 1893 and the Election of 1896.—The decade of the 'nineties opened with the nation approaching the end of another business cycle. Railroad building during the 'eighties had been accompanied by inordinate speculation, which had undermined supposedly strong organizations. Corporations on the verge of bankruptcy declared stock dividends and paid regular dividends out of capital. The failure of the Philadelphia and Reading and the National Cordage Company early in 1893 aroused the nation to the unhealthy industrial situation, which had already been foretold by financial conditions in Europe. A reaction from the highly speculative years through which the country had just passed was inevitable, but the panic of 1893 was precipitated because of apprehension that the government would not be able to maintain the gold standard.

The precarious situation in which the government found itself was due to a number of causes. The amount of silver purchased under the Sherman Act was too large to be readily absorbed, and gold began to be crowded out of circulation. The financial crisis in England in 1890 brought about liquidation there, which resulted in net loss of $68,000,000 in gold exported from the United States. The bumper wheat crop of 1891 coincident with a failure of European crops gave a temporary favorable balance of trade; but in 1893 the situation was reversed, with a net loss of $87,000,000 in gold exported. To complicate the difficulty of the Treasury Department, the surplus of the 'eighties had been wiped out by the extravagances of the Harrison administration and by the McKinley tariff of 1890 [1] and a Treasury deficit was impending in 1893. An act of 1882 which authorized the Secre-

---
[1] See p. 579.

tary of the Treasury to suspend the issue of gold certificates whenever the amount of gold coin or bullion in the Treasury reserved for the redemption of United States notes fell below $100,000,000, tacitly recognized the existence of a reserve and set a minimum safety point. Subsequent Treasurers had not allowed the reserve to fall below this point and it had so far been sufficient to maintain the gold standard even after the added strain imposed by the Sherman Silver Purchase Act of 1890. A wiping out of the gold reserve would mean the suspension of specie payments or the substitution of silver for gold in the payment of paper presented under the act of 1890. Either case would mean the elimination of the gold standard and the cheapening of money. While this would have brought joy to the inflationists, the mere possibility paralyzed with fear the holders of fixed capital and business in general.

When Cleveland was inaugurated the reserve was $100,982,410 and, on April 22, 1893, it fell below the $100,000,000 mark, recovered temporarily in July, and then declined until in November it reached $59,000,000. Failures of well-known concerns had already shaken public confidence in the business structure, and the decline of the reserve set in motion a period of liquidation the most severe yet experienced. During 1893 over 600 banking institutions failed, while during the summer 74 railroad corporations owning 30,000 miles of road passed into the hands of receivers. By the end of the next year 194 roads operating 39,000 miles had failed, including the Philadelphia and Reading, the Erie, the Northern Pacific, and Union Pacific. More than 15,000 commercial failures involving liabilities of $346,000,000 were recorded for 1893. The production of iron and coal declined, and to add to the general distress there was a poor corn crop in 1894 and a decreased demand on the part of Europe for wheat. Unemployment, strikes, discontent, and much actual suffering characterized the winters of 1893 and 1894, a period which encompassed the Pullman strike in Chicago and the marching of "Coxey's army."

Cleveland, a hard-money man, was determined at all costs to maintain the gold standard. Believing rightly that the distress of the Treasury was partially due to the Sherman Silver Purchase Act, he called Congress in special session on August 1 and demanded its repeal. A bill to this effect passed the House with little delay, but the Senate held it up until October 30, when it was granted by a sectional vote, with the west and south aligned against the north and east. The repeal of the Sherman Act came too late to stave off the panic and the exhaustion of the gold reserve. In January, 1894, the Treasury sold $50,000,000 of 5-per-cent ten-year bonds to obtain gold, and in November resorted to an additional loan of $50,000,000. The gold obtained in this way soon drained out, for there was nothing to prevent the man who loaned gold one day from presenting paper on the next and demanding it back. Borrowing on these conditions seemed useless, and when in February, 1895, the Treasury found itself with a reserve of only $41,000,000 and that declining at the rate of $2,000,000 a day, Cleveland negotiated with J. P. Morgan and a group of bankers for a loan of 3,500,000 ounces of gold to be paid for in 4-per-cent United States bonds. It was agreed that half the gold should be obtained abroad and that the bankers would exert every influence to prevent its withdrawal until the contract had been fulfilled. The action of Cleveland in borrowing privately from the bankers brought down upon him a storm of abuse, and when, a year later (January, 1896), a fourth loan was resorted to, it was offered to the public. Liquidation by this time had run its course; the loan was several times oversubscribed, and during the year the gold reserve in the Treasury continued to mount.

The efforts of the inflationists to expand the currency, continuously evident since the close of the Civil War, now reached their climax in the campaign of 1896. Bitterly disappointed over the stand taken by Cleveland during the panic of 1893, the Populists and free-silver Democrats of the west and south rallied to William J. Bryan and his platform of free and unlimited coinage of silver at a ratio of 16-1. In one of the hardest-fought and most important campaigns in our history, a campaign which hinged on the question of free silver, bimetallism was repudiated.[1]

---

[1] See chap. xviii for further discussion of this campaign.

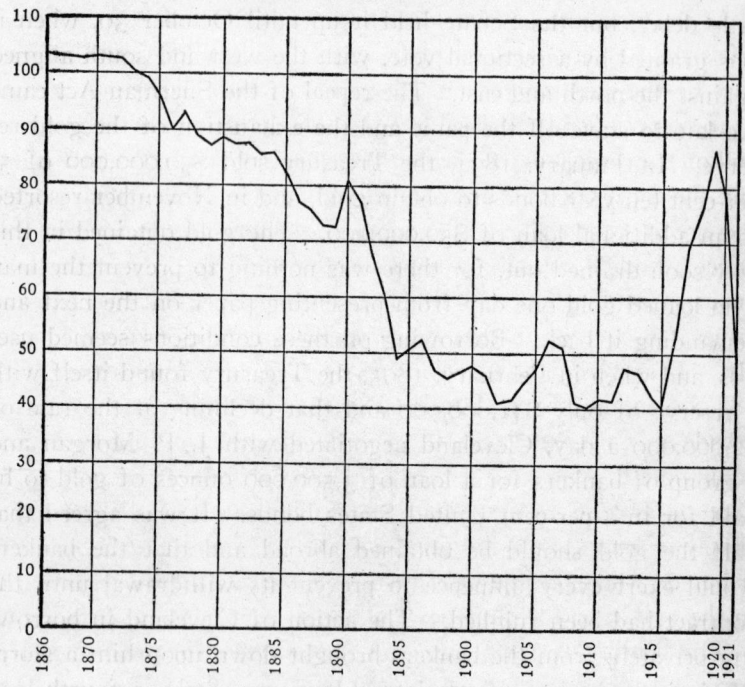

BULLION VALUE OF 371¼ GRAINS OF SILVER (CONTENTS OF ONE UNITED STATES SILVER DOLLAR) AT THE ANNUAL AVERAGE PRICE OF SILVER EACH YEAR, 1866-1921 [1]

The Currency Act of 1900.—Although the result of the campaign of 1896 settled the question of bimetallism, the silver advocates were still so strong in the Senate that it was not until four years later that the gold standard was adopted. In the Currency Act of 1900 it was provided (1) that the gold dollar of 25.8 grains be established as the standard unit of value, and that all other forms of currency be maintained at a parity with this; (2) that the gold reserve in the Treasury be kept at $150,000,000; and (3) that in a crisis this reserve be maintained by the sale of short-term bonds and by the retention of legal tender or Treasury notes once presented until sufficient gold had been accumulated to release them. Other provisions called for the

---

[1] *Statistical Abstract*, 1921, p. 619.

retirement of the Treasury notes of 1890 and the substitution for them of silver certificates based on coined silver dollars, and the liberalizing of the laws governing national banks.

At the time the act was passed it was doubtful if in a severe crisis the $150,000,000 of gold reserve could withstand the pressure of the existing redeemable money, which then amounted to $346,000,000 of greenbacks, $76,000,000 of silver (coined or bullion), every dollar of which was worth but forty-seven cents, and the $331,000,000 of national bank notes based entirely upon the credit of the United States. The machinery for maintenance, though clumsy, has proved efficient and the revival of business after 1898 and the influx of gold from the Alaskan mines (discovered in that year) speedily increased the gold supply. The average annual coinage of gold, which had been $67,185,000 in the years 1891-1900, increased during the following decade to $101,022,000. This with additions to the supply of bank notes increased the per-capita circulation from $23.85 in 1893 to $33.86 in 1907 and $34.20 in 1911. Both the extra supply of money and the prosperity of the years after 1896 contributed to bring a rise in prices and a decline, for the time being, in the agitation for currency inflation.

The Panic of 1907 and the Movement for Banking Reform.—In the campaign of 1896 McKinley had been heralded as the "advance agent of prosperity." In truth, liquidation had about run its course and he entered office on a returning wave of prosperity which advanced with but few interruptions until halted in 1907. The discovery of gold in Alaska, a succession of good harvests, the greater activity of American exporters in foreign fields, coincident with our embarkation upon a career of colonial power, all served to stimulate business. Rising prices and confidence in the administration helped to promote rapid expansion and a great movement toward consolidation.

By 1907 the demand for capital had outstripped the available supply. The nation had overspeculated and overinvested. Investors became more cautious and the banks began to contract their loans. The action of the Knickerbocker Trust Company in closing its doors on October 22, to prevent a run on the bank, precipitated a panic which ruined many of the more speculative ventures and caused the partial suspension of specie payments by the banks. As an extraordinary measure sound banks were allowed to resort to clearing-house loan certificates based upon approved securities to tide them over the crisis. The panic was largely limited to business in the cities, and its effects were not widespread, a fact which has given it the name of the "rich man's panic." Recovery was rapid and the nation continued to enjoy further prosperity until the temporary depression of 1914. The panic of 1907 was laid by capital at the doors of the Roosevelt administration, whose meddling with business, it was claimed, had brought on the catastrophe. The real causes, of course, were deeper and were those normally indicating the end of a business cycle. The panic of 1907 at least served one good purpose in that it brought out clearly the defects of the national-bank system founded in 1863. While the national banks had marked a long step in advance by providing safe banking facilities and a standard bank note based on the credit of the national government, some further improvements were necessary. Probably the chief criticism brought against the system was its lack of elasticity. The Currency Act of 1900 had extended the issue of bank notes from 90 per cent to the full face value of the bonds upon which they were issued, but in times of emergency this did not provide sufficient currency. More could be obtained only by purchasing additional bonds, a fact which tied up the whole question of paper money with the public debt and meant that a decrease in the debt would contract the bank notes. It was further maintained that the rigid reserve limits fixed by law made the credit facilities needlessly inelastic. The smaller communities urged that the minimum requirement of $25,000 capital for banks in towns of 3,000 or less and $50,000 for towns of 6,000 or less was so large that they were debarred from the benefits of the national banks. The agricultural interests complained that the system encouraged the flow of funds out of communities where they had been accumulated to the large financial centers, where they were used for speculative purposes rather than in the more legitimate needs of agriculture and industry. The efficiency of the banking system was hampered by the cumbersome and expensive exchange and transfer system and the decentralization of the gold supply.

Profiting from the lessons of the panic, an emergency measure, known as the Aldrich-Vreeland Act, was passed in 1908 providing temporarily for the issue of bank notes upon approved securities of states, cities, towns, or municipalities, and upon commercial paper, and providing for the formation of associations of national banks for the purpose of issuing notes, the act to be in force until June 30, 1914 (later extended another year). The same bill also called for the appointment of a National Monetary Commission to study banking conditions and report to Congress. Much popular interest was now aroused and both major parties were pledged to some kind of reform. The report of the Commission was submitted in January, 1912, with specific recommendations known as the "Aldrich plan." This plan did not suit either political group, but President Wilson, having disposed of the tariff, next pressed for banking legislation which took form in the Federal Reserve Act of December 23, 1913.

The Federal Reserve Act.—The chief defects of the national-bank system were eliminated under the Federal Reserve Act. The new law divided the country into twelve districts, and in the principal banking city of each a Federal Reserve bank was to be placed. The Federal Reserve cities as decided upon were Boston, New York, Philadelphia, Cleveland, Richmond, Atlanta, Chicago, St. Louis, Minneapolis, Kansas City, Dallas, and San Francisco. Every national bank is required, and other banks are encouraged, to become members of the system by subscribing to the capital stock of the Federal Reserve bank in their district an amount equal to 6 per cent of their capital stock and surplus. In this manner each Federal Reserve bank is owned by the various member banks scattered throughout the district. Each Reserve bank is governed by a board of nine directors, divided into three classes, A, B, and C. The members of the first two classes are chosen by the stockholding banks, the three Class A directors representing the bankers, while the three Class B directors are chosen from those actively engaged in agriculture and commerce. The three Class C directors are appointed by the Federal Reserve

Board, which designates one of them as chairman of the board with the title of Federal Reserve Agent.

Directing the whole system is the Federal Reserve Board of Washington, consisting of eight members, including the Secretary of the Treasury, the Comptroller of the Currency, and six members appointed by the President. This board exercises supervisory powers and determines the larger questions of policy. A Federal Advisory Council, composed of one representative from each Federal Reserve bank, was created to consult with the board and help in unifying and carrying out the policies decided upon.

The Federal Reserve banks do no direct banking with individuals or business houses. They are simply bankers' banks, central agents for performing certain duties for the member banks. These duties include the rediscount of commercial paper for the member banks of the district, the purchase and sale of bills of exchange, the granting of loans to member banks upon government securities as collateral, and other similar banking operations; and in addition the issuing of the Federal Reserve notes. Furthermore the Reserve banks act as the fiscal agents of the government. The Federal Reserve bank notes are issued, like the national bank notes, upon the deposit of government bonds, and were designed to supplant eventually the earlier notes. To provide a type of money which would expand and contract as needed, the Reserve banks are empowered to issue Federal Reserve notes on the security of commercial paper. No reserve bank may pay out the notes of another reserve bank, but must send them in to be retired. As these notes may not be counted as reserves, the currency should automatically contract when it is not needed. The Federal Reserve notes are receivable for taxes, customs, and all public dues, are obligations of the United States, and are redeemable on demand in gold at the Treasury Department.

Federal Reserve banks are required to carry a reserve of 40 per cent in gold against Federal Reserve notes outstanding, and 35 per cent in gold and lawful money against deposits. The member banks must maintain with the Federal Reserve banks a reserve of 3 per cent of their time deposits and from 7 to 13 per cent of demand deposits, depending on the location of the bank.

CIRCULATION STATEMENT OF UNITED STATES MONEY—JANUARY 1, 1924

Kind of money	Stock of money	Money held in the treasury	Money outside of the treasury		
		Total	Total	Held by Federal Reserve Banks and Agents	In circulation
					Amount	Per capita
Gold Coin and Bullion........	$4,247,200,861	$3,553,932,238	$693,268,623	$277,949,206	$415,319,417	$3.71
Gold Certificates.............	(976,605,729)	....	976,605,729	394,576,520	582,029,209	5.20
Standard Silver Dollars.......	498,382,769	421,484,478	76,898,291	18,194,251	58,704,040	.53
Silver Certificates............	(409,726,165)	....	409,726,165	34,360,907	375,365,258	3.35
Treasury Notes of 1890.......	(1,442,926)	....	1,442,926	....	1,442,926	.01
Subsidiary Silver.............	276,887,941	7,169,115	269,718,826	9,092,362	260,626,464	2.33
United States Notes..........	346,681,016	3,510,856	343,170,160	36,347,526	306,822,634	2.74
Federal Reserve Notes........	2,822,326,620	1,092,164	2,821,234,456	597,560,054	2,223,674,402	19.86
Federal Reserve Bank Notes...	14,420,170	331,230	14,088,940	478,189	13,610,751	.12
National Bank Notes..........	771,566,979	17,543,198	754,023,781	40,533,499	713,490,282	6.37
Total January 1, 1924....	$8,977,466,356	$4,005,063,279	$6,360,177,897	$1,409,092,514	$4,951,085,383	$44.22

Amendments to the original act have made it profitable for the state banks and trust companies to enter, their number amounting, on January 28, 1924, to 1,618. At the same time this rapid extension has necessitated another amendment establishing branches of the Federal Reserve banks in a number of the larger cities.

It is perhaps too early to estimate the achievements of the Federal Reserve system, and it has not fully met expectations, especially in the matter of a central discount market. "On the other hand, an elastic currency has been created which has met the requirements of the nation. With its aid, the country was financed through the Great War without paper money inflation or relinquishment of the gold standard."[1]

Money Now in Circulation.—The present currency is an interesting mixture of past and present systems, as a critical examination of the bills that find their way into (and out of) your pocket will show. The paper money now issued directly by the government, United States notes (greenbacks), gold certificates, and silver certificates, is in denominations of $1 to $10,000. In addition, Treasury notes of 1890 are still outstanding in denominations of $1 to $1,000, to the amount of about $1,500,000, for the redemption of which silver dollars of 1890 are kept in the vaults of the Treasury. Metal for the redemption of gold and silver certificates is also on hand. Silver certificates since 1900 are issued mostly in denominations of $1, $2, $5, and $10, though 10 per cent of the whole issue may be in amounts up to $50; while gold certificates run from $10 to $10,000. Federal Reserve notes in denominations of $5 to $10,000 and Federal Reserve bank notes in denominations of $1 to $50, are replacing to a great extent the national bank notes, issued in denominations of $1 to $1,000. The coins now minted consist of gold pieces from $5 to $20; silver from ten cents to a dollar; the "nickel," or five-cent piece, and the bronze cent.

According to a note appended to the circulation statements issued on the first of the month by the Treasury Department:

---

[1] *Harvard Business Review*, October, 1923, p. 125, in a review of H. Parker Willis, *The Federal Reserve System* (1923).

Gold certificates are secured dollar for dollar by gold held in the Treasury for their redemption; silver certificates are secured dollar for dollar by standard silver dollars held in the Treasury for their redemption; United States notes are secured by a gold reserve of $152,979,025.63 held in the Treasury. This reserve fund may also be used for the redemption of Treasury notes of 1890, which are also secured dollar for dollar by standard silver dollars, held in the Treasury. Federal reserve notes are obligations of the United States and a first lien on all the assets of the issuing Federal reserve bank. Federal reserve notes are secured by the deposit with Federal reserve agents of a like amount of gold or of gold and such discounted or purchased paper as is eligible under the terms of the Federal Reserve Act. Federal reserve banks must maintain a gold reserve of a least 40 per cent, including the gold redemption fund which must be deposited with the United States Treasurer, against Federal reserve notes in actual circulation. Federal reserve bank notes and National Bank notes are secured by United States Government obligations, and a 5 per cent fund for their redemption is required to be maintained with the Treasurer of the United States in gold or lawful money.

The statement itself sums up the amount and location of the currency of the country.

NOTES FOR FURTHER REFERENCE

The *Annual Report on the Finances* by the Secretary of the Treasury forms an essential part of the material for detailed study on our finances. Special reports of value are that of the Special Commissioner of the Revenue, 1869, giving the results of David A. Wells's investigation of the money cost of the Civil War; and the *Monetary Commission Report,* 1898. Extracts from both of these are given in Bogart and Thompson.

The most satisfactory financial history is the standard work of Davis R. Dewey, *Financial History of the United States* (8th ed., 1922). Other volumes valuable in various phases of the subject are: A. B. Hepburn, *History of Coinage and Currency in the United States* (rev. ed., 1915); J. F. Johnson, *Money and Currency* (new ed., 1921); A. D. Noyes, *Forty Years of American Finance* (1909); William T. Foster and Wallace Catchings, *Money* (1923), Pollak Foundation for Economic Research, No. 2.

On the struggle for the gold standard, see W. C. Mitchell, *A History of the Greenbacks* (1903), which contains useful tables; and J. Laurence Laughlin, *History of Bimetallism in the United States* (4th ed., 1897).

Of the many books on banking, the following are selected for reference: William A. Scott, *Money and Banking* (4th ed., 1910); C. F. Dunbar, *Theory and History of Banking* (5th ed., rev., 1916); John T. Holdsworth, *Money and Banking* (3d ed., 1921); H. P. Willis and G. R. Edwards, *Banking and Business*

(1922); D. R. Dewey and M. J. Shugrue, *Banking and Credit* (1922); E. W. Kemmerer, *The A B C of the Federal Reserve System* (1916); H. Parker Willis, *The Federal Reserve System* (1923); Gustav Nord, *American Financial Methods* (1912); Chester A. Phillips, *Readings in Money and Banking* (1916); Charles A. Conant, *A History of Modern Banks of Issue* (5th ed., enlarged, 1915) and J. D. Magee, *Materials for the Study of Banking* (1923).

On commercial crises and business depressions much has recently been written: O. M. W. Sprague, *History of Crises under the National Banking System*, Senate Document No. 538, Sixty-first Congress, Second Session; Otto C. Lightner, *History of Business Depressions* (1922); Lionel D. Edie (ed.), *The Stabilization of Business* (1923); Alvin H. Hansen, *Cycles of Prosperity and Depression 1902-1908* (1921), University of Wisconsin Studies in the Social Sciences and History, No. 5; Hudson B. Hastings, *Costs and Profits: Their Relation to Business Cycles* (1923), Pollak Foundation for Economic Research, No. 3; G. H. Hull, *Industrial Depressions* (1911); Henry L. Moore, *Generating Economic Cycles* (1923), and *Economic Cycles: Their Law and Cause* (1914); W. C. Mitchell, *Business Cycles* (1913), Memoirs of the University of California, Vol. 3; Warren M. Persons, *Construction of a Business Barometer*, American Economic Review, Dec., 1916; Thorstein B. Veblen, *Theory of Business Enterprise* (1904); National Bureau of Economic Research, Inc., *Business Cycles and Unemployment* (1923), pp. xxvii-xl, and 1-405 in Conference on Unemployment, Washington, 1921; Report of Joint Commission of Agricultural Inquiry, *The Agricultural Crisis and Its Causes* (1921), House of Rep., Sixty-seventh Congress, First Session; Report No. 408.

Much of interest is contained in the autobiographies of two men who were in the thick of the currency controversies of the period: Hugh McCulloch, *Men and Measures of Half a Century* (1889), by a Secretary of the Treasury who served under Lincoln, Johnson, and Arthur; and John S. Sherman, *Recollections of Forty Years* (1895), by Hayes's Secretary of the Treasury.

SELECTED READINGS

DEWEY, D. R., *Financial History of the United States*, Chaps. XII-XXII.
JOHNSON, J. F., *Money and Currency*, Chap. XVI.
HEPBURN, A. B., *A History of Currency in the United States*, Chaps. XI-XXV.
WILLIS, H. P., *The Federal Reserve*, Chaps. I-VIII.
LAUCK, W. J., *The Causes for the Panic of 1893*.
BOGART, E. L., and THOMPSON, C. M., *Readings in the Economic History of the United States*, Chap. XX.

CHAPTER XXII

BUSINESS CONSOLIDATION

The Culmination of Laissez-Faire.—All of the processes of the American Industrial Revolution were immensely speeded up by the Civil War; but in the rapidly growing industrial and agricultural life unbridled freedom and competition reigned supreme. The *laissez-faire* doctrine of Adam Smith and his successors had been accepted as final by the great majority of Americans in the years immediately following the war, and a fitting capstone had been put upon the theories by the first section of the Fourteenth Amendment.[1] Although this had supposedly been incorporated in the Constitution to protect the negro, the increasing pressure of corporations upon the courts eventually led to an interpretation which went far to restrain the interference of the state legislatures in the operation of business. To the rising capitalist and, in fact, to the average citizen, it seemed not only unnecessary, but bad economics, to regulate private capital. Capital should be aided, not impeded, in the development of the vast natural resources of which, it was believed, there was a sufficiency for all. Furthermore, the pioneer individualism of a frontier people demanded the utmost freedom of action. As a consequence, competition and *laissez-faire* were the order of the day. These were the years when millions of acres were given to the railroads and charters bestowed with a free hand. The most valuable of the oil, lumber, and metal lands were occupied under federal land acts, bought in, or obtained by fraud. Although there was indeed plenty for all, victory went to the strongest and the most unscrupulous. The same was true in the struggle for markets. The home market, which had been freely supplied by

---

[1] "No State shall make or enforce any law which shall abridge the privileges or immunities of citizens of the United States; nor shall any State deprive any person of life, liberty, or property without due process of law; nor deny to any person within its jurisdiction the equal protection of the laws." See previous discussion, p. 463 ff.

manufactured goods from abroad during the early decades of the century, by 1860 bought 89 per cent of its manufactured commodities from domestic producers, and 97 per cent by 1900. In the struggle for these resources and markets legislatures were bribed, the people robbed,[1] all sorts of illegal methods used, and even armed force resorted to upon occasion. But the evils of unrestrained competition and *laissez-faire* to a certain extent brought their own curb. The public reacted against the wasteful appropriation of the country's resources and the illegal methods so commonly used, while cut-throat competition was so disastrous that some way out had to be found. Business consolidation and government regulation have been the outcome.

Consolidation of Business.—The years previous to the Civil War saw the golden age of the small manufacturing business. Since the war the tendency has been to consolidate. Economists at opposite ends of the scale of economic thought have agreed that the consolidation of smaller units into larger is an inevitable result of the conditions brought about by the Industrial Revolution. Whether this trend is inevitable or not, it has without doubt been very marked in our economic life. Most of the witnesses appearing before the Industrial Commission in 1899 believed that "competition so vigorous that profits of nearly all competing establishments were destroyed,"[2] was the chief motivating force for the business combinations. This was the immediate cause which led many to unite to escape being driven to the wall. The bitter rate wars of the railroads during the early 'seventies had driven fares and rates between competitive points below the costs of transportation. Competition was so excessive in the refining of sugar that eighteen out of about forty refineries had failed before consolidation set in. Added to the losses from price cutting were the inherent losses of competition due to costs of advertising and salesmen, and the many disadvantages which a small industry must suffer in comparison with a large one in the

---

[1] Orfield, M. N., *Federal Land Grants to the States With Special Reference to Minnesota*, Bulletin of the University of Minnesota (1915), shows how the public lands, forests, and mineral wealth of one richly endowed state passed into private hands.

[2] *Preliminary Report of the Industrial Commission*, p. 9.

utilization of by-products, the securing of the best management, and the bargaining with labor, bankers, and transportation companies. The desire to eliminate needless costs went hand in hand with eagerness to reap greater profits, which were especially obtainable when the business, as in the case of the Standard Oil, was large enough to effect a monopoly.

While these were the immediate causes, certain results of the Industrial Revolution made big business possible. The invention of labor-saving machinery made large-scale production profitable, and the development of transportation and communication made the distribution practical. Gradually smaller inventions, such as the typewriter, adding machine, and many other appliances contributed necessary factors to the age of big business. This development was greatly aided by the adoption of the corporate form, under which most large industrial units were carried on. In a survey of manufacturing since 1860, it will be seen that the average product per establishment has increased over sixteenfold, while the average of workers has more than trebled.

AVERAGE PRODUCTS AND WAGE-EARNERS IN ESTABLISHMENTS SINCE 1860 [1]

Per establishment	1859	1869	1879	1889	1899	1904	1909	1914	1919
Average product.	$13,429	$16,785	$21,152	$26,371	$54,969	$68,433	$76,993	$87,916	$215,157
Average wage-earners........	9.34	8.15	10.76	11.96	20.49	25.30	24.64	25.51	31.36

The appended table gives the distribution of wage-earners and value of products, according to the value of the annual output. The per-cent distribution columns are worthy of careful scrutiny, for periods of five, ten, fifteen, or twenty years. The tendency to further consolidation is evident in every period. Over the whole twenty years the per-cent distribution of the number of establishments with an annual output up to $20,000 decreased from 33.3 to 26.4; those from $20,000 to $100,000 increased slightly (4.7); those up to $1,000,000 increased 6.6 per cent; and those over $1,000,000 2.5 per cent. The last named in 1919

---

[1] Computed from figures in *Statistical Abstract*, 1921, table 482, p. 868.

numbered 3.6 per cent of the whole number of establishments, but employed 56.9 per cent of the wage-earners and manufactured 67.8 per cent of the products according to value. When the last two groups are combined, it is found that the concerns doing a business of over $100,000 number barely one-fifth of the total, but they employ 88 per cent of the workers, and the value of their products is 92.6 per cent of the total.

The Corporate Form—Advantages and Disadvantages.— As the size of the business unit increased and competition became more reckless and exacting under the impetus of transportation facilities and the prevalent economic doctrine of *laissez-faire,* the old-fashioned methods of conducting a business by means of individual ownership or partnership became inadequate. The amounts of capital needed were too great for individuals to supply, and the risk was too great to be undertaken singly. As a consequence the corporate form of business has been adopted since the Civil War to suit the new needs. Before that time it was used chiefly in the formation of banks or the building of turnpikes and railroads under conditions where it was advisable to distribute the risks widely. It was generally looked upon as a dangerous and undemocratic form associated with the idea of monopoly, a form to be carefully supervised. In New York state incorporation under general laws was not permitted until the constitution of 1846.

A corporation according to an excellent definition is "a voluntary autonomous association formed for the private advantage of its members, which acts with compulsory unity and is authorized by the state for the accomplishment of some public good." [1] In other words, a corporation is an organization or association created by law under a charter which authorizes it to do certain

---

[1] Haney, L. H., *Business Organization and Combination,* p. 82. A more famous definition is that of Chief-Justice Marshall in the Dartmouth College case—"A corporation is an artificial being, invisible, intangible, and existing only in contemplation of law. Being the mere creature of law, it possesses only those properties which the charter of its creation confers upon it, either expressly, or as incidental to its very existence. . . . Among the most important are immortality, and, if the expression may be allowed, individuality; properties, by which a perpetual succession of many persons are considered as the same, and may act as a single individual." Dartmouth College *v.* Woodward, vol. iv, Wheaton's Reports, 518, p. 636.

ESTABLISHMENTS CLASSIFIED ACCORDING TO VALUE OF PRODUCTS FROM 1904 TO 1919 [1]

Annual output	Establishments		Wage-earners		Value of Products	
	Number	Per cent distribution	Average number	Per cent distribution	Amount	Per cent distribution
Less than $5,000:						
1904	71,147	32.9	106,353	1.9	$176,128,212	1.2
1909	93,349	34.8	142,430	2.2	222,463,847	1.1
1914	97,060	35.2	129,623	1.8	233,381,081	1.0
1919	65,485	22.6	45,813	0.5	167,085,044	0.3
$5,000 to $20,000:						
1904	72,791	33.7	419,466	7.7	751,047,759	5.1
1909	86,988	32.4	470,006	7.1	904,645,664	4.4
1914	87,931	31.9	429,037	6.1	905,693,168	3.7
1919	87,440	30.1	249,722	2.7	945,602,857	1.5
$20,000 to $100,000:						
1904	48,096	22.2	1,027,047	18.8	2,129,257,883	14.4
1909	57,270	21.3	1,090,449	16.5	2,544,426,711	12.3
1914	56,814	20.6	999,510	14.2	2,550,229,411	10.5
1919	77,911	26.9	793,528	8.7	3,571,283,301	5.7
$100,000 to $1,000,000:						
1904	22,246	10.3	2,515,064	46.0	6,109,012,538	41.3
1909	27,824	10.4	2,896,532	43.8	7,946,935,255	38.4
1914	30,167	11.0	3,002,071	42.7	8,763,070,135	36.1
1919	48,855	16.9	2,832,598	31.1	15,433,003,131	24.8
$1,000,000 and over:						
1904	1,900	0.9	1,400,453	25.6	5,628,456,171	38.0
1909	3,060	1.1	2,015,629	30.5	9,053,580,393	43.8
1914	3,819	1.4	2,476,006	35.2	11,794,060,929	48.7
1919	10,414	3.6	5,172,712	56.9	42,301,103,617	67.8
All classes:						
1904	216,180	100.0	5,468,383	100.0	14,793,902,563	100.0
1909	268,491	100.0	6,615,046	100.0	20,672,051,870	100.0
1914	275,791	100.0	7,036,247	100.0	24,246,434,724	100.0
1919	290,105	100.0	9,096,372	100.0	62,418,078,773	100.0

things. Although not a person, a corporation is an artificial being which like a person may carry on business, break the law, sue and be sued. A corporation has many advantages which explain its

[1] From table 199, Abstract of the Census of Manufactures, 1919, p. 354.

almost universal adoption. (1) It makes easier the raising of large amounts of capital. Under the terms of the charter corporations are allowed to capitalize their holdings and issue stock. This stock may be bought by many persons who often contribute comparatively small amounts to build up a great business. Thus both the American Telephone and Telegraph Company and the United States Steel Corporation boast of over 175,000 stockholders. A corporation may also borrow money and issue bonds, thus giving it access to large resources of capital. (2) By owning corporate stocks many people may share in the development of the country and in the profits of the largest concerns managed by men of great ability, without themselves contributing anything but money. (3) The risk of the stockholders is limited by the law of the state. (4) The shares may usually be bought and sold, thus allowing a person voluntarily to enter or leave a concern as his private interests dictate. (5) The corporation has great advantages in that it is not disrupted by the death or retirement of members.

On the other hand, the corporate form has disadvantages. Where the number of stockholders is large and scattered, it is impossible for them to exercise any real control over their delegated agents, the directors elected at the annual meetings. The irresponsibility of directors is accentuated by the legal attitude that a corporation is a separate legal person, and that the directors are the agents of the corporation and not of the stockholders, thus making it useless for a stockholder and a minority to sue a director or his agents for loss incurred through fraud or negligence. The lack of control which the stockholders exert upon their directors has often encouraged the latter to use their position to promote personal interests, to indulge in speculative management, fraudulent promotions, and overcapitalization, which have in the end worked havoc to the stockholders, who are not inclined to inquire too closely while dividends are unimpaired. From the view of the investor the numerous stock and bond issues so common to corporations is confusing and only an expert can work out their true valuation. From the broader view of public policy, corporations seem to promote monopoly; for stock ownership facilitates interlocking directorates and interlocking ownerships. Whatever its disadvantages may be, however, the corporation has become the dominant form of business organization to-day. Although in 1919 corporations numbered only 31.5 per cent of the establishments, they employed 86 per cent of the wage-earners and produced 87.7 per cent of the total value of the products. The following table gives the situation from 1904 to 1919:

ESTABLISHMENTS IN THE UNITED STATES, 1904 TO 1919 [1]

Character of ownership	Establishments		Wage-earners		Value of Products	
	Number	Per cent distribution	Average number	Average per establishment	Amount	Per cent distribution
Individuals:						
1904	113,946	52.7	755,923	7.0	$1,702,830,624	11.5
1909	140,605	52.4	804,883	6.0	2,042,061,500	9.9
1914	142,436	51.6	707,568	5.0	1,925,518,298	7.9
1919	138,112	47.6	623,469	4.0	3,536,321,836	5.7
Corporations:						
1904	51,097	23.6	3,862,698	76.0	10,904,069,307	73.7
1909	69,501	25.9	5,002,393	72.0	16,341,116,634	79.0
1914	78,152	28.3	5,649,891	72.0	20,183,147,103	83.2
1919	91,517	31.5	7,875,132	86.0	54,744,392,855	87.7
All others:						
1904	51,137	23.7	849,762	21.0	2,187,002,652	14.8
1909	58,385	21.7	807,770	18.0	2,288,873,736	11.1
1914	55,203	20.0	678,788	12.0	2,137,769,323	8.8
1919	60,476	20.8	597,771	10.0	4,137,364,082	6.6
All classes:						
1904	216,180	100.0	5,468,383	25.0	14,793,902,563	100.0
1909	268,491	100.0	6,615,046	25.0	20,672,051,870	100.0
1914	275,791	100.0	7,036,247	26.0	24,246,434,724	100.0
1919	290,105	100.0	9,096,372	31.0	62,418,078,773	100.0

The Evolution of Concentration.—While some large concerns have achieved their size by internal growth and natural expansion, many more have come to their present greatness through a consolidation of industries engaged in the production of similar commodities. Attempts like that of the salt producers in western Virginia (after 1830) to restrict output and thus con-

[1] From table 195, Abstract of the Census of Manufactures, 1919, p. 340.

trol prices had been made before the Civil War, but it was not until after the panic of 1873 that the movement toward consolidation became noticeable. It is possible to divide roughly the periods through which the consolidation movement has passed according to the forms which it has taken: (1) pools, (2) trusts, (3) holding companies, (4) amalgamations and mergers, and (5) "community of interest."[1]

Pools.—The period of pools which arose after the panic of 1873 continued until about 1887. A pool is an organization of business units whose members seek to control prices by apportioning in some method the available business. This form was especially popular among the railroads where the bitter rivalry between competitive points was fast leading to ruin. Although forbidden in the Interstate Commerce Act of 1887, the practice was continued, especially in the south, where the transportation of cotton was for a long time apportioned and the freight rates fixed by common consent. In addition to traffic pools there have been "output" pools illustrated by the agreement between the powder manufacturers in 1886, which sought to eliminate "ill-regulated and unauthorized competition" by mutual understanding in regard to output and price. Informal apportionment of the business among different units of the same industry undoubtedly still persists to some extent. Another form of pool is that of allotment of territory and market. A typical example was the agreement entered into in 1902 between the Imperial Tobacco Company of Great Britain and the American Tobacco Company, giving the former the exclusive control of the British Isles, and the latter control of the United States, its colonies, and Cuba; while a new corporation, the British-American Tobacco Company, Limited, was to handle the business of the rest of the world. Still another type of pool was the depositing of a certain part of profits or income with a central body, to be later redistributed.

Trusts.—Pools in railroads were declared illegal in 1887, again in 1897 in the case against the Trans-Missouri Freight Association. Beginning with the former date, pooling was de-

---

[1] These divisions follow Professor Ripley's introduction to *Trusts, Pools, and Corporations* (rev. ed., 1916), pp. xi-xii.

serted in favor of a new form of understanding which appeared to be legal and at the same time much more efficient. From 1887 until 1897 the trust was the most favored form of combination. A trust is a form of organization in which the stockholders under a trust agreement deposit with a board of trustees a controlling portion of their stock and receive in return trust certificates. It was a case of using the old legal idea of a trusteeship to create a monopoly and was introduced as early as 1879 and 1883 by the Standard Oil Company. It was followed by the formation in 1887 of the "Whisky Trust" (Distillers' and Cattle Feeders' Trust) and "Sugar Trust" (Sugar Refineries Company), the "Lead Trust," the "Cotton-oil Trust" and in the succeeding years by many others. The trust form, which gave absolute power to the trustees, effected a monopoly, opposition to which produced anti-trust laws on the part of various states in 1889 and later, and the Sherman Anti-trust Act on the part of the national government in 1890. The early prosecutions of the federal courts under the Sherman Act were generally unsuccessful, but the dissolution of the Standard Oil Trust by the Ohio courts in 1892 put a decided damper upon the trust method of consolidation. The panic in 1893 and the succeeding years of depression held up aggressive moves on the part of business, but at the same time helped to prepare for the greatest period of consolidation, the years 1897-1904.

Holding Companies.—The anti-trust legislation led to the adoption of a new form of consolidation, namely the holding company; and during the period of 1897 to 1904 this was the popular form. A holding company is an organization created to dominate other corporations by owning or controlling a portion of their stocks. Although the device of the holding company had been employed before this time by the Pennsylvania Company and by the American Bell Telephone Company, it was now adopted rapidly, the Standard Oil taking the lead, with certain states, notably New Jersey and Delaware, passing laws to make incorporation easy. During these years over $6,000,000,000 worth of securities were marketed, the year 1899 showing new combinations organized with a nominal capital of $3,512,000,000, of which at least one-fourth was water. Most of the great combinations of to-day were formed during this period, including perhaps the greatest of all, the United States Steel Corporation, organized in 1901 with a capital of $1,100,000,000, exclusive of a bonded indebtedness of $304,000,000.[1] The United States Steel Corporation is a typical holding company, its property consisting of the securities of eleven constituent companies, which in turn own a controlling interest in 170 subsidiary concerns. In a similar manner railroads have widened their influence. Since 1904 the position of the holding company when it controls a monopoly has been unsafe. In that year the government secured a conviction and dissolution of the Northern Securities Company, an organization formed to hold the stock of the three great railroads tapping the northwest. The courts affirmed that while a holding company was legal under the laws of the incorporating states, it was illegal when the obvious intent was to effect a monopoly. In the same tenor subsequent decisions dissolved the Standard Oil and American Tobacco holding companies.

Mergers and "Community of Interest."—Since 1904 the holding company has usually given way to new forms of consolidation—either amalgamation and merger or the outright purchase by one organization of the property of related industries. Furthermore, laws against monopolies and the prosecutions under these laws have made it desirable to effect the same end without forming an actual merger. Thus there has developed the principle of control through "community of interest." One company can buy stock in another of sufficient quantity to make its influence felt, and directors of one company in this manner may sit on the board of another. True, the Clayton Act has forbidden interlocking directorates in competitive companies engaged in interstate business whose capital, surplus, and undivided profits aggregate more than $1,000,000; but even here, as stockholders, the same persons may exercise great influence. So extensive was the "community of interest" in the oil companies that the dissolution of the trust in 1892 and of the holding company in 1911

---

[1] Cotter, Arundel: *Authentic History of the United States Steel Corporation*, 1916, p. 26.

made practically no difference. The railroads for so long have been under the questioning eye of the people that they especially have resorted to consolidation through "community of interest." To such an extent has the purchase of stock been consummated between the railroads that it is comparatively easy to divide them into eight or ten different groups according to their controlling financial interests, a situation which has almost eliminated competition in the sections served by these systems. Indeed, this concentration of capital has gone on until automatically a "community of interest" has been built up not only between units in the same line of business, but between all sorts of organizations, banks, and bond houses.

The Standard Oil Company.—The history of the oil business is of particular significance in the studying of industrial combinations, for the rise and progress of the Standard Oil Company illustrates practically every phase in the development and methods of monopoly under American conditions. Successful drilling for oil commenced in 1859 in the vicinity of Titusville, Pennsylvania, after the discovery of the Drake well. While the business of drilling wells and refining oil expanded rapidly during the war, the production in 1865 was behind the demands and the whole industry was severely handicapped by lack of transportation facilities and efficient refining machinery. The fact that transportation was the great problem and the chief expense of the expanding oil industry made it quite evident to the most able men in the business that success would come to the large concern with capital enough to install the best machinery for large-scale production and sufficient output to force favorable railroad rates. In 1867, while the industry was still in its infancy, John D. Rockefeller united the refineries of William Rockefeller & Co., Rockefeller and Andrews, Rockefeller & Co., S. V. Harkness, and H. M. Flagler, into the firm of Rockefeller, Andrews & Flagler. "The cause leading to its formation," he said, "was the desire to unite our skill and capital in order to carry on a business of some magnitude and importance in place of the small business that each separately had theretofore carried on."[1] Further capital was needed and in 1870 the

---

[1] Preliminary Report of the Industrial Commission, p. 95.

company was reorganized into the Standard Oil Company of Ohio, with a capital of $1,000,000 and a refining capacity in its Cleveland plant of about 600 barrels a day. This amounted, however, to only 4 per cent of the oil refineries in the United States, and the Standard plant was not even then the largest in the country.

Up to 1870 competition between oil men had been largely in production. In the succeeding years it was a competition for transportation facilities and favorable rates, a bitter war which left the Standard Oil Company in complete control. The victory of the Standard Oil may be attributed largely (1) to the business acumen of Rockefeller and his associates, (2) to the securing of favorable freight rates, and (3) to the unscrupulous and illegal methods to which they resorted to destroy competition and win favorable concessions from railroads and legislatures. The desire of the oil men to gain cheap transportation rates was aided by the railroads (chiefly the Erie, the New York Central, and the Pennsylvania, which were in competition for the oil business), and, in keeping with the policy of the time, the roads lowered their rates at competitive points and to promising concerns. In all the dickering with the railroads, no group of refiners was so successful as the Standard Oil. The favorable location of the company at Cleveland was, to be sure, a factor in this success, since it freed the concern from complete dependence on the railroads by affording water transportation to the seaboard by way of the Great Lakes.

The most notorious of the rate agreements was made through the South Improvement Company chartered by the Pennsylvania legislature in 1871, with the widest powers, including authority "to construct and operate any work, or works, public or private, designed to include, increase, facilitate, or develop trade, travel, or the transportation of freight, livestock, passengers, or any traffic by land or water, from or to any part of the United States." [1] This company, of which 900 of the 2,000 shares were held by Rockefeller and his close associates, made contracts with

---

[1] Preliminary Report of the Industrial Commission, p. 608. The charter of the South Improvement Company is given on p. 607, and the contract with the Pennsylvania on p. 610.

the Pennsylvania, the New York Central, and the Erie whereby the company agreed to ship 45 per cent of all the oil transported by it over the first-named railroad and to divide the remainder between the other two roads. In return the railroads agreed to allow rebates on all petroleum shipped by the company, but to charge all others the full rates and in addition to furnish to the South Improvement Company waybills of all petroleum and its products transported over their lines. It was also agreed by each road "at all times to coöperate, as far as it legally may, with the party hereto of the first part against loss by injury or competition, to the end that the party hereto of the first part may keep up a remunerative, and so a full and regular business, and to that end shall lower or raise the gross rates of transportation over its railroads and connections, as far as it legally may, for such times and to such extent as may be necessary to overcome such competition." The South Improvement Company aroused such a storm of opposition that its charter was revoked after three months, but nevertheless, rebates and favorable discriminations were continued. The Standard Oil gradually extended its operations to include the ownership of pipe lines and by 1879 controlled from 90 to 95 per cent of the oil refined and was able in turn to dictate its rates to the roads. The "Hepburn Committee," reporting in January, 1880, to the New York legislature, said:

It owns and controls the pipe lines of the producing regions that connect with the railroads. It controls both ends of these roads. It ships 95 per cent of all oil. . . . It dictates terms and rates to the railroads. It has bought out and frozen out refiners all over the country. By means of the superior facilities for transportation which it thus possessed, it could overbid in the producing regions and undersell in the markets of the world. Thus it has gone on buying out and freezing out all opposition, until it has absorbed and monopolized this great traffic, this great production which ranks second on the list of exports of our country. The parties whom they have driven to the wall have had ample capital, and equal ability in the prosecution of their business in all things save their ability to acquire facilities for transportation.[1]

In order the more completely to dominate the situation, the Standard Oil Company of Ohio worked out a scheme by which

---
[1] New York Assembly Document, No. 38, 1880.

the stock holdings of fourteen companies and the majority holdings in twenty-six other companies were placed in the hands of nine trustees having irrevocable powers of attorney. The stockholders received in return trust certificates. The par value of the trust certificates amounted to $70,000,000, of which $46,000,000 were owned by the nine trustees who dictated the policies of the constituent companies. The trust form was looked on askance by the people as a whole, and the passing of the Sherman Anti-trust Act in 1890 enabled the state of Ohio to secure in 1892 the dissolution of the Standard Oil Trust, which now broke up and was reorganized into twenty constituent companies having a capitalization of $102,233,700. The trust certificates were replaced by proportionate shares of stock in the new companies.

In 1899 a second attempt was made to bring the entire properties under a single control by the formation of the Standard Oil Company of New Jersey, a holding company, as well as an operating company, formed with the intention of transferring to it the stock of the different corporations so that in time one concern might own and direct the whole industry. The position of the Standard Oil of New Jersey as a holding company was gravely imperiled by the decision in the Northern Securities case (1904) and finally made untenable by the Supreme Court order of dissolution in 1911. The business is now carried on by corporations chartered by the several states, which act harmoniously and exercise a virtual monopoly through a "community of interest" brought about by the ownership by certain individuals of controlling stock in the several companies. By 1904 the Standard Oil controlled about 85 per cent of the domestic and 90 per cent of the export trade. The earning capacity had increased from $8,000,000 in 1882 to $57,459,356 in 1905 and dividends from 5¼ per cent in 1882 to 30 per cent in 1898. In recent years the company has stretched into foreign fields, notably in the Mexican oil districts, in Rumania, and in the Baku regions of Russia. Increased demand for oil for motor traffic has added impetus to production and stimulated the formation of many new companies; but new enterprises have so far effected little diminution in the power of this mighty combination.

Combinations and Monopolies.—The discussion so far has been largely concerned with the combination movement and the various forms it has taken. It should be remembered, however, that a pool, a holding company, or even a trust may be organized without effecting a monopoly. But the desire for a monopoly with its advantages has ordinarily been in the minds of the organizers, for a virtual monopoly may be brought about by controlling hardly more than 50 per cent of the product. So obvious was it that the trusts were organized to eliminate competition and to control products that the term "trust" has been commonly used in America to designate any large combination which approaches a monopoly and is sometimes even applied indiscriminately to any big business.

Moreover, there are a number of different kinds of monopolies with which we are familiar. There may be (1) personal monopolies in which an individual possessing special talent or knowledge may be able to drive out competitors. There are (2) legal monopolies: public, as the case of the post office in America, or private, such as those based on patents, copyrights, or franchises. An important group is that of (3) the natural monopoly of situation or organization, as illustrated by a street railway, gasworks, or anthracite coal. (4) Labor monopolies resulting from combinations of skilled laborers often control the labor supply. But of special interest to us here are (5) the capitalistic monopolies or monopolies of organization which by the concentration of large aggregations of capital and the unification of a sufficient number of production units, have been able to exercise a monopoly.

Even a cursory consideration of these types brings home the fact that certain forms of monopoly are inevitable, and that others are encouraged for the sake of public welfare. Personal talent or a steam railway are often inevitable monopolies. A government post-office system and a franchise creating a street railway may be monopolies created for the public good; broad social welfare is considered in the granting of patents. On the other hand, capitalistic monopolies and monopolies of labor lead at once into controversial fields. But even here modern conditions prevent us from taking too dogmatic an attitude. The cost of erecting a sugar refinery or a steel mill is so great that free competition is almost automatically cut off, while in the case of labor the perfectly laudable determination to secure better conditions through a stronger bargaining organization cannot be condemned too hastily.

Advantages and Disadvantages of Capitalistic Monopolies.—Large-scale monopolistic production, it is claimed, effects savings both in production and in marketing. As to production, it is urged (1) that the large resources make it possible to use only the best located plants and the most efficient machinery, especially in slack times; (2) that large-scale production allows more complete utilization of by-products and economies in the division of labor; (3) that it permits the specialization of production at the different plants; (4) that administrative expenses can be saved by the elimination of duplicated offices of high-salaried officials and at the same time the best talent in the field may be secured; (5) that research may be pursued on a larger scale; (6) that waste and ineffective methods may be more easily detected through the careful comparison of different plants producing the same article; and (7) that there is greater strength in dealing with labor. As to marketing, it is maintained that expenses are reduced (1) by the elimination of salesmen and advertising; (2) by the elimination of cross freights, as orders may be filled from the nearest plants; and (3) by the development of greater strength in the export business. The argument is also advanced that a control of both the market price of a raw commodity and of the finished article helps to stabilize prices and production and thus exerts a healthy influence upon economic life. During the periods of rapid monopolistic development, the evils of competition were always emphasized and the motto "competition is the death of trade" kept well to the front.

On the other hand, it is argued that while a monopoly may manufacture more cheaply, the savings are not passed on to the consumer, for it is usually formed to enhance profits, and there is conclusive evidence that in many cases the public has been gouged by unwarranted charges. It was the belief of the Industrial Commission in 1902, after a most exhaustive study, "that

BUSINESS CONSOLIDATION

in most cases the combination has exerted an appreciable power over prices and in practically all cases it has increased the margin between raw materials and finished products. Since there is reason to believe that the cost of production over a period of years has lessened, the conclusion is inevitable that the combinations have been able to increase their profits." [1] At about the same time, Professor Jenks came to the conclusion that "the fact that the power to increase the margin temporarily at least, somewhat arbitrarily, and the fact that the margin has been increased in specific cases, seems to be clearly established." [2] While the price has often been raised to the consumer, the producer of the raw materials, such as cattlemen, sugar raisers, and others have suffered from the lack of competition among buyers. Furthermore, monopoly has often resulted in inefficient and careless service to the consumer, who has been forced to accept what was given him.

In a comparison of the advantages and disadvantages, it should be pointed out that many of the alleged advantages of monopolies are similarly applicable to any large-scale industry where there is no monopoly. Steady consolidation of business has gone far and there is every reason to believe that the process will continue. Nevertheless, monopoly as such has generally been distrusted as both an economic and a social evil, and persistent efforts either to restore competition or to control the inevitable monopolies have been undertaken through legislative means.

Growth of Opposition to the Trusts.—Notwithstanding the dominance of *laissez-faire* and the enthusiasm with which business consolidation proceeded, there developed a strong opposition to the movement. This came first from a deep-seated antipathy to monopoly inherited from the old English common-law conception, a dislike which was undoubtedly stimulated by the misfortunes of those whose means of livelihood was injured by the new consolidations. Secondly, there was a fear that the natural resources of the country would be brought under the control of a few irresponsible men. By 1873 six corporations owned the

---

[1] Final Report of the Industrial Commission, vol. xix, 621.
[2] Jenks, Jeremiah W., *Trusts and Industrial Combinations*, a Bulletin of the Department of Labor, No. 29, July, 1900, p. 765.

anthracite coal deposits of Pennsylvania and the transportation facilities to carry the coal out, and in the succeeding years much of the bituminous field was appropriated. By 1882 thirty-nine refineries of the Standard Oil controlled 90 per cent of the product. "A small number of men," said Henry Demarest Lloyd (1894), "are obtaining the power to forbid any but themselves to supply the people with fire in nearly every form known to modern life and industry, from matches to locomotives and electricity. They control our hard coal and much of the soft, and stoves, furnaces, and steam and hot-water heaters; the governors on steam boilers and the boilers; gas and gas-fixtures, natural gas and gas-pipes, electric lighting, and all the appurtenances. You cannot free yourself by changing from electricity to gas, or from the gas of the city to the gas of the fields. If you fly from kerosene to candles, you are still under the ban."[1] By 1904 most of the great products of the country were in the control of big combinations, so large as to constitute monopolies.

Not only were the people disturbed over the appropriation and consolidation of the resources of the country, but they were thoroughly aroused over the dishonest methods of competition which had, in many cases by open evasion of the law, brought success. The concern which did not want to join the trust was throttled by every unfair means known, among the least vicious of which was the obtaining of special railroad rebates, a factor which more than all else made possible the success of the Standard Oil. Not only was there evasion of the law, but tampering with the government; the unwholesome influence of big business upon politics is evidenced by the free distribution of railroad passes and still more by activities at election time. The Supreme Court of Michigan undoubtedly expressed the current feeling when it said in a case involving the Diamond Match Company, one of the most notorious of the trusts of the period: "Indeed, it is doubtful if free government can long exist in a country where such enormous amounts of money are allowed to be accumulated in the vaults of corporations, to be used at discretion in controlling the

---

[1] Lloyd, H. D., *Wealth Against Commonwealth*, pp. 9-10.

BUSINESS CONSOLIDATION

property and business of the country against the interest of the public and that of the people, for the personal gain and aggrandizement of a few individuals." [1]

Moreover, the financial practices incident to consolidation, the watering of stocks, the paying of enormous commissions to lawyers and banking houses, had helped to fleece the general public. And finally labor has found it more difficult to deal with the increased power of consolidated capital, notably in the case of the United States Steel, and has been among the severest critics of the trust.

This rapid growth of monopoly and the irresponsible use of the power which went with it were viewed with concern by many of the most thoughtful. Among the literature calling attention to defects in the economic life of the time three widely read books stand out preëminently. Henry George in 1880 published his *Progress and Poverty,* in which he advocated a single tax on land values as one solution for the problem of monopoly. Edward Bellamy's *Looking Backward* (1887), by glorifying the socialistic state, pointed to another solution; and Henry Demarest Lloyd's *Wealth Against Commonwealth* (1894) was the ablest and most effective attack ever delivered against the trust. The opposition which had developed had already made itself felt in political channels. Further grants to railroad corporations and monopolies had been opposed by both the major parties in 1872. The Greenbackers in 1880 and the Anti-Monopolists in 1884 had called for government action to prevent or control monopolies, and the same was true in 1888 of the platforms of the Union Labor, the Prohibitionists, and the Republicans. By the close of 1890 twenty-seven states and territories had passed laws intended to prevent and destroy monopolies and fifteen states had incorporated provisions in their constitutions for the same purpose. In that year the federal government also took action.

The Sherman Anti-Trust Act.—By 1890 public opinion had become so aroused over the subject of monopolies that federal legislation was demanded to supplement the state laws. Investi-

---

[1] Richardson *v.* Buhl *et al.*, 77 Michigan State Reports, p. 658.

gations undertaken in 1888 by a committee of the House of Representatives [1] and by a committee of the Senate of the State of New York,[2] while offering little in the shape of constructive suggestion, had confirmed the current beliefs as to the evils of monopolies. President Harrison in his message of December, 1889, had urged legislation against trusts which partook of the nature of conspiracies.[3] A number of anti-trust bills were introduced in the Senate in 1888, but two years of discussion ensued before a bill was eventually passed.

The Sherman Anti-trust Act of 1890 [4] contained eight sections, the principle and theory of the act, however, appearing in the following:

Sec. 1.—Every contract, combination in the form of trust or otherwise, or conspiracy, in restraint of trade or commerce among the several States, or with foreign nations, is hereby declared to be illegal. . . .

Sec. 2.—Every person who shall monopolize or attempt to monopolize or combine or conspire with any other person or persons to monopolize any part of the trade or commerce among the several States, or with foreign nations, shall be deemed guilty of a misdemeanor. . . .

Fines and imprisonment were provided for violation and the injured person might recover three times the damages sustained. The several circuit courts of the United States were invested with jurisdiction to prevent or restrain violations of the act, and the Attorney-General directed to institute proceedings in equity against such violations.

The Sherman Act was looked upon by many as an unnecessary blow at legitimate business and a futile opposition to an inevitable economic development. The committee who framed it main-

---

[1] *Report of Investigation of Trusts*, House Reports, First Session, Fiftieth Congress, 1887-88, vol. 9, Serial Number 3112.
[2] *Report of the Senate Committee of General Laws on Investigation Relative to Trusts*, March 6, 1888.
[3] Richardson, *Messages and Papers of the Presidents*, ix, 43.
[4] 26 Stat., 209. The texts of this act and the other federal Anti-trust Acts mentioned in this chapter are to be found in Jenks and Clark: *The Trust Problem*, in Appendix F: Federal Trust Legislation in the United States. The various sections of the Appendix contain much valuable source material.

tained truly, however, that the bill was simply a restating of the usual English common-law principles and the extension of them to America. The act did not attempt to define "contract, combination, or conspiracy in restraint of trade." It was purposely drawn in general terms for the courts to interpret, the intention being that no business legitimately carried on need fear interference.

Senator Cullom called the Sherman Act "one of the most important enactments ever passed by Congress," but it was decidedly ineffective for a long while. This was due chiefly to two reasons—first, the economic depression in the succeeding years deferred for some time further large-scale consolidation; and second, the general terms in which the bill was stated required much legal interpretation. The panic of 1893 temporarily crippled business and made both national and state governments loath to increase its burdens. The political weakness of the Harrison administration, followed by the necessary affiliations of Cleveland with eastern capitalists during his second term, prevented aggressive legislation; while under McKinley the combination movement went on merrily, with no apparent desire on the part of the administration to interfere. Down to 1901 the government had instituted eighteen suits, but with a discouraging lack of success. The spirit of *laissez-faire* and the economic tendency of the period toward consolidation, combined with the difficulties of handling the technical questions involved in the trust and corporate form, hindered decisive and clear-cut judicial action. The Supreme Court in 1893 refused to dissolve the American Sugar Refining Company, although it had purchased plants to enable it to control 98 per cent of the refining business of the country, on the ground that the purchase was not an act of interstate commerce. After this decision any hope of accomplishing much from the Sherman Act seemed futile. The act was further weakened in 1895 in the case of V. E. C. Knight Company, when the court voted that the law was applicable only to monopoly in restraint of trade and not to monopoly in manufacture. In case a suit was successful, the monopoly was ordinarily continued under some other form. In general we must agree with Professor Jenks when he says, "a study of these statutes and of the decisions of our courts of last resort which have been made under them will show that they have had comparatively little, practically no, effect, as regards the trend of our industrial development." [1]

The "Muck-Rakers" and the Revival of Anti-Trust Activity.—The tremendous revival of the combination movement in the prosperous years immediately following the Spanish-American War, coincident with the abuses and the high-handed disregard of public welfare as evidenced by the large corporate interests brought a logical reaction. Beginning with the publishing in 1903-04 of Ida M. Tarbell's "History of the Standard Oil Company" in *McClure's Magazine,* there ensued a period in which many of the worst features of our economic and social life were aired before the public. Lincoln Steffens's *Shame of the Cities* (1904) exposed the rottenness of many of the local governments; Thomas Lawson's "Frenzied Finance," published in *Everybody's Magazine* (1905-06) showed Wall Street at its worst; Upton Sinclair in *The Jungle* (1906) revealed the horrible filth and misery of the workers in the meat-packing industry; Winston Churchill in *Coniston* (1906) drew a picture of the subserviency of state legislatures to the railroads; while in other books and numerous magazine articles the lawlessness and greed of big business and the venality of politicians were enlarged upon. The Democrats in the campaigns of 1896, 1900, and 1904 directed part of their artillery against the trusts.

This exposure eventually degenerated into "muck-racking," but it inaugurated a healthy reaction for reform, a movement in which President Roosevelt took the lead. On a campaign speaking tour in 1902 he attacked the trusts, and in the next year Congress passed three acts to control big business more effectively. The first of these, known as the Expediting Act,[2] gave preference to federal suits brought under the Interstate Commerce Act and the Sherman Anti-trust Act. The second was the Elkins Anti-rebate Act,[3] which aimed to clarify the law and eliminate

---

[1] Jenks, Jeremiah W., *The Trust Problem* (rev. ed., 1905), p. 218.
[2] 32 Stat., 823.
[3] 32 Stat., 847.

one of the worst practices of the railroads. The third was the creation of a Department of Commerce and Labor with a subsidiary Bureau of Corporations to make "diligent investigation into the organization, conduct, and management of corporations." [1] In the same year the President directed his Attorney-General to institute proceedings against the Northern Securities Company, a New Jersey holding corporation designed to create a transportation monopoly in the northwest by controlling the stock of the Great Northern, the Northern Pacific, and the Chicago, Burlington, and Quincy. The successful issue of this suit [2] in 1904 showed that the Sherman Act might not be a useless reed in the hands of an aggressive administration. Under Roosevelt sixteen civil suits and eighteen criminal suits were prosecuted with considerable success. The Pure Food Law of 1906 [3] marked a distinct step forward in the policy of government intervention to protect the welfare of the public, while the more comprehensive act of 1907 [4] aimed especially to bring under supervision the meat-packing business.

Dissolution of the Standard Oil Company and the American Tobacco Company.—The Taft administration believed that legitimate business might go on undisturbed and a solution be found to the trust problem by the voluntary federal incorporation of concerns whose charters were to be approved by a projected corporation commission, with the power reserved to Congress to revoke such charters. A bill to this effect was introduced, but public interest was never sufficient to push it through. As a consequence, the government continued to press the prosecutions already commenced, and succeeded in obtaining two notable decisions in 1911. The first of these, against the Standard Oil Company of New Jersey, [5] had been in the courts more than four years. The defendant argued that the Standard Oil com-

---

[1] 32 Stat., 825.
[2] U. S. *v.* Northern Securities Company *et al.*, 120 Fed. Rep., 721; 193 U. S., 197.
[3] 34 Stat., 768.
[4] 34 Stat., 1256, 1260.
[5] U. S. *v.* Standard Oil Company of New Jersey *et al.*, 152 Fed. Rep. 290; 173 Fed. Rep. 177; 221 U. S. 1.

panies were the natural products of the growth of a single business, that they had never competed with one another and consequently could not have conspired or combined in restraint of trade. Both the circuit and Supreme courts, however, affirmed the government's contention that the concerns had so conspired by many and devious methods to build up a monopoly. The dissolution was carried out by apportioning pro rata to the stockholders of the holding company shares in the various constituent concerns.

The case of the American Tobacco Company [1] was more complicated because the organization was not merely a holding company, but an actual manufacturing concern and one which was engaged in making a number of products, including chewing and smoking tobacco, snuff, little cigars, cigarettes, and tin foil. The court attempted to restore competition by creating separate companies in each line; for example, the manufacture of smoking tobacco was divided between four companies, cigarettes among three concerns, plug tobacco among four, and tin foil among two. A proportionate distribution of stock was made in the new companies corresponding to the holdings in the old. Each new company was enjoined from coöperating with, or holding stock in, another company.

Two interesting facts stand out in regard to these decisions. The first is that the dissolutions failed in their purpose. In form there was competition, in fact there was little. The distribution of stock created simply a community of interest among the various concerns which appeared to work as harmoniously together as when under a single management. The increase in value of Standard Oil stocks after the dissolution showed that no detrimental results were feared. After more than thirty years of operation and numerous court dissolutions and interpretations, the Sherman Anti-Trust Act appeared to have failed utterly in its purpose of preventing monopoly and restraint. The second point to be noted was the interpretation given to the act by the two decisions. The Trans-Missouri Freight case decision (1897) [2]

---

[1] U. S. v. American Tobacco Co. et al., 164 Fed. Rep. 700; 221 U. S. 106.
[2] U. S. v. Trans-Missouri Freight Association, 53 Fed. Rep. 440; 58 Fed. Rep. 58; 166 U. S. 290.

had refused to see any difference between reasonable and unreasonable combinations in restraint of trade, but the judges in the two decisions in 1911 professed to see a difference and maintained that the only restraint of trade which was intended by the law was that which monopolizes or attempts to monopolize. In other words they introduced the so-called "rule of reason" and tried to differentiate between "good trusts" and "bad trusts." Many believed the "rule of reason" was an unwarranted interpretation and that it simply weakened the act. It certainly made more complicated further consideration of trust cases by the courts.

Democrats and the Trusts—The Clayton Act and the Federal Trade Commission.—For years the Democratic party had assailed the Republicans as the friends and allies of the trusts. In their platform of 1912 the Democrats had demanded that the Sherman Act be made more stringent in order to restore free competition. Their candidate, Woodrow Wilson, in a remarkable series of campaign speeches, had emphasized what he called "The New Freedom." While claiming not to be one of those who think that competition can be established by law against a world-wide economic tendency, he still believed that much of our old, free coöperative life could be restored. Without condemning big business as such, he laid the destruction of competition to the trusts. "American industry is not free, as it once was free," he said. "American enterprise is not free; the man with only a little capital is finding it harder to get into the field, more and more impossible to compete with the big fellow. Why? Because the laws of this country do not prevent the strong from crushing the weak." [1] To restore, if possible, some of the old competition appeared to be the purpose of the new administration and it was obvious that after years of criticism some legislation would be passed.

Wilson had further affirmed that the trouble with the Sherman Act was that it was not definite enough and needed a more careful statement of unlawful practices, so that legitimate business might

---

[1] Wilson, Woodrow, *The New Freedom*, p. 15.

better know when it was within the law. These ideas Congress sought to embody in the Clayton Anti-Trust Act of 1914.[1] The following are the chief provisions:

1. The act forbids (a) any person to discriminate in price, either directly or indirectly, between purchasers of commodities whenever such discrimination lessens competition or tends to create monopoly, (b) a manufacturer to sell his goods to a dealer under conditions requiring the latter not to handle the products of competitors—a hit at the so-called "tying" agreements.

2. Corporations were forbidden to acquire stock in another concern where the effect was substantially to lessen competition. The holding of stock solely for investment was allowed.

3. Interlocking directorates were forbidden in concerns engaged in interstate commerce whose capital, surplus, and undivided profits aggregate more than $1,000,000, if such concerns are competitors.

4. It was made unlawful in the case of banks for one person to serve as director or officer in another if the deposits, capital surplus, and undivided profits of any of the institutions exceeded $5,000,000.

5. Labor unions and farmers' organizations were specifically declared not to be conspiracies in restraint of trade.

A few days earlier a Federal Trade Commission of five members had been created [2] whose business it was to investigate persons or corporations (except interstate carriers and banks) subject to the anti-trust laws, and present reports of its activities. It was also granted power to issue orders requiring the cessation of illegal practices, and if these were not obeyed it was to apply for federal action to the circuit court of appeals in the district where the alleged offense was committed. The commission took over the work of the old Bureau of Corporations and was designed to act for corporations along somewhat the same line that the Interstate Commerce Commission has done for interstate carriers.

In order that Americans might compete on more equal terms with great foreign concerns, the anti-trust laws were modified in

---

[1] 38 Stat., 730.
[2] 38 Stat., 717.

1918. The Webb Export Act[1] stated that nothing in the Sherman Act was to be construed as making "illegal an association entered into for the sole purpose of engaging in export trade and actually engaged solely in such trade," providing this association is not party to any attempt to restrain competition or control prices within the country. Furthermore, the Clayton Act under the same condition was not to be construed as forbidding the "acquisition or ownership by any corporation of the whole or any part of the stock or other capital of any corporation organized solely for the purpose of engaging in export trade."

The "Money Trust."—No discussion of business consolidation would be complete without reference to the concentration of banking power. Parallel with the rapid but extensive consolidation of business has gone that of the banking interests. The increasing wealth of the country had naturally enlarged the size of the banks, while the greater demands of their customers necessitated growth and consolidation in order to meet them. This concentration by the opening of the twentieth century, however, had been so vast that there was a firmly grounded conviction among many that a small group controlled the financial resources of the land, loaning and withholding funds where they pleased, thus holding in the hollow of their hands the fate of many a business.

We have already seen how the important lines of railroads through interlocking directorates and stockholdings were in the power of six influential groups dominated by a score of men. It was now asserted that the same men controlled the banking facilities. Around the Morgan-Rockefeller interests, wrote John Moody in 1904,[2]

or what must ultimately become one greater group, all other smaller groups of capitalists congregate. They are all allied and intertwined by their various mutual interests. For instance, the Pennsylvania Railroad interests are on the one hand allied with the Vanderbilts and on the other with the Rockefellers. The Vanderbilts are closely allied with the Morgan group, and both the Pennsylvania and the

---
[1] 40 Stat., 516, 518.
[2] Moody, John, *The Truth About the Trusts*, p. 493.

Vanderbilt interests have recently become the dominating factors in the Reading system, a former Morgan road and the most important part of the anthracite coal combine which has always been dominated by the Morgan people. . . . Viewed as a whole, we find the dominating influences in the Trust to be made up of an intricate network of large and small capitalists, many allied to one another by ties of more or less importance, but all being appendages to or parts of the greater groups, which are themselves dependent on and allied with the two mammoth, or Rockefeller and Morgan groups. These two mammoth groups jointly . . . constitute the heart of the business and commercial life of the nation.

The concentration of capital was promoted by the fact that the Rockefeller and the Morgan interests worked through banks which they controlled; thus the National City Bank, the greatest of American banking institutions, became the Rockefeller bank, while the Morgans controlled the First National, The Bankers' Trust, and others. Wall Street and the insurance companies formed a community of interest in the joint direction of the great trust companies and thus the influence of Wall Street became dominant in the vast loaning operations of the insurance companies.

The general belief was fully confirmed in the report of the Pujo Committee (1913), which pointed out that the concentration of control of money and credit had been effected chiefly through consolidations of competitive or potentially competitive banks and trust companies; through interlocking directorates and stockholdings; through the influence of the powerful banking houses, banks, and trust companies brought to bear on insurance companies, railroads, producing and trading companies; and finally through partnership arrangements between a few of the leading banking houses in the purchase of security issues, which has had the effect of virtually destroying competition. The committee named J. P. Morgan & Co., the First National Bank of New York, and the National City Bank as the most powerful banking units, placing their combined assets in New York City, as controlled through seven subsidiary banks, at over $2,000,000,000. In addition to the interests named, the committee believed that Lee Higginson & Co., Kidder, Peabody & Co., and Kuhn, Loeb & Co.

BUSINESS CONSOLIDATION 543

were the principal banking agencies through which the corporate enterprises of the United States obtain capital for their operations. Four allied financial institutions in New York City, it affirmed, held 341 directorships in banks, transportation, public utility, and insurance companies, whose aggregate resources were $22,245,-000,000.

If by a "money trust" is meant an established and well defined identity and community of interest between a few leaders of finance which has been created and is held together through stock holdings, interlocking directorates, and other forms of domination over banks, trust companies, railroads, public service, and industrial corporations, and which has resulted in a vast and growing concentration of control of money and credit in the hands of a comparatively few men—your committee has no hesitation in asserting as a result of its investigation that this condition, largely developed within the past five years, exists in this country today.[1]

While it was true that to a considerable extent this growth and consolidation had followed natural economic laws—as is illustrated by consolidation of financial power in other countries—at the same time there was a real danger in a situation in which the economic life blood of the nation was controlled by a small group of men using their power for private ends. The report of the committee contained a number of recommendations in regard to bettering the banking facilities, breaking up concentration, and supervising the stock exchange. Some of the best of these have been incorporated in the law creating the Federal Reserve System, the adoption of which was undoubtedly furthered by this investigation; in the Clayton Act, which forbade interlocking directorates in the large banks, and in the Esch-Cummins bill, which empowered the Interstate Commerce Commission to supervise plans and security issues in the reorganization of interstate railroads.

NOTES FOR FURTHER REFERENCE

Some of the most important source material on industrial and financial concentration is to be found in the various investigations made by the state and

---

[1] Report of the Committee Appointed to Investigate the Concentration of Control of Money and Credit, Sixty-second Congress, Third Session, p. 130, quoted by Phillips, Chester A.: *Readings in Money and Banking*, 1916, p. 606.

federal legislatures. Of these the most valuable are: *Preliminary Report of the Industrial Commission on Trusts and Industrial Combinations*, Vol. I of the Commission's Report (1900); *Final Report of the Industrial Commission*, Vol. XIX of the Commission's Report (1902); *Report of the Special Committee on Railroads Appointed Under a Resolution of the Assembly of February 28, 1879, to Investigate Alleged Abuses in the Management of Railroads Chartered by the State of New York*, Assembly Doc. No. 38, 1880 (Hepburn Committee), especially informing on rebates; *Report of the Committee Pursuant to House Resolutions 429 and 504 to Investigate the Concentration of the Control of Money and Credit* (1913) (Pujo Committee). See also the Thirteenth Census Abstract, Chap. X, and the Abstract of the Census of Manufactures, 1914, Chaps. VI and VII.

General studies of the trust movement include J. W. Jenks and W. E. Clark, *The Trust Problem* (4th ed., 1917), a standard work, scholarly and kept up to date; Eliot Jones, *The Trust Problem in the United States* (1921); C. R. Van Hise, *Concentration and Control* (rev. ed., 1914); B. J. Hendrick, *The Age of Big Business* (1919), in the Chronicles of America, interestingly written, with good bibliography; R. T. Ely, *Monopolies and Trusts* (1900); John Moody, *The Truth About the Trusts* (1904), with valuable statistical information by an expert; and John Moody, *The Masters of Capital* (1919), in the Chronicles of America. Helpful source material is collected in W. Z. Ripley, *Trusts, Pools, and Corporations* (rev. ed., 1916).

The heaviest guns in the early anti-trust agitation were fired by Henry Demarest Lloyd in his unsparing denunciation of monopoly, *Wealth Against Commonwealth* (1894). See also C. Lloyd, *Life of Henry Demarest Lloyd* (2 vols., 1912), for a survey of the growth of anti-trust feeling. In later years Woodrow Wilson lifted his voice in favor of competition and the small business in his *New Freedom* (1913), a collection of campaign speeches. A short résumé of the literature of protest is that of C. C. Regier, *The Muck-Raking Campaign* in Vol. XV, No. 1 (January, 1924), Historical Outlook.

Special industries may be studied in Ida M. Tarbell, *History of the Standard Oil Company* (2 vols., 1904), a pioneer work ably and unbiasedly written; in G. H. Montague, *Rise and Progress of the Standard Oil Company* (1903), a defense of the oil monopoly; in H. R. Mussey, *Combination in the Mining Industry* (1905); in Abraham Berglund, *The United States Steel Corporation* (1907), Vol. XVIII, No. 3, Columbia University Studies in History, Economics and Public Law; in H. L. Wilgus, *A Study of the United States Steel Corporation* (1901); in Arundel Cotter, *The United States Steel—a Corporation with a Soul* (1921); in the *Report of the Commissioner of Corporations on the Steel Industry* (3 parts, 1911); in Eliot Jones, *The Anthracite Coal Combination*, Harvard Economic Studies, Vol. II (1914); and in Scott Nearing, *Anthracite: An Instance of Natural Resource Monopoly* (1915). See also *Report to the President on the Anthracite Coal Strike of May-October, 1902*, by the Anthracite Coal Strike Commission (1903), especially Appendix J.

On the legal aspect consult W. H. Taft, *The Anti-Trust Act and the Supreme Court* (1914). For a summary of monopoly advantages and disadvantages, see L. H. Haney, *Business Organization and Combination* (rev. ed., 1913).

SELECTED READINGS

JENKS, J. W., and CLARK, W. E., *The Trust Problem*, Chaps. III-V, IX, XIII-XV.
JONES, ELIOT, *The Trust Problem in the United States*, Chaps. I-IV.
HANEY, L. H., *Business Organization and Combination*, Chaps. VI-XVI, XXIII-XXVII.
RIPLEY, W. Z., *Trusts, Pools, and Corporations,* Introduction.

CHAPTER XXIII

MANUFACTURING SINCE 1860

ALTHOUGH the factory system in America obtained its first real foothold during the period of the Embargo and the War of 1812 and in the mills which sprang up in the succeeding years, nevertheless it seems safe to say that "until about the year 1850, the bulk of general manufacturing done in the United States was carried on in the shop and the household, by the labor of the family or individual proprietors, with apprentice assistants, as contrasted with the present system of factory labor, compensated by wages, and assisted by power."[1] Since 1850 our economic life has been revolutionized as we followed in the wake of western Europe in substituting factory-made products for those of hand labor. The development of the factory system has been continuous, but the process was immensely hastened by the demands of the Civil War. If the War of 1812 introduced the factory system, the Civil War effected an industrial revolution. More and more America turned to factory-made goods and to large-scale industry, and Americans occupied themselves increasingly with manufacturing. Until the decade of the 'eighties agriculture was the principal source of wealth, but the census of 1890 showed that manufacturing has forged to the front and ten years later the value of manufactured products was more than double that of agricultural.

COMPARISON OF AGRICULTURAL AND MANUFACTURED WEALTH [2]

Value of products	1889	1899	1909	1919
Agricultural	$2,460,107,000	$4,717,076,000	$8,498,311,000	$23,783,200,000
Manufactured (including those based on agriculture)	9,372,379,000	11,406,927,000	20,672,052,000	62,418,079,000

[1] Twelfth Census. vol. vii, pp. liii.
[2] *Statistical Abstract*, 1921, pp. 862, 868.

The increasing value of manufactured over agricultural products is reflected in the concentration of population. The census of 1920 reported, for the first time in our history, the urban population (those living in towns of 2,500 or over) as more than the rural, the percentage being 51.4 and 48.6, respectively. The per cent of people living in towns of over 8,000 increased from 16.1 in 1860 to 43.8 in 1920. While the population from 1850 to 1900 trebled (from 23,192,000 to 76,129,000), and the products of agriculture nearly trebled ($1,600,000,000 to $4,717,070,000), the value of manufacturers increased eleven fold ($1,019,107,000 to $11,406,927,000). From 1859 to 1914 the value of American manufactures increased eighteen fold, and from 1859 to 1919 thirty-three fold. Naturally this increase stands out when a comparison is made with foreign nations. M. G. Mulhall in his *Industries and Wealth of Nations* (1896) shows how the United States, which had ranked fifth in the value of manufactured products in 1840 and fourth in 1860, had taken first place in 1894. At that time we produced twice as much as Great Britain and half as much as all Europe together.

The census of manufactures in 1909 gave the value of manufactured products at over $20,000,000,000. This figure may be compared with the census of 1907 in Great Britain, which returned $8,000,000,000 for that nation, and with the estimate of Germany in 1913, which reported between $11,000,000,000 and $12,000,000,000. The United States, however, consumed a larger proportion of its manufactured goods at home than either of these nations, although there was a great actual increase in exports. In the normal years before the recent war, while Great Britain exported one-fourth, this nation exported less than one-tenth of her manufactured products, a proportion changed temporarily by the war.

Manufacturing was progressing favorably in the 'fifties when the panic of 1857 halted development. The impetus to production given by the Civil War increased the number of establishments during the decade of the 'sixties 79.6 per cent, and the number of wage-earners 56.6—the largest relative advances made in any decade in our history. The severe panic of 1873 again retarded development, especially in the founding of new establishments, but before the decade had run its course recovery set in and a healthy progress was evident. The 'eighties showed the largest increase in our manufacturing up to 1909 in capital invested and in wages paid, the growth during the decade amounting to 133.8 per cent for the former, and 99.5 for the latter. Despite the depressing effect on industry of the panic of 1893, figures for 1899 evidenced substantial gains. They pale into insignificance, however, before the enormous strides made in the twentieth century, the second decade of which, including as it did the period of the Great War, surpassed all others in our history in percentage of increase in capital, wages, and value of products. It is worthy of note that while these three advanced 150 to 200 per cent, the number of wage-earners increased only 37.5, and the number of establishments only 8.1.

GROWTH OF MANUFACTURES, 1849–1919 [1]

Year	Number of establishments	Average number of wage-earners	Capital	Wages	Value of products
1849.....	123,025	957,059	$ 533,245,000	$ 236,755,000	$1,019,107,000
1859.....	140,435	1,311,246	1,009,856,000	378,879,000	1,885,862,000
1869.....	252,148	2,053,996	1,694,567,000	620,467,000	3,385,860,000
1879.....	253,852	2,732,595	2,790,273,000	947,954,000	5,369,579,000
1889.....	355,405	4,251,535	6,525,051,000	1,891,220,000	9,372,379,000
1899.....	512,191	5,306,143	9,813,834,000	2,320,938,000	13,000,149,000
1899.....	207,514	4,712,763	8,975,256,000	2,008,361,000	11,406,927,000
1909.....	268,491	6,615,046	18,428,270,000	3,427,038,000	20,672,052,000
1919.....	290,105	9,096,372	44,688,094,000	10,533,400,000	62,418,079,000

The appended list of leading manufacturing industries according to rank in 1860, 1914, and 1919 illustrates a number of interesting facts. In the earlier year the first four groups were dependent upon either agriculture or lumbering; while in 1914 four and in 1919 three groups (including the first) out of the first six were still dependent upon these sources. Although

[1] From table 3, p. 13, Abstract of the Census of Manufactures for 1919.
Until 1899 estimates included hand and neighborhood industries, but thereafter were limited to factory industries. Figures on each basis were compiled for 1899.

manufacturing from agricultural raw materials predominates, iron and steel manufacture had advanced by 1914 from fifth place to second, while foundry and machine-shop products, which in 1860 were included with crude iron and steel, ranked fourth as a separate division. Two new groups have appeared among the first ten, car construction and repairs, and automobiles, both transportation products. Even among the great manufacturing groups the rank is not static, and no list long remains the same. For instance, between 1900 and 1914 automobile manufacture became so important as to be separated from machine-shop products and to assume a position of eighth rank, and by 1919 of third rank. The industry standing twenty-fifth in rank in 1914 (silk goods, including throwsters) had a higher value of products ($254,011,000) than the industry standing first in 1860.

This remarkable expansion in manufacturing has been attended by a corresponding growth in economic independence. During the colonial period and the years before the Civil War we had been largely dependent upon Europe for much of the better class of manufactured goods; our exports had consisted in our surplus of food and raw materials. The high tariff walls and the influx of immigration from Europe, providing cheap labor, made it possible to exploit the unsurpassed mineral and agricultural wealth. Practically anything which we need can now be manufactured at home; the recent war has broken even the great chemical monopoly of the Germans. Imports now consist to a large extent of luxuries, tropical fruits, rubber, and manufactured goods involving hand labor. An indefinite blockade which cut us off from the rest of the world would not seriously interfere with our ability to live and carry on the economic functions.

Causes for the Growth of Manufactures.—Undoubtedly a strong impetus was given to the growth of manufacturing by the imperative needs of the Civil War and by the stimulation of high prices caused by war demands and the printing of fiat money. The real causes of the long upward swing are more fundamental. The United States had become a great manufacturing nation first of all because of her unsurpassed natural resources. Rich agricultural products, such as livestock and cotton, have formed the

RANK OF LEADING INDUSTRIES, 1860, 1914, AND 1919 [1]

Rank	1860 Industry	1860 Value of products (in thousands)	1914 Industry	1914 Value of products (in thousands)	1919 Industry	1919 Value of products (in thousands)
1	Flour and meal	$248,580	Slaughtering and meat packing	$1,651,965	Slaughtering and meat packing	$4,246,291
2	Cotton goods	115,726	Iron and steel, steel works and rolling mills	918,665	Iron and steel, steel works and rolling mills	2,828,902
3	Lumber planed and sawed	104,928	Flour-mill and gristmill products	877,680	Automobiles	2,387,903
4	Boots and shoes	91,889	Foundry and machine-shop products	866,545	Foundry and machine-shop products	2,289,251
5	Iron founding and machinery	88,648	Lumber and timber products	715,310	Cotton goods	2,125,272
6	Clothing, including furnishing	88,095	Cotton goods	676,569	Flour-mill and gristmill products	2,052,434
7	Leather, including morocco and patent leather	75,598	Cars and general shop construction and repairs by steam-railroad companies	510,041	Petroleum	1,632,533
8	Woolen goods, including yarn, etc.	65,706	Automobiles	503,230	Ship building	1,456,490
9	Liquors	56,588	Boots and shoes	501,760	Lumber and timber products	1,387,471
10	Steam engines	46,757	Printing and publishing, newspapers and periodicals	495,906	Cars and general shop construction and repairs by steam railroad companies	1,279,235
11	Iron, cast	36,638	Bread and other products	491,893	Clothing, women's	1,208,543
12	Iron, forged, rolled and wrought	36,537	Clothing, women's	437,888	Clothing, men's	1,162,986
13	Provisions (beef, pork, etc.)	31,986	Clothing, men's	458,211	Boots and shoes	1,155,041
14	Printing (book, job, etc.)	31,063	Smelting and refining copper	444,022	Bread and other bakery products	1,151,896
15	Carriages	26,849	Liquors, malt	442,149	Woolen and worsted goods	1,065,434

[1] Figures for 1860 compiled from Census of 1860, Volume on Manufactures; for 1914 from Abstract of Census of Manufactures, 1914, table 220, pp. 516 ff.; for 1919 from Abstract of Census of Manufactures, 1919, table 9, pp. 19-20.

MANUFACTURING SINCE 1860

basis of some of the most important manufacturing industries; while iron, coal, oil, copper, and other minerals have been obtainable in large quantities. In addition to raw materials, manufacturing is dependent upon labor and a market. Labor was obtained by the natural rapid increase of population in an undeveloped country and by millions of immigrants, many of whom were unfitted by training and environment for other than factory work. American manufacturers could not look to the older countries for a large market, but had to build one up at home in competition with foreign products; such a market was partially supplied by the continued accretion of immigrants, but more especially by the large agricultural population of the south and west.

The high tariffs which the Civil War inaugurated have become a fixture in our system and have greatly stimulated manufacturing, both by the high profits which they have allowed to well-established industries and by the protection given to infant enterprises. Under Republican and Democratic administrations alike the great aim of the government has been ordinarily to promote industrial prosperity. In contrast to the system of high protective tariffs, the internal policy of the government has been largely *laissez-faire*. The lack of government interference during the period of great growth undoubtedly imparted a spirit of confidence amounting sometimes to recklessness on the part of the organizer. Even after attempts were made in 1887 and in 1890 to exercise control such efforts for a long time met with slight success. Manufacturing has also been aided by the freedom of interstate commerce, a necessity which was instrumental in the acceptance of the Constitution and amply justified it. "The mainland of the United States is the largest area in the civilized world which is thus unrestricted by customs, excises, or national prejudice, and its population possesses, because of its great collective wealth, a larger consuming capacity than that of any other nation."[1] "It is the enjoyment of free-trade and protection at the same time," said James G. Blaine, "which has contributed to the unexampled development and marvelous prosperity of the

---
[1] Twelfth Census of the United States, vol. vii, p. lvii.

United States."[1] The newness and freedom of the country has reacted upon the character of both capital and labor. The former has been inventive, resourceful, ready to take risks and seize whatever advantages offered; and labor has developed a mobility unknown elsewhere and has been free to desert the old hand processes for the new machinery.

Without transportation facilities, manufacturing, other than purely local, would be well-nigh impossible. The 26,000 miles of navigable rivers, the Great Lakes, the roads and the canals, helped in the early years of the introduction of the factory system, but it was not until the construction of a network of railways that large-scale manufacturing was practicable. More recently the railroad facilities have been augmented by the enlarging of some of of the old canals, by the invention of the automobile and resulting highway construction. Manufacturing and transportation have helped to create wealth, which has been constantly available for reinvestment in similar projects.

As the Industrial Revolution has progressed, new inventions have made possible many new manufacturing industries which have been stimulated by the purchasing power of the American consumer. Among these important industries may be mentioned the manufacture of transportation equipment; electrical supplies used for telephone, telegraph, radio and lighting purposes; bicycles and automobiles. The distribution of these and many other products has been aided by the development on a large scale of advertising and salesmanship.

Characteristics of American Manufacture.—The manufacturing industry in America has been influenced in its development by several factors contributing to make it differ from the European system. Foremost among these is the scarcity of labor, prevalent during most of our history, and inevitably directing the inventive genius of the nation to the creation of labor-saving machinery. American products are preëminently machine made, not hand made, a characteristic not wholly favorable as to quality, since the substitution of machine for hand products has

---

[1] Blaine, James G., *Twenty Years of Congress*, vol. i, p. 211.

MANUFACTURING SINCE 1860

meant a certain sacrifice of the artistic, the delicate, and the beautiful for the sake of large-scale production. While our commodities are cheaper and more rapidly made than those of Europe, it is all too true that they are not so fine nor so artistic. An excellent example of this is apparent in an examination of European and American inexpensive jewelry.

Scarcity of labor in America also helped to develop earlier here than in any other country the standardization of machinery and parts, permitting the rapid production of complicated mechanisms in large quantities, each part of which is made separately and the whole assembled later. This enables easy replacement of parts, and keeps down expenses in running machinery.

Another characteristic partly attributable to the scarcity of labor and partly to the character of the raw materials, is that much of our manufacturing produces small changes, and the value added by manufacture forms a relatively small proportion of the total value of the product. "Thus the slaughtering and meat packing industry, which ranks first in gross value of products, and the flour-mill and gristmill industry, which ranks third in that respect, both hold a comparatively low rank in regard to number of wage earners and value added by manufacture." [1] This form of manufacturing is likely to be located close to the supply, whereas more complicated forms, like the metal and textile industries of New England, are often located far from the raw material. The dependence of many of our leading manufactured products upon agriculture should again be emphasized. Of the twelve great groups of manufactures summarized in the Statistical Abstract, seven are directly dependent upon agriculture— food, textiles, lumber, leather, paper, liquors, tobacco.

Closely allied with the scarcity of labor has been the enthusiastic adoption of any power other than hand, and most of the factories are operated by steam, water, electricity, or gasoline. Since 1870 mechanical power in the United States has increased from 2,346,000 horse-power to 29,567,117 in 1919. Whereas in 1870 the horse-power contributed by waterfalls and steam was

---

[1] Abstract of Census of Manufactures, 1914, p. 27.

about equal, in 1919 the distribution was as follows: steam, 57.8 per cent; electricity, 31.6; water 6.0; internal-combustion engines, 4.3; other horse-power, .3. New England in 1914 reported 42 per cent of her power generated from water and the Middle Atlantic states 24 per cent. With the increasing expense of coal there has been recently a growing interest in the development of water power, especially evidenced in the expansion of hydro-electric power plants.

Some note has already been made of an outstanding characteristic of American manufacturing—freedom from tradition. Our rapid development may be traced to some extent to the freedom from inherited ideas, leaving our industries unhampered to seek the best and quickest way. Guild regulation and the medieval legislation of town and nation were not felt. Furthermore, American labor is intelligent, quick to comprehend and to adopt new methods. Environment has made of the American a jack-of-all-trades and nurtured the inventive genius. Nowhere have new methods of machine production been more enthusiastically sought. Liberal patent laws have aided; the 276 patents granted in the decade 1790-1800 grew to 6,480 for the decade 1840-50, to 25,200 for the decade 1850-60, to 71,800 for the ten years 1860-70, and to 221,500 for the decade 1890-1900. In 1911 the total number of patents issued since 1790 reached the million mark. In the single year of 1921 the number issued and reissued amounted to 41,404. Over 80,000 applications are now made annually and about 40,000 annually granted, a number equal to many times the combined patents granted by the rest of the world. The greatest manufacturing plants at present maintain laboratories devoted exclusively to the development of new devices and improvements. A mere recitation of some of the great patents issued in the field of electricity or gas engines would provide a thumb-nail sketch of the development of recent American manufacture and of technical advancement.

Westward Movement of Manufacturing.—Like population and agriculture, the movement of manufactures has been steadily westward. A map published with the Twelfth Census [1] shows

---

[1] Twelfth Census of the United States, Statistical Atlas, plate No. 179.

the center of manufacturing in 1850 (computed upon the gross value of the products) near the center of Pennsylvania, 41 miles northwest of Harrisburg. In 1860, 1870, and 1880 the center had moved to western Pennsylvania, and by 1890 had moved nearly to the center of Ohio, a few miles southwest of Canton, with a further progress westward in the next census to a point southeast of Mansfield. During the half century from 1850 to 1890 the westward movement of the center of manufacturing was 225 miles and the westward movement of the center of population 243 miles, indicating both the close relationship of the two movements and the more rapid advance of population.

The westward movement of manufacture has been caused primarily by the filling up of the west, which has provided labor and a market, and secondarily by a desire to be close to raw materials. It has been retarded by the scarcity of labor in the new communities and by the concentration of capital in the older ones. It has likewise been forced to wait upon the development of transportation facilities. Although it has lagged behind both agriculture and population in the westward advance, manufacturing has usually followed the raw material as fast as labor could be obtained. Thus the milling of flour moved west from the coast rivers to Rochester on the Erie Canal, then to Chicago, and finally to Minneapolis and Kansas City. The meat-packing industry had its trans-Alleghany beginnings about 1816 at Cincinnati, but has moved to Chicago and Kansas City. Lumbering is an excellent example of an industry forced to follow the source of supply. The northeast originally furnished most of the lumber, but at present it is obtained largely from the northwest and south. As in the case of milling and meat packing, the manufacture of agricultural machinery moved westward. The factories largely shifted from their original home in central New York to Illinois and Wisconsin, following both the hickory forests and the farmer. In a similar manner the last thirty years have shown a tendency for the cotton industry to shift toward the source. The Carolinas and Georgia in 1880 produced only 6.2 per cent of the cotton manufactured, but in 1919 their output had risen to 33.7; the consumption of raw cotton in the mills of these states during the same years increased from 205,000 to 2,500,000 bales and the number of spindles over twenty-six fold in the cotton-growing states. In many cases capital has set up its industry close to the raw materials and has imported labor to handle it. Thus slaves were formerly imported to raise tobacco and cotton, and each year now lumberers during the cutting season are shipped to the lumber camps.

Localization of Industry.—Although the center of manufactures has moved westward, this tendency has been hampered as well as aided by the many influences making for localization. The Twelfth Census has ably summed up the general causes for the localization of industry as follows: (1) nearness to materials, (2) nearness to markets, (3) nearness to water power, (4) a favorable climate, (5) supply of labor, (6) capital available for investment in manufactures, and (7) the momentum of an early start. Any one of these, or a combination of several, explains the location of most of our manufacturing.

The nearness to materials explains the concentration of milling in the Twin Cities and Kansas City; of meat packing in Chicago, Omaha, and Kansas City; of furniture at Grand Rapids; of fruit and vegetable canning in central New York and at Baltimore; of fish canning in Oregon and on the New England coast; and of tobacco in Virginia and St. Louis. It also explains, in part, the recent migration of many cotton mills to the south.

The nearness to market is an influential factor in the localization of industry, especially in the production of bulky and heavy articles. Four of the six states ranking highest in value of manufactured products—New York, Pennsylvania, Illinois, Ohio, Massachusetts, and New Jersey—are in the northeast, and contribute (1919) more than half of the total for the United States. By sections, the Middle Atlantic (New York, Pennsylvania, and New Jersey) produced 31.9 per cent of the value of the manufactures; the East North Central (Wisconsin, Michigan, Illinois, Indiana, and Ohio), 28.5; and New England, 11.6. The localization of the manufacture of such luxuries as jewelry at Providence, of silk at Paterson, of furs at New York, or the localization of factories producing certain high-grade necessities in the northeast, is due either to the fact that originally the only market was east of the Alleghanies or that the principal market remains there. Transportation costs, especially before the days of railroads and automobiles, were naturally a powerful factor in locating industries in thickly populated communities or on rivers and highways leading directly thereto.

Before the introduction of the steam engine, manufacturing was largely dependent upon water power or hand. The preëminence of New England and eastern New York as manufacturing communities is in no small degree due to the water power furnished by the Hudson, the Mohawk, the Connecticut, the Housatonic, the Merrimac, and scores of other streams. Industries like cotton and wool, founded in early days on water power, have continued to depend upon that source. In New England this has been partially due to the distance from the coal supply. The utilization of coal has made manufacturing in many districts independent of water power, and has caused the centering of many industries near the coal fields; as a result the horse-power generated by coal has increased with greater rapidity than that obtained from water. The possibility of running many factories by electricity, which may be produced by water power (in view of the increased cost of coal), has stimulated in recent years an interest in hydro-electric power. The future may see inventions making practical the storage of energy from the sun's rays, the tides, or the winds.

Along with water power, a favorable climate has helped to determine the geographical position of the textiles. High humidity and even temperature have fitted Fall River and New Bedford for cotton manufacturing. The invigorating air of the north is infinitely more suitable for labor than the enervating climate of many parts of the south, and seems to be a permanent factor tending toward the industrial development of the north.

Industries naturally tend to establish themselves where there is a supply of labor. While American labor is probably more mobile than that of foreign nations, the expense of moving and the attachment of home and friends tend to keep it relatively fixed. The decline of the merchant marine and the meager profits from agriculture freed much labor in New England for manufacturing, a supply augmented by women and children coming in from the farms. Immigrants who were factory workers at home drifted to the factory towns to swell the available labor force. An industry of a certain type draws to it skilled labor in that particular line, thus giving a further impetus to the establishment of new factories. As an industry becomes concentrated, future skilled labor must be trained in this center, an influence which keeps it from spreading. Nine-tenths of the collars and cuffs are made in New York State, three-fourths of the plated ware in Connecticut, and almost all of the carpets in Philadelphia, Yonkers, Hartford, and Amsterdam, New York. Industries employing women and children often follow those employing men; thus textile mills are often set up in foundry and mining towns. It is easier to move capital than labor, and the human factor must always remain a vital one in the establishment of an industry.

In recent years most large enterprises secure their capital from the great financial centers. This, however, is usually a second stage. Before financiers step in to reorganize or enlarge a manufacturing plant, the industry has ordinarily been established by the enterprise and capital of local business men, illustrating "the tendency of a town to own itself in the early stages of its industrial life." Outside capital is more easily attracted to a prosperous town and to an industry in which local people have invested. The rapid rise of textiles in New Bedford and other New England cities is partially due to the capital set free by the decline of whaling and the merchant marine. Fall River is an excellent illustration of a town which has specialized in one industry, the control of which has largely been retained in the community.

The momentum of an early start can be given as a leading cause for the localization of industry. It has been said that if the population of New England was suddenly wiped out it is doubtful if its future would be other than that of a summer resort, its natural advantages for manufacturing as against its disadvantages being so slight. Johnstown and Gloversville, New York, the greatest glove center in America, originally drew glovers because it was advantageously located to make use of deer skins. As skilled labor gravitated there it has continued to produce leather, kid, and cloth gloves. An early carpet factory in Amsterdam, a few miles to the south, has drawn to it skilled weavers which have made it the second largest carpet city in the country. The chance settlement of a skilled shoemaker at Lynn in 1750 has made it the leading shoe town, and kept it so notwithstanding its distance from the source of raw materials. A similar early settlement of jewelers in Providence has made it a center in that industry. The habit of industrial imitation is great, for the average man has not the courage to be an industrial pioneer; with skilled labor at hand and successful industries already in operation, the line of least resistance is likely to be followed. What has been said of industry in general is true to a limited extent of the concentration in certain districts of the great cities of the middlemen, such as the stock brokers, the textile wholesalers, the fur merchants, and others.

Manufacturing in the Northeast—(New England, New York, Pennsylvania and New Jersey).—The northeast from the beginning of our industrial history has been the most important manufacturing region, the value of its products in 1919 amounting to 43.5 per cent of the total. Almost all of the factors which influence localization have been operative here—markets, labor, capital, transportation facilities, and the impetus of an early start. The streams of New England and the coal of Pennsylvania provided ample power. In certain raw materials alone was this section handicapped. In the small-scale industries of the early years a sufficiency of raw materials such as iron, wool, and hides could be found at home, but later these had to some extent to be imported. The other advantages, however, not only preserved to this region its supremacy in many types of metal, leather, and textile manufacturing, but created cotton mills and sugar refineries, the materials for which are entirely imported. The northeast surpasses all other sections in the value of its output per person, the percentage of people engaged in manufacturing, and the number and variety of such enterprises.

SUMMARY OF MANUFACTURES BY GEOGRAPHIC SECTIONS, 1919[1]

	Number of establishments	Wage-earners (average number)	Wages	Capital	Value of products	Value added by manufacture	Per cent of total value of products
			Expressed in thousands				
New England...	25,528	1,351,389	$1,436,437	$5,780,410	$7,183,071	$3,231,163	12
Middle Atlantic.	88,360	2,872,653	3,464,931	15,072,300	19,854,773	8,430,677	32
East North Central	61,332	2,396,618	2,990,931	12,163,595	17,737,480	7,115,793	28
West North Central	29,166	499,635	546,373	2,690,626	5,187,065	1,408,940	8
South Atlantic...	29,976	817,212	778,027	3,332,332	4,455,152	1,858,887	7
East South Central	14,655	329,226	298,710	1,296,449	1,642,391	664,567	3
West South Central	13,909	285,244	293,022	1,463,838	2,277,861	729,868	4
Mountain	7,612	109,216	141,901	833,984	922,676	312,437	1
Pacific	19,567	435,179	581,269	2,054,470	3,157,610	1,289,368	5
Total for United States	290,105	9,096,372	10,533,600	44,688,094	62,418,079	25,041,699	100

The leading five states in the value of their manufactured products in 1919 were New York, Pennsylvania, Illinois, Ohio, and Massachusetts, three of which are located in the northeast. In the Thirteenth Census (1909) New York is reported as ranking first in 104 industries and second in 49; of the twenty leading industries in the country she then held the first position in seven and second in five. New York's most valuable five products in 1919 were (1) women's clothing; (2) men's clothing; (3) sugar refining; (4) slaughtering and meat packing; (5) printing and publishing, newspapers and periodicals. Pennsylvania held first rank in 35 industries and second in 37 in 1909; in 1919 her leading manufactures were (1) iron and steel, steel works and rolling mills; (2) foundry and machine-shop products; (3) iron and steel, blast furnaces; (4) ship building, including boat building; (5) silk goods, including throwsters; (6) cars and general shop construction and repairs by steam-railroad companies. New Jersey in 1909 took leading rank in 21 industries and second in 19. Her principal industries (1919) are (1) petroleum refining; (2) smelting and refining, copper; (3) ship

---

[1] *Statistical Abstract*, 1921, pp. 254-255. Per cent distributions compiled by author.

building, steel; (4) silk goods, including throwsters; (5) foundry and machine-shop products; (6) electrical machinery, apparatus, and supplies; (7) slaughtering and meat packing. In petroleum refining and silk goods she leads all of the states.

Turning to New England, we find Massachusetts, fifth among all states in the value of her products in both 1909 and 1919, ranking first in 1909 in 19 industries, including boots and shoes, woolens, cotton goods, cutlery and tools, and second in 22 industries. The remainder of the New England states, like Massachusetts, excel in textiles and the manufacture of the smaller metal products. Connecticut leads in the production of brass, bronze, and copper objects, clocks, firearms, screws, and silverware; and Rhode Island until recently in jewelry. The distance from raw materials has turned the northeast to the manufacture of smaller commodities in which the labor cost is higher, transportation cost lower, and the value added by manufacture greater.

Manufacturing in the Middle West (Including the East North Central and the West North Central States).—This region had to wait upon immigration and settlement before sufficient labor and capital could be accumulated for manufacturing. The wealth of raw materials both agricultural and mineral destined this section to an industrial future. As the greatest corn country in the world it has naturally created a slaughtering and meat-packing industry and has drawn to itself the manufacturers of vehicles and farm machinery. The bituminous coal of Ohio, Illinois, and Michigan provided power; the iron of Ohio, Missouri, and the Lake Superior region of Minnesota furnished raw materials for foundries, while petroleum formed the foundation of other industries. In this section are located Illinois and Ohio, the third and fourth states, respectively, in the value of manufactured products in 1919.

The principal products of the most important manufacturing states of the middle west were in 1919 as follows:

Illinois: (1) slaughtering and meat packing, (2) foundry and machine-shop products, (3) men's clothing, (4) iron and steel, steel works and rolling mills, (5) agricultural implements.

Ohio: (1) iron and steel, steel works and rolling mills, (2) rubber tires, tubes, etc., (3) foundry and machine-shop products, (4) automobiles, (5) iron and steel, blast furnaces, (6) slaughtering and meat packing.

Michigan: (1) automobiles, (2) automobile bodies and parts, (3) foundry and machine-shop products, (4) engines, steam, gas and water, (5) ship building, steel, new vessels and small boats, (6) furniture.

Wisconsin: (1) slaughtering and meat packing, (2) automobiles, (3) leather, tanned, curried, and finished, (4) cheese, (5) engines, steam, gas, and water, (6) foundry and machine-shop products, (7) paper and wood pulp, (8) condensed milk.

Indiana: (1) iron and steel, steel works and rolling mills, (2) slaughtering and meat packing, (3) automobiles, (4) cars, steam railroad, not including operations of railroad companies, (5) foundry and machine-shop products, (6) flour-mill and gristmill products, (7) automobile bodies and parts.

Missouri: (1) slaughtering and meat packing, (2) flour-mill and gristmill products, (3) boots and shoes, (4) automobiles, (5) foundry and machine-shop products, (6) cars, steam railroad, not including operations of railroad companies.

Iowa: (1) slaughtering and meat packing, (2) butter, (3) food preparations, (4) cars and general shop construction and repairs by steam-railroad companies, (5) glucose and starch, (6) flour-mill and gristmill products.

The Far West (Including the Pacific and Mountain States).—The economic life and the manufactures of the Pacific states are founded primarily upon agriculture. In Washington, Oregon, and California three of the five industries ranking highest depend upon agriculture or lumber. Washington is first among all the states in forest products, with California third and Oregon fourth. California leads the Union in canning and preserving fruits and vegetables, although her leading industry in point of value is petroleum refining, a recent development. The prominence of steel-ship building in these states in 1919 was an aftermath of the World War.

MANUFACTURING SINCE 1860

LEADING INDUSTRIES OF THE PACIFIC STATES, 1919, RANKED BY VALUE OF PRODUCTS

Rank	Washington	Oregon	California
1	Lumber and timber products	Lumber and timber products	Petroleum refining
2	Ship building, steel	Flour-mill and gristmill products	Canning and preserving fruits and vegetables
3	Flour-mill and gristmill products	Foundry and machine-shop products	Ship building, steel
4	Slaughtering and meat packing	Slaughtering and meat packing	Slaughtering and meat packing
5	Ship building, wooden	Ship building, wooden	Flour-mill and gristmill products
6	Foundry and machine-shop products	Canning and preserving fruits and vegetables	Foundry and machine-shop products

It is customary to think of the eight mountain states as interested fundamentally in the extraction of minerals, and of their manufactures as closely connected with smelting and refining. This impression is not borne out by the facts, for the agricultural products of these states are double those of the mineral. Furthermore, manufactures based on lumber and agriculture total three out of the first five industries in seven states. The chief industry in value of products in Montana is flour-mill and gristmill products; in Idaho, lumber and timber products; in Colorado, slaughtering and meat packing. It is interesting to note that the chief industry in the three states of Wyoming, New Mexico, and Nevada is that of cars and general shop construction and repairs by steam-railroad companies. In Arizona, however, the industry first in value of products is the smelting and refining of copper; and in Utah the smelting and refining of lead, an industry in which no other state is even listed by the census.

Manufacturing in the South (Including all of the States Designated by the Census Under the Terms South Atlantic,

East South Central and West South Central States).—The Civil War ended forever the popular southern belief that the future of that section lay wholly in agriculture. Without capital the great plantations were broken up to be tilled by tenant farmers, while the remnants of the planter class pursued their fortunes in the cities. It was not until the decade of the 'eighties that manufacturing had made much headway. By that time the south had sufficiently recovered to accumulate local capital while northern investors began to pause from their exploitation of the west to see the latent possibilities in the south. This section was still and probably always will be primarily agricultural; but the great crops of cotton and tobacco needed preparation before marketing, and the cheap labor available from the negroes and "poor whites" made it likely that factories would be set up close to the raw materials. Until recent years the lumber and mineral resources had scarcely been tapped, but both have provided inviting fields for outside capital.

The most spectacular development which any southern industry has experienced is lumbering. The interest of the south in forest products goes back to colonial days, but its premier position dates but recently with the partial depletion of the northern forests. In 1870 the northeastern group of states produced nearly 40 per cent and the Lake states 25 per cent of the lumber cut in the United States, and ten years later these two regions still produced 60 per cent of the total, although the Lake states were now in the lead. By 1900 the southern states took the lead and have continued to hold it, pushed hard by the states of the northwest. Their production now amounts to about two-fifths of the total cut. In value, lumber and timber products ranked high in 1919 in each of the southern states; in eight of these states, this industry employed more workers than did any other kind of manufacturing, and in number of establishments it led in every southern state. In addition to hard pine and other coniferous woods, turpentine and rosin are manufactured in Florida and Georgia. The following table interestingly illustrates the migratory nature of the lumber industry as well as its southern location:

LUMBER PRODUCTION 1869-1919. PERCENTAGE BY SECTIONS

	1869	1879	1889	1899	1909	1919
Northeast	35.7	25.8	19.8	16.3	11.7	7.5
Central	17.9	18.4	13.1	16.1	12.3	8.7
Southern	10.1	13.8	20.3	31.7	44.9	46.6
Lake	28.2	34.7	34.6	24.9	12.3	7.8
Pacific and Mountain	4.9	4.5	9.6	9.9	18.4	29.3
All others	3.3	2.8	2.6	1.1	0.4	0.2

Since 1880 the monopoly of New England in the manufacture of cotton has been broken by the southern states, in particular by North and South Carolina. Nearness to the source of raw materials and cheap labor, combined with increased taxation and labor difficulties in New England, have been the chief causes. This southern movement had forced the New England manufacturer to turn his attention to the production of the finer grades of cloth and has left the coarser grades of sheeting and ducks to the south. In 1880 there were less than 500,000 spindles in North Carolina, South Carolina, and Georgia, with a product valued at scarcely $13,000,000, while there were over 8,500,000 spindles in New England. By 1910 over half of the raw cotton was manufactured in the south. As notable as the increase in cotton manufactures has been the development in the utilization of the by-products. Cotton seed, formerly thrown away, is now fed to cattle or manufactured into cooking oil. The value of cotton by-products has risen from $18,000,000 in 1892 to $352,000,000 in 1919.

Abundant supplies of coal, iron, and copper in South Carolina, Tennessee, and Alabama have originated, especially in Alabama, extensive iron and steel works. The production of pig iron increased from 347,000 tons in 1880 to 2,130,000 in 1919. The city of Birmingham had grown from 3,000 to 179,000 in 1920 and is one of the steel centers of the United States. More recently the discovery of oil wells in Texas and Louisiana has laid the foundations of a new industry and a new prosperity.

The value of southern manufactured products has increased

from $338,791,898 in 1880 to $8,375,404,000 in 1919, and the capital invested from $192,949,654 to $6,092,619,000. Nevertheless, the southern states in 1919 produced but 13.5 per cent of the total manufactures of the nation, measured in terms of value. The future must necessarily see a further industrial development in the south, but it will be handicapped by unskilled labor, an enervating climate, and the superior possibilities of agriculture.

The Manufacture of Food.—When the position of the United States as a food-producing nation is taken into account, it is not at all surprising to find that slaughtering and meat packing and the output of flour- and gristmills rank first and sixth in the list of manufactured products. The slaughtering of meat, the preserving of food, and even the grinding of grain were at one time largely household industries. The entrance of women into industry and professions, the multifarious activities of the modern housewife, the growing scarcity of household servants, the growth of urban life, and the cheapness of manufactured products have all tended to take the preparation of food, at least in the preserving stages, out of the home. Baker's bread, prepared breakfast food, and canned meat, vegetables, fruit, fish, and milk are all indications of this tendency.

Much of this manufacturing, it is true, is of a simple kind and has depended for its growth rather upon increased production of foodstuffs than upon new inventions. Nevertheless, the invention of modern methods of refrigeration in the 'seventies stimulated immensely cattle raising and the transportation of fresh meats, while the roller process of making flour permitted the utilization of spring wheat and so put under cultivation the wheat fields of the Dakotas, Montana, and Minnesota. Innumerable patents have made possible the use of the old staples in new form. In like manner manufacturing processes have introduced new but valuable foods, such as cotton-seed oil and peanut butter. The total value of foodstuffs altered by manufacture amounted in 1919 to $12,438,941,000, of which $2,327,344,000 were added by manufacture. Food products contributed nearly 20 per cent of the total value of the manufactures of the country.

Americans consume more meat per capita than any other

people. Slaughtering and meat packing ranks first among industries, with a production valuation in 1919 of nearly $4,250,000,000. The industry, as before suggested, has followed the westward movement of the corn belt. Its center from 1816 to 1860 was in Cincinnati and the cities of Ohio, but has now shifted west to Chicago, St. Louis, Omaha, and Kansas City. In addition to the introduction of refrigeration and canning the industry has profited by the utilization of an ever-increasing number of by-products which are now the chief source of profit. No part of the animal is wasted; fertilizer, leather, glue, wool, and many other products are derived therefrom. The packing industry has grown from 259 establishments in 1859, the value of whose output was $29,441,000, to 1,304 in 1919, with an output of $4,246,291,000. Besides meat, the chief animal products in 1919 were butter, valued at $583,163,000, condensed milk and milk products other than cheese, valued at $339,507,000; cheese at $143,456,000; and canned and preserved fish at $77,284,000.

The lead in the manufacturing of vegetable food is taken by flour-mill and gristmill products, whose value in 1919 was $2,052,434,000. The manufacture of flour is carried on in every state in the Union, and widely dispersed in establishments whose invested capital averaged only $74,797, and whose wage earners averaged only four in 1919; but notwithstanding the preponderance of small flour-mills, the 274 mills reporting a consumption of 100,000 (or over) bushels of wheat turned out over two-thirds of the total product in the United States. Although the industry is so widely distributed, the center has shifted west as the great wheat fields of the prairies have been opened up. Minnesota takes first place in flour-mill and gristmill products, followed by Kansas, New York, Illinois, and Missouri. The value of these products has increased almost ninefold since 1860, due to the opening up of the west, the growth of transportation facilities, an increasing home market, and the introduction of the roller process. Bread and bakery products, valued in 1919 at $1,151,896,000, are manufactured chiefly where the population is most dense. The manufacture of confectionery and ice cream, with an output valued in 1919 at $637,209,000, is centered chiefly in New York, Illinois, Pennsylvania, and Massachusetts. The canning and preserving industry is well distributed throughout the United States, Nevada and Wyoming being the only states not represented by one or more establishments in 1919. Nevertheless, the Pacific states led with 42 per cent of the total value of the products, followed by the Middle Atlantic group with 18.1, and the East North Central with 17.4. The leading states are California, New York, and Maryland. The manufacture of most of the other food preparations is greatest in New York and Illinois.

Clothing, Textiles and Shoes.—The textile industry is largely concerned with the manufacture of cloth and clothing, and ranked in 1919 as the third great group of industries, with a total product valued at $9,216,103,000. At the head of the textiles stands cotton manufacturing, an industry ranking fifth among the specific industries of the country in 1919. The value of the product of all of the textile mills and the industries dependent upon them is equal to nearly one-seventh of that of all our manufactures combined. The textiles increased in value in the last half of the nineteenth century more than sixfold, and in the amount of the product tenfold. The general causes for this do not differ from those which affected other manufactures—an increasing market and labor supply, an abundance of raw material in the case of cotton, and a high protective tariff.

GROWTH OF THE TEXTILE FABRIC INDUSTRY, 1859-1919 [1]

Year	Number of establishments	Average number of wage earners	Capital	Value of products
1859	3,104	191,152	$ 148,440,000	$ 211,707,000
1869	4,709	267,321	285,175,000	418,527,000
1879	4,290	387,554	406,337,000	534,674,000
1889	4,056	497,822	729,333,000	730,567,000
1899	4,999	631,979	982,559,000	886,882,000
1909	4,825	834,087	1,717,795,000	1,591,736,000
1914	4,991	874,702	1,921,925,000	1,761,711,000
1919	6,087	969,260	4,102,969,000	5,006,641,000

[1] Abstract of Census of Manufactures, 1919, table 34, p. 50.

While the present value of manufactured textiles is enormous we still import considerable quantities of high-grade silks and woolens.

Before 1860 cotton was the chief textile manufactured and was the first American industry to be brought into the factory. In that year there were 572 mills in New England, 540 in the middle states, 159 in the south and 22 in the west, the total production valued at $115,600,000. The abnormal demand for woolens during the Civil War and the interruption in the supply of raw materials pushed back cotton temporarily. It was not until 1900 that cotton again attained first rank, when its manufactured product was valued at $339,200,320 as against $296,990,484 for woolen manufactures; the value of cotton manufactures is now nearly double that of woolens. As a matter of fact cotton manufactures still rank as the most important branch of the textile industry, with a value in 1919 of over $2,000,000,000 or nearly 40 per cent of the total value of products for all kinds of textiles combined. The United States ranks second among the nations in the manufacture of textiles, being surpassed only by Great Britain, a position achieved chiefly by the fact that we are the only great cotton-manufacturing nation whose raw material is found at home.

Mention has already been made of the movement of cotton manufacturing toward the source of supply. New England's 570 establishments of 1860 had decreased to 459 in 1919, chiefly through consolidations, although the number of spindles had increased from 3,859,000 to 17,542,926, but the 159 mills of the south in 1860 had increased to 711 in 1919 and the spindles from 561,000 to 14,568,272. The cotton-growing states consume over three-fifths of the cotton manufactured in the United States, although the better grades and the greater profits are still made in New England.

The great competition in cotton manufacturing has brought astounding progress since 1860. Although the fundamental patents had been taken out before the Civil War, notable achievements, particularly in labor-saving devices, have been made since then so revolutionary that the large concerns have been forced to replace their machinery two or three times during this period. Among the most famous of the improvements are the ring spinner, the Northrop loom, the Barber warp-tying machine, and the automatic seamless knitting machine. The first three of these inventions have doubled the production per operative. The ring spinner does not turn out as even or soft a thread as the old mule spinner, but its greater production speed has won it an assured place in the American mills, where labor is costlier and scarcer. The Northrop loom is almost human in its ability to stop if the warp breaks or the shuttle gets out of place, and in inserting the colored warps at the proper time. The Barber machine ties the ends of the threads together, and the Bronson knitting machine will knit stockings complete in every detail.

If the Civil War temporarily set back cotton manufacturing, it gave the woolen industry an impetus which has established it firmly. The value of woolens manufactured in 1850 was less than $50,000,000, while the combined manufactured value of woolens, worsteds, carpets, and felts in 1919 amounted to $1,234,657,000, divided as follows: worsteds, $700,537,000; woolens, $364,897,000; carpets and rugs, $123,254,000; felt goods, $39,230,000; and felt hats, $6,740,000. These are manufactured in 1,016 mills, of which 259 are in Pennsylvania, 207 in Massachusetts, 88 in Rhode Island, and 68 in New York. The tendency toward consolidation has been marked in the woolen industry, as is shown by the fact that the number of establishments has declined from 3,208 in operation in 1869 to 1,016 in 1919.

The American manufacturer because of special conditions has until recently carried out all the processes of manufacture in a single plant, carding the short fiber for woolens, combing the long fiber for worsteds, spinning the thread, weaving the cloth, and dyeing it. The European, on the other hand, has specialized these processes in different plants and thus concentrated skilled labor on certain phases. This partially accounts for the frequent superiority of the British woolens. The American manufacturer leads the field in the grade and variety of flannels

MANUFACTURING SINCE 1860

and blankets but in quantity his greatest output consists of the cheaper woolens which enter into the consumption of the wholesale clothing houses. The manufacture of worsteds dates from about 1870 and is rapidly growing at the expense of other woolens, but its growth has been retarded by the greater amount of skill and labor necessary for its production. The development of woolen machinery has been less rapid than that of cotton, and the inventions have been mostly English in their origin, as is the machinery used.

The most spectacular advance in the woolen branch of textiles in America has been in the manufacture of carpets. Until after the Civil War carpets were largely hand woven. Most of the inventions have been the work of Americans, who have made the United States the greatest carpet-producing nation in the world and the products the finest obtainable excepting the handloom rugs of the Orient. The prosperity and wealth of the nation have enabled the consumption to approach 100,000,000 yards annually. The factory production of carpets commenced with the adaptation in 1841 by Erastus B. Bigelow of Boston of the power loom to the weaving of ingrain carpets, and a few years later to the weaving of Wiltons and Brussels, a work supplemented in 1864 by Smith & Skinner of Yonkers, who applied the power loom to Axminsters. The value of manufactured carpets and rugs rose from $7,857,000 in 1860 to $123,254,000 in 1919. The 75 mills operating in 1919 were located chiefly in New York and Pennsylvania, and their product is not only supreme at home, but is exported in considerable amounts.

In hosiery and knit goods the United States also leads the world. Hosiery was knit in the home until, in 1832, Egbert Egberts at Cohoes, New York, successfully applied the principles of knitting by power, and thereafter the manufacture of hosiery gradually shifted to the factory. Owing to the variety of products, the tendency in this industry to build comparatively small mills with moderate capital has been marked. The number of establishments has increased from 197 in 1860 to 2,050 in 1919, and the value of the products from $7,280,000 to $713,-

140,000. The industry is centered in New York and Pennsylvania.

The most remarkable advance in textiles in America has been in the silk industry. This is the more astonishing in that it has been built up upon raw material which must be imported from a great distance and in competition with the long established and well equipped mills of France. The success has been due to tariff protection, to the enterprise of manufacturers, and to the nation's wealth, which has created a market. With these favoring factors, but against great odds, the industry has grown from $6,608,000 in 1860 to $688,470,000 in 1919, an amount which equaled nearly 93 per cent of the domestic consumption, and a product larger than that of any other nation. John Ryle of Paterson, New Jersey, and the Cheneys of South Manchester, Connecticut, founded the industry by the manufacture of thread and ribbons, branching out eventually into piece goods. Paterson, New Jersey, is now the center of the industry, with Pennsylvania, New Jersey, New York, Connecticut, and Massachusetts the principal producing states, in the order named.

Following closely upon the heels of textile development came that of ready-made clothing. It was stimulated by the demands of the Civil War and made possible by the invention of the sewing machine. Labor was provided by women and children and by the steady influx of foreigners, first the Irish, and after 1876 the Russian Jews. The application of factory methods to ready-made clothing is a comparatively recent development, and during most of our history the industry has been cursed by sweat-shop methods under which operatives worked for a miserable pittance in their own homes often with the most unhealthy surroundings. Considerations of humanity, of public health, and of business are gradually eliminating sweat-shop clothing through both voluntary and legal means. Some modern clothing factories are as up to date as can be found in any industry, though much clothing is still manufactured in small factories hardly better than the sweat-shop. The value of men's ready-made clothing increased from $80,830,000 in 1860 to $1,162,986,000 in 1919, and of women's from $7,181,000 to $1,208,543,000.

New York City is far in the lead in both branches, but its progress is augmented by that of other metropolitan centers such as Chicago, Philadelphia, Rochester, and Baltimore.

Until 1845 shoemaking was a hand trade and usually carried on in the home. The change to the factory commenced in that year by the invention of the leather roller which hastened the preparation of hides. Then came the Blanchard lasting machine which could turn out lasts of standardized sizes. Soon J. B. Nichols, a Lynn shoemaker, adapted the sewing machine of Howe to the sewing of uppers, and just before the Civil War Lyman B. Blake invented the McKay machine for sewing the uppers to the soles. Shoe machinery is among the most ingenious in the world, and in an average factory sixty kinds of machines are used. A peculiarity of the industry is the control of the most important patents by a single company, the United Shoe Machinery Company, who lease the principal machines and are paid a royalty upon each pair of shoes manufactured.[1]

"The American shoe of today is the standard production of the world," says the Census of 1910. The Census of Manufactures of 1919 credits the industry with 1,449 establishments, the product valued at $1,155,641,000, and employing 211,049 persons. It is estimated that the American people spend $1,000,000,000 a year for footwear and average three pairs per capita annually. The centers of the manufacturing are in the vicinity of Boston and St. Louis. The boots and shoes produced in the three New England states, Maine, Massachusetts, and New Hampshire, represented 48.9 per cent of the total for the United States. The five leading shoe cities in the order named are Brockton, New York, Lynn, St. Louis, and Haverhill.

Iron and Steel.—It is impossible to overestimate the importance of iron and steel in modern life. We live in an age of machinery which rests upon iron and steel, as does our transportation system and our large structures. Through steel rails, railroad

---

[1] The United Shoe Machinery Company was prosecuted by the government on the grounds that their "tying clause," which requires that one machine shall be used in conjunction with another, was in violation of the Clayton Act, a contention upheld in 1922 by the Supreme Court.

coaches, and an infinite variety of tools and machinery, the basic metal enters intimately into our everyday life. The rapid rise in output is accounted for chiefly by the increased demand, by the substitution of coke for anthracite coal in the smelting, and by revolutionary inventions which have facilitated and cheapened production.

Iron and steel and their products rank second among the great groups of American manufactures, with 20,120 establishments and total products valued at $9,403,634,000, over one-seventh of the value of manufactures of the nation. This group includes many industries, such as blast furnaces, steel works and rolling mills, and foundry and machine-shop products; but it excludes important industries founded mainly on iron and steel, such as vehicles for land transportation, steel-ship building, agricultural implements, and railroad repair shops.

The manufacture of iron and steel falls into two divisions, the production of pig iron and the conversion of pig iron into commercial iron and steel. Until after the Civil War, steel was a rare commodity used chiefly in cutlery and in the finer grade of tools. Iron, however, was the chief metal in use, and up to 1839, when anthracite was introduced, was chiefly smelted with charcoal. In the years immediately after the war bituminous coal, chiefly in the form of coke, was introduced in smelting, a factor enabling the industry to distribute itself so widely that in 1880 iron was manufactured in thirty states. The greatest single event in the history of iron and steel was the invention in 1856 by two Englishmen, Henry Bessemer and Robert Musket, of a method, known as the Bessemer process, by which a blast of cold air is forced through the molten pig iron, oxydizing the foreign substances, after which such quantities of carbon and other elements may be introduced as will make the desired quality of steel. The Bessemer process was first used in this country at Wyandotte, Michigan, in 1864. By cheapening its production it made universal the use of steel and relegated iron to a position of comparative unimportance.

The Bessemer process has its limitations. It is not suitable for ore high in phosphorus, and this factor has led to its being supplanted by the open-hearth, or Martin, process. Owing to the high-grade ore of the Lake Superior region the Bessemer method was widely used until 1906, when 12,000,000 tons were manufactured by that process. Since then it has been superseded by the open-hearth method, which makes available lower-grade ores. Of the 34,026,979 tons of steel produced in 1919, the open-hearth process turned out 26,726,036 and the Bessemer only 6,946,939.

With greater production and better smelting methods there have appeared other improvements. The furnaces have doubled in size and tripled in heating capacity since 1850; the average output per furnace of 50 tons a day in 1870 was increased to 400 tons a day. The mechanical progress of the industry is illustrated by the following description:

The bulk of the domestic iron ore used in the United States is not touched by the hand of man from the time it leaves the mine until it is converted into forms for sale, as rails, structural shapes, wire rods, sheets, merchant bars, etc., and in many of the operations the ore is dug by power shovels, or "milled" through chutes, so that even in mining but little hand labor is necessary. Lifted from its bed by dippers on power shovels, ore is loaded into railway cars or directed through "mills" into chutes, or mined by pick and shovels and shot into mine tram cars. The cars run to pockets feeding skips, which are raised by power and automatically dumped into waiting railway cars, except when climatic conditions, stagnant trade, or the desirability of discontinuing operations, require that ore be stocked for future shipment. Where necessary the ore passes through intermediate crushers to reduce the size of the pieces; but in crushing, stocking or reloading, man directs the movement of machinery and touches but little of the ore.

Railway cars, with their loads of iron ore, convey it to blast furnaces, or, as is the custom with most Lake Superior ores, to docks provided with pockets, into which the ore drops through the opened bottoms of the cars, while spouts connecting with the dock pockets deliver the ore by gravity into holds of the vessels specially constructed for the iron ore trade. At the end of the vessel's trip, mechanical devices remove the ore and deposit it in cars or on stock piles. . . .

When the furnace is reached to which the ore is consigned, the railway cars are run onto "car dumpers" which turn the cars over to discharge their contents into pits from which the ore is carried by cable or tramways to stock piles; or the cars drop their contents into the furnace supply bins. From the bins, whether fed direct from cars or from stock piles, the ore is chuted to scale charging cars which feed to skip cars. These skip cars, which also receive the fuel and fluxing material, are elevated to the furnace top and automatically discharge their contents.[1]

The steel, once made, is strengthened and cast into the desired forms by forging, pressing, and rolling.

The chief centers of pig iron production are (1) Pennsylvania and Ohio; (2) Illinois, and (3) Tennessee and Alabama, all located close to the Lake Superior or Alleghany fields. An abundance of raw material and an expanding market have enabled the United States to produce over three times as much pig iron and twice as much steel as Germany, her nearest competitor, and over four times as much of both as Great Britain. Statistics of the industry follow:

IRON AND STEEL INDUSTRY, 1859-1919

Year	Number of establishments	Average number of wage earners	Capital	Value of products
1859	402	22,014	$ 23,343,000	$ 36,537,000
1869	808	77,555	121,772,000	207,209,000
1879	792	140,798	209,905,000	296,558,000
1889	719	171,181	414,045,000	478,688,000
1899	668	222,490	513,392,000	803,968,000
1909	654	278,505	1,492,316,000	1,377,152,000
1919	695	416,748	3,458,935,000	3,623,369,000

The value of all non-ferrous metals mined in the United States in 1919 was nearly $1,000,000,000 and the mining and smelting of these metals constitute an important part of our industrial system. The total value of the products of smelting and refining establishments in the United States in 1919 was $972,094,000. New Jersey, Arizona, Pennsylvania, Illinois, and Colorado ranked highest in the order named. Utah led the Union in the smelting and refining of lead, New Jersey and Arizona in that of copper,

---

[1] *The A B C of Iron and Steel,* edited by A. O. Backert (4th ed., 1921), chap. i, by John Birkinbine, "Iron Ore and Mining Operations," p. 1.

and Illinois, Oklahoma, and Kansas in that of zinc. Gold and silver are among the products of copper and lead smelting and refining, and constitute a large part of the value of the products of these industries.

The manufacture of electric supplies is another industry which depends on the metals—copper, lead, zinc, and aluminum, as well as steel. Although the telegraph was in use in the 'forties, the general development of the industry has taken place during the past thirty-five years, during which time 17,500 patents have been taken out. The value of electric supplies in 1919 amounted to almost $1,000,000,000 and has since undoubtedly risen rapidly due to the development of the radio. The application of electricity to telegraphy, telephony, wireless communication, lighting, and motors has influenced economic and social life more potently than any group of mechanical inventions since those which brought into being the Industrial Revolution.

Tariff History Since 1860.—The American system of high protective tariffs may be said to date from the Civil War. There had been protective tariffs, of course, before that time; but they had not been exorbitantly high and the tendency in the years immediately before the war had been downward. The act of 1857 had reduced the maximum protection to 24 per cent, and the general level of duties was reduced to the lowest point since 1815. In 1861, just before the opening of the war, the Morrill Act had been passed, which aimed to restore the duties of 1846, and by adding to the duties on iron and wool, hoped to attach to the Republican party Pennsylvania and some of the western states. Hardly had the measure been enacted when the south broke away and some source of revenue had to be found immediately to help prosecute the war. As the tariff up to that time had been the chief source of revenue, Congress naturally turned to it in this dilemma. Scarcely a month of any congressional session passed during the struggle without some increase of duties on imports. This action was the more easy because of the growing protectionist sentiment as manufacturing was stimulated during the war, and also because it was felt that manufacturers should receive additional protection to compensate them for the high internal duties which were imposed on all manufactured articles. "Probably no country," said Professor Taussig, "has seen in so short a time, so extraordinary a mass of financial legislation." [1] The most important of the war tariffs were the acts of 1862 and 1864. An internal-revenue act earlier in 1862 had levied in a wholesale manner taxes on manufactured products and the tariff of 1862 was framed partially to offset this former bill. Again in 1864 there were wholesale advances in both internal and tariff duties. Under the guidance of strong protectionists and under the spur of war needs this tariff of 1864, which raised the duties "greatly and indiscriminatingly," was jammed through; "five days in all were given by the two houses to this act, which was in its effects one of the most important financial measures ever passed in the United States." [2] The average level of the tariffs had been increased from 19 per cent in 1851 to 47 per cent by the act of 1864.

These hastily drawn and imposed measures were obviously so unscientific that many of the war-time supporters were convinced that a revision downward was in order. An attempt in 1867 to reduce the tariff was a failure, and further advances were made in 1869 and 1870. The growth of the liberal Republican movement in 1872, which advocated reduction, caused a 10-per-cent lowering of the rates immediately before the election, but the depression of 1873 and the loss therefrom of revenue gave an excuse for restoring the duties to their former level. In the clamor for lower taxes after the war it was much easier to remove the internal revenues. Furthermore, the political parties were not yet aligned definitely upon the issue, and as always the lobby of the protectionists was more efficient and successful than the influence of the people as a whole. The returning prosperity after the war was ascribed to the high tariffs and the latter became a fixed part of our financial structure.

No further attempt of importance to deal with this matter was made until the Treasury surplus of $100,000,000 in 1881 and 1882 brought from President Arthur a request for tariff

---

[1] Taussig, F. W., *Tariff History of the United States*, 6th ed., p. 160.
[2] *Ibid.*, p. 168.

reduction. Although a strongly protective commission advised a reduction averaging 25 per cent, the protectionists so manipulated the bill of 1883, that when it was eventually passed the general level was reduced only 5 per cent. Up to this time neither of the political parties had taken a very definite stand on the tariff. President Cleveland, however, influenced again by a surplus in the Treasury, devoted his entire message of 1887 to the question. Affirming that industries must have necessary protection, he demanded that consideration should be given to the people at large by reducing the cost of living and declared the existing tariff to be a "vicious, inequitable, and illogical source of unnecessary taxation." "Our progress toward a wise conclusion," he said, "will not be improved by dwelling upon the theories of protection and free trade. This savors too much of bandying epithets. It is a condition which confronts us, not a theory." Led by John G. Carlisle, the Democratic House prepared and passed the Mills bill. In the meantime the Republican Senate prepared, under the direction of Senator Aldrich, a protectionist measure. As neither bill could pass the other House, the matter was deadlocked until after the election.

Cleveland forced the election of 1888 on the tariff issue and the people as a whole received their first instruction on the question during the course of the campaign. Although Cleveland received more popular votes, his opponent, Harrison, captured the electoral college and the Republican party interpreted the victory as an indorsement of a high-tariff policy. By revising the House rules to prevent dilatory methods, the Republicans were able in 1890 to pass the McKinley bill, which has been described as the "climax of protection." High duties were placed on the finer grades of woolens, cottons, linens and clothing, and on iron, steel, glass, tin plate, and other commodities. To propitiate the farmer, tariffs were laid on agricultural products; and to take care of the surplus, the duty on sugar was removed and a bounty of two cents a pound placed on the domestic product in order to protect the Louisiana producers. The McKinley tariff raised the average level of duties to 49.5 per cent. Retail prices so quickly reflected the increased tariffs and the reaction of the public was so great that the Democrats in 1890 were swept into possession of the House and in 1892 into the Presidency.

With the return of the Democrats to power in 1892 Cleveland pressed Congress for action on the tariff. His efforts, however, toward a healthy reduction were foiled by an active lobby supported by Democratic protectionists, and Cleveland, after denouncing the Wilson-Gorman tariff of 1894 as an example of party perfidy and dishonor, allowed it to become law without his signature. It did, nevertheless, put wool, copper, and lumber on the free list and lowered the average level to 39.9 per cent. The decreased revenue was to be supplied by an income tax of 2 per cent on incomes over $4,000, a feature of the act which was declared unconstitutional in 1895.[1] The panic of 1893 and the business depression which ensued were partially attributed to the anticipated tariff, and the discontent was sufficient to bring back the Republicans to power in 1896. A special session of Congress thereupon wiped the Wilson-Gorman Act from the statute books and substituted the Dingley Act. The latter, modeled closely on the McKinley bill, raised the average level to 57 per cent, the highest in our history, put prohibitive rates on wool, woolen goods, cutlery, and pottery, replaced lumber on the dutiable list, and taxed hides, heretofore free, to insure the support of certain western senators.

The reaction from the trust making of the years 1897-1903 had brought the progressive movement of the Roosevelt administration. Many believed that the high tariff fostered monopoly, and the Republican platform of 1908 declared "unequivocally for a revision of the tariff by a special session of Congress immediately following the inauguration of the next President," a revision which was generally interpreted as meaning downward. Protectionist senators, however, prevented any serious reduction, and the Payne-Aldrich bill of 1909 produced no essential change in the tariff system, with the exception that hides were placed on the free list.

Failure in the Payne-Aldrich bill to revise the tariff downward

---

[1] Pollock v. Farmers' Loan and Trust Company, 158 U. S. 429.

was a factor in the formation of the Progressive party in 1912 and the election of the Democrats in that year. The last-named party was pledged to reduction and had emphasized the idea of a "competitive tariff"—that is, the lowering of duties sufficiently to place domestic goods on a competitive basis with foreign. The Underwood tariff of 1913 put iron and steel and raw wool on the free list, and sugar in 1916; certain agricultural products were to come in free and big reductions were made on cotton and woolen goods. Some few advances were made, chiefly in chemicals. Although the Underwood Act left the tariff still highly protective, a real attempt had been made to reduce the cost of living. Any deficiency in revenue was to be supplied by receipts from an income tax, incorporated in the bill and now made constitutional by the Sixteenth Amendment.

American tariffs in the making have been characterized by the most unscientific procedure of political give and take. The need of turning the question over to a group of experts had been recognized in the Tariff Commission of 1882 and the Tariff Board of 1909. Neither of these had been taken seriously by Congress and the latter board was legislated out of existence. In 1916 a new Tariff Commission of six members was created to investigate all questions with reference to tariffs and to submit reports to Congress.

With the conclusion of the war and the return to power of the Republican party in 1920, a radical revision of the tariff was to be expected. Two factors were dominant in its making: first, the agricultural distress following the collapse of the war boom; and second, the voice of economic interests, using the nationalism kindled during the war to give force to their demands, and clamoring for protection of the industries stimulated by the war. To prevent post-war dumping and to meet the demands of the farmers an "emergency" tariff was rushed through a special session of Congress (May 27, 1921); it imposed duties on wheat, corn, meat, wool, and sugar, and was to be kept on the statute books until a more detailed act could be framed. The latter, known as the Fordney-McCumber tariff (passed September 19, 1922), not only returned to the high duties of 1909 and earlier tariffs, but surpassed them in the high protection given. Agricultural products were protected to a high degree, although the protection was hardly needed and did not affect perceptibly the decline in prices. Hides, however, remained on the free list to offset the effects of an absence of tariff on boots and shoes, an omission insisted upon by the farmers. The duties on manufactured goods were fully as high as those on agricultural products and of more significance. The duties on iron and steel, omitted in 1912, were re-imposed, and those on textiles, especially silk, were increased. In response to the demands mentioned above, the act was particularly concerned with the so-called "war babies," especially with the chemical and dyestuff industries, and gave them ample protection. There was much talk during the passage of the act about equalizing "the differences in costs of production in the United States and the competing foreign countries," and for fear that foreign competition might possibly injure the American producer, the President was given power to raise or lower duties not exceeding 50 per cent upon recommendation of the Tariff Commission. This power given to the President, and the extreme protection conferred by the tariff, are the outstanding features of the act.

NOTES FOR FURTHER REFERENCE

The bases for the study of manufacturing since the Civil War are the census reports, especially Vol. VII of the *Twelfth Census of the United States* and the *Abstract of the Thirteenth Census*. As companion volumes of equal value are the *Reports on Manufactures* for 1905 and 1914. Essential for recent industrial developments, but unwieldy to handle is the *Report of the Industrial Commission* (19 vols., 1902), which took testimony for two years on general industrial conditions. Extracts from this report are included in the *Readings* of Bogart and Thompson. A valuable piece of work, beautifully done, is *A Graphic Analysis of the Census of Manufacturing 1849–1919*, by the National Industrial Conference Board (1923). Excellent chapters of résumé are to be found in E. L. Bogart, *Economic History of the United States* (new ed., 1922); in Isaac Lippincott, *Economic Development of the United States* (1922); in L. R. Wells, *Industrial History of the United States* (1922), and in the old book of C. D. Wright, *The Industrial Evolution of the United States* (1897). Instructive are the observations of a foreigner, Pierre Leroy-Beaulieu, *The United States in the Twentieth Century* (2d ed., 1907). On the location of industry see Malcolm Kier, *Manufacturing Industries in America* (1920).

The history of special industries may be traced in encyclopedias, especially the *Americana*, and in the following books: T. M. Young, *The American Cotton Industry* (1903); P. H. Nystrom, *Textiles* (1916); M. T. Copeland, *The Cotton Manufacturing Industry in the United States* (1912); S. N. D. North, *A Century of American Wool Manufacture, 1790-1890* (1895); W. C. Wyckoff, *American Silk Manufacture* (1880); *Flour Milling* in the semi-centennial issue of the *Northwestern Miller* (1923); F. J. Allen, *The Shoe Industry* (1916); B. E. Hazard, *The Organization of the Boot and Shoe Industry in Massachusetts Before 1875* (1921); M. D. Swank, *History of the Manufacture of Iron in All Ages* (2d ed., 1892); H. N. Casson, *The Romance of Steel* (1907); J. R. Smith, *Story of Iron and Steel* (1913), and J. V. Woodworth, *American Tool Making and Interchangeable Manufacturing* (1905).

On the tariff, consult P. Ashley, *Modern Tariff History* (3d ed., 1920); Edward Stanwood, *American Tariff Controversies in the Nineteenth Century* (1903); F. W. Taussig, *Some Aspects of the Tariff Question* (1915) and *Tariff History of the United States* (7th ed., 1923). Also C. W. Wright, *Wool Growing and the Tariff; a Study in the Economic History of the United States* (1910), Harvard Economic Studies, Vol. V. On the Fordney-McCumber tariff, see F. W. Taussig, *The Tariff Act of 1922*, and A. H. Cole, *The Textile Schedules in the Tariff of 1922*, both articles in the Quarterly Journal of Economics, Vol. XXXVII, No. 1 (November, 1922), and W. S. Culbertson on *The Making of Tariffs*, in Yale Review (January, 1923).

SELECTED READINGS

TAUSSIG, F. W., *Tariff History of the United States* (7th ed.), Part II.
LIPPINCOTT, ISAAC, *Economic Development of the United States*, Chaps. XIX, XX.
Twelfth Census of the United States, Vol. VII, Chap. II.
Abstract of the Thirteenth Census, pp. 435-469.
BOGART, E. L., and THOMPSON, C. M., *Readings in the Economic History of the United States*, Chaps. XXI, XXIII.

CHAPTER XXIV

THE LABOR MOVEMENT

Growth of the Wage Earning Class.—Our economic history from the earliest colonial days has been characterized by a "labor problem"; but a "labor movement"—that is, an organized continued effort on the part of wage earners to better their standard of living—necessarily waited upon conditions arising from the growth in population, the rapid increase in manufacturing, and the concentration of population in cities. These effects of the Industrial Revolution were delayed in this country, owing to many causes—scarcity of labor, lack of liquid capital, abundance of rich unoccupied farming land—all tending to direct the energies of the people into rural occupations and delay the era of manufacturing and urban life.

Nevertheless, population grew rapidly, almost doubling every twenty years. The percentage of the total population living in cities of 8,000 or over increased slowly before 1840 and then more rapidly; only 8.5 per cent of the people lived in such cities in 1841, but by 1860 the percentage had risen to 16.1 and by 1920 to 43.8. The years of most rapid growth as shown in the following table correspond closely with the period of greatest activity on the part of labor.

GROWTH OF CITY POPULATION [1]

Year	Total population	Places of 8,000 inhabitants or over		
		Population	Number of places	Per cent of population
1790	3,929,214	131,472	6	3.3
1800	5,308,483	210,873	6	4.0
1820	9,638,453	475,135	13	4.9
1840	17,069,453	1,453,994	44	8.5
1860	31,443,321	5,072,256	141	16.1
1880	50,155,783	11,365,698	285	22.7
1900	75,994,575	25,018,335	547	32.0
1920	105,710,620	46,307,640	924	43.8

[1] *U. S. Census* of 1920, vol. i, table 27, p. 43.

If the word urban is used, as it is in the Fourteenth Census, to designate places of 2,500 inhabitants or over, it is found that in 1920 51.4 per cent of the population lived in urban territory as compared with 28.6 per cent in 1880.

Although the growth of urban population has been nation-wide, it has been most notable in the manufacturing sections; in 1920 more than two-thirds was contained in three geographic sections—the New England, the Middle Atlantic, and the East North Central states. Rhode Island and Massachusetts each showed over 90 per cent living in towns, New York over 80 per cent, and New Jersey over 70 per cent. The three sections noted above turned out almost three-quarters of the manufactured products of the nation, reckoned in terms of value. The Census of 1850 recorded 957,059 wage earners producing commodities to the value of $1,019,106,616; that of 1889 showed 4,251,535 whose products were valued at $9,372,437,283; the figures of 1914 placed the wage earners at 7,036,247, and the value of the product at $24,246,434,724; and those of 1919 at 9,096,372, and $62,418,078,773.

Other Causes for the Development of the Labor Movement.—The rise of a class of wage earners and their concentration in urban communities are the fundamental factors leading to the growth of the labor movement. The increase in manufacturing after the Civil War developed larger business units, usually under the corporate form. This accretion in the power of capital stirred the wage earners to action, especially those skilled workers whose occupations were imperiled by the invention of new machinery. The passing of the small industry in which it was possible to maintain a close personal relationship between employer and employee, and in its place the coming of the corporation with its thousands of owners scattered throughout the country, tended to a lack of understanding between labor and capital and a seeming diversity of interest.

If the new machinery and the growth of mighty business units affected detrimentally the wage earner, they also contained the elements of his salvation. Big factories brought the workers together in cities where they could mingle with their fellows, exchange more readily their ideas, and combine more easily for resistance. Improvements in paper manufacture and printing made it possible to spread their program, develop loyalty, and weld more firmly together the local organizations. An aggressive labor press, as it developed, contributed not only to the education of the worker in the problems involved, but in like manner made capital conversant with the aims of labor. The whole movement was accelerated by the diffusion of knowledge through our democratic system of education and was integrated by the development of railways and more rapid methods of communication.

Labor Movement Before the Civil War.—Although manufacturing grew steadily during the first half of the century, as did the number of wage earners dependent upon it, the organization of labor proceeded slowly. As long as public land could be had at nominal cost, "wage slavery," in the sense that there was no escape from the system, did not exist. If times were hard and wages low, the worker could always go west. To women the factory system during its early years was looked upon as an escape from dependence and an otherwise colorless life. Nevertheless, the first quarter of the century may be considered, says Professor Ely, "as a germinal period, preceding the modern labor movement, and preparing the way for it."[1] It was characterized by the formation of an increasing number of local unions, especially in the cities of the eastern seaboard. As early as 1803 there was incorporated the "House Carpenters of the City of New York." Twenty-four such associations, partaking of the nature of benevolent or friendly societies, rather than unions, were incorporated in New York between 1800 and 1810. Strikes were not unknown, one of the first being conducted in 1795 by the tailors of Baltimore, the first group to organize in that city.

Lack of facilities for easy transportation and communication kept these efforts local until the decade of 1827 to 1837. These years were marked by such activity that the period may be considered as containing the first real labor movement. The impetus

---

[1] Ely, Richard T., *The Labor Movement in America*, p. 39 (1886).

originated from many causes: a rise in prices; the hostility of the courts toward labor; a general moral awakening throughout the nation, as illustrated by the numerous social schemes and new religions offered during these years for the salvation of mankind; and a feeling of increased power resulting from the extension of the suffrage.

The first trades union in the world was formed in Philadelphia in 1827 by various labor groups which rushed to the support of the carpenters, who had been defeated in a strike for a ten-hour day. Their example was followed in 1833 by the organization of the General Trades Union of New York City and by similar societies in Boston, Albany, and elsewhere. An outgrowth of these city central bodies was the formation of a short-lived National Trades Union (1834-37), which boasted of over 26,000 members. In the middle 'thirties there were probably 300,000 enrolled in unions in the seaboard cities. The new consciousness of power and the general optimism of the period contributed to the belief that the quickest results could be obtained through political action. In 1828 the Mechanics' Union of Philadelphia presented candidates for city and state offices; in 1829 the Workingmen's Party of New York was formed, and at the same time the New England Association of Farmers, Mechanics, and Other Workingmen sought to bring together all manual laborers. The strongest political move of labor during the period was the formation in New York State in 1835 of the Equal Rights Party, dubbed the Loco Focos, which proved sufficiently strong in the elections of the following year to defeat the regular Democratic organization, Tammany Hall, and to force the latter in the future to rely mainly upon the labor vote.

The principal demand of the early labor movement was the ten-hour day. Western land to some extent kept wages up. "All of the strikes I heard of were on the question of hours, not of wages," testified De Tocqueville. Although bitterly opposed by capital, this demand found wide sympathy and made some progress, notably the promulgation of an order in 1841 by President Van Buren inaugurating the ten-hour day in all government establishments. Other demands were made—those for restriction of child labor; for abolition of imprisonment for debt and of the practice of hiring out convicts to contractors in competition with other workmen; for exemption of wages and tools from seizure for debt, and at the same time the right of mechanics to place liens on property to secure payment of their wages; and for the abolition of sweat-shops. To these demands were generally added other reforms, which were at that time in the air—the abolition of monopolies, of private banks, of capital punishment and compulsory militia service; or, on the other hand, the adoption of free trade, temperance, and woman suffrage.

This remarkable period of union growth and political activity on the part of the wage earner was brought to a sudden stop by the panic of 1837, which caused widespread unemployment. Out of work and with their funds exhausted, the unions disintegrated, and the labor press, which had commenced to function, disappeared. Gradually, as trade revived, there came, during the 'fifties, a new awakening on the part of labor. In 1852 the National Typographical Union,[1] the oldest existing trade union in America, was organized. It was followed in 1854 by the National Trade Association of Hat Finishers, and in 1858 by the "Sons of Vulcan," which later (1876) grew into the Amalgamated Association of Iron and Steel Workers. The Iron Moulders' Union and the Machinists' and Blacksmiths' Union of North America were both founded in 1859. By 1860 more than twenty trades had national organizations.

Effect of the Civil War on Labor.—A distinct impetus was given to the labor movement by the Civil War. The struggle resulted in a deeper consideration of economic and social matters, for the question of the liberation of the slaves could hardly be discussed without also involving the status of northern labor, especially when economists of the south maintained that the condition of a southern slave was preferable to that of a northern wage earner; a contention which, considering hours and factory conditions, was not without point. The increasing cost of living

---

[1] This name was changed in 1869 to the "International Typographical Union" in order to include Canadian printers. The word "international" as applied to labor unions in America is ordinarily used in this sense.

brought on by the war was not met by a parallel rise in wages, for a steady influx of immigrants and the adoption of labor-saving machinery helped to meet the scarcity of labor. Especially irritating to labor was an act of July 4, 1864, which enabled agents of employers to engage foreign laborers under a contract in which their transportation was to be paid by future wages. War tariffs and war contracts brought a sudden accumulation of wealth in the hands of a few men and at the same time accentuated the sharp contrast between the rich and poor. On the one hand was a growing power in the hands of capital, and on the other a misgiving concerning the likelihood of a glutted labor market when the soldiers returned; both factors strengthened the determination of labor to keep wages up after they had commenced to ascend during the last two years of the war. Numerous local unions and at last ten national unions sprang into existence between 1863 and 1866. The first of the great railroad brotherhoods, namely the Brotherhood of Locomotive Engineers, was organized at Detroit in 1863 as the "Brotherhood of the Footboard"; their example was followed in 1869 by the founding of the Brotherhood of Locomotive Firemen. By 1870 there were in existence no less than thirty-two national trade unions, while each important city had its trade assembly, its labor press and workingmen's library.

Knights of Labor and Its Forerunners.—Before the impetus given by the Civil War had spent itself, at least one notable attempt had been made to bring all labor together in a single organization. Under the leadership of W. H. Sylvis, and on the basis of the city assemblies of trade unions, a National Labor Union was organized which held seven annual conventions, beginning in 1866, and at the height of its power had a membership of 600,000. The idea that the future of the wage earner lay in coöperative enterprises rather than in militant trade unionism was strongly held and numerous experiments made. The National Labor Union, among other things, advocated Chinese exclusion, the eight-hour day, and the establishment of a government bureau of labor. Another notable development immediately after the war was the growth of the Knights of St. Crispin,[1] founded in Milwaukee in 1867 and especially strong in the shoe-trade centers of Massachusetts. It throve from 1868 to 1870, when it became the "undoubted foremost trade organization of the world."[2] Participation in politics and the depression following the panic of 1873 smashed both of these groups as it did many others. The period 1873-80 was characterized by business demoralization, unemployment, desperate and usually unsuccessful strikes sometimes accompanied by violence and crime,[3] all of which left the labor movement disintegrated and to some extent discredited. Only 18 per cent of the national trade unions survived these years.

During this discouraging period labor turned whole-heartedly to political action and secret organizations. One of these was destined to play an important rôle in the recuperative years of the early 'eighties. In 1869 Uriah S. Stevens, a Philadelphia garment maker, and six fellow craftsmen organized the Noble Order of the Knights of Labor. The high idealism of Stevens was written into the constitution and upheld by Terence V. Powderly, who succeeded him as Grand Master. Taking their motto from Solon, they affirmed, "That is the most perfect government in which an injury to one is the concern of all." The set of instructions given to every initiate into the order read: "Labor is noble and holy. To defend it from degradation; to divest it of the evils to body, mind, and estate which ignorance and greed have imposed; to rescue the toiler from the grip of the selfish—is a work worthy of the noblest and best of our race. . . . We mean no conflict with legitimate enterprise, no antagonism to necessary capital; but men, in their haste and greed, blinded by self interests, overlook the interests of others, and sometimes violate the rights of those they deem helpless. We mean to uphold the dignity of labor, to affirm the nobility of all

---

[1] So-called after St. Crispin, the patron saint of shoemakers.
[2] McNeill, G. E., *The Labor Movement*, p. 200.
[3] The deeds of the "Mollie Maguires" in the anthracite-coal regions of Pennsylvania illustrate the extreme of labor lawlessness during this period. See Rhodes, James F., *History of the United States from Hayes to McKinley, 1877-1896*, chap. ii, pp. 52-87.

who earn their bread by the sweat of their brows. We mean to create a healthy public opinion on the subject of labor (the only creator of values) and the justice of its receiving a full, just share of the values or capital it has created. We shall, with all our strength, support laws made to harmonize the interests of labor and capital, and also those laws which tend to lighten the exhaustiveness of toil. To pause in his toil, to devote [himself] to his own interests, to gather a knowledge of the world's commerce, to unite, combine and co-operate in the great army of peace and industry, to nourish and cherish, build and develop, the temple he lives in, is the highest and noblest duty of man to himself, to his fellow-man and to his Creator."[1] In other words, their aim was to secure to the wage earner the fullest enjoyment of the wealth he creates, and leisure for the development of his intellectual, moral, and social faculties. They favored the eight-hour day, a tax on incomes and inheritances, postal savings banks, workingmen's compensation for injuries received through lack of necessary safeguard, and the appropriation by the community of the unearned increment on land. That the socialism, strong in Europe at that time, had made some progress here is seen in their advocacy of the public ownership of such utilities as railways, gas plants, and waterworks. In addition they urged private coöperative organizations of workingmen to handle the production and distribution of goods. The leaders and a minority of the order felt "that strikes are deplorable in their effect and contrary to the best interests of the order," and that success lay in "agitation, education and organization." "Without organization," said Powderly, "we cannot accomplish anything; through it, we hope to forever banish that curse of modern civilization—wage slavery."[2]

The order was secret at first; even the name was unknown. It was designated by five asterisks and usually spoken of as the "Five Stars." Growth was slow in the beginning. In 1869 only eleven tailors comprised the membership of Assembly No. 1,

---

[1] Wright, C. D., *Historical Sketch of the Knights of Labor*, Quarterly Journal of Economics, January, 1887, pp. 142-143.
[2] Speech before the annual convention in Pittsburgh, 1880.

and in 1873 there were only six assemblies, all in Philadelphia. Two years later the organization had grown to eighty assemblies in the city and vicinity and in 1875 a national convention was called at Tyrone, Pennsylvania, and an invitation extended to other labor organizations to join them. By 1883 the membership was 52,000, but within three years had jumped to 700,000, and at the height of its career numbered close to a million. Widespread, though unwarranted, distrust of the organization led the Knights in 1881 to abolish its secret character. In make-up it resembled a "grand national union of industrial workers" rather than a federation of craft unions as exemplified later in the American Federation of Labor. Composed both of national trade unions and local assemblies, its composition was heterogeneous. Only three-fourths of its members need be wage earners, a rule which gave access to all types of reformers, while many of the assemblies were composed of women and unskilled laborers. Under such conditions strong divergence of opinion as to policies was bound to arise. Two main factions, one favoring reform through political channels, and the other advocating direct action, appeared. The latter group proved the more powerful and embarked the Knights in an aggressive campaign to raise the standard of living, thus involving them in many severe strikes. The most notable and successful of these was directed against the Gould railway system in 1885, and wrung concessions from the most powerful capitalist of the day. Writing in 1886 of the Knights of Labor, Professor Ely described them as "the most powerful and the most remarkable labor organization of modern times . . . established on truly scientific principles which involved either an intuitive perception of the nature of industrial progress, or a wonderful acquaintance with the laws of economic society."[1]

But the year 1886, which marked the height of their power, marked also the beginning of their downfall. Unsuccessful strikes in that year undermined their prestige and alienated public sympathy;[2] factional differences prevented united action; po-

---

[1] Ely, R. T., *The Labor Movement in America*, p. 75.

[2] The Haymarket riot of 1886, with the attendant bomb throwing, also

litical activity hurt them; overcentralization of power created suspicion; but more than all else, the rising opposition of a new organization, the American Federation of Labor, proved their undoing. The decline of the order after 1888 was as rapid as had been its growth. Its brief but spectacular career, however, had not been without results. Many weak unions had been reorganized and put on their feet through affiliation with the Knights, while others had been founded. A standing Committee of Labor was established by the House of Representatives in 1883 and in the following year a national Bureau of Labor was created to gather expert information. Even President Cleveland, who little understood the significance of the labor disputes which filled his first administration, sent in 1886 to Congress the first presidential message devoted to labor, in which he advocated the creation of a board of labor commissioners to act as official arbiters in labor disputes. Congress half-heartedly in 1888 followed his suggestion by enacting a law for the settlement of railway disputes by arbitration, provided both parties were willing.

American Federation of Labor.—The origin of the American Federation of Labor dates from 1881, when a joint call for a convention was issued by the Knights of Industry, an organization strong in the middle west, and the Amalgamated Labor Union, an offshoot of the Knights of Labor. This convention called a second, which met at Pittsburgh in the same year and formed a union which was reorganized at Columbus in 1886 as the American Federation of Labor. Although the early platforms of the Federation embodied such demands as a protective tariff, anti-contract immigration, the abolition of conspiracy laws as applied to trade unions, and compulsory education, the trend of its policy has been quite consistently away from direct political action to unionism pure and simple. In this they differed radically from the Knights of Labor.

The American Federation, as its name implies, was distinctly a federation of craft and industrial unions rather than a "one

---

alienated public sympathy from the labor movement, although neither the Knights of Labor nor the American Federation of Labor had any connection with the anarchists who were responsible.

big union" affair, and its policy has been one of distinct liberality toward the autonomy of its constituent groups. In 1920 it was composed of the following elements:

1. National and International Unions, of which there were 110, comprising 36,741 local unions.
2. Local Trade and Federal Unions, composed of seven or more wage earners whose trade and calling is not organized and who are not members of any body affiliated with the Federation. There were 1,286 such unions in 1920.
3. Forty-six State Federations, with which the labor groups inside the several states are directed to join.
4. City Central Bodies, numbering 926, to which the locals are required to ally themselves, but whose powers are limited by the American Federation and the various national trade unions.
5. The National and International Unions are grouped more or less roughly into departments, according to the line of work followed, as the Building Trades Department, Metal Trades Department, Mining Department, Railroad Employees Department, and Union Label Trades Department. Local Department Councils to the number of 682 supervise more locally the work of the departments. Each of the five departments has its separate set of officers.

The officers of the Federation consist of a president, eight vice-presidents, a secretary, and a treasurer, who form a very powerful executive body and are elected at the annual convention. The National and International unions have virtual self-government, but the power of the local union is distinctly circumscribed. Ample funds are obtained by a per-capita tax upon all of the members to carry on the work of the Federation whose executive offices are in the organization's own building in Washington.

The growth of the American Federation for a number of years was slow, its membership in 1890 numbering only 100,000. By 1900 this had grown to 548,000; by 1904 to 1,676,000 and by 1914 to 2,000,000. The period of war prosperity reacted favorably upon the Federation, which boasted in 1920 of 4,076,740 members, a figure which declined in the succeeding years. Although built primarily upon the basis of craft unions, certain large industrial unions like the United Mine Workers, the Western Federation of Miners, and the International Union of Brewery Workers have affiliated without changing materially the complexion of the bigger organization. On the other hand, certain important groups, such as the National Association of Letter Carriers and the National Association of Steamfitters, have grown and prospered without the aid of the American Federation. The Railway Brotherhoods, probably the most powerful, firmly knit, and wealthy of the labor groups, have constantly refused to affiliate with the Federation.

Policies of the American Federation of Labor.—Broadly speaking, the purpose of the American Federation of Labor has been: (1) to agitate all questions looking toward the benefit of the working classes, in order to bring about the enactment of favorable measures and the repeal of oppressive laws in both state and national legislatures; (2) to use all possible means to remedy abuses under which the wage earner works and to uphold him in his just rights and privileges; and (3) to promote close and thorough organization to insure such results. More definitely they have attempted to raise the standard of living by fighting for shorter hours, higher wages, and better working conditions. At the same time they have sought to protect themselves by benefit and insurance schemes and by pushing union labor products.

The policies of the labor unions and the measure of success they have won during the past forty years has been to no small degree due to the leaders. As a whole, American labor has been most fortunate in its leadership, and the long tenure of office which the best officials have held demonstrates a realization of this fact on the part of the rank and file. Perhaps the most brilliant was John Mitchell (1870-1919). Starting to work in the coal mines at thirteen, he joined the United Mine Workers of America at its organization in 1890 and nine years later, at the age of twenty-nine, was its president. Although there were only

43,000 members when he rose to power, so skillfully did he direct their uphill fight that he lived to see a membership of 400,000, probably the largest trades union of his time. At the age of thirty-two he led the miners through the spectacular and successful coal strike of 1902 in such superb fashion that his prestige became national and the confidence of the labor world in him absolute. The most famous of the labor leaders and the most valuable in his services to the wage earner is Samuel Gompers. Born in London in 1850 of Dutch-Jewish parentage, he emigrated to America at the age of thirteen and soon after joined as an apprentice the first cigar makers' union organized in that city. An active worker in the founding of both the American Federation and the Cigar Makers' International Union, he became president of the former in 1882, and with the exception of one year (1895), has been ever since annually reëlected. Scores of other able leaders could be mentioned—the Railway Brotherhoods have been particularly productive of them—such men as P. M. Arthur and Warren S. Stone of the Brotherhood of Locomotive Engineers, Edgar E. Clark of the Railway Conductors, William Carter of the Firemen, and William G. Stone of the Trainmen.

These men, trained in the rough school of experience, have developed a hard-headed and practical, but at the same time aggressive, policy. Whatever may be their attitude as to ultimate ends, they refuse to allow any dreams of a millennium to stand in the way of fighting for what small gains may be attained at the moment. "We are all practical men," said Adolph Strasser, president of the Cigar Makers' Union, before a Senate committee in 1883. "We have no ultimate ends. We are going on from day to day. We are fighting only for immediate objects—objects that can be realized in a few years." Samuel Gompers's insistence upon a rather strict adherence to a policy of organization based on national craft or trade unions, upon frugality in money matters, and upon avoidance of radical economic theories, has enabled him with considerable success to bring the pressure of organized labor to bear on such practical demands as the eight-hour day, the Saturday half holiday, federal child-labor legislation, the restriction of immigration and alien contract labor, and workingmen's compensation.

While glad of whatever may be obtained peaceably, the American Federation of Labor has not hesitated to back its constituent unions in the fierce warfare of strikes and boycotts. Ordinarily a local is forbidden to strike without the consent of the national, but, that consent once given, the national is responsible for the successful outcome. With the growth of the Federation has come a corresponding growth of strikes and lockouts, rather than a diminution, notwithstanding the improved methods for settling disputes adopted by many of the leading unions. The two chief ends hoped for in strikes in recent years have been raising of wages and recognition of the union; it is interesting to note that the proportion of strikes attributable to the last-named cause has constantly increased. By means of the boycott, the unions have attempted to put their stamp of disapproval upon the products of certain employers hostile to organized labor, as in the case of the Buck Stove and Range Company of St. Louis and of Daniel Lowe, hat manufacturer of Danbury, Connecticut. What amounts to an indirect boycott is the appeal to all friends of labor to use only goods bearing a union label. Other methods by which labor has sought to protect itself are infinite. Regulations are demanded regarding hours of work, relations of union to non-union men in the shop, use of non-union materials, number of helpers and apprentices, and many other matters of daily importance about which it seems advisable to have a definite understanding.

The most fundamental desire of the wage earner is some sense of security as regards his work and wages. This, along with the growing strength of trade unionism, has brought a rapid development in collective bargaining. John Mitchell believed that "the hope of future peace in the industrial world lies in the trade agreement," and it must be admitted that where given a fair trial collective bargaining has proved the most hopeful factor in lessening the possibility of serious labor strife. The trade agreement ranges from the simplest type to the more complicated forms taken by the International Typographical Union in its dealing with the American Newspaper Publishers' Association or in the Protocol of the New York Garment Trade. Some employers have become so enthusiastic over the stabilizing influence of these agreements that they are actually reconciled to the closed shop.

Industrial Unionism and the I. W. W.—Although trade unionism is undoubtedly in the ascendant, it does not occupy the entire stage. Industrial unionism has persisted among such groups as the United Mine Workers of America, where every workman from slate picker to engineer in one industry belongs to the same union. Certain unions, like the Cigar Makers' International, have in recent years shown a distinct tendency toward this type, while the International Longshoremen's Association illustrates an advanced stage of industrialization. For practical reasons, industrial unionism will survive, for it is a logical development in an industry where there are a small number of workmen in many trades, or in an industry like mining, which may be isolated from great centers.

Militant industrial unionism never quite died out after the disintegration of the Knights of Labor, but it was not until 1905 that it seriously challenged the trade unions. A convention held in that year in Chicago under the influence of the Western Federation of Miners and the socialistically inclined American Labor Union, and dominated by such radicals as Eugene V. Debs and William D. Haywood, founded the Industrial Workers of the World. Declaring that the "universal economic evils affecting the working class can be eradicated only by a universal working class movement," they demanded the formation of "one great industrial union, embracing all industries, providing for craft autonomy locally, industrial autonomy internationally and wage class unity generally." "It must be founded on the class struggle," said the manifesto, "and its general administration must be conducted in harmony with the recognition of the irrepressible conflict between the capitalist class and the working class."[1] In a new preamble affixed to their constitution in 1908 it was as-

---

[1] Proceedings of the First Convention of the I. W. W., pp. 5, 6.

serted that "a struggle must go on until the workers of the world organize as a class, take possession of the earth and the machinery of production and abolish the wage system."[1] Believing in the class struggle, they advocate direct action as the means to victory. Direct action includes such tactics as the general strike, boycott, and sabotage. Sabotage may be peaceful, simply soldiering on the job, or it may involve such violent tactics as destruction of property. Enmity to the present order is fundamental in their philosophy, and their methods of warfare are those best suited to the moment.

Their doctrines were appealing to the great class of unorganized, unskilled workers generally, especially to certain groups of eastern factory operatives and to the migratory workers of the west who follow the harvest and cut the lumber. From 1909 to 1917 the I. W. W. was an aggressive organization. Its agitators handled the Lawrence strike of 1912, the Paterson strike of 1913, and kept the northwest in a state of unrest. Their revolutionary language and violent methods eventually aroused the hostility of the public and inclined it to condone the lawless and extra-legal methods employed by communities in their efforts to rid themselves of this group. The opposition of the I. W. W. to the war brought them into direct hostility to the government, which further curtailed their operations. Although membership at the height of its activity in 1917 probably did not number more than 75,000, the lack of numbers was counterbalanced by the enthusiasm and revolutionary ardor of its members.

Before going farther it may be well at this point to note briefly the current criticisms of the labor unions. The hostility of the public to a revolutionary organization of the I. W. W. type is easily understandable. The American Federation of Labor, however, professes to be neither radical nor socialistic, so that the attack proceeds from a different angle. A recent writer[2] has attacked labor unions on the grounds of "authorized practices that destroy efficiency, limit output, increase costs enormously, pro-

---

[1] Quoted in Brissenden, P. F., *The I. W. W.*, Appendix ii.
[2] Bullard, F. L., "Labor Unions at the Danger Line," *Atlantic Monthly*, vol. 126, No. 6 (December, 1923).

duce a labor monopoly," and this arraignment undoubtedly includes the chief counts brought against the unions. The unions are accused of so minutely prescribing the amount of work to be done by their members and the manner in which the job shall be carried on, and of so arbitrarily limiting membership, as to show an utter indifference to other workmen and to the public welfare, amounting almost to a conspiracy against society. The classic examples of this evil, it is pointed out, exist in the building trade, one of the great basic industries, where regional disputes, limitation of output, and arbitrary rules of all kinds are carried to absurd and dangerous extremities.

Labor and the Courts.—While the wage earner has been able to influence state and national legislatures, and to make distinct progress in his dealings with the employer, his experience with the judiciary has not been so fortunate. Under our governmental system of checks and balances, in which a judiciary (by its very nature conservative and not representative of the working class) passes on the constitutionality of legislation, it is not at all surprising that progress has been slow. Labor has had to struggle not only against a conservative judiciary, but also against legal theories and economic philosophies whose origin antedated the Industrial Revolution.

The formation of labor unions had scarcely commenced before they were haled into court on the ground that, in absence of a statute or legislation on the point, the old common law of England applied in America, and a combination of workmen to raise wages was a conspiracy against the public and as such illegal. Decisions during the first two decades of the century were generally against the workmen, but gradually the attitude of the courts shifted. Mere combining was no longer taken as conspiracy, the judiciary now directing their attention rather to the methods employed by the unions to gain their ends. For years, however, the right to strike, to boycott, and to picket was questioned in the courts.

The Fifth Amendment to the Constitution asserted that no one could be "deprived of life, liberty, or property without due process of law," while in most of the state constitutions were to be found similar statements. The idea was incorporated again in the first section of the Fourteenth Amendment. Although Justice Holmes asserted that this amendment did not write *laissez-faire* into the Constitution, his belief was not generally held by the sturdy exponents of that economic doctrine, who interpreted many labor laws as an infringement of liberty, an abridgment of contract, or as class legislation. With such a background it is not surprising that many labor laws have been declared unconstitutional. Legislation which has come under the ban of the courts at various times includes laws fixing the hours of labor engaged on public works, laws limiting the hours of labor in private industries designed to protect the health of adult male workers, laws prohibiting the payment of wages in scrip or the enforced dealing at company stores, laws prohibiting the manufacture of such commodities as cigars in tenement houses, laws forbidding employers from holding back wages, workingmen's compensation laws, and minimum-wage laws.

Notwithstanding the long list of adverse decisions, labor has been insistent in affirming the constitutionality of labor legislation under the police power given to the state to look after the health and safety of the people. This power has been generally upheld by the courts in the case of laws governing the hours and conditions of work of women;[1] while the position of guardian which the state maintains towards children has allowed protective legislation for minors. In the case of adult males there has been much judicial interpretation. Where the laws have obviously been designed to protect the health and safety of the community, such as those limiting the hours of employment on public carriers, they have generally been upheld. In the case of laws governing

---

[1] A discouraging blow to labor legislation, particularly with reference to women, was dealt by the Supreme Court on April 11, 1923, when it declared a minimum-wage law in the District of Columbia unconstitutional. Florence Kelley, general secretary of the National Consumers' League, speaking of the decision, said: "Under the Fifth and Fourteenth Amendments of the federal Constitution as now interpreted by the court, it is idle to seek to assure by orderly processes of legislation, to wage-earning men, women or children, life, liberty, or the pursuit of happiness. This decision fills those words with the bitterest and most cruel mockery." *Proceedings of the National Conference of Social Work* (1923), p. 114.

hours of labor in private industries intended to protect not the safety of the public, but the health and safety of the workers, even the Supreme Court in Lochner v. New York as recently as 1905 declared null and void a law which fixed the hours of work in bakeshops at ten a day on the contention that it was a violation of the right of individuals to contract as to hours and labor, a liberty enjoyed under the Fourteenth Amendment. The court, however, has reversed itself in later decisions by upholding similar laws.[1] Legislation respecting safety and sanitation in factories is now generally held to be constitutional, and the courts have come to be more liberal in rewriting the old common law and in placing more responsibility upon employers in case of accident. Less success has been experienced with laws aimed to control methods of wage payment.

The control of the national government over interstate commerce has aroused hope that something might be accomplished here. The Adamson Act of 1916, which provided for a basic eight-hour day on interstate carriers, was a great victory and was upheld by the Supreme Court (Wilson v. New, 243 U. S. 332). On the other hand, however, the Keating-Owen bill, passed in 1916 to prohibit the interstate commerce in the products of children under sixteen, was declared unconstitutional in 1918 (Hammer v. Dagenhart, 247 U. S. 251). An attempt in 1919 to accomplish the same object by the imposition of a 10-per-cent tax on the net profits of factories employing children under fourteen years of age met the same fate (May 18, 1922).

No judicial activity has been so bitterly opposed by labor as the use of the injunction.[2] Designed originally as a powerful

---

[1] Holden v. Hardy, 169 U. S. 366; and Bunting v. Oregon, 243 U. S. 246.

[2] "An injunction," says Professor Watkins (*An Introduction to the Study of Labor Problems*, p. 324), "is an order issued by a court of equity for the purpose of preventing injury to a person or property or of preserving the existing conditions until the final determination of rights." In theory it is an extraordinary expedient to be used when property and personal rights are imperiled, and when there are no other remedies at law adequate to meet the emergency. A violation of an injunction is punished as contempt of court without jury trial and may involve fine or imprisonment. In earlier years the injunction was an instrument used by the courts only in extraordinary situations. In recent years it has become a common instrument in labor cases. The development of the

weapon in the hand of the crown to be employed most rarely against threatened lawlessness and riotous outbreaks, it has come to be used in recent years in America quite commonly by the courts to limit the activity of labor during strikes. The imprisonment of Eugene V. Debs for violating a federal injunction during the Pullman strike of 1894, and the sentencing to prison of Gompers, Mitchell, and Morrison for ignoring the Buck Stove and Range injunction, are two striking incidents in the more or less free use made of this instrument. "Government by injunction" has been denounced as a one-sided and unjust use of power, and labor succeeded in writing into the Clayton Act of 1914 a clause which prohibits the use of restraining injunctions in cases between employers and employees "unless necessary to prevent irreparable injury to property, or to a property right, of the party making the application, for which injury there is no adequate remedy at law." As might be expected from the looseness of the phraseology of this prohibition, little actual change has been made in the status of the use of the injunction.

Labor and Politics.—It seemed obvious that as soon as the franchise was extended downward far enough to include the wage earners, the demands of labor would become intertwined with politics. This was true of the first labor movement of the 'twenties and 'thirties, but inadequate facilities for communication, as well as the localization of manufacturing, prevented this first political effort from becoming national in scope. Workingmen's parties, however, were formed in New York State and candidates of workingmen presented themselves in Philadelphia, New England, and elsewhere. The depression following the panic of 1837 interrupted the political activities of labor, and their efforts until after the Civil War were directed along other channels.

Although a Labor Reform party entered a presidential candidate in 1872, the radical labor vote during the next few years

---

free use of the injunction has a two-fold significance. In the first place, it has enormously increased the power of the judiciary. In the second place, the original theory of the injunction has been strained enormously to fit labor cases, and has been developed chiefly as a weapon to be used against labor.

was absorbed in the Greenback party, which coalesced in 1878 with the Labor Reform group. The Greenback platform, in addition to its views on currency reform, included demands for the regulation of interstate commerce, a graduated income tax, prohibition of the importation of contract labor, and labor legislation. The support of the Greenbackers in 1880 and 1884 was drawn chiefly from the western farmers and eastern labor; the party disappeared in the election of 1888 and its place was taken by the Union Labor party which drew its vote chiefly from the west and south. The strongest and the most radical of the early third parties was the Populist, or People's party, which polled over a million votes in 1892 on a platform which included free coinage of silver, a graduated income tax, postal savings banks, and the government ownership of railways, telegraphs, and telephones. Their convention had declared itself in sympathy "with organized workingmen to shorten hours of labor" and maintained that "the interests of rural and city labor are the same, their enemies identical." This effort to tie up the political interests of the farmer and city laborer, which has extended even to the Farmer-Labor party of 1920, is one of the most interesting developments in the history of American politics.

With the passing of the Populist party and the growing strength of the American Federation of Labor, organized labor has been less ready to embark officially in politics. Samuel Gompers during his thirty-five years of control has persistently and successfully opposed the formation of a distinct labor party. Nevertheless, this policy of the Federation has been under the continued fire of the radicals, some of whom insist upon independent political action, while others would secure from that body the indorsement of socialism. That the radical labor vote is by no means insignificant is seen from the 901,873 ballots cast for Debs in 1912 and the 919,799 votes in 1920.[1]

The failure to work through a distinct labor-party movement does not mean that the labor unions hesitate to throw their political influence where it may be of the most benefit. Holding in

---

[1] The votes in 1920 were cast for Debs, notwithstanding the fact that he was in prison at the time, for alleged violation of the Espionage Act.

numerous instances the balance of power, they have been able to elect candidates friendly to labor, and by aggressive lobbying have influenced much labor legislation. Furthermore, the American Federation has not hesitated to take a stand officially on issues involving political action; it has indorsed proposals for the initiative, the referendum, and the recall, the direct election of senators, woman suffrage, government ownership or regulation of public utilities, restriction of immigration, the establishment of state and national labor bureaus and a national department of education, abolition of child labor, and all manner of legislation protecting the life, health, and future of the worker.

Although the "full dinner pail" argument which the Republicans have advanced to reconcile the workmen to the Republican party has been powerful, labor in general has in the last twenty years been inclined to divide its vote among opposing parties. When the Democratic organization in 1896 absorbed the Populist party and at the same time severely criticized the Supreme Court decision on the income tax and the use of the injunction, the appeal to labor was strong. Mr. Gompers and other leaders have unofficially but openly worked for both Bryan and Wilson. Through the medium of this party two of their greatest legislative victories were obtained—the Adamson Act and the Clayton Act. Spurred by post-war conditions, they have joined with the farmers in certain states of the northwest, and a Farmer-Labor party in Minnesota has recently elected both United States senators from that state.

Progress of the Wage Earners.—Although the income of the great majority of wage earners is still inadequate to maintain a proper standard of living, it is undoubtedly true that distinct progress has been made along certain lines vital to labor. During the Civil War prices rose faster than wages, but at its conclusion real wages were not far below the level of 1861. After the war, prices declined, but wages tended to remain near the point of 1865; between that year and 1890 real wages rose more than 100 per cent in industry and 70 per cent in agriculture. Since 1897 wages have continued to advance, but hardly as rapidly as prices, with the result that in the last thirty years labor has either barely held its own or actually lost ground, as far as real wages are concerned. This is very significant, for it brings up immediately the question whether labor is receiving its fair share in the increase of wealth due to the unprecedented productivity of the nation's industries during this period.

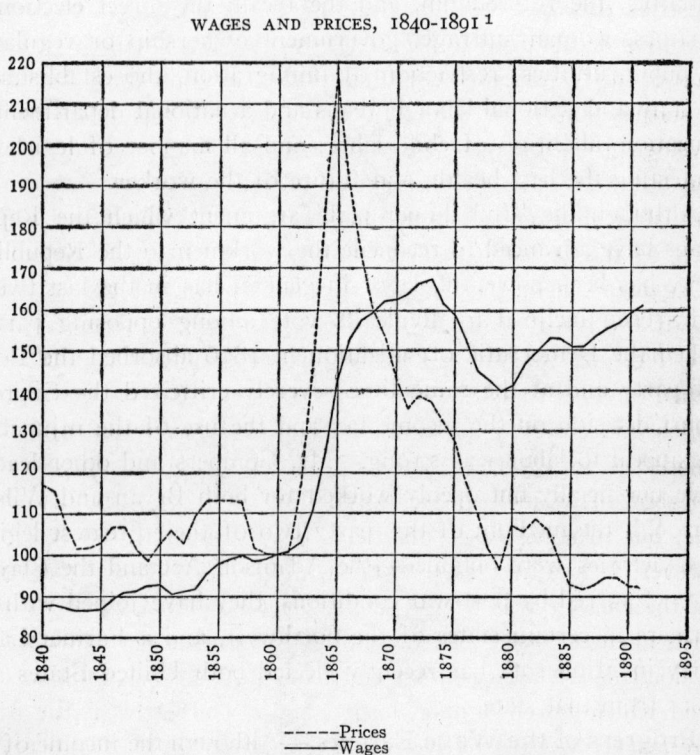

WAGES AND PRICES, 1840-1891 [1]

-----Prices
———Wages

On the other hand, real progress has been made in decreasing the length of the working day. Operatives in the cotton mills of the 'forties sometimes labored thirteen and fourteen hours. By 1860 the average day for all labor was eleven hours, and this after an agitation for a ten-hour day which had extended over thirty years. Though inadequate, the best figures obtainable for the period 1840-90 are from the Aldrich Report; this gives the average in 1844 at eleven and one-half hours a day, in 1865

---

[1] *Statistical Abstract*, 1899, p. 92, and *Statistical Abstract*, 1921, p. 854.

eleven, with a gradual reduction until 1890, when the average was ten hours. Since its organization the American Federation of Labor has steadily advocated the eight-hour day. This with a Saturday half holiday, making a forty-four-hour week, has been the great demand of labor. In some of the highly organized industries this has been achieved by the unions, and in certain hazardous occupations by legal enactment. Beginning with Utah in 1896, thirty states by 1916 had placed an eight-hour day upon the statute books for miners. Some impetus was given to the movement by a law passed by Congress in 1892 providing for an eight-hour day for government employees. During the World War labor shortage gave an impetus to this demand and a majority of trades using skilled labor worked under this schedule.

Organized labor has fought strenuously against the exploitation of the labor of women and children, and in this way they have received the aid of many outside their own ranks. It seemed obvious that the effects of excessive labor on the part of women and children must cause physical degeneration and thus menace the future of the race and nation. To the male worker the labor of women and children imperiled not only his wage scale, but in some cases his job. Before much could be done, however, it was necessary to undermine the old-fashioned belief that the factory was a God-sent protector against the evils into which idleness might lead the children; to break the influence of the doctrine of *laissez-faire;* and to counteract the influences of greed which fattened upon such labor. The early textile mills were largely worked by women and children, as they still are in certain sections. Although the number of children working has increased, relatively there has been a decline. The Census of 1870 reported 739,164 children between the ages of ten and fifteen engaged in gainful occupations, while the Census of 1910 reported 1,990,225, not quite half of whom were girls. This amounted to 5.2 per cent of all those gainfully employed and was a decrease of 0.8 since 1900. The proportion of children gainfully employed to the whole number of children, however, increased from 16.8 in 1880 to 18.8 in 1910. While a majority of the children listed were engaged in agriculture, it is still discouragingly true that thousands are to be found in factories, especially in southern states where northern capital has built large textile mills. It is true that agricultural labor up to a certain extent may not be physically deleterious to a child of immature years, but the same can hardly be said of factory or sweatshop work. This stunting of physical and mental development cannot be too strongly deplored.

Attempts to do away with child labor have taken the form of laws limiting the working hours, setting an age limit below which children may not be gainfully employed, prohibiting night work, and providing for compulsory education. Legislation in regard to child labor started in Massachusetts in 1836, when a law was passed regulating the instruction of children employed in manufacturing establishments. In 1842 the working day for children under twelve was limited to ten hours (considered a great advance!), and acts of 1866 and 1867 forbade the employment of any child under sixteen more than sixty hours a week. In 1873 the length of the school year was extended to twenty weeks and the age of attendance to twelve years, and ten years later all towns of more than 10,000 population were compelled to establish evening schools. A law of 1888 excluded children under thirteen from factories, workshops, and mercantile establishments, and those under fourteen except during vacation; while other indoor work was forbidden children under thirteen unless they had attended school twenty weeks. The age of compulsory attendance was raised in 1889 to fourteen years and the school year to thirty weeks. In this hesitating manner has Massachusetts, a state always in the forefront of labor legislation, tackled the evil of child labor, and by methods much the same other states have taken up the problem. All of the states now have some sort of child-labor legislation on the statute books, although six of them (all in the south) as late as 1914 had no compulsory education laws. The minimum age of lawful employment varies from twelve to fourteen years, depending on the economic background and the social consciousness of the states. In New Mexico, for example, there is no law except one prohibiting children under fourteen working in mines; in Utah and

Wyoming only mines and dangerous occupations are forbidden to children; and in four of the southern states there is no minimum age for employment in stores. With few exceptions, the states now limit the working day of a child under sixteen to eight hours, while twenty-four states and the District of Columbia prohibit night work. Laws of this nature, combined with legislation for compulsory education vigorously enforced and with appropriate penalties for infringement, are necessary to eliminate this curse. Child labor is most common in the families of newly arrived immigrants and among the poor whites of certain sections of the south, where economic pressure is sufficient to cause the parents in many cases to coöperate with the employer in ignoring existing laws. Two federal child-labor laws have been declared unconstitutional.[1]

The factory system and other results of the Industrial Revolution have thrown open to women innumerable new opportunities to earn a livelihood. Many of the old home occupations, such as cloth making, soap making, fruit and meat preserving, have passed to the factory and here women have followed. Usually unorganized, often living at home and looking upon their work as temporary, they have been subject, like children, to economic exploitation. They have been particularly the victims of the "sweated industries," where work is contracted out to be done in the home. In recent years the number of women employed outside the home has increased faster than the total population, but the advance has come in middle-class occupational groups rather than in factory work; it is also noticeable that there has been a decided falling off in the number of women employed in such traditionally feminine occupations as waitresses, general servants, and seamstresses. The number of women engaged in manufacturing industries in 1914 was 1,500,000 and in 1920 about 1,930,000.

The two great evils attendant upon the employment of women which legislation has attempted to mitigate are insufficient pay and physical injury, both of which may have deleterious effects

---

[1] A vigorous agitation has been started for the adoption of a constitutional amendment which will permit federal legislation controlling child labor.

extending to society as a whole. Gradually the feeling has spread that women, like children, need the protection of the state, and especially that their physical well-being as mothers of future citizens is a concern of society. In certain of the states commissions have been instituted to study the cost of living and to decide upon minimum-wage scales, which in some cases are compulsory. The first minimum-wage law in the country was passed in Massachusetts in 1912. Although the World War drew thousands of women into industry under supposedly advantageous circumstances, it is the belief of many students [1] in close touch with labor conditions that the wages of women relative to those received by men were hardly bettered. Of 117 plants investigated in 1919 in New York State, 29 paid women less than $12 a week, and 69 less than $14.[2] The Consumers' League in 1919 claimed on the basis of statistics compiled by the United States Bureau of Labor, that only one out of fourteen industries in New York City which employed large numbers of women paid a living wage.[3] While it is true that women are still underpaid in most occupations, the conditions under which they work have been bettered. Numerous laws improving factory conditions have been passed, while forty-two states (1920) had laws limiting the hours of labor for women and several had laws forbidding night work.

Although the percentage of accidents among workmen is larger in the United States than in any of the great industrial nations, we have been the last to recognize that these accidents should be borne by the industry rather than wholly by the workman. Until within the last few years the theory under which the law worked was that the responsibility for an accident could be placed upon some person who must bear the loss. This responsibility was almost always placed upon labor rather than capital, on the ground that the workman knew the risk that he must run if he accepted

---

[1] Carlton, F. T., *History and Problems of Organized Labor*, p. 483.

[2] *The Industrial Replacement of Men by Women.* Bulletin issued by the Industrial Commission of New York, March, 1919.

[3] The *Survey* for April 17, 1920. The percentage for the entire state was even more striking. "Nineteen per cent of the workers received less than $11 a week, 71 per cent received less than $14, and 88 per cent received less than $16."

a job, or else that the accident was caused by contributory negligence upon his part or that of a fellow worker. If the employer could prove that he had exercised reasonable precautions, he was ordinarily relieved of responsibility.

Eventually the point of view of society changed. It was realized that the old common law which might have fitted conditions before the Industrial Revolution was no longer fair. Under modern conditions, it became impossible in the case of many accidents to prove the negligence of anyone; obviously it was the fault of the inevitable risks of industry. In reality, the industry was the guilty party, not the workman. Following in the footsteps of Germany (1884) and England (1897), various American states, beginning with Maryland in 1902, began to pass workingmen's compensation acts. The first compensation laws of Maryland (1902), Montana (1909), and New York (1910) were declared unconstitutional, but after 1911 laws were framed which stood the tests of the courts. By 1920 forty-two states and the territories of Alaska, Hawaii, and Porto Rico had workingmen's compensation laws, as did the United States government for its civilian employees. Most of these laws, besides cutting away the old common-law defenses of the employer, provide (1) for the payment in case of death or permanent disability of a maximum amount in weekly allotments extending over a period of from 300 to 500 weeks, (2) in the case of temporary disability the payment of doctors' bills and for a certain period a percentage of the regular wages, and (3) in the case of certain specified industries the payment of a fixed lump sum. Usually agricultural laborers and domestic servants are excluded, as are commonly those employed in establishments hiring from three to five men. Provisions for the payments required by the law are usually met by some sort of insurance either through state or private companies. It is now decidedly to the advantage of the employer as well as to the employee to avoid accidents, and the salutary effect of these laws is increasingly evident. As a rule, their operation is supervised by special tribunals which pass on the claims for compensation.

Immigration.—Closely tied up with the labor problem, but containing elements of significance to all phases of our political and economic life, is the matter of immigration. Between 1860 and 1920 close to 28,500,000 foreigners have sought our shores to enter permanently or temporarily into the nation's labor force, a number almost equal to the total population of the country in 1850. This incoming tide of labor as it rose and fell corresponded quite closely with the periods of prosperity and depression. Thus we find the peak years in 1873 with 459,803 arrivals,

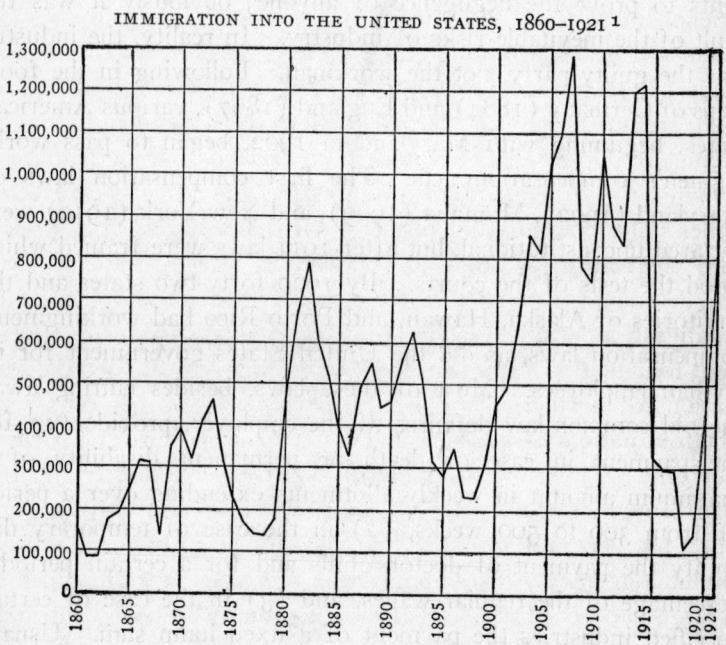

IMMIGRATION INTO THE UNITED STATES, 1860–1921 [1]

in 1882 with 788,922, in 1892 with 579,663, in 1907 with 1,285,349, and in 1914 with 1,218,420. While in actual numbers immigration increased by each decade up to the opening of the World War, emigration and the normal growth of population have kept the percentage of foreign born to the whole population at about 14 per cent, slightly under that figure in 1860, and slightly over in 1910.

As before the Civil War, the hope of economic betterment or

---

[1] *Statistical Abstract*, 1921, p. 883.

the desire for greater political and religious freedom has been the compelling motive in the minds of the immigrants themselves. Nevertheless, the impetus has been partially supplied from without. Professor Commons is of the opinion that "the desire to get cheap labor, to take in passenger fares, and to sell land have probably brought more immigrants than the hard conditions of Europe, Asia, and Africa have sent."[1] Capital seeking renewed supplies of cheap labor has coöperated with steamship companies in scouring Europe for prospective immigrants. Relatives and friends in at least one-fourth of the cases now send over the means of transportation; emigration is no longer a hazardous undertaking limited to the strong and self-reliant. As emigration from Europe has become easier, the type coming in has changed. Up to 1880 Great Britain, Ireland, and Germany contributed the larger part, men aggressive and forceful and oftentimes skilled artisans and farmers, not radically different in blood and characteristics from the population already here. During the decade 1851-60 these three countries sent 88 per cent of the immigration, while Austria-Hungary, Italy, Russia, and Poland sent four-tenths of one per cent. In the years 1891-1910 the three nations of northern Europe mentioned sent 31.6 per cent, while the four nations of southern and eastern Europe furnished over 50 per cent. This flow of immigration from southeastern Europe has brought a different type, hard working and thrifty, to be sure, but generally unskilled and accustomed to political and economic autocracy. The immigration of recent years has included a large number of the Jewish race, a group who have contributed much to our civilization, but have in some respects accentuated the problems arising from this influx from abroad.

Opposition to immigration comes from two sources: (1) a large part of organized and unorganized labor, who hold that the continued inflow of cheap labor keeps wages low and prevents a rise in the standard of living; and (2) many ardent Americans who believe that ideals and standards are jeopardized by a too rapid addition to the "melting pot" of those who do not

---

[1] Commons, John R., *Races and Immigrants in America*, p. 108.

readily "melt." On the other hand, the advocates of comparatively easy immigration laws are represented by capital, which argues that a continued supply of cheap labor is necessary to develop the resources of the nation and to fill the jobs which the native American avoids. For this contention there is much to be said, for without doubt our economic structure has been reared to no small extent upon the rough labor of newly arrived immigrants. Yet it is a question whether this tireless hunt for cheaper and cheaper labor is not at last bringing to our shores undesirable accessions more rapidly than they can be absorbed. That it is possible to maintain industry without fresh supplies of labor from abroad, even during periods of great demand, was proved during the war; while the widespread unemployment in times of depression hardly bespeaks the need of a greater labor force. Cheap labor is usually the most expensive in the long run, and it is quite probable that the nation might profit more by being forced to develop greater efficiency upon the part of the labor that is here rather than by the importation of unskilled and consequently low paid wage earners. Space will not permit further discussion here, but the question of immigration restriction is one of our greatest economic and social problems.

Until recently whatever restrictions were placed on immigration were due to the demands of labor. A number of acts culminating in 1882 finally prohibited Chinese immigration, and subsequent agreements with Japan have aimed at a similar exclusion of her citizens. By the act of 1882 the first step in federal control of immigration was taken. It placed a head tax of fifty cents on those entering, excluded certain undesirable classes of aliens, and provided for coöperation between the states and federal government in the enforcement of the act. Under the influence of the Knights of Labor, laws were enacted in 1885 and 1889 prohibiting the bringing over of immigrants under contract to labor, laws which were generally evaded. The office which corresponds to the present Commissioner-General of Immigration was created in 1891. Under acts of 1891, 1893, 1907 and 1917 the policy has been developed to exclude those morally, mentally, and physically unfit; those afflicted with physical and mental diseases, vagrants, paupers, anarchists, and contract laborers are debarred. For illegally bringing in immigrants steamship companies are liable to fine and to the necessity of returning them: nor may they encourage or solicit immigrants. The 1917 act created a literacy test which was passed over the President's veto. As a result of war conditions and from an apprehension that we might be deluged by an inflow from the war-torn nations of Europe, further legislation was secured, this time backed by both labor and the public generally. This act, which went into force in May, 1921 (operative until 1924), restricted the yearly immigration of any nationality to 3 per cent of its total in the United States in 1910. A new act passed in May, 1924, provided further limitations by changing the apportionment to 2 per cent of any nationality residing here in 1890, and by forbidding the immigration of Japanese (except certain specified classes), thus abrogating the existing "gentlemen's agreement." The national government supervises only the entrance of the immigrant. His future touch with American life and ideals is a problem of the state, the community, and the individual American.

NOTES FOR FURTHER REFERENCE

For recent statistics consult the Abstracts of the Census of Manufactures. Excellent histories of the labor movement in America are: Mary Beard, *A Short History of the American Labor Movement* (1920), a brief summary; G. C. Groat, *Organized Labor in America* (1919); F. T. Carlton, *History and Problems of Organized Labor* (rev. ed., 1920); and *Organized Labor in American History* (1920); S. P. Orth, *Armies of Labor* (1919), in Chronicles of America Series; G. S. Watkins, *An Introduction to the Study of Labor Problems* (1922); Selig Perlman, *A History of Trade Unionism in the United States* (1922); and R. T. Ely, *The Labor Movement in America* (1905), good on the early period. A more detailed account is that of J. R. Commons and others, *History of Labour in the United States* (2 vols., 1918). A tremendous collection of source materials gathered and edited by J. R. Commons and associates, but dealing chiefly with the period before 1860, is the *Documentary History of American Industrial Society* (10 vols., 1910-11). Paul H. Douglas, Curtice N. Hitchcock and Willard E. Atkins, *The Worker in Modern Economic Society*, University of Chicago Press (1923), is an excellent collection of readings designed to supplement a text-book. Carroll D. Wright, *Historical Sketch of the Knights of Labor*, in the quarterly *Journal of Economics*, January, 1887, pp. 127-168, is an excellent contemporary account. The best of the histories and interpretations by labor leaders are those of T. V. Powderly, *Thirty Years of Labor,*

1859-1889 (1890), by the Grand Master of the Knights of Labor and particularly interesting on that organization; G. E. McNeill (ed.), *The Labor Movement, the Problem of Today* (1887), by one of the earliest state labor officials; John Mitchell, *Organized Labor* (1903), by the one-time president of the United Mine Workers of America; and Helen Marot, *American Labor Unions* (1914).

There are numerous studies by experts, many of which have received their inspiration from the researches and training of J. R. Commons. Among the best are J. R. Commons, *Trade Unionism and Labor Problems*, First Series (1905) and Second Series (1921), collections of readings; J. R. Commons, *Labor and Administration* (1913); T. S. Adams and H. L. Sumner, *Labor Problems* (1905); J. R. Commons and J. B. Andrews, *Principles of Labor Legislation* (1916); W. Jett Lauck and Edgar Syndenstricker, *Conditions of Labor in American Industries* (1917); D. D. Lescohier, *The Labor Market* (1919); Hayes Robins, *The Labor Movement and the Farmer* (1922); and Daniel Bloomfield, *Employment Management* (1920), a source book.

On the labor of women and children, there is now a large amount of material. Pioneer government work was done in the *Report on Condition of Woman and Child Wage-Earners in the United States* (19 vols., 1910-12), published by the United States Department of Labor, Sixty-first Congress, Second Session, Senate Documents, vols. 86-104, and subsequent reports by the Women's Bureau and Children's Bureau of the Department of Labor. A recent study is that of Adelaide M. Anderson, *Women in the Factory* (1922). On the horrible conditions of child labor in the early mills, see Helen L. Sumner's work in J. R. Commons et al., *History of Labour in the United States* (1918), Vol. I, p. 169 ff. An earlier work by John Spargo, *The Bitter Cry of the Children* (1906), is still valuable.

On the more radical developments, the books of P. F. Brissenden, *The I. W. W., a Study of American Syndicalism* (1919), Vol. LXXXIII, Columbia University Studies, and J. G. Brooks, *American Syndicalism: The I. W. W.* (1913), the latter emphasizing the philosophy of the movement and its international aspect, will be found valuable. Carleton H. Parker, *The Casual Laborer and other Essays* (1920), and John Spargo, *Syndicalism, Industrial Unionism and Socialism* (1913), are illuminating. A notion of European labor movements may be gained from Paul U. Kellogg and Arthur H. Gleason, *British Labor and the War: Reconstructors for a New World* (1919); and Louis Levine, *Syndicalism in France*, Columbia University Studies in History, Economics and Public Law, Vol. 46, No. 3 (1914).

The immigration problem may be studied in J. R. Commons, *Races and Immigrants in America* (1907), a succulent survey; J. W. Jenks and W. J. Lauck, *The Immigration Problem* (1917), a scholarly presentation; I. A. Hourwich, *Immigration and Labor* (1922); Philip Davis and Bertha Schwartz, compilers and editors, *Immigration and Americanization* (1920), a book of selected readings; John P. Gavit, *Americans by Choice* (1922); F. J. Warne, *The Tide of Immigration* (1916); Grace Abbott, *The Immigrant and the Community* (1917); J. Drachsler, *Democracy and Assimilation* (1920); S. P. Orth, *Our Foreigners*, Chronicles of America Series (1920); National Industrial Conference Board, *Immigration Problems in the United States* (1923). For statistics, see the Report of the Immigration Committee, containing *Statistical Review of Immi-*

THE LABOR MOVEMENT

gration, 1820-1910, and *Distribution of Immigrants, 1850-1900*, Sixty-first Congress, Third Session, Senate Document No. 756, Vol. 20 (1911).

Three very significant reports on specific strikes are the *Report to the President on the Anthracite Coal Strike of May-October, 1902*, by the Anthracite Coal Commission (1903); *Report of Strike of Textile Workers in Lawrence, Massachusetts* (1912), Sixty-second Congress, Second Session, No. 870; *Report of the Steel Strike of 1919* by the Commission of Inquiry of the Interchurch World Movement.

SELECTED READINGS

CARLTON, F. T., *History and Problems of Organized Labor*, Chaps. IV-VII.
WATKINS, G. S., *An Introduction to the Study of Labor Problems*, Chaps. III-XIV, et al.
BRISSENDEN, P. F., *The I. W. W., a Study in American Syndicalism*, Chaps. I-V.
COMMONS, J. R., *Races and Immigrants in America*.
COMMONS, J. R. (ed.), *Trade Unionism and Labor Problems*, 2d series, Chap. II.
COMMONS, J. R., and ANDREWS, J. B., *Principles of Labor Legislation*, Chaps. I-III, et al.

CHAPTER XXV

WORLD TRADE AND THE NEW IMPERIALISM

The Old Imperialism.—Imperialism, that national policy which tends toward the extension of political, economic, and intellectual dominion over regions geographically situated beyond the national boundaries, is a phenomenon discernible from the earliest times in nations which have progressed to a position of wealth and power. In modern times the world has witnessed two distinct waves or outbursts of imperialism. The first of these, which we may designate as the Old Imperialism, commenced with the discovery of new trade routes to the East at the close of the fifteenth century and lasted until 1815, the end of the Second Hundred Years' War between France and England. Then ensued a lull in imperialistic activity, during which statesmen were little interested in extending their foreign dominions. The last third of the century, however, witnessed a renewed interest and activity in foreign expansion on the part of many nations, a period inaugurated by Disraeli (1874-80) in England and by the embarkation of France and Germany after 1880 upon new imperial efforts.

The Old Imperialism was influenced by mercantilism. It looked toward the founding of foreign settlements of colonists from the home country, who were to set up little Spains, little Englands, and little Hollands throughout the world to serve as sources of raw materials and markets for home products. Under the impetus of the Old Imperialism, North and South America and Siberia were conquered and peopled by Europeans, while settlements and trading posts were established in South Africa, in India, in the East Indies, in Australia and elsewhere.

If the Old Imperialism is defined as the acquisition of land which is actually settled by those who acquire it, American imperialism up to 1898 may be largely considered as such. The area of the United States in 1800 was 892,135 square miles, sufficient in the belief of most men to accommodate for an indefinite period the needs of our population. But the restless, land-hungry pioneer spirit had so entered into the blood of large groups of the people that it was only three years later that the Louisiana Purchase of 885,000 square miles was consummated. Florida, containing 59,600 square miles, was purchased from Spain in 1819; Texas, a region of 389,000 square miles, was annexed in 1845; and the Oregon territory (285,000 square miles) was secured by treaty in 1845. The Mexican War, instigated for apparently no purpose but to confirm the annexation of Texas and to extend our boundaries to the Pacific, added 529,000 square miles, augmented in 1853 by the Gadsden Purchase of 30,000 square miles. In the case of Florida, Texas, and Oregon, settlers had gone ahead of acquisition, but in general these accumulations of large stretches of land were made without any immediate expectation of use. The same was true of Alaska purchased in 1867 for $7,200,000; but in all cases white settlers speedily entered to dominate and occupy. The habit of imperialism was too easily formed. The prices paid were trivial in comparison to the value of the land; and where wars were fought they were neither sanguinary nor costly. The anti-imperialists were easily overrun by the frontiersmen or by the southern slaveholders so generally in control of the national government before the Civil War. The Indian inhabitants were ruthlessly brushed aside or were overpowered by a superior civilization. For such an imperialism a strong case can be presented. An inferior civilization in a sparsely occupied region had to give way to an aggressive people possessed with energy and numbers to conquer, and resources to develop the land.

The New Imperialism.—The new wave of imperialism which has swept over the world since 1870 has brought in its wake results more far-reaching than any other human event since the Industrial Revolution ushered in modern times. In fact, the New Imperialism is a direct result of the Industrial Revolution. Its causes are principally economic. (1) The new inventions in machinery increased production so enormously that new markets had to be developed to dispose of the surplus production and the vast population of Africa and Asia came to be considered as potential customers. Improvements in transportation and communication by land and sea were of inestimable value in speeding up this search for new markets. (2) As the Industrial Revolution increased the population and hence the markets at home, and new markets were discovered abroad, it was necessary to develop new sources of raw material. Those interested in manufacturing and commerce professed to believe that such sources of supply were safer when controlled by the home government. (3) In addition to these economic factors came a third equally important. With the tremendous increase in manufacturing and transportation, there followed accumulations of capital seeking investment. As the surplus of capital increased in Europe, interest rates declined and financiers were forced to go far afield for profitable investments. As a result we find European capital invested heavily abroad. British investments abroad in 1914 were estimated by Sir George Paish at about $20,000,000,000, approximately 23 per cent of the total capital investments of the nation.[1] British investments in India amounted to nearly £378,776,000 and before the recent war, close to £754,617,000 in the United States. In 1912 France was estimated to have loaned abroad over $8,000,000,000 amounting to 37 per cent of the total personal securities of the French, chiefly in the Near East and Russia; while Germany had between $7,500,000,000 and $8,500,000,000 invested abroad at the beginning of the war. This money invested in factories, mines, oil wells, railroads, and traction companies, or loaned to foreign governments, must be protected; it continually directed the eyes of capitalists and governments to foreign fields and served to weaken the independence of smaller powers as the more wealthy nations gained economic control. It gave the tone to the New Imperialism, which was in reality financial imperialism. As in the sixteenth and seventeenth centuries, the homeland sent out settlers to conquer and occupy, so now the capitalists of the nineteenth century sent out manufactured products and money. They were not interested in settle-

---

[1] Bogart, E. L., *War Costs and Their Financing*, pp. 14-16.

ment, for the lands now exploited were often quite densely occupied.

In addition to the economic there were of course other motives; first, a sincere desire on the part of Christians to convert the followers of other religions—a course favored by imperialists of all kinds who realized that the missionaries were blazing the trails which the soldier and merchant were only too ready to follow. The last century has seen a remarkable effort on the part of both Catholic and Protestant missionaries, who sometimes have sought the protection of their government to facilitate their work in foreign fields. Second, the argument that the colonies might absorb the surplus population and products of Europe was frequently advanced. Between 1870 and 1900 Great Britain added to her possessions (exclusive of spheres of influence) about 5,000,000 square miles, with an estimated population of 88,000,000; France added 3,500,000 square miles with a population of 37,000,000 and Germany 1,000,000 square miles with an estimated population of 14,000,000. Third, the whole movement was stimulated and condoned by the desire for national power and prestige. It is this intimate relation to national policy and to diplomacy, and to European competition in armies and navies, that forms the background of the World War.

The United States and the New Imperialism.—Into this competition for extra-territorial possessions the United States entered; late, to be sure, but with vigor. Until 1898 there had been little reason for overseas expansion. Up to 1890 there had been free land suitable for settlement, and abundant opportunities for whatever free capital might be seeking investments. In fact, the usual scarcity of capital in a new country was so acute that European wealth to the extent of over $6,000,000,000 (more than half of which was British) was still invested here in 1910. But the Spanish-American War marked a turning point. From a position of inferiority in 1860 we had advanced to a position of great economic importance. Population had increased 97 per cent between 1870 and 1900, and during the same period the annual production of wheat had increased from 236,000,000 bushels to 522,000,000 bushels, corn from 1,094,000,000 to

2,105,000,000 bushels; cotton from 4,352,000 to 10,100,000 bales; the annual production of petroleum from 221,000,000 to 2,672,000,000 gallons; of coal from 29,000,000 to 241,000,000 tons, and pig iron from 1,665,000 to 13,789,000 tons.

The nation was rapidly demonstrating that its resources were the greatest of any single nation and that they were well in hand for exploitation. In amount of productive land the United States is second only to Russia, and first in lands under actual cultivation. With her forest reserves of over 400,000,000 acres, second only to Russia, with 30 per cent of the world's iron; with more than half of the world's coal, and with ample copper, petroleum and water power, she was well supplied with essential raw materials.[1] While approximately nine-tenths of the production was consumed at home, the tenth exported, amounting in 1898 to $1,210,291,913, had become sufficient to make the matter of foreign markets important. By that year the necessity of calling upon Europe for continual loans to develop transportation and manufacturing was beginning to pass. Although the flow of capital was inward until the opening of the World War, for the most part we financed our own operations, and even accumulated a surplus for foreign investment. In comparison with present investments abroad the amount so placed in 1898 was small, yet even at that time it was undoubtedly a strong factor in hastening the war with Spain. Americans had invested more than $50,000,000 in Cuban business before 1898, and our commerce with the island amounted to $100,000,000 annually, a fact which President Cleveland noted in his last message to Congress. Economically as well as geographically, Cuba was closer to the United States than to Europe.

From the Promulgation of the Monroe Doctrine to the Spanish-American War.—Although the Spanish-American War is generally considered as marking the definite embarkation of the United States upon a career of imperialism, a gradual development can be traced from a much earlier period.[2] The

---

[1] See chap. i.

[2] "It is true," says John Bassett Moore, "that the expansion of 1898 involved, so far as concerns the Philippine Islands, the taking of a step geographically in

Monroe Doctrine in its stated intention—(1) to refrain from interference in European affairs, (2) to consider any attempt by the allied powers to "extend their political system to any portion of either continent of America as endangering our peace and happiness," and (3) that the era of colonization in the Americas was over—seemed rather to be a reaffirmation of Washington's policy of isolation than any move toward foreign power. If any imperialism was furthered by it, not American, but English economic imperialism was the gainer. Yet the Monroe Doctrine has so grown in importance and so broadened in its interpretation that it has eventually become a strong factor in American imperialism. It was an early indication of the region in which our interests would become keen and our financial imperialism make its start. In a sense it preserved "America for Americans." In combination with England's desertion of the Quadruple Alliance, it prevented further European conquests.

That the United States had every intention of upholding the Monroe Doctrine was demonstrated by the immediate ousting of France from Mexico at the conclusion of the Civil War. On the other hand, whatever foundations the Monroe Doctrine had laid in the Latin-American republics for confidence in our friendliness and unselfishness was largely sacrificed by our wholesale annexations at the end of the Mexican War. The Latin American came to look upon the Monroe Doctrine as a policy of keeping out Europeans from regions which we intended to dominate. Nevertheless, the ideal of closer relations between the American Republics first conceived by Adams and Clay was kept alive, notably through the efforts of James G. Blaine and the Pan-American Congress of 1889. Cleveland's aggressive championship of the South American republic in the boundary line dispute between Great Britain and Venezuela was a continuation of the same policy.

In the meantime, political developments in the Pacific were

---

advance of any that has been taken before; but so far as concerns the acquisition of new territory we were merely following a habit which had characterized our entire national existence." *Four Phases of American Development* (1912), pp. 147-148.

following economic penetration. The conquest of California and the purchase of Alaska had definitely put us in the Pacific. Long before American settlers had reached the western coast, Yankee sailors had built up a brisk trade with the Orient. Commodore Perry's famous voyage to Japan in 1854 was a natural sequel to our commercial interests, as was eventually our appearance in the Samoan Islands. Rival German, English, and American interests led to the establishment of a joint protectorate in 1889 and to the annexation of Tutuila in 1899. American control of Hawaii was presaged as early as 1875, when a treaty of reciprocity was arranged with the stipulation that none of its territory should be leased or sold to any other power. Eventually American economic interests in Hawaii became so powerful that they were able to instigate a rebellion against the autocratic Queen Liliuokalani, which was consummated with the active cooperation of the American minister, and the presence of an American naval force. A provincial government was established and annexation sought. Cleveland refused to sanction acquisition by such methods, but the Americans in Hawaii would not return to the old régime. The matter hung fire until 1898, when the administration of McKinley under the stress of war pushed through annexation.

The eyes of Americans had turned toward the control of Cuba for a half century before the Spanish-American War. Cuba occupied the strategic position controlling the entrance to the Gulf of Mexico. It would provide a natural extension for southern slavery, and its accession was a logical continuation of the policy which brought us Florida and New Mexico. American control of Cuba was a leading question from 1850 to 1861. "It is our destiny to have Cuba," said Stephen A. Douglas in 1858, "and it is folly to debate the question. It naturally belongs to the American continent."[1] With the overthrow of the slave-owning aristocracy in the Civil War, the matter of Cuban annexation became quiescent until the nation was sufficiently advanced for financial imperialism. This situation had arrived by 1898

---

[1] Speech in New Orleans, December 6, 1858, in *Life of Stephen A. Douglas*, by "A Member of the Western Bar" (1860), p. 184.

and it provided the great cause for the war. Among the subsidiary and contributing causes were sympathy for the Cubans, and disgust with Spanish methods, the agitation of the "yellow" press, and the blowing up of the *Maine*. It appeared that Spain had come to her senses and was prepared to make any concessions before hostilities commenced, but war was nevertheless declared. The war itself was easily won; the results, however, are important. The declaration of April 19th, which preceded the formal declaration of war on April 21st, declared that "the people of the Island of Cuba are, and of right ought to be, free and independent," and that "the United States hereby disclaims any disposition or intention to exercise sovereignty, jurisdiction, or control over said Island, except for the pacification thereof, and asserts its determination, when that is accomplished, to leave the government and control of the Island to its people."[1]

The war left us with the Philippines, Porto Rico, Guam, and the destiny of Cuba on our hands. The question of what to do with these regions was troublesome, for an active minority opposed their retention. Mr. Bryan forced the campaign of 1900 on the question of imperialism, and the return of the Republicans was looked upon as an answer to the question. Already, however, the future had been foretold. "The Philippines, like Cuba and Porto Rico, were intrusted to our hands by the war, and to that great trust, under the providence of God and in the name of human progress and civilization, we are committed. . . . We could not discharge the responsibilities upon us until these colonies became ours, either by conquest or treaty," said President McKinley in 1899.[2]

"Our concern was not for territory or trade or empire," he continued, "but for the people whose interests and destiny, without our willing it, had been put in our hands. . . .

"No imperial designs lurk in the American mind. They are alien to American sentiment, thought and purpose. Our price-

---
[1] Text of House Resolution 233, Senate Resolution 149, approved by the President April 20th, pp. 738-739, vol. xxx, U. S. Statutes at Large.
[2] Speech in Boston, February 16, 1899. In Boston *Herald*, February 17, 1899, pp. 2-3.

less principles undergo no change under a tropical sun. They go with the flag."

A year later Senator Beveridge expressed the attitude of the victorious imperialists: "The Philippines are ours forever, 'territory belonging to the United States,' as the Constitution calls them. And just beyond the Philippines are China's illimitable markets. We will not retreat from either. We will not repudiate our duty in the archipelago. We will not abandon our opportunity in the Orient. We will not renounce our part in the mission of our race, trustee, under God, of the civilization of the world." [1]

The promise of 1898 not to annex Cuba was kept, but Cuban independence was restricted and the domination of the United States in the affairs of the island republic assured by the "Platt Amendment" which our Senate forced the Cuban convention to incorporate in their constitution before military occupation of the island was relinquished. The clauses of the "Platt Amendment" provided, among other things, that (1) the Cuban government should never enter into agreements with other powers which might impair the independence of the island or grant foreign powers military or naval bases; (2) that it should not contract public debts beyond an amount the interest and sinking fund of which could be carried by the ordinary revenues; (3) that it should sell or lease necessary coaling stations to the United States when necessary to maintain the independence and adequate government of Cuba. Under this provision an army was dispatched to put down an insurrection in 1906, remaining until 1909; and active interference was again resorted to in 1922 for the purpose of effecting political and financial reforms. Porto Rico was taken over with scarcely any opposition on the part of the people, who were allowed after 1900 participation in the government. Since 1901 free trade has existed between the island and the United States.

The situation in the Philippines was more complicated. The islands had been taken from Spain with the active coöperation

---

[1] *Congressional Record*, January 9, 1900, Fifty-sixth Congress, First Session, vol. xxxiii, part i, p. 704.

of Philippine insurrectos, who were led to believe that they might expect independence at the end of the war. The retention of the Philippines, especially in the face of armed native resistance, seemed to be a definite embarkation on the road to imperialism, and the opposition at home was strong. Nevertheless, the United States decided to remain, and after the revolutionists had been put down provision was made in the Philippine Act of 1902 for a government in which the islanders might coöperate through elections to the lower house. Philippine participation was furthered in the Jones Act of 1916, which increased the powers of the Philippine government and gave the natives control of the upper house. Continued hope of independence was held out by the Wilson administration, but any further developments in that direction seem to be indefinitely postponed. An act of March 8, 1902, allowed goods grown or produced in the Philippines to enter the United States under a 25-per-cent reduction. Free importation was allowed in 1909 except that only a specified amount of sugar and tobacco could be brought in; and since 1913 complete free trade has been operative between the United States and her colony.

From the Inauguration of the "Open Door" Policy to the Acquisition of the Virgin Islands.—Now that we had a commercial outpost and a rapidly growing market in the Far East, our next move was to be expected in that quarter. The European nations since 1840, and Japan since 1894, had cast covetous eyes upon the existing and potential resources of China, and had intrigued, jockeyed, and fought to increase their respective powers. The climax to this came almost simultaneously with the Spanish-American War. On the pretext of the murder of two missionaries, Germany in 1897 seized the bay of Kiao-chau in the province of Shantung, extorted from the Chinese government a ninety-nine-year lease of Kiao-chau, and proceeded to fortify it. In the following year France secured a similar lease on Kwang-chow Wan, while Russia obtained a lease of Port Arthur and the neighboring harbor of Talien-wan which she intended to use as a terminal for the trans-Siberian Railway. England followed (1898) by occupying the harbor of Wei-hai-wei, from which, situated midway between the Russians and Germans, she could keep an eye on both. Japan, having captured Formosa (1895), was aggressively pushing her interests in Korea. All of these nations (except Germany) had already appropriated valuable Chinese territories, and now without exception hoped to make these new leases points from which extensive spheres of influence might be built up. To prevent the absolute disintegration of China and to protect America's growing commerce, John Hay, on September 6, 1899, addressed a circular note to London, Berlin, and St. Petersburg which enunciated the now famous "open door" policy. It requested that in its sphere of influence each nation give assurances that (1) all existing treaty ports and established interests in each sphere of influence would be unmolested, (2) that the Chinese tariffs and no others would be enforced and collected by Chinese officials, and (3) that no differentiation in port and railroad charges be made between the citizens of any nation carrying on business. Upon the acceptance of these principles by England, and the expressed sympathy in them on the part of Russia and Germany, Hay notified each of the powers of the others' approval and stated that the United States considered their assent as "final and definitive."

In 1900 the United States showed her intention of playing an active part in the Far East by coöperating in the joint expedition to put down the Boxer Rebellion, and her friendly interest in China by later remitting $11,000,000 of her share of the indemnity. The Boxer Rebellion gave Secretary Hay another chance to affirm his policy when in his circular note of July 3, 1900, to seven powers he announced that "the policy of the Government of the United States is to seek a solution which may bring about permanent safety and peace to China, preserve Chinese territorial and administrative entity, protect all rights guaranteed to friendly powers by treaty and international law, and safeguard for the world the principle of equal and impartial trade with all parts of the Chinese Empire."[1] The principles of this note were followed in the Boxer settlement.

---

[1] House Documents, Fifty-sixth Congress, Second Session, 1900-1901, vol. i, p. 299 (serial number 4069).

During the succeeding years, economic equality was outwardly maintained in China for all nations, but beneath the surface each of the mighty powers was seeking to strengthen her own political and economic gains. So solicitous was our own government in fostering business with the Orient that our activities as directed by Secretary of State Knox (1909-13) were popularly known as "dollar diplomacy." The gradual internal awakening of China was proving a hindrance to Western diplomacy before 1914, and the World War further complicated the Far Eastern question. In 1921 Secretary Hughes took occasion at the Washington Conference to reaffirm to the world in dramatic and unmistakable fashion the principle of equality of economic opportunity in China, and in particular to call a halt to the attempted monopoly by the Japanese. The Washington Conference in its Far Eastern aspect was simply an attempt to continue John Hay's policy of the "open door."

With the close of the Boxer uprising, the scene shifts closer home. The old project of an interoceanic canal now assumed a new importance. It was looked upon as necessary in maintaining the American empire, and its control by our government was considered an essential preliminary. The Clayton-Bulwer Treaty (1850) with England was brushed aside by the new Hay-Pauncefote Treaty (1901), in which Great Britain removed all former restrictions on our building and fortifying the canal on condition that it be regulated by certain rules, and that it be "free and open to the vessels of commerce and war of all nations observing these Rules, on terms of entire equality." President Roosevelt favored the Panama route, rather than that through Nicaragua, and reached an understanding with the French Canal Company, which had been bought up by an American syndicate, to purchase their rights and equipment for $40,000,000, if negotiations with Columbia were successful in obtaining control of a zone around the proposed canal and the right to fortify it. This proposition was embodied in the Hay-Herran Treaty (1903), which was ratified by the United States Senate, but failed of ratification by the Columbian Congress, notwithstanding pressure from the Province of Panama. Roosevelt, exasperated over the delay in his plans, which he looked upon as simply an attempt to extract a higher price for the concession, listened willingly to rumors of a rebellion against Colombia by the discontented Panamanians, and when the revolution (fostered aggressively by the agents of the French Canal Company) actually took place, he was careful to have battleships at hand to prevent Colombia from landing troops to put down the rebellion. Before the insurrection, which occurred on November 3, 1903, could be crushed, he recognized the new republic and by November 18th the Hay-Varilla Treaty had been signed in Washington, by which the United States guaranteed the independence of Panama, agreed to pay $10,000,000 outright, and an annuity of $250,000, beginning nine years later, in return for a strip ten miles wide upon which to build the canal. Construction was finally started in 1906 and the canal completed in 1914. The Panama Canal not only put in the hands of the United States the shortest water route from the Atlantic to the Pacific,[1] but the conditions under which it was built and obtained made Panama virtually a protectorate of the United States.

With the building of the Panama Canal the preservation of United States' interests in Central America and the Caribbean became a matter of even greater concern. The unstable governments, continually changing through forcible political upheavals, endangered the constantly increasing investments of American financiers. Inability to pay or repudiation of foreign debts made the interference of foreign creditor nations a constant danger. To meet the situation our government has evolved a policy somewhat as follows. First, the Monroe Doctrine must be maintained, a position demonstrated in 1902 when Roosevelt forced Germany to give up her blockade and submit the Venezuelan debts to arbitration. Second, while our policy is opposed to European nations forcibly collecting claims it admits that the recalcitrant tropical countries must be required to meet their just debts. Third, where this is necessary or where our own financial interests are at stake, the policy of intervention and supervision has been adopted.

---

[1] In 1855 a group of American capitalists had built a railroad across the isthmus.

This policy has been pursued notably with three countries. Upon a number of occasions United States marines had been landed in Nicaragua at the request of the Nicaraguan government, even before our participation in the revolution of 1912. At this time (figures for 1913) the United States handled 35 per cent of the imports and purchased 56 per cent of the exports of the country. Following this interference, American bankers, with the unofficial sanction of the Department of State (after a treaty providing for intervention had failed to pass the Senate), reorganized the finances of the little country, established a national bank, and assumed control of the principal industries. A treaty ratified by Nicaragua in 1914, and by the United States in 1916,[1] provided that the United States should pay $3,000,000, receiving in return the exclusive right to build a canal on the Nicaraguan route, a lease of ninety-nine years on three small islands, and a naval base on the Gulf of Fonseca. Economic control and the presence of marines have brought peace and made Nicaragua a virtual protectorate. The same is true of the Dominican Republic and Haiti. In the former case the condition of the country became so chaotic after the death of President Heureaux in 1899, that in 1904 the nation was bankrupt and unable to meet the interest on its debt. An executive arrangement was made in 1905 (later passed in treaty form in 1907) whereby the United States was to take over the administration of the customs houses, and to pay the Dominican government 45 per cent of the income for current expenses and 55 per cent to pay foreign claims. In 1908 Kuhn, Loeb & Company of New York refunded the debt of $20,000,000, under a treaty by which the United States was to collect the customs for fifty years.

Haiti, the sister republic of Santo Domingo and located on the same island, was next brought under the control of the United States. With its 2,000,000 inhabitants, Haiti is the most thickly populated of the West Indies and probably the richest in natural resources. For a number of years American bankers had been interested in Haitian finances, and the Great

---

[1] Treaty Series No. 624, Washington, 1916.

War gave an opportunity for government intervention. Professing to believe that Germany had designs upon Haiti, and determined that no European nation should assume control of Haitian customs in the hour of financial difficulty, our country invaded the island, forced through a treaty, and, in spite of the armed opposition of the natives, has continued occupation. The treaty of 1915 [1] imposed on Haiti provides for (1) American aid in development of natural and commercial resources; (2) an American receivership of customs and supervision of expenditures, and an American financial adviser; (3) a native Haitian rural and urban constabulary commanded by American officers; and (4) the cession of no Haitian territory to any "foreign" nation. The reports of the Senate committee investigating conditions in Haiti and Santo Domingo under United States occupation, submitted in 1921 and 1922, advised against the withdrawal of American troops from either country.

Although President Wilson in January, 1916, stated that "there is not a foot of territory belonging to any nation which this nation covets or desires," in the summer of that year sufficient inducements were offered to the Danish government to make it sell its possessions in the West Indies, known as the Virgin Islands.

Exceedingly critical have been our relations with Mexico since the Madero Revolution of 1910. Encouraged by thirty years of strong rule under Porfirio Diaz, United States oil drillers, silver miners, railroad builders, ranchers, and others had invested close to a billion dollars, while Europeans had interests aggregating half that amount. During this period our government had been under pressure to intervene for the protection of United States capital, and at the same time European nations have almost forced us either to look after their interests or to allow intervention on their part, an action hardly in keeping with the new interpretation of the Monroe Doctrine. The Mexican revolution, itself largely an agrarian revolt against the inordinately large accumulations of land, was accompanied by much banditry

---

[1] U. S. State Department, Treaty Series No. 623.

and considerable loss to investments of citizens of the United States. Upon two occasions (1914 and 1917) armed forces entered Mexico, but withdrew, and probably only the European war has prevented further outside interference. Certain clauses in the Mexican Constitution of May, 1917, have been held to discriminate against foreign capital, and recognition was withheld until August 31, 1923, when a satisfactory understanding was reached concerning American economic interests. The

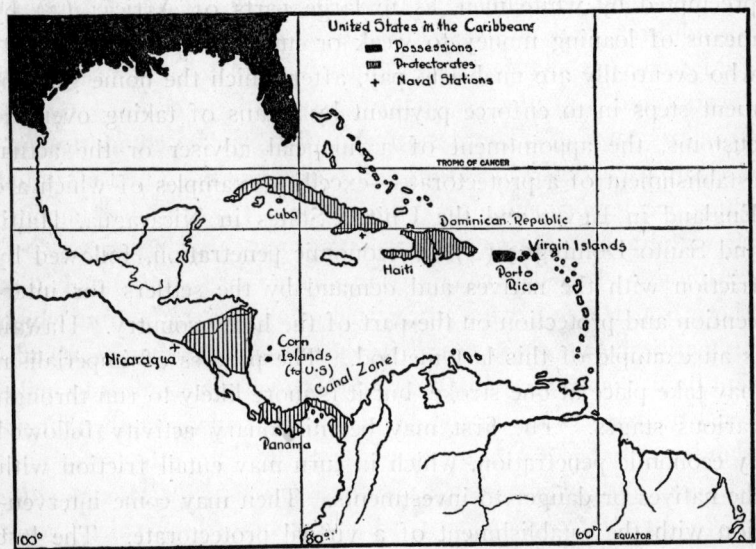

policy of the Taft, Wilson, and Harding administrations was essentially the same—namely, to let the Mexicans work out their own salvation, but at the same time to exert increasing pressure from the outside for more settled conditions and for the protection of foreign interests.[1] Notwithstanding the chaotic condition of the country, considerable business has been carried on by American firms; the vast Tampico oil fields, especially, have functioned with little interruption.[2]

Technique of Imperialism.—Enough has been said in a

---

[1] A new step was inaugurated in 1924 by the Coolidge administration in selling munitions to the Obregon government, but refusing to sell to the revolutionists.

[2] The map on this page is drawn on a McKinley Outline map, copyrighted by the McKinley Publishing Company, and used with their permission.

general way to give some idea of the manner in which imperialism is pursued and it may be wise to pause for a moment to emphasize this aspect. The New Imperialism of the last forty years has generally been carried out by one of the following methods, or variations and combinations of them: (1) by means of military conquest, as in the case of the Boer War or the Spanish-American War; (2) by the appropriation, frequently by treaty consent from the natives, of certain regions not yet preëmpted by white men, as in large parts of Africa; (3) by means of loaning money to weak or impoverished governments who eventually are unable to pay, after which the home government steps in to enforce payment by means of taking over the customs, the appointment of a financial adviser or the actual establishment of a protectorate—excellent examples of which are England in Egypt and the United States in Nicaragua, Haiti, and Santo Domingo; (4) by economic penetration, followed by friction with the natives and demand by the settlers for intervention and protection on the part of the home country. Hawaii is an example of this last method. The process of imperialism may take place at one stroke, but it is more likely to run through various stages. The first may be missionary activity followed by economic penetration, which in turn may entail friction with the natives or danger to investments. Then may come intervention with the establishment of a virtual protectorate. The last stage is actual annexation.

American Economic Expansion Since 1898.—Five reasons may be given for the remarkable economic expansion outside of our own boundary lines since the Spanish-American War. In the first place, by 1898 the United States had become a great industrial country as compared with earlier agricultural predominance. It is the industrial nations that are the great investing nations. Though an agricultural community may support comfortably a fair-sized population, it produces little surplus wealth. Surplus liquid wealth comes from an industrial society, in which machinery increases man's capacity for production, and where a thickly populated region affords cheap labor and better opportunities to amass wealth. Likewise, industrial regions must have raw products and markets. Thus the conditions are present which may lead to imperialism.

Secondly, the rapid increase in production and wealth during these years has put America in a position not only to finance her own economic expansion, but to invest heavily abroad. The percentage of increase in population during the first two decades of the twentieth century has been far surpassed by the percentage of increase of production of almost all industrial products. By 1920, with 6 per cent of the world's population and 7 per cent of the world's land, we were producing 20 per cent of the world's supply of gold, 25 per cent of the wheat, 40 per cent of the silver, 50 per cent of the zinc, 52 per cent of the coal, 60 per cent of the aluminum, 70 per cent of the copper, 60 per cent of the cotton, 66 per cent of the oil, 70 per cent of the corn, and 80 per cent of the automobiles. Estimates of wealth in terms of purchasing power in 1920 puts that of the United States at 500 billion dollars, the whole British Empire at 230, France at 100, Russia at 60, and Italy at 20 billions. As the richest nation in the world, with the production of agricultural and industrial products mounting more rapidly than population, we may expect an increasing interest in foreign commerce and investment.

In the third place, the rapid concentration of capital and consolidation of business which has been a marked feature of our economic life since 1898 has aided in financial imperialism. An industrial organization of the size and power of the Standard Oil or a banking concern of the importance of the National City Bank with resources of over a billion, is obviously in a better position than a small company to push its interests in foreign fields, in competition with European investors. Fourthly, the immense influence of the large corporations upon government policies was bound to make itself felt in foreign fields. If, as Woodrow Wilson pointed out, "the masters of the government of the United States are the combined capitalists and manufacturers of the United States,"[1] embarkation upon a policy of imperialism was a normal evolution.

In the fifth place, the World War has contributed to this ex-

---
[1] Wilson, Woodrow, *The New Freedom*, p. 57.

pansion. The tremendous cost of carrying on the war forced the withdrawal of much European capital invested abroad, thus leaving the field open to our own financiers. By the end of the conflict we had not only taken over many European holdings in other countries, but had loaned heavily in Europe. Before the war the British were the chief investors in "foreign" fields, with almost twenty billion dollars loaned in the international markets as compared with nine billions for France and six for Germany. At the conclusion of the war the foreign investments of Great Britain and France had decreased to but a fraction of their former amounts, while Germany had been practically eliminated. On the other hand, the United States found herself the leading investing nation of the world, with nine and a half billion dollars loaned to the allied nations, other foreign investments amounting to eight billions, and with notice loans and goods on consignment amounting to several billions more.

Some indication of the increasing commerce since 1898 with our own dependencies is given in the following table:

COMMERCE BETWEEN THE UNITED STATES AND HER DEPENDENCIES
1898 AND 1920 [1]

Country	Year	Exports from United States	Imports into United States
United States (total foreign commerce)	1898	$1,231,482,330	$ 616,049,654
	1920	8,228,016,307	5,278,481,490
Cuba	1898	9,561,656	15,232,477
	1920	515,208,731	721,693,880
Danish West Indies (Virgin Islands)	1898	707,622	327,759
	1920	3,993,478	4,540,386
Haiti	1898	2,968,579	876,582
	1920	19,900,380	8,973,534
Dominican Republic	1898	1,151,258	2,382,139
	1920	45,522,750	33,878,099
Nicaragua	1898	1,049,505	1,095,865
	1920	9,542,964	7,971,426
Panama	1904	979,724	440,747
	1920	33,333,155	8,272,586
Porto Rico	1898	1,505,946	2,414,356
	1920	121,561,574	158,275,729
Philippines	1898	127,804	3,830,415
	1920	99,396,564	112,951,409
Hawaii	1898	5,907,155	17,187,380
	1920	74,052,452	192,308,454

[1] Compiled from the *Statistical Abstracts* of 1900, 1905, and 1920.

Notwithstanding the enormous cost of the war and the very doubtful security of the money loaned to our allies, the war undoubtedly left this country much stronger financially than any other nation. Overnight, as it were, we became the great creditor nation, and the interests this position implies will ultimately be reflected in changes in both domestic and foreign policies.

Causes for Growth of Foreign Commerce Since the Civil War.—The foreign commerce of the United States has grown tremendously since 1860. Exports have increased twenty-four fold and imports fourteen fold. In 1850 imports exceeded exports by $20,040,062, but in 1920 the excess of exports over imports was $2,870,636,549. Per-capita imports have increased three fold, while per-capita exports have increased seven fold. The principal causes for this increase are (1) the opening up to settlement and cultivation of immense stretches of agricultural country between the Mississippi and the Pacific, (2) the development of the Lake Superior iron regions, as well as the various mineral resources of the Cordillera range, (3) the enormous increase of industrial life and the differentiation of industrial products as the nation grew older, (4) the growth of an internal transportation system capable of handling the commerce as it expanded, (5) the rapid increase in wealth which went hand in hand with these developments, (6) technical improvements in the manufacture of cables, wireless, and nautical apparatus, as well as the expansion of trade facilities such as banking and credit, and (7) government aid. These causes are essentially the same as those which have prompted internal commerce and those which spurred on imperialism.

EXPORTS AND IMPORTS, 1860-1920 [1]

Year	Exports of merchandise	Exports of merchandise and specie	Imports of merchandise	Imports of merchandise and specie	Excess of total exports
1860	$ 333,576,000	$ 400,122,000	$ 353,616,000	$ 362,166,000	$ 37,956,000
1900	1,394,483,000	1,499,462,000	849,941,000	929,771,000	569,691,000
1920	8,228,016,000	8,663,723,000	5,278,481,000	5,783,609,000	2,880,114,000

[1] Compiled from *Statistical Abstract*, 1921, table 482, pp. 840-841, 854-855.

The opening up of the far west has been discussed elsewhere.[1] It is significant in the history of foreign commerce, for until the close of the century the chief exports were agricultural products; as, in fact, they had been since early colonial times. As the nation grew older our unequaled minerals naturally induced manufacturing and the export of finished products. At the same time agricultural products were gradually supplemented and eventually surpassed in value by manufactured commodities, as the importance of the mineral resources in stimulating foreign commerce became increasingly potent. Yankee ingenuity, protected by patents, had evolved numerous types of labor-saving machinery so epoch-making in their significance that they found a ready sale abroad. Of these may be mentioned agricultural implements, sewing machines, typewriters, and cash registers. Foreign commerce, of course, originates and is stimulated by the needs of one part of the world for the products of another. As we have repeatedly emphasized, no nation has been more fortunately situated in respect to resources than the United States, and we have been in a position to furnish raw materials and eventually manufactured products to other parts of the world on most advantageous terms.

Importation has increased along with exportation, although the excess of exports over imports has grown since 1860 out of proportion to the increase of either exports or imports. Importation has primarily been caused by the necessity of taking in exchange some commodities classed as luxuries and many products of minute and skilled workmanship which the cheaper labor costs and more artistic training of foreign artisans have enabled them to produce. It should be remembered also that the excess of exports in value over imports is partially compensated for (1) by large sums paid to Europeans (especially before the war) for marine freight and insurance, (2) by dividends and interest paid to foreign stock and bond holders, (3) by money sent home by immigrants, and (4) by American travelers, who were reported to spend in the pre-war years about $350,000,000 annually.

---

[1] See chap. xviii.

The growth of the internal transportation system has also been discussed elsewhere.[1] The fact that railroad mileage increased from 30,000 miles in 1860 to 266,000 in 1916 demonstrates the fact that internal transportation has more than kept pace with foreign commerce and, in fact, has made the latter possible. In the same group of influences should be placed the better facilities for conveying information relative to commercial transactions and business conditions. The principle of the telegraph was put into operation in undersea cables after 1852, the year in which a short line was laid from Dover to Ostend. Chiefly through the efforts of Cyrus Field, a successful transatlantic cable was laid in 1866. By 1918 at least nine active cables connected North America with Europe, while the principal ports of the world were joined with 281,000 nautical miles of cable, enough to girdle the world thirteen times. Experiments by Heinrich Hertz in Germany after 1888, by Marconi in Italy, and by numerous inventors throughout the world, perfected the wireless to such an extent that messages were sent across the Atlantic in 1903. At the present time it is possible at small expense for anyone to obtain by radio in his own home the stock quotations and commercial news of the day.

Along with the growth of foreign commerce has gone that of extended financial facilities. Up until the last decade European, South American, and African banking was largely taken care of by branch offices abroad which are able to furnish credit information, negotiate loans, issue bills of exchange, and give other aids to commerce.

Government Aid.—Through innumerable channels the hand of our government is ever at work rendering aid to commercial activity. This may be described under three general heads: (1) construction and improvement of harbors, and measures to increase the safety of navigation; (2) efforts to make commerce more profitable; and (3) aids to the ship-building industry. Such aid originates in the first place from congressional legislation and is carried on either by bureaus of the executive de-

---

[1] Chap. xx.

partments or through independent agencies. This activity has been greatly extended since the rapid expansion of American commerce, after the Spanish-American War, and especially since the division in 1913 of the Department of Commerce and Labor into two executive divisions. A mere citation of the bureaus at present under the Department of Commerce will give some idea of the extensive interests of that branch of the government; Bureau of Foreign and Domestic Commerce, Bureau of Lighthouses, Coast and Geodetic Survey, Steamboat Inspection Service, Bureau of Navigation, Bureau of Census, Bureau of Standards, Bureau of Fisheries. Other departments through their special bureaus are able to render service and information.[1]

Since 1802, when the first federal appropriation was made for rivers and harbors, over a billion dollars has been expended. The appropriation for 1921 was $19,452,700. This work is recommended and supervised by the Corps of Engineers of the United States army. Owing to the fact that the appropriations are influenced to a considerable extent by political log rolling, not all of this money has been wisely spent. Safety in navigation has been promoted by the Bureau of Lighthouses, which constructs, inspects, and superintends lighthouses, beacons, buoys, and other aids to navigation; by the Coast and Geodetic Survey, which prepares charts and maps of the seacoast and adjacent ocean; by the Steamboat Inspection Service, which enforces the navigation laws, thus promoting the safety of navigation; by the Coast Guard of the Treasury Department, whose activities are divided between the Life Saving Service, which maintains 325 stations (1916) to warn ships of impending danger and help those in distress, and the Revenue Cutter Service, which aims not only to enforce the revenue laws, but to supplement the work of the Life Saving Service. Much of the work of the Coast and Geodetic Survey is supplemented by the Hydrographic Office of the Bureau of Navigation of the Navy Department, whose duty it is to provide accurate nautical charts, sailing directions, and

---

[1] *Service Monographs of the United States Government,* published by the Institute for Government Research, are valuable accounts of the actual working of these bureaus.

manuals of instruction. To the Weather Bureau of the Department of Agriculture is charged the forecasting of weather; the issue of storm warnings; the display of weather and flood signals for the benefit of agriculture, commerce and navigation; the gauging and reporting of height of water in rivers; the maintenance and operation of seacoast telegraph lines; and the collection and transmission of marine intelligence for the benefit of commerce and navigation.

Efforts on the part of the government to make commerce more profitable have been exerted from the beginning of our history. Tonnage acts have been passed favoring American ships and treaties negotiated to secure favorable treatment for American products. In addition to activities of the State, Treasury, and Post Office departments and various subsidiary bureaus in providing valuable information, there was instituted in 1912 the Bureau of Foreign and Domestic Commerce under the Department of Commerce, whose business it is "to develop the various manufacturing industries of the United States and markets for their products at home and abroad, by gathering and publishing useful information, or by any other available method." Through special agents and commercial attachés in foreign countries and by means of coöperation with the consular service of the State Department, data covering conditions and commercial opportunities abroad are gathered and distributed. This Bureau publishes a weekly journal known as *Commerce Reports*, a monthly statement known as *Monthly Summary of Foreign Commerce of the United States*, an annual *Statistical Abstract of the United States*, an annual summary of our foreign trade known as *Commerce and Navigation, a World Trade Directory*, a biennial statement of *Trade of the United States*, and many special bulletins of trade conditions, opportunities in foreign markets, tariff laws, and other matters of interest to traders.

Special encouragement to foreign commerce was afforded in the Webb Export Act of April 10, 1918, which exempted associations entered into "for the sole purpose of engaging in export trade and actually so engaged," from the provisions of the Sherman Anti-Trust Act, on condition that such associations did not enter into a conspiracy to control prices or restrain competition. Corporations desiring to acquire part or entire ownership of such foreign trading corporations were exempted from the provisions of the Clayton Act.

The Merchant Marine.—A merchant marine owned at home is not absolutely essential to an extensive foreign commerce, but the advantages which accrue from it may be so great as to warrant active government aid. The Civil War dealt our once famous merchant marine a blow from which it never recovered. Destruction by Confederate privateers and large sales abroad decreased the amount of tonnage. Delay in adopting iron steam-driven ships gave British builders an advantage which they continued to hold. More profitable investments in internal transportation and the exploitation of raw materials in the great industrial age which dawned after the war, drew capital away from the sea. Lack of government interest helped complete the downfall of American shipping.

The five years following the Civil War showed a slight revival, but the forces tending to a decline continued operative. American shipping engaged in foreign trade and the fisheries, which amounted to 2,642,628 tons in 1870, had dropped to 826,694 tons in 1900. In 1860 the percentage of imports and exports carried in American ships was 66.5, but this dropped in 1870 to 35.6, in 1880 to 13, in 1890 to 9.4, in 1900 to 7.1. The merchant marine had a loyal friend in Senator Frye of Maine, who in 1891 introduced bills to subsidize mail steamers, freight steamers, and sailing vessels; but the encouragement was insufficient and a subsequent bill introduced by him in 1901 providing for a more liberal subsidy of $9,000,000 for thirty years was defeated by the agricultural and manufacturing interests.

The La Follette Seamen's Act [1] of 1915 was unjustly denounced by the shipping interests as the final blow at a declining merchant marine, but its backers claimed that it was a protection to American shipping and a simple act of justice to American seamen. Among other things it provided (1) that 75 per cent

---

[1] 36 Stat., 1164.

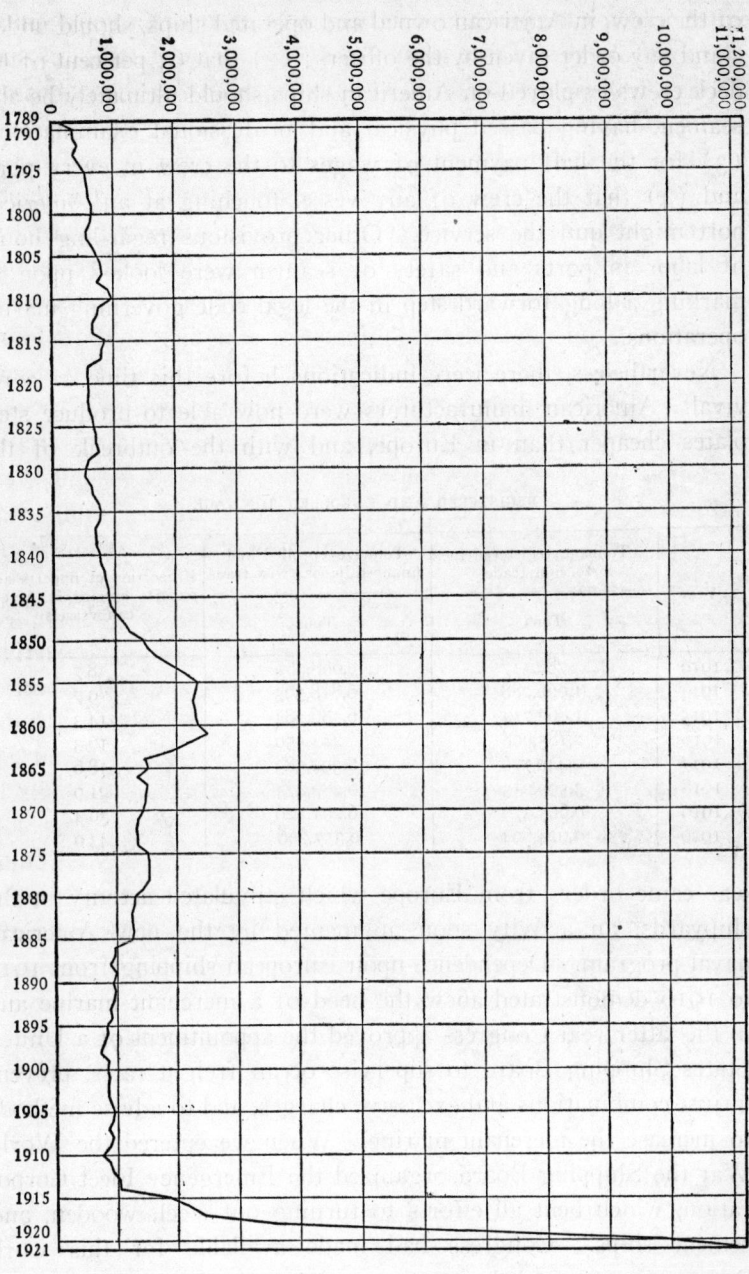

TONNAGE OF THE UNITED STATES MERCHANT MARINE ENGAGED IN FOREIGN COMMERCE, 1789-1921[1]

[1] *Statistical Abstract*, 1921, and W. Bates, *American Marine*, p. 462 ff.

of the crew, in American owned and operated ships, should understand any order given by the officers; (2) that 64 per cent of the deck crews employed on American ships should ultimately be able seamen, having passed physical and professional examinations; (3) for the half payment of wages to the crew in every port; and (4) that the crew of any vessel touching at an American port might quit the service. Other provisions regarding hours of labor in ports and safety of seamen were looked upon as marking a long forward step in the legal code governing marine operations.

Nevertheless, there were indications before this time of a revival. American manufacturers were now able to produce steel plates cheaper than in Europe, and with the outbreak of the

REGISTERED AND ENROLLED TONNAGE

Year	Tonnage registered in foreign trade Tons	Tonnage enrolled and financed in coastwise trade Tons	Proportion of imports and exports carried in American vessels
1910	782,517	6,668,966	8.7
1914	1,066,288	6,818,363	9.7
1915	1,862,714	6,486,384	14.3
1916	2,185,008	6,244,550	16.3
1917	2,440,776	6,392,583	18.6
1918	3,599,213	6,282,474	21.9
1919	6,665,376	6,201,426	36.4
1920	9,924,694	6,357,706	43.0

war came orders from Europe which stimulated activity in the shipyards, an activity soon augmented by the new American naval program. Dependence upon European shipping from 1914 to 1916 demonstrated anew the need of a merchant marine and in the latter year Congress approved the appointment of a United States Shipping Board to supervise ocean freight rates, prevent unjust combinations and excessive charges, and to advise methods to increase the merchant marine. When we entered the World War the Shipping Board organized the Emergency Fleet Corporation, which bent all efforts to turning out steel, wooden, and cement ships. Congress had made available for this work $4,000,000,000, and during the nineteen months of the war 875 vessels of 2,941,845 gross tons were built. The 61 shipyards of 1917 with their 235 ways had increased by November, 1918, to 341 shipyards and 1,284 launching ways, while the number of workmen had grown from 45,000 to 380,000.

A large proportion of the gross tonnage in 1920 was owned by the government and the question of the future was one which Congress was called upon to decide. The bitter hostility of capital to government-owned transportation presaged the retirement of the government from the ocean carrying trade, an end partially achieved in the Merchant Marine Act of 1920.[1] This bill, known as the Jones Act, continued the United States Shipping Board, gave partial recognition to the principle of subsidies to American shipping by exempting companies from excess profits tax up to a certain amount, and by providing for government loans to the amount of $25,000,000 a year for five years, the sum to be obtained from the sale of government ships, and in a number of other ways sought to stabilize and aid the new-born merchant marine. The effects, however, of the Jones Act in promoting American shipping were very slight, and the advocates of further government aid have pressed for more efficacious legislation. Since private capital has not been willing to assume the risk of operating the giant liners still in government hands as a result of the war, the United States Shipping Board has reconditioned them for regular passenger and freight service, and continues to run them.

Trend of Commerce Since 1860.—The Civil War profoundly affected foreign commerce. The high tariffs inaugurated by the war stimulated manufacturing. The cutting off of the southern trade forced northern producers to seek other markets, with the result that between 1860 and 1865 there was an actual increase of exports other than cotton. The war temporarily ruined the cotton business of the south, and the amount of cotton exported did not reach the pre-war level until 1875. This in turn cut down imports into the south. While the war ruined the merchant marine, it stimulated the building of trans-

---

[1] 41 Stat., 988, 1008.

continental railroads, and further aided commerce by the concentration of capital and the better banking system which was introduced in 1863.

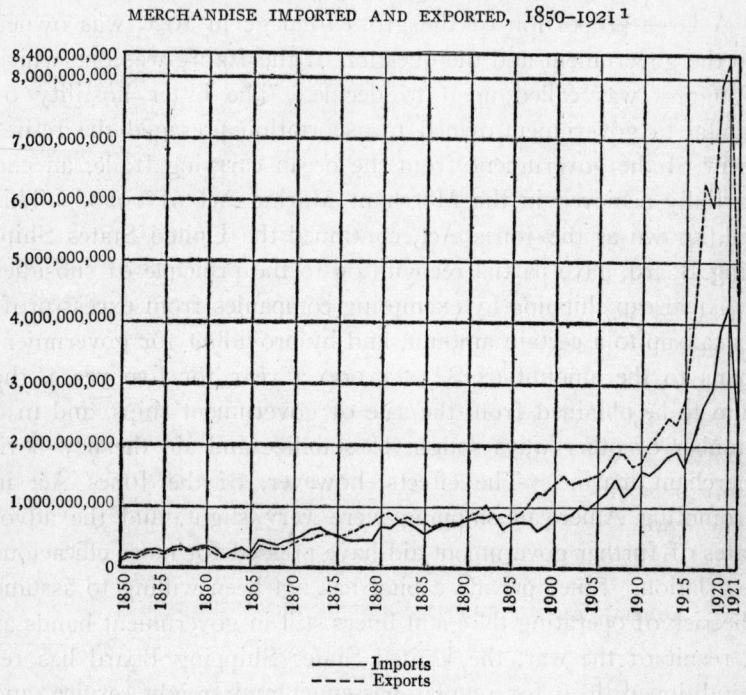

MERCHANDISE IMPORTED AND EXPORTED, 1850–1921 [1]

—— Imports
- - - - Exports

Since 1865 the increase in imports and exports has been irregular, but large in the long run. Between 1870 and 1890 the value of the annual imports of merchandise increased 95 per cent and the value of the annual exports 255 per cent. The first decade of the century saw an increase of 25 per cent in the value of the exports and 83 per cent in the value of the imports. The figures of the next decade are even more striking, but the increases in value must be discounted to a certain extent by the rise in prices. To obtain a true conception of the growth of foreign commerce the following figures should be compared with the imports and exports of certain commodities in amount rather than value.

---

[1] *Statistical Abstract,* 1899, p. 92, and *Statistical Abstract,* 1921, p. 854.

WORLD TRADE AND IMPERIALISM

EXPORTS AND IMPORTS BY DECADES [1]

Year	Exports	Imports
1860	$ 333,576,157	$ 353,616,119
1870	392,771,768	453,958,408
1880	835,638,658	667,954,746
1890	857,828,684	789,310,409
1900	1,394,483,082	849,941,184
1910	1,744,984,720	1,556,947,430
1920	8,228,016,307	5,278,481,490

Up to 1890 the leading exports were agricultural, including wheat, corn, and meat products and such semi-agricultural products as flour, glucose, cotton and vegetable oils, butter and cheese. Agricultural products amounted to 79.4 per cent of the country's exports in 1870, 83.3 per cent in 1880, 74.5 per cent in 1890, and 60.9 per cent in 1900. Manufactures ranked in importance next to agriculture, followed by minerals, food products, and fisheries. Of individual commodities, exported grains took the lead, followed by cotton, meat and meat products, iron and steel, and mineral oil.

DOMESTIC EXPORTS GROUPED ACCORDING TO SOURCES OF PRODUCTION [2]

	1870		1880		1890		1900	
	Value*	Per cent	Value*	Per cent	Value*	Per cent	Value*	Per cent
Agriculture	$361,188	79.35	$685,961	83.25	$629,821	74.51	$835,858	60.98
Manufactures	68,280	15.00	102,856	12.48	151,102	17.87	433,852	31.65
Forests	14,898	3.27	17,321	2.11	29,473	3.49	52,218	3.81
Mining	5,026	1.10	5,863	.71	22,298	2.64	37,844	2.76
Fisheries	2,836	.62	5,255	.64	7,453	.88	6,327	.46
Miscellaneous	2,981	.66	6,689	.81	5,141	.61	4,665	.34
Total	$455,208	100.00	$823,946	100.00	$845,294	100.00	$1,370,764	100.00

* In thousands of dollars.

The percentages taken in connection with the figures of valuation show a gradual decline in the relative importance of agricultural products in exports and an increase in that of partly or

---

[1] Fiscal years until 1920. Compiled from *Statistical Abstract*, 1921, table 482, pp. 840-841, 854,855.

[2] From *Monthly Summary of Commerce and Finance*, April, 1903, p. 3249.

wholly manufactured commodities. Particularly is this to be noted during the last decade of the century. By far the greatest market was Great Britain; but there was a growing demand in all of the countries of northern Europe except Russia. Trade with southern Europe was stagnant, while with South America, although the amount almost doubled between 1870 and 1900, the percentage of the total was less. Healthy progress was made in the Asian, Australian, and African markets. In 1870 the percentage of the total of our exports which went to Europe was 79.35 per cent, to North America 13.03, to South America 4.09, to Asia 2.07, to Oceanica .82, and to Africa .64. In 1900 Europe received 76.60 per cent, North America 13.45, South America 2.79, Asia 4.66, Oceanica 3.11, and Africa 1.79.

The figures for imports 1870-90 show foodstuffs, both in crude condition and partly manufactured, holding up well, and in the latter case actually increasing, with sugar as the leading single import. The group which advanced the most rapidly was that of crude materials for use in manufacturing, including rubber, hides and skins, raw silks and fibers. While Europe in 1900 took 74.6 per cent of our exports, she sent us but 51.8 per cent of our imports, most of which were in manufactured products. Relatively the import trade with North America declined greatly and that with South America increased slightly, while the imports from Asia, which had been 6.8 per cent of the total in 1870, advanced to 16.5 per cent in 1900.

By 1900 industrial development in the United States had advanced to a stage where there was an excess of manufactured products and minerals to export, and this, furthered by the impetus furnished by the Spanish-American War, drove American capital and products into foreign markets. The same tendencies that were apparent in the last decade of the preceding century became more evident. Foodstuffs, which had led in exportation, gave place to manufactures. In 1900 all classes of foodstuffs constituted 39.80 per cent of our total exports; in 1910 but 21.58, with a decline in value of nearly $177,000,000. This decline in absolute value is all the more marked in that prices on the whole were rising. It will be noticed that while the value of foodstuffs exported increased nearly four fold from 1900 to 1920, their percentage of total exports decreased from 39.80 to 25.18. During these twenty years manufactured exports increased nearly ten times in value, and amounted from 35.38 to 51.52 in percentage of total exports. Crude materials for further use in manufacturing were relatively in the same position in 1920 as at the opening of the century. Cotton, of which about two-thirds was annually exported, up to 1914 made up the bulk of this group and retained its position as the leading single American export. The gradual superseding of agricultural products by manufactures is shown in the following table:

EXPORTED DOMESTIC MERCHANDISE, CLASSIFIED BY GREAT GROUPS [1]

Year	Crude materials for use in manufacturing		Manufactures ready for consumption and for further use in manufacturing		Foodstuffs, crude and manufactured, and food animals		Miscellaneous	
	Value	Per cent *	Value	Per cent *	Value	Per cent *	Value	Per cent *
1900	$ 325,244,296	23.73	$ 485,022,156	35.38	$ 545,602,580	39.80	$14,894,539	1.09
1910	565,934,957	33.10	766,981,245	44.85	369,087,974	21.58	8,079,822	.47
1915	510,455,540	18.80	1,163,327,840	42.83	961,568,583	35.40	80,826,502	2.97
1920	1,870,767,054	23.15	4,163,354,637	51.52	2,034,596,001	25.18	11,763,129	.15

* Per cent of total exports.

As American manufactured articles were exported in larger amounts, other nations by keen competition and high tariff duties have endeavored to keep them out. Although in 1914 our European exports amounted to 63.37 per cent of the total, they consisted chiefly of cotton, wheat, flour, meat products, and tobacco. An outlet for manufactured products had to be found in other quarters. North American countries, especially Canada, have furnished the chief new markets, and exports to them have shown both absolute and relative gains. The first two decades saw an intensified rivalry in the Far East, complicated now by the entrance of Japan, but our nation has been fortunate in obtaining her share of Asiatic trade. Particularly noticeable, also, has been the advance in South and Central American commerce, part of which was a normal development and part promoted by the

[1] Compiled from *Statistical Abstract* for 1921, table 482, pp. 848, 849.

World War. The relative importance of our export markets as shown by the percentage of exports sent out in 1920 is: to Europe, 54.27; to North America, 23.45; to Asia, 9.38; to South America, 7.58; to Oceanica, 3.30; and to Africa, 2.02. In contrast to our exports, more than half of our imports came from non-European countries. In 1920 only one-fourth of our imports were from Europe and about the same amount from Asia; from North America came 31.5 per cent, and from South America 14. From the latter we receive the two important commodities we are unable to raise at home, rubber and coffee; while our domestic supply of minerals, hides, chemicals, and tropical foodstuffs is considerably augmented from this source.

NOTES FOR FURTHER REFERENCE

General discussions are those of C. R. Fish, *The Path of Empire* (1919), in Chronicles of America and *American Diplomacy* (1915), and A. C. Coolidge, *United States as a World Power* (1916). The historic background of the period is sketched by J. F. Rhodes, *The McKinley and Roosevelt Administrations* (1922), weaker than his earlier volumes and inclined to hero worship; by C. A. Beard, *Contemporary American History* (1914); by C. R. Lingley, *Since the Civil War* (1920); by F. L. Paxson, *Recent American History* (1921); and by J. H. Latané, *America as a World Power, 1897-1907* (1907). An illuminating résumé of American expansion is to be found in John Bassett Moore, *Four Phases of American Development, Federalism—Democracy—Imperialism—Expansion* (1912). See also his *American Diplomacy, Its Spirit and Achievements* (1905). An excellent biography of an arch-imperialist is that of H. Croly, *Marcus A. Hanna* (1912), while strong arguments against imperialism are to be found in G. F. Hoar, *Autobiography of Seventy Years* (1913). The reminiscences and speeches of a bitter anti-imperialist senator, interesting and containing information not easily obtainable, are found in R. F. Pettigrew, *The Course of Empire* (1920). Much of value on this subject may be obtained from the lives of the chief executives of the period: C. S. Olcott, *William McKinley* (2 vols., 1916); R. M. McElroy, *Grover Cleveland; the Man and the Statesman* (2 vols., 1924); Theodore Roosevelt, *An Autobiography* (1913); Lord Charnwood, *Theodore Roosevelt* (1923); W. E. Dodd, *Woodrow Wilson and His Work* (1920); and C. Seymour, *Woodrow Wilson and the World War* (1921), in Chronicles of America.

A penetrating and stimulating survey of American imperialism is that of H. H. Powers, *America Among the Nations* (1919). A radical but brilliant economic interpretation with special emphasis on the development of financial imperialism as affected by the World War is that of Scott Nearing, *The American Empire* (1921). Illuminating surveys are the lectures of Achille Viallate, *Economic Imperialism and International Relations During the Last Fifty Years* (1923). On special phases, see J. H. Latané, *The United States and Latin*

America (1920); C. L. Jones, *Caribbean Interests of the United States* (1916); A. B. Hart, *The Monroe Doctrine, an Interpretation* (1916).

Reliable and detailed on the war with Spain is F. E. Chadwick, *Relations of the United States and Spain*, Vol. I, *Diplomacy* (1909), and Vols. II and III, *The Spanish War* (1911). There are excellent volumes on the Philippines: D. C. Worcester, *The Philippines: Past and Present* (2 vols., 1914); J. A. Leroy, *The Americans in the Philippines* (1914); C. B. Elliott, *The Philippines* (2 vols., 1917); F. B. Harrison, *The Corner-Stone of Philippine Independence* (1922), by the anti-imperialistic governor-general during the years 1913-21; and J. S. Reyes, *Legislative History of America's Economic Policy Toward the Philippines* (1923). Other books on our insular possessions are L. S. Rowe, *The United States and Porto Rico* (1904); W. F. Willoughby, *Territories and Dependencies of the United States* (1905); and L. K. Zabriskie, *The Virgin Islands of the United States of America* (1918). See also the *Report of the Philippine Commission* (1917), by W. H. Taft, Senate Doc. No. 200, Sixtieth Congress, First Session, Vol. 7.

On the history and significance of the Panama Canal consult Arthur Bullard, *Panama, the Canal, the Country and the People* (2d ed., 1914); J. B. Bishop, *The Panama Gateway* (1913); I. E. Bennett (ed.), *History of the Panama Canal* (1915); and E. R. Johnson, *The Panama Canal and Commerce* (1916).

Studies of various economic phases are in the following: F. M. Halsey, *Investments in Latin America and the British West Indies,* Department of Commerce Special Agent Series, No. 169 (1918); Francis W. Hirst and George Paish, *The Credit of Nations and the Trade Balance of the United States* (1910), National Monetary Commission Publications, No. 2, Senate Doc. No. 579, Sixty-first Congress, Second Session; and H. R. Mussey, *The New Normal in Foreign Trade*, Political Science Quarterly, Vol. XXXVII, No. 3. The best histories of American commerce during this period are those of E. R. Johnson, T. W. Van Metre et al., *History of Domestic and Foreign Commerce of the United States* (2 vols., 1915), reprinted in 1922 in one volume; and A. L. Bishop, *Outlines of American Foreign Commerce* (1923).

SELECTED READINGS

MOORE, J. B., *Four Phases of American Development,* Lectures III and IV.
FISH, C. R., *American Diplomacy*, Chaps. XXVI-XXXVI.
COOLIDGE, A. C., *The United States as a World Power*, Chaps. V-IX, XV-XIX.
FISH, C. R., *The Path of Empire.*
POWERS, H. H., *America Among the Nations*, Chaps. I-XII.
JOHNSON, E. R., and HUEBNER, G. G., *Principles of Ocean Transportation*, Chaps. XXIII-XXXII.
BISHOP, A. L., *Outlines of American Foreign Commerce*, Chaps. IX, XI-XIV.

CHAPTER XXVI
RECENT ECONOMIC TENDENCIES

The Passing of Laissez-Faire.—As in Europe, the Industrial Revolution was followed in the United States by a period of industrial development unrestricted and free from interference on the part of the government. The economic doctrine of *laissez-faire* was gladly welcomed by industrial leaders and represented for many years the dominant belief and policy of the government. In one respect only, that of the protective tariff, was the principle seriously violated; here *laissez-faire* was not considered advantageous. Although the farmer, in certain cases, felt the need of government assistance, nevertheless his frontier life developed the independence of spirit that made him naturally fall in with the idea of *laissez-faire*. Opposition of the agricultural group was never formidable so long as there was an abundance of unoccupied land.

Economic distress after the Civil War, arising from currency deflation and the resulting fall in prices, as well as from other factors, caused a serious outbreak against the theory of *laissez-faire*, an outbreak which found vent in the "granger movement" of the 'seventies. Nevertheless, the 'eighties and 'nineties saw the golden age of *laissez-faire*. A change, however, was bound to come and was caused chiefly (1) by the grievances of the farmers against the railroads and other representatives of capital, (2) by the distrust of the average citizen toward the consolidation of industry and capital, (3) by the advance of state control and social experiment in Europe, and (4) by the rapidly increasing population of newly arrived immigrants reared in such an atmosphere of state socialism. There was a growing feeling that America was no longer the land of opportunity, and that uncontrolled *laissez-faire* had benefited only a small group who had amassed in their hands the major share of the wealth and resources of the nation. The culmination of the discontent came in the first decade of the century when Roosevelt, catching the prevailing feeling of unrest, led in a vigorous campaign toward government regulation.

Preliminary shots in the campaign had already been fired in the Interstate Commerce Act of 1887 and in the Sherman Antitrust Act of 1890, but neither of these laws proved of much avail. Further railroad legislation as embodied in the Elkins (1903), Hepburn (1906), and Mann-Elkins (1910) acts increased the power of the Interstate Commerce Commission and more closely circumscribed the activities of the railroads. Federal regulation had made rapid strides before the opening of the World War, and the taking over by the government of the railroads in 1918 was but a logical step in a long development. The reaction after the war returned the railroads to private hands, but with enlarged powers of control retained by the federal authorities. This story has already been recounted, as has that of the efforts of the government through judicial prosecution and such subsequent legislation as the Clayton Act (1914) to define more carefully and restrict combinations in restraint of trade. The passing in 1906 of the federal Pure Food Law marked the entrance of the government into a new phase of activity in the protection of the whole people from the unscrupulous few.

The extension of federal activities has not been confined simply to restraining harmful practices. The government has definitely and successfully enlarged its own business interests by embarking more aggressively in the transportation business through the parcels post (1912) and in the banking business through the postal savings banks (1910). In its efforts to encourage, to understand more clearly, and to coöperate more intelligently with the economic life of the nation a Department of Agriculture was established in 1889, a Department of Commerce and Labor in 1903, and the latter separated (1913) into two departments, all of which through their numerous bureaus are centers of scientific research and a clearing house for information. More recently (1913) the banking and currency system of the nation has been radically altered and, in so doing, centralization and federal control have been strengthened.

Not only has the movement toward government regulation and

aid been marked, but steps have been taken in the direction of wealth equalization, or at least toward the amelioration of certain economic inequalities. The lead was taken by the state of Wisconsin during the years 1900 to 1905, when it became a veritable laboratory for social legislation. One of the most important laws of this nature was one providing for an income tax, put on the Wisconsin statute books in 1911. An amendment making possible a federal income tax was added to the Constitution of the United States in 1913 and an income tax section added to the Simmons-Underwood tariff.[1] To-day sixteen states [2] collect income taxes in addition to the federal tax. Inheritance taxes by which the state appropriates a portion of the estate of deceased citizens, the rate usually increasing with the size of the property left, have been enacted in practically every state. These new taxes, it should be pointed out, are a result of the mounting expenses of the government as well as a change in the attitude of the people, but the increased expenses in turn are a result of the enlarged government activities.

Along the same line is the enactment in a number of states of workingmen's compensation laws, of laws prohibiting child labor, those providing for minimum hours for dangerous trades, factory acts of various kinds, and minimum hour and wage acts for women and children. A federal act designating eight hours as the standard day on railroads was passed in 1916, but two federal child labor laws have been declared unconstitutional. Where the national government has failed, state governments have in many cases succeeded; and the state laws have generally gone much farther than the federal. But for the conflict of authority between federal and state constitutions probably social legislation

---

[1] An income tax had been passed to provide revenue for the Civil War and had been continued until 1872 (see chap. xxi). A second tax, imposed in 1894, however, had been declared unconstitutional.

[2] Income tax laws:
  A. Combined personal and corporation income tax laws; 8 states.
  B. Personal income tax laws and distinct corporation income tax laws; 2 states.
  C. Personal income tax laws, but no corporation income tax; 3 states.
  D. Corporation income tax laws, but no personal income tax; 3 states.
—Bulletin of the National Tax Association, vol. ix, no. 2, November, 1923.

on the part of both might have made greater progress. In this connection, however, it may be said that the attitude of both state and federal courts had become more tolerant before the war of acts looking toward social betterment, but that reaction seems to have set in since 1918.

Coincident with the breakdown of *laissez-faire* have appeared political and economic groups advocating more complete socialization. The discontent of the farmers of the northwest has made itself felt recently in the activities of the Nonpartisan League and that of the city proletariat in the Socialist party, an organization which polled 901,873 votes in 1912, 590,579 in 1916, and 919,799 in 1920. Born of discontent with economic conditions, these various radical movements have undoubtedly contributed in a constructive way toward the movement for social legislation. "It was apparent," says Dr. Beard, "from an examination of the first decade of the twentieth century, that they [the states] were well in the paths of nations like Germany, England and Australia. . . . Eminent economists turned from free trade and *laissez-faire* to consider some of the grievances of the working class, and many abandoned the time honored discussions of 'economic theories,' in favor of legislative programs embracing the principles of state socialism to which Germany and England were already committed."[1] The extent to which socialism, as exemplified by enlarged government activity, may go, is difficult to predict. Just as there was a reaction from *laissez-faire* to government regulation, so the pendulum has swung back again since the close of the late war. Nevertheless, the war, through its heritage of debt and new obligations assumed, has unavoidably extended the regulating activities of the government. Whatever the future of state socialism may be, the movement toward greater social justice will surely continue.

Conservation.—No recent development more pregnant of beneficial possibilities has appeared than the awakening interest in conservation. Blessed for three centuries with an apparently exhaustless supply of land and raw materials, the American people

---

[1] Beard, Charles A., *Contemporary American History*, pp. 304-305.

became prodigal of their heritage and wasteful in their habits. As population began to press seriously upon the land and the cost of raw materials to rise, attention was drawn to the conservation of remaining resources. Investigations disclosed that our supplies of wood and minerals were not inexhaustible, but that their duration was limited. It was discovered that the waste in mining bituminous coal was 50 per cent and that of mining anthracite was 100, while yearly millions of gallons of mineral oils either evaporated or were lost in pumping. Unscientific methods of lumbering were both denuding the nation of wood and releasing the floods to spread havoc in the valleys. Soil was being robbed of its fertility without sufficient return of fertilizer. Not alone in the processes of production was waste to be found, but in the use of the finished material. Furthermore, in no highly developed industrial nation is human life more lightly regarded than in the United States. Of the 29,000,000 workers it is estimated that 500,000 are killed or crippled each year, a larger number than the casualties suffered by the United States in the World War. For every 100,000 tons of coal mined it is claimed that one man is killed and several injured. The Interstate Commerce Commission reported for the year 1919 over 2,000 railroad employees killed and over 131,000 injured.

The untiring labors of Gifford Pinchot were responsible for first arousing widespread interest in national conservation. He enlisted the interest of President Roosevelt, who called the famous conference of governors in 1908, a conference followed by the appointment of forty state conservation committees and a national conservation committee. The three-volume report of the last named, submitted in 1909, was the first scientific survey of the wasteful methods employed in the exploitation of our resources and of the possibilities of conservation. The impetus given by Roosevelt has undoubtedly directed the state and national governments to such a policy, but the progress of both in matters of conservation has been slow. Nevertheless, the sum total of public and private efforts has been encouraging.

Conservation may take three forms: (1) the insuring to the people as a whole the wealth of the forests, mines, and rivers by preserving them from exploitation through private hands; (2) actual saving through more efficient and thrifty methods in production, distribution, and consumption of the raw materials of the nation; and (3) the greater saving of the country's man power. Under the liberal land acts it was found that large corporations had acquired, often by fraudulent methods, a large part of the mineral resources of the country. That a similar fate might not befall the water-power resources, the great power of the future, Roosevelt hastened to withdraw 148,346,925 acres from public entry, almost six times as many acres as had Grover Cleveland, his nearest rival in attempting to save for the nation as a whole some parts of the public domain. Executive efforts to preserve intact some parts of the public lands and to care for them have been continually hampered by congressional lack of interest and by the opposition of those who would exploit them. Nevertheless, there are at present in the United States still some 190,000,000 acres under the control of the federal government, besides about 24,000,000 in Alaska.

Conservation through more efficient production and more thrifty consumption has taken many forms. The increase in the cost of living and the thrift propaganda of the World War have helped to a more economical manner of life, although there is still much to strive for in this respect. To preserve the supply of lumber, reforestation, fire prevention, the study of insect pests and the better utilization of by-products are all being urged and in many places attempted. Although 65 per cent of the original lumber is still standing, the rapidity with which it is cut or destroyed by fire means its utter depletion before many decades unless heroic efforts are made along the lines indicated to avert such a calamity.[1] The preservation of forests will also help re-

---

[1] The consumption figures for a single issue of a great metropolitan newspaper furnish an excellent illustration of the rapid use of our natural resources. The Public Service Bureau of the Chicago *Tribune* give "the approximate figures on the materials that are required to publish the Sunday edition of the Chicago *Tribune*" as follows: "Standing timber, 54 acres; sulphur, 21 tons; coal, 665 tons; electric horsepower, 63,000; water, 18,200,000 gallons; limestone, 28 tons; paper 800 tons." Mr. S. M. Williams of the New York *World* suggests 400 to 450 tons of newsprint paper as the average of the New York Sunday *World* with coal consumption varying from 1,400 to 1,700 pounds of coal per ton

tain the water for irrigation and power purposes.[1] More efficient methods of mining and the greater utilization of inferior ores, which science is making increasingly profitable, may save much of the minerals now wasted. The soil must be continually replenished by natural and artificial fertilizers, and new land, hitherto believed unarable, must be brought under cultivation to provide food for the increasing population. Science is doing much for conservation, but the determination of the public as expressed individually and through legislatures is essential.

Strange as it may seem, the conservation of human life has ordinarily been a matter of less interest to industry than the conservation of materials, and this has been true even in America, where the supply of labor has not been excessive. The eight-hour day has barely been achieved where unionization has been strongest, and is still unknown in many industries. It is but recently that laws requiring sanitary surroundings and guarded machinery have made certain kinds of factory work even tolerable. Child labor is still too often a curse which mortgages the nation's future. Nevertheless, gradually through humanitarian efforts, and through trade union strength the conditions of work are improving. Medical science has made rapid advances in both preventive and curative medicine; together with our greater knowledge and practice of sanitation it has appreciably increased the span of life. If to the conquest of disease future generations add some progress in the application of eugenics, the debilitating effects of our high-speed civilization may be counteracted.

Efficiency—Scientific Management.—The subject of conservation leads naturally into a more detailed discussion of that phase of it which deals with the more adequate and efficient handling of the raw materials and labor during the process of manufacturing and distribution. Close on the heels of the inauguration of the conservation movement came a campaign for greater efficiency, with particular attention placed upon "scientific

---

of paper and approximately fifty acres of pulp wood (chiefly spruce and balsam) forests cut over.

[1] See essay by R. Zon in Caldwell and Slosson, *Science Remaking the World* (1923).

management." For years Frederick W. Taylor had experimented and preached the doctrine of greater conservation through the application of scientific methods to shop management and production. His disciples, H. L. Gantt, Carl G. Barth, F. B. Gilbreth, Harrington Emerson, and others have by their own studies carried on the work with the spirit of the master, although with differences in detail and method.

It was Taylor's belief that by careful studies of time and methods the quickest and best manner of doing a piece of work could be discovered, after which the job should be pitched to the standard of the most efficient. By eliminating the incompetent and stimulating the best workers through wage systems believed to be psychologically sound, greater production might be obtained and at the same time higher wages paid. Taylor's own wage system was that of the "differentiated piece rate" which consists of two rates; the first, or ordinary piece rate, which is paid for production up to a possible standard; and second, a higher rate paid per piece for the entire production if a man turns out more than the ordinary amount. Others, like Halsey, Gantt, and Emerson, have devised different methods, usually a combination of wages paid for a set task with a bonus for more than the average production, the basic idea being to share with the men the profits accruing from the surplus produced by extra efforts. The details are not so important as the fact that science is at last touching in industry the human factor.[1]

The question of efficiency in industry goes beyond the matter of efficiency in wage systems, better machinery, cost accounting, or even more scientific organization of management. It includes the solving of such problems as waste from poor sanitation and ill health, from inadequate factory training, and from labor turnover. It has been estimated that there are 30,000,000 workers in the United States whose average annual loss through sickness

---

[1] It should be pointed out that labor, as a rule, opposes scientific management on the ground that it speeds up the workmen to an abnormal pace and that it reduces men to mere machines or automatons. A discussion of these charges and the answers made by the advocates of scientific management may be found in H. B. Drury, *Scientific Management* (1915), p. 188 *ff*.

amounts to nine days, or an equivalent of 739,736 years; and, if the daily wage is estimated at $2.50 and medical attendance at $1 a day the annual loss would be $945,000,000. In reality the loss to the employer would increase this figure by a large amount. To counteract this loss the more intelligent concerns have installed sanitary and safety engineers, medical directors, hospitals, dental clinics, visiting nurses, gymnasiums, recreation rooms and parks, sanitary lunch rooms, and other means of health preservation both in and out of the factory. Such complete application can be tried only in the larger establishments, but the experiments have amply demonstrated the value of these efforts.

In addition to the efforts to promote the health of the employees it has been realized that efficiency would be stimulated and losses reduced if proper training systems could be established. These training systems have taken two forms: (1) the education of foreigners in English and in general information, and (2) special technical training in factory operations or in the preparation of executives. In both cases it has been found that efficiency and skill have been increased and waste reduced. It is at last becoming apparent that the wisest and most economical plan to pursue is to make the most by training and otherwise of the labor at hand rather than the old plan of consuming human labor as rapidly and wastefully as possible in the expectation of throwing aside the useless remnant for the fresh supply in the market.

No factor has been of greater importance in stimulating interest in labor conservation than the enormous losses incurred each year because of labor turnover. To the employer frequent change in employees is costly, involving as it does the expense of hiring, training, extra wear and tear on machinery, work spoiled, more accidents, as well as the general impairment of the efficiency and morale of the whole force. To the employee there is the loss of earnings, the development of lazy habits, demoralization through frequent idleness, impairment of skill, a lowering in the standard of living for the whole family, and a decrease in buying power. The cost of taking on a new man has been estimated by many concerns at from $25 to $50 without counting many secondary expenses. The United States Department of Labor believes the annual cost of labor turnover to be close to $1,125,000,000. Without entering into the many causes for the termination of employment, it may be said that a more enlightened labor policy on the part of employers and a more scientific management of seasonal trades can and is increasingly cutting down labor turnover. A certain amount of labor turnover is normal and beneficial as men fit into their proper places and seek to better themselves, while the restless nature and quick temper of others will always be responsible for some changes, but there is much avoidable waste from this cause.

Industrial Democracy.—Political philosophers from Aristotle to Madison have truthfully emphasized the economic basis of politics. It was the economic as well as numerical strength of the bourgeoisie and working classes that won for them the franchise. But political democracy did not allay the economic unrest which has grown as the effects of capitalism have drawn a sharp distinction between the holders of capital and the wage earners. It has become more and more apparent, however, that political democracy without greater social or economic democracy is an anachronism which cannot permanently endure. The development, in consequence, of greater coöperation between labor and capital in the management and prosecution of our industrial life, while expected, is one of the most significant factors of recent economics and bids fair to continue with increasing momentum in the future. Undoubtedly the strength of labor is the primary cause for this trend toward industrial democracy, for in many industries organized labor has reached a position of such power that the continuation of production is practically within its hands. In such a situation the prolongation of the bitter struggle between labor and capital would mean destruction of both and loss to the public. With destruction or coöperation as the alternative it is hardly surprising that the more intelligent representatives of both sides read the writing on the wall. A second cause has been the growth of a more sensitive social conscience which has made the employer react more quickly to the rights of the employee, especially when the latter has won the backing of the public. Furthermore, the success of experiments in industrial democracy in England has been an incentive to similar attempts here. The whole movement is a phase in the decline of autocracy whether in politics or elsewhere.

Industrial democracy comprehends more than a mere participation in extra profits by means of a bonus system or even of stock ownership. It implies an actual coöperation in the management and the determination of policies.[1] Quite naturally it has so far been chiefly confined to exercising joint control over problems of immediate interest to the employee—working conditions, hours, wages. The experiments along this line have been classified under the following heads:[2] (1) the conference system, (2) the works committee plan, and (3) the house and senate plan. They are typical of the efforts so far made.

Under the conference system representatives of the employers and employees meet together to discuss matters of mutual interest. The representatives of the employers are appointed, and those of the employees are elected on the basis of one representative for a certain number of workers regardless of crafts or departments. Examples of this type are to be found in the International Harvester Company and the Bethlehem Steel Corporation. In the steel company joint committees composed of representatives of management and employees are formed on (1) rules, procedures and elections; (2) ways and means; (3) safety and medical service; (4) works, practice, improvement in method and production economics; (5) employees' transportation; (6) wages, piece work, bonus and tonnage schedules; (7) employment and working conditions; (8) housing, domestic economics, and living conditions; (9) health and works sanitation; (10) education and publications; (11) pensions, insurance, and relief; (12) recreation, athletics, and entertainment; (13) continuous employment and trade fluctuations. Inability to arrive at satisfactory conclusions in one of the joint committees throws the matter to a

---

[1] Rockefeller, J. D., Jr., *Representation in Industry*, a reprint of an address before the United States Chamber of Commerce in 1918.
[2] Tipper, Harry, *Human Factors in Industry*, p. 135 ff.

general joint committee, and if that is not productive of results to arbitration by mutual consent.

The works committee plan differs from the conference system in that the representatives of labor comprise appointees of the various shop unions or occupational groups within the factory selected by the groups. Representing not the entire establishment, but only the wage earners and those comprising craft groups, the work of the committees has largely been confined to adjusting grievances which might lead to strikes.

The third plan, that of an organization modeled somewhat on the plan of the American government, has been tried in a number of industries, the best examples of which are to be found in the American Multigraph Company and the William Demuth Company of New York City. In the latter case there is a cabinet consisting of the head of the firm and his executives, a senate consisting of thirty foremen from the different departments, and a house of representatives made up of employees elected from each department. An issue which cannot be settled by agreement of the house, senate, and cabinet goes to a board of conciliation, composed of one man selected by the employer and one by the employees and a third agreed upon by both. Under the house and senate plan joint committees of the two are chiefly instrumental in ironing out difficulties and determining policies.

Although some of the early attempts have ended in failure, others have demonstrated the soundness of intrusting labor with a share in the responsibility of management and in formulating policies. An immense impetus was given to the whole movement by the World War, when it was absolutely necessary to eliminate friction and maintain continuous production. It was then that the sound psychology of industrial democracy was urged by the War Labor Board and through its efforts shop committees of varying types were installed in some of the nation's greatest industries.[1]

Economic Outlook of the Church.—One of the most interesting recent phenomena is the tendency of the churches, both

---
[1] *How the Government Handled Its Labor Problem During the War* (1919), a pamphlet issued by the Bureau of Industrial Research.

Catholic and Protestant, to take a rather definite stand with respect to future economic and social development. Although many sincere church members oppose any practical interference in economic matters on the ground that such action is not a function of the church, several programs, official and semiofficial, have been presented. It is fruitful to speculate on the causes for this new attitude. Does it arise from the growing strength of labor, from a greater social consciousness on the part of the churches, from a changing belief in the functions of organized Christianity, or from the feeling that there may be fundamental weaknesses in the present economic order incompatible with the doctrine of Christ? All of these elements have probably contributed.

The most notable emanation from the Catholic source is the document popularly known as the "Bishops' Program of Social Reconstruction." In the post-war period the bishops propose the following program:[1] (1) the general level of wages attained during the war should not be lowered; (2) the right of labor to organize and deal with employers through chosen representatives should never again be called in question by any considerable number of employers; (3) labor ought gradually to receive greater representation in the industrial part of business management; (4) minimum wage laws providing for decent maintenance should be passed by the states; (5) women should receive equal wages with men for the same task and quality of work, they should be kept out of occupations harmful to health and morals, and their participation in industry should be kept within the smallest possible limits; (6) the cost of living should be reduced, and coöperative stores are suggested as a possible method, as well as government competition with monopolies that cannot be restrained by the ordinary anti-trust laws; (7) a national system of labor exchanges acting in harmony with state, municipal, and private employment bureaus should be set up; (8) the problem of congestion and other forms of bad

[1] Written by Bishops P. J. Muldoon, Joseph Schrembs, Patrick J. Hayes, and W. T. Russell, who comprised the Administrative Committee of the National Catholic War Council.

housing ought to be tackled by the cities; (9) the state should make provision for insurance against illness, invalidity, unemployment, and old age, but as far as possible by a levy on industry; (10) there should be improvement in public health by health inspection, municipal clinics, etc.; (11) vocational training should be enlarged with at least the elements of cultural education; (12) child labor should be taxed out of existence; (13) land colonization should be promoted by bringing under cultivation arid, swamp, and cut-over timber lands. These reforms, it is believed, will decrease the inefficiency in production and distribution, will promote coöperation and bring about increased incomes for labor. That the "present system stands in grievous need of considerable modifications and improvement" is maintained, but the reforms must come not through socialism, but through the Christian view of work and wealth.

Not quite so definite, but covering substantially the same ground, are the social ideals of the majority of Protestant churches adopted by the Federal Council of the Churches of Christ in America in 1919. The Council stands officially for: (1) equal rights and justice for all men in all stations of life; (2) protection of the standard of the family by the single standard of purity, uniform divorce laws, proper regulation of marriage, proper housing; (3) the fullest possible development of every child, especially by the provision of education and recreation; (4) the abolition of child labor; (5) such regulation of the conditions of toil for women as shall safeguard the physical and moral health of the community; (6) abatement and prevention of poverty; (7) protection of society from the social, economic, and moral waste of the liquor traffic; (8) conservation of health; (9) protection of the worker from dangerous machinery, occupational diseases, and mortality; (10) the right of all men to the opportunity for self-maintenance, for safeguarding this right against encroachments of every kind, for the protection of workers from the hardships of enforced unemployment; (11) suitable provision for the old age of the workers, and for those incapacitated by injury; (12) the right of employees and employers alike to organize, and for adequate means of conciliation and arbitration in industrial disputes; (13) release from employment one day in seven; (14) gradual and reasonable reduction of hours of labor to the lowest practical point, and for that degree of leisure for all which is a condition of the highest human life; (15) a living wage as a minimum in every industry, and for the highest wage that each industry can afford; (16) a new emphasis upon the application of Christian principles to the acquisition and use of property, and for the most equitable division of the product of industry that can ultimately be devised.

In addition to these two programs individual churches have presented social creeds pointing in the same direction. Most of these demands have been advocated by labor for a long period. If the aspirations of labor are now to be backed by a powerful organization such as the church, with mainly a middle-class constituency, a significant hint may possibly be found here as to future economic and social evolution.

Control of the Business Cycle.—The last two decades have seen an extraordinary growth in the facilities for scientific economic research and in the actual amount of such work accomplished. Government agencies, endowed bureaus, and private individuals have contributed enormously to the scientific data available. Of the various economic phenomena under observation, an understanding of none is more important than that of the business cycle. Even a cursory review of our economic history reveals it in one phase as a "constant recurrence of irregularly separated booms and slumps,"[1] and the crisis years in these cycles (1812, 1818, 1825, 1837, 1847, 1857, 1873, 1884, 1890, 1893, 1903, 1907, 1910, 1913, 1920) stand out as key dates. "Of our one hundred and thirty years as a nation," says O. C. Lightner, "thirty-three years have been wasted in disastrous and ruinous depression and perhaps an equal number have been marked by over-spending, extravagance and waste of wealth. The remainder have been years of normalcy."[2] W. C. Mitchell,

---

[1] From a definition by Herbert Hoover in his foreword to *Business Cycles and Unemployment, Report and Recommendation of a Committee of the President's Conference on Unemployment*, including an investigation made under the auspices of the National Bureau of Economic Research, part i, p. vi.

[2] Lightner, O. C., *The History of Business Depressions*, p. 7.

a leading American expert, states that " . . . the modern view is that crises are but one feature of recurrent 'business cycles.' . . . A crisis is expected to be followed by a depression, the depression by a revival, the revival by prosperity, and prosperity by a new crisis. Cycles of this sort can be traced for at least a century in America." [1]

If our economic history has been chiefly a succession of periods of prosperity and depression, a more intimate view of business cycles is worth while. According to Herbert Hoover:

Analyses of past cycles of business show certain common tendencies. If we begin the analysis when business is reviving, in general the characteristic features are increased volume of manufacturing, rising stock exchange prices followed by rising commodity prices, then by business expansion and increased demand for credit from both business men and speculators. As the result of the advance of commodity prices, money rates stiffen and credit gradually becomes strained, and these conditions may be accompanied by a curtailment of credit for speculative purposes. Then stock exchange prices fall; for a while longer general business continues to increase unevenly, transportation facilities are overburdened and deliveries are delayed, the apparent shortage of goods is intensified by speculative buying and duplication of orders by merchants and other buyers until credit expansion nears its limit. Public confidence is then shaken, resulting in widespread cancellation of orders if the cycle is extreme. This is always followed by liquidation of inventories and sharp and irregular fall of prices. During the period of depression, there is always more or less widespread unemployment.[2]

In other words, during the period of hard times following a panic, capital becomes more cautious and the nation as a whole retrenches and saves. As the renewed activity and thrift of the people gradually build up fresh resources, the purse strings are slowly loosened to start new enterprises. Under the stimulus of competition and the increased liberality of investors, business

---

[1] Mitchell, W. C., in *Business Cycles and Unemployment*, p. 5.
[2] Hoover, Herbert, in foreword to *Business Cycles and Unemployment*, pp. xii, xiii.

booms, and wages and prices are on the upgrade. At the height of prosperity money is spent recklessly and invested often in the most risky and unwarranted ventures. Not only is the available capital used up, but the future is mortgaged. When the bubble inevitably bursts liquidation wipes out much of the former profits and leaves in a safe position only those strongly intrenched financially. The latter, indeed, are often so placed as to take advantage of the weaker, and consolidation usually ensues, while business returns to saner methods.

Although the causes for these recurrent cycles of prosperity and depression differ at different times, all of them rest in the end, according to Mitchell and other experts, upon "this crucial factor—the prospects of profits." [1] While it may be impossible in a capitalistic society to remove this primary cause and we may be destined in the future to a continued alternation of good and bad times, it may be possible to soften the effects by increased knowledge of how the cycles function and when they may be expected. A distinct advance has been made in controlling the business cycles by the adoption of the Federal Reserve banking system, a system which has enabled the nation to weather a financial crisis and period of depression without a panic, so that the last crisis, when it came in 1920, was "regulated." After the panics of the 'nineties, business sought to save itself by consolidation, but to-day the emphasis is upon increased scientific economic knowledge.

National Income and the Distribution of Wealth.—The most thorough investigation into the extent and distribution of the nation's income has been made by the National Bureau of Economic Research. Pursuing their studies by the double method of estimating income both by sources of production and by incomes received, they obtained results strikingly similar. Omitting the money value of the work done by housewives for their own families, an item which would add several billions to the figures, their findings are as follows:

---

[1] *Business Cycles and Unemployment*, p. 6.

TWO ESTIMATES OF NATIONAL INCOME, 1909-1919 [1]

Year	National income (in billions)			Population (in millions)	Income per capita (in dollars)		
	By sources of production	By incomes received	Final estimate		By sources of production	By incomes received	Final estimate
1909	$28.8	$....	$28.8	90.37	$318	$...	$319
1910	31.8	31.1	31.4	92.23	344	337	340
1911	31.2	31.2	31.2	93.81	332	333	333
1912	33.6	32.4	33.0	95.34	352	340	346
1913	35.6	33.3	34.4	97.28	366	342	354
1914	33.9	32.5	33.2	99.19	342	328	335
1915	36.1	35.9	36.0	100.43	360	357	358
1916	45.4	45.5	45.4	101.72	446	447	446
1917	53.9	53.9	53.9	103.06	523	523	523
1918	60.4	61.7	61.0	104.18	579	592	586
1919	....	66.0	....	104.85	...	629	...

In studying the above table, due consideration should be given to the inflated war prices after 1914. When reduced to the price level of 1913, the purchasing power of the $61,000,000,000 national income in 1918 was only 38.8 billion dollars, and that of the per-capita income of $586 was only $372.

An attempt to distribute the national income by industries gives first rank to manufacturing and second to agriculture. Unclassified industries and miscellaneous income in the table below includes the incomes credited to merchandising, to the professions, to the rental value of houses occupied by their owners, and to interest on consumption goods owned by families.

PERCENTAGE OF THE NATIONAL INCOME CONTRIBUTED BY THE VARIOUS INDUSTRIES, 1909-1918 [2]

	1909	1914	1918	Average, 1909-1918
Agriculture	16.29	17.80	21.01	17.43
Minerals	3.14	3.06	3.33	3.24
Manufacturing	30.32	27.27	31.47	29.97
Transportation	9.60	9.34	8.67	9.28
Banking	1.51	1.52	1.27	1.45
Government	5.00	5.72	8.87	5.61
Unclassified industries, and miscellaneous income	34.14	35.29	25.38	33.02

[1] *Income in the United States, Its Amount and Distribution, 1909-1919*, vol. i, by the Staff of the National Bureau of Economic Research, Inc., 1921. This table is adapted from tables 1, 9, and 11, pp. 13, 64, and 68.

[2] *Income in the United States, 1909-1919*, by the National Bureau of Economic Research, Inc. (1921), adapted from the table in vol. i, p. 23.

Turning from the amount to the distribution, it is interesting to note the proportion of income paid to employees of all grades as compensation for their services (including wages, salaries, pensions, compensations for accidents, etc.). This amounted in 1909 to 53 per cent, and in 1918 to 54 per cent, of the total; it varies from about one-eighth of the total in agriculture to three-quarters in the case of mining, manufacturing, water transportation, and government work. In the highly organized industries from 28 to 31 per cent goes to "management and capital," in the less highly organized considerably more. Of the total payment to employees in highly skilled industries the manual laborers and clerical staff receive 92 per cent and the officers 8 per cent.

AVERAGE ANNUAL EARNINGS OF EMPLOYEES NORMALLY ENGAGED IN VARIOUS INDUSTRIES, 1909, 1913, and 1918 [1]

	Current money			Value at prices of 1913			Indices of the purchasing power of annual earnings, base, 1913		
	1909	1913	1918	1909	1913	1918	1909	1913	1918
All industries	$626	$723	$1,078	$656	$723	$682	90.7	100.	94.3
Agriculture	302	328	590	316	328	373	96.3	100.	113.7
Minerals	599	755	1,283	627	755	812	83.0	100.	107.5
Manufactures (Factory workers)	571	705	1,148	597	705	726	84.7	100.	103.0
Transportation	657	762	1,286	688	762	814	86.8	100.	106.8
Banking	770	930	1,461	807	930	925	90.3	100.	99.5
Government	739	823	895	774	823	567	94.0	100.	68.9
Unclassified industries	716	779	1,054	750	779	667	96.3	100.	85.6

When the average earnings are examined it is found that, while in most cases there have been substantial increases in wages during the period 1909-1918, when compared with the increase in the cost of living the actual improvement has been slight. In fact, labor statisticians present estimates to show that real wages have actually declined if reckoned over a period of two

[1] Adapted from table 20, pp. 102-103, *Income in the United States.*

decades. The table on page 670 gives a résumé of the annual earnings in leading industries as estimated by the National Bureau of Economic Research. The indices of the purchasing power based on prices for 1913 should be particularly noted.

The amount of wages as given in current money in the preceding table helps to clarify the income-tax statistics as published by the government since the tax was imposed in 1913. The figures for the year 1920 are fairly indicative of recent incomes, for that year included the last months of the war inflation and the beginning of the subsequent depression. Even in this year of partially inflated wages 83 per cent of those over ten years of age gainfully employed did not receive an income amounting to $1,000. Although it is extremely difficult, even with the force of government pressure, to get accurate information on incomes, it appeared that in 1920 less than 1 per cent of the income receivers had close to 12 per cent of the national income and that 10 per cent had 34 per cent. At the same time, whereas the incomes of those between $1,000 and $5,000 amounted to 64.35 per cent of the total income reported, they paid but 15.43 of the tax.

INCOME TAX, 1920[1]

Income classes	Number in each class	Per cent of total	Amount of net income in each class	Per cent of total	Amount of tax in each class	Per cent of total
1,000–2,000	2,671,950	36.80	$4,050,066,618	17.06	$ 36,859,732	3.43
2,000–3,000	2,569,316	35.39	6,184,543,368	26.06	45,507,821	4.23
3,000–5,000	1,337,116	18.42	5,039,607,239	21.23	83,496,116	7.77
5,000–10,000	455,442	6.27	3,068,330,963	12.93	97,886,033	9.11
10,000–25,000	171,830	2.37	2,547,904,786	10.73	172,259,321	16.02
25,000–50,000	38,548	.53	1,307,785,113	5.51	154,265,276	14.35
50,000–100,000	12,093	.17	810,386,333	3.41	163,717,719	15.23
100,000–150,000	2,191	.03	265,511,505	1.12	86,587,694	8.05
150,000–300,000	1,063	.014	215,138,673	.91	92,604,423	8.61
300,000–500,000	239	.003	89,313,552	.38	47,043,461	4.38
500,000–1,000,000	123	.002	79,962,894	.34	45,641,005	4.25
1,000,000 and over	33	.001	77,078,139	.32	49,185,085	4.57

Although the per-capita income and the per-capita wealth are greater in the United States than in European countries, the distribution of the general wealth is not radically different. This is a striking and sobering fact when the greater resources, the newness of the country and its democratic institutions are remembered.

[1] Adapted from tables on p. 715 ff. of *Statistics of Income from Returns of Net Income for 1920.* Published by the Treasury Department (1922).

Some idea of the distribution of wealth can be obtained from the income-tax statistics already cited, where 7,259,944 wage earners over ten years of age out of 41,614,248 reported incomes of $1,000 or more, and 72 per cent of those filing returns received less than $3,000. But the distribution of wealth is far more unequal than income. Possibly the best estimates of the distribution of wealth are those made by W. I. King [1] from studies of estates probated in Massachusetts during four different three-year periods—1829-31, 1859-61, 1879-81, and 1889-91—and in five Wisconsin counties during the year 1900. Mr. King for purposes of comparison divided the decedents into four classes—the poor, comprising 65 per cent and consisting of those possessing little or no property except furniture, clothing, and personal belongings; the lower middle class, composed of 15 per cent and consisting of those who had a little property, possibly a thousand dollars on the average; the upper middle class, comprising 18 per cent and consisting of those possessing property valued at from $2,000 to $50,000; and finally the rich, comprising 2 per cent. In both states, and during all periods in Massachusetts, it was found that the poorest two-thirds of the population owned but 5 or 6 per cent of the wealth, and the lower middle class even less. The poorest four-fifths of the people owned scarcely 10 per cent of the total wealth. The upper middle class possessed about one-third of the wealth, while the rich owned about three-fifths. In fact, the richest 1 per cent owned almost one-half of the property probated while one four-hundredth owned one-fourth. Although Massachusetts is a manufacturing state and the counties studied in Wisconsin contained several cities, including Milwaukee, there is good reason to believe that they are typical of conditions throughout the country.

Prices and the Cost of Living.—The general movement of prices from 1840 to 1890 may be readily studied from data in the Aldrich Report,[2] and after that date from studies made by the Bureau of Labor Statistics.[3]

---

[1] King, W. I., *Wealth and Income of the People of the United States* (1917).
[2] Senate Doc. 1394, Fifty-second Congress, Second Session, March 3, 1893.
[3] Monthly Review of the U. S. Bureau of Labor Statistics, vol. ix, no. 3, September, 1919, p. 98.

RECENT ECONOMIC TENDENCIES

The Massachusetts Commission on the Cost of Living[1] attributes the drop in prices 1840-43 to the industrial depression; the rise from then until 1847 to the high tariff; the fall from 1847 to 1849 to the reduction in the tariff; the advance to 1857 to the increased gold supply from the California discoveries and to the general prosperity; the drop in 1857 and the succeeding years to the panic of 1857 and the opening of the Civil War; the sharp rise to the high point of 1866 to the disruption of prices caused by the war and the disturbance of the monetary system; the fall in prices at the conclusion to deflation, greater production, and a less active market, a tendency arrested by the extraordinary speculation 1870-73, but continued as a consequence of the panic of 1873 and the demonetization of silver; the rise 1880-82 to the resumption of specie payments and business improvement; the drop 1882-86 to the railroad collapse; the gradual decline from 1886 to 1897 to the appreciation of gold, currency agitation, and depression; the rise from 1897 to 1907 to the depreciation of gold, with temporary declines of 1900 due to the Boer War and in 1907 to the panic of that year.[2] Since this report was made there was a slight decline in 1912, after which prices resumed their upward movement until 1920, when they began to slump. Speaking broadly, it may be said that from 1850 to 1865 there was a substantial increase in price levels; from 1865 to 1896 a great decrease; from 1896 to 1914 another rise, and from 1914 to 1920 a sudden jump of over 100 per cent.

In examining the broad price swings, the jump of over 100 per cent from 1860 to 1866 may be attributed to increased demand for commodities and excessive currency inflation. The long decline from 1865 to 1896 was due primarily to deflation and to the fact that the rapid exploitation of natural resources was producing food and manufactured products in excess of the demand. The unsteady but sure rise from 1897 to 1914 may be attributed to the increase in the gold supply and to the fact that the population and its needs were catching up to production,

---

[1] *Report of the Massachusetts Commission on the Cost of Living,* House Doc. 1750 (1910).

[2] See graph on page 606 for the trend of wages and prices 1840-1891.

especially agricultural. The World War with its inflation and its excess of demand over supply was a repetition to some extent of the Civil War situation. To the general causes already mentioned to account for the long upward swing of prices since 1897 the following contributory or subsidiary influences have been stressed by various economists: (1) the rising standard of living, (2) exhaustion of natural resources, (3)

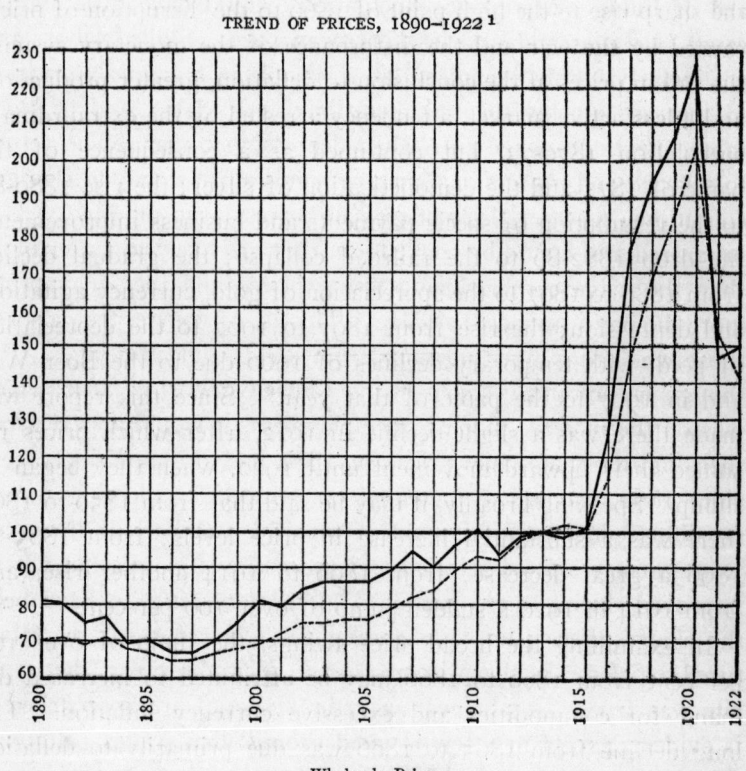

TREND OF PRICES, 1890–1922 [1]

——— Wholesale Prices
– – – Retail Cost of Food

withdrawal of population from agriculture and the growth of cities, (4) arbitrary raising of prices by trusts, (5) labor unions, (6) the high tariffs, (7) wasteful habits of the American people combined with uneconomical and unscientific production and dis-

---

[1] Based on statistics in the Monthly Labor Review, p. 23, February, 1921, and p. 41, March, 1923.

tribution, (8) increasing burden upon society of disease, accident, crime, and pauperism. Eliminating the controversial items of trusts, tariffs, and unions, for it should be noted that the tendency of rising prices has been world wide and in countries where these influences are not so potent, it seems quite certain that the other factors mentioned have played a part in the movement.

More significant than the matter of prices is that of the standard of living. Mere price fluctuations would have little significance if wages and salaries kept even pace. Skilled labor, and particularly highly organized labor, is able in a favorable market to increase quickly the remuneration for its labor and to block more effectually wage cuts when depression comes. The return to salaried and professional men responds more slowly to changed conditions, whether the tendency of prices is up or down. Unorganized and bound by convention, they are likely to suffer more than the wage earner during rising prices and they may profit more during a period of decline. This is, of course, true also of those who enjoy fixed incomes from bonds and mortgages. Although no account is taken in the Aldrich Report of rent, which was on the increase, nor of unemployment, it would appear from their figures that while prices in 1880 had dropped 6 per cent since 1860, wages had increased 60 per cent. Allowing for all possible deductions, it is evident that the standard of living of wage earners and of salaried and professional classes must have improved during these years. The farmer seems to have been the chief sufferer, notwithstanding the lower cost of manufactured articles. In the price rise after 1897, however, the opposite was true. In whatever way the statistics are approached it is evident, except in a few isolated groups, that wages (although they have more than doubled in many trades) have not kept pace with the cost of living. The standard of living of workingmen has declined relatively during the past twenty-five years. This does not mean that they do not have conveniences now which they did not have then, but that relative to the rest of the population they are worse off. What is true of wage earners is even more true of salaried men and those in certain of the professions. The recent war period simply accentuated a tendency which seemed to the superficial observer temporarily stayed. Although real wages have declined, in some respects the standard of living has been bettered. To this, according to the belief of many economists, the Eighteenth Amendment has contributed. The gradual adoption of the eight-hour day, and the passing of legislation providing for more sanitary working conditions, the adoption by a number of states of workingmen's compensation, and other similar measures, as well as an awakened social conscience upon the part of employers has certainly helped to soften the decline in real wages and to remedy some of the glaring faults in our industrial system.[1]

NOTES FOR FURTHER REFERENCE

A standard volume on conservation is that of C. R. Van Hise, *Conservation of Natural Resources in the United States* (1910). Other studies are C. G. Gilbert and J. E. Pogue, *America's Power Resources* (1921); Benton MacKaye, *Employment and Natural Resources* (1919), Publications of the Department of Labor; F. K. Lane, *Conservation Through Engineering* (1920, Bulletin 705, Department of the Interior, United States Geological Survey). A wealth of material is to be found in the *Proceedings of the Conference of Governors* (1909); in the *Annals of the American Academy of Political and Social Science* (1909, Vol. XXXIII, No. 3); in the *Proceedings of the Second Pan-American Congress*, December 27-January 8, 1916, Section III; and in John Ise, *The United States Forest Policy* (1920).

For discussions of scientific management, see C. W. Gerstenberg, *Principles of Business* (1919); Harrington Emerson, *Efficiency as a Basis for Operation and Wages* (1911); H. B. Drury, *Scientific Management*, Vol. LXV, No. 2, Columbia University Studies in History, Economics and Public Law (1918); the *Addresses and Discourses on Scientific Management Held at the Amos Tuck School, Dartmouth College, October, 1911* (1912); and F. B. Copley, *Frederick W. Taylor* (2 vols., 1923).

Material on industrial democracy may be found in David and Meyer Bloomfield, *Selected Articles on Modern Industrial Movements* (1919); J. R. Commons, *Trade Unionism and Labor Problems* (2d Series, 1921); J. R. Commons et al., *Industrial Government* (1921); W. L. Stoddard, *The Shop Committee* (1919); and Harry Tipper, *Human Factors in Industry* (1922). See also the Whitley Committee report on *The Industrial Council Plan in Great Britain*, published by the Bureau of Industrial Research.

Excellent studies of income have been made by the Staff of the National Bureau of Economic Research, Inc., *Income in the United States, Its Amount and Distribution, 1909-1919*, Vol. I, *Summary* (1921); Vol. II, *Detailed Report* (1922); also their *Distribution of Income by States in 1919*, Publication No. 3 (1922). See also W. I. King, *The Wealth and Income of the People of the*

---

[1] For a complete survey of the wage question in the United States on the eve of the World War see Lauck, W. J., and Syndenstricker, E., *Conditions of Labor in American Industries* (1917).

United States (1917); David Friday, *War, Profits and Prices* (1920); and the publications of the Treasury Department on income statistics. Important government investigations are the *Aldrich Report on Prices, Wages, and Transportation*, Senate Document 1394, Fifty-second Congress, Second Session, March, 1893; *Investigation Relative to Wages and Prices of Commodities*, Sixty-first Congress, Third Session, Senate Document 847 (1911), and the *Report of the Massachusetts Commission on the Cost of Living* (1910). The Department of Labor studies the cost of living and reports regularly in the *Labor Review*. Significant studies of the business cycle are those of Alvin H. Hansen, *Cycles of Prosperity and Depression in the United States, Great Britain and Germany*, a study of monthly data, 1902-08, University of Wisconsin Studies in the Social Sciences and History, No. 5 (1921); Lionel D. Edie (ed.), *The Stabilization of Business* (1923), containing articles by W. C. Mitchell, Irving Fisher, F. H. Dixon, *et al.*, with an introduction by Herbert Hoover; O. C. Lightner, *History of Business Depressions* (1922); and *Business Cycles and Unemployment* (1921), *Report and Recommendations of a Committee of the President's Conference on Unemployment*, including an investigation made under the auspices of the National Bureau of Economic Research.

SELECTED READINGS

VAN HISE, C. R., *Conservation of Natural Resources in the United States*, Introduction and Part I.
DRURY, H. B., *Scientific Management*, Chaps. I-III and Part II.
KING, W. I., *The Wealth and Income of the People of the United States*, Chaps. III, IV and IX.
STODDARD, W. L., *The Shop Committee*.
Business Cycles and Unemployment, Chap. I on "Business Cycles," by W. C. Mitchell.
Social Reconstruction, Reconstruction Pamphlets, No. 1, January, 1919. Published by the Committee on Special War Activities, National Catholic War Council.

CHAPTER XXVII

THE WORLD WAR AND RECONSTRUCTION

Early Effect of the War Upon the United States.—The opening of the World War was followed in the United States by a period of uncertainty and depression, an accentuation of a business condition which had existed during the latter part of 1913 and the early months of 1914. Producers of raw materials and foodstuffs found their European market temporarily cut off. As one great nation after another was drawn into the conflict, however, the United States became the most important neutral, and orders for war materials and foodstuffs began to pour in. Then ensued, beginning with 1915, a period of enormous industrial and agricultural expansion which lasted until 1920. Up to our entrance into the war in 1917, our increased foreign trade and the current high prices drew vast amounts of securities as well as gold from Europe. The lack of transportation facilities upon the ocean during the early years of the war was the only factor seriously interfering with our war prosperity.

Financing the World War.—The United States entered the World War on April 6, 1917, and continued an increasingly active participant until the armistice of November 11, 1918. That the financial cost of our participation would be stupendous was obvious from the experience of the allied nations, and plans were at once formulated to meet the increased expenditures. Two general policies were advocated: one, that the cost of the war be paid immediately by taxation, thus laying the entire cost upon the generation engaged in the conflict; the other, that loans be chiefly relied upon to meet the increasing burden. A compromise between these two policies seemed the most practical way out, as well as being in line with precedent, and was adopted. About one-third of the direct cost of the war was met by immediate taxation, and about two-thirds by loans.

The expenses of the federal government had not been materially increasing in the years before the war. The normal net expenses in 1916 ($674,230,020) were less than $35,000,000 in excess of those of 1910 ($639,502,470). But succeeding years saw a vast increase. In two years the interest charge alone of the federal government had become greater than the entire cost of running the government before the war. The total direct cost of the war to the United States, including the nine and a half billions loaned to the Allies, was about $35,500,000,000, an amount three times the total expenditures of the federal government during the first hundred years of its existence and close to $2,000,000 an hour during the duration of the war. The national debt, which amounted to only $1,000,000,000 before the war, jumped by the end of August, 1919, to the unprecedented total of $26,596,701,648.

AGGREGATE EXPENDITURES AND FOREIGN LOANS OF UNITED STATES GOVERNMENT, FISCAL YEARS 1910-1920.[1]

Year	Normal net expense	Net war cost (Excess above estimated normal expenses)	
		Excess army and navy	Excess interest, pensions, etc.
1917	$ 659,860,650	$ 393,852,949	$ 2,690,164
1918	682,458,285	6,770,295,897	120,952,611
1919	691,858,252	10,917,817,469	379,367,891
1920	826,550,410	1,073,892,747	1,073,392,874
	$7,442,039,118	$19,155,859,062	$1,576,403,540

Year	Net war cost (Excess above estimated normal expenses)		Loans to European governments (less repayments)
	Special war activities	Total war cost	
1917	$ 33,060,510	$ 429,603,623	$ 885,000,000
1918	1,094,994,128	7,986,242,636	4,739,434,750
1919	2,487,710,885	13,784,896,245	3,470,280,265
1920	1,634,695,094	3,781,980,715	350,291,840
	$5,250,460,617	$25,982,723,219	$9,445,006,855

[1] Part of table 3, p. 21, of Rosa, Edward B., *Expenditures and Revenues of the Federal Government* in Annals of the American Academy of Political and Social Science, vol. xcv, no. 184, May, 1921.

To obtain these vast amounts spent at home and lent to the Allies, the government relied chiefly upon loans. Five bond issues were subscribed to by the people in units as low as $50, the first four issues known as Liberty Loans, and the fifth, floated after the armistice, as the Victory Liberty Loan. At the height of war enthusiasm, almost $7,000,000,000 was subscribed in a single loan by over 27,000,000 people. In addition to the Victory and Liberty loans, war saving certificates of five dollars and war saving stamps for twenty-five cents were sold to a total of $1,000,000,000. The grand total of the loans floated in these two ways was close to $22,500,000,000.

UNITED STATES WAR LOANS [1]

Loan	Billions asked	Subscribed	Allotted	Subscribers (approximately)
First Liberty Loan	2	$ 3,035,226,850	$ 1,989,455,550	4,500,000
Second Liberty Loan	3	4,617,532,300	3,807,865,000	9,420,000
Third Liberty Loan	3	4,176,516,850	4,175,650,050	18,376,815
Fourth Liberty Loan	6	6,993,073,250	6,964,581,250	22,177,680
Victory Liberty Loan	4.5	5,249,980,300	4,497,818,750	12,000,000
Total	18.5	$24,072,257,550	$21,435,370,600	

Not alone was large-scale borrowing resorted to, but also the imposition of new and heavier taxes. Contrary to the method pursued during the Civil War, the Democratic Congress, which had reduced the tariff in 1913, refused to consider the import duties as an important source of revenue, and scarcely 5 per cent of the taxes for the war year of 1918 were derived from this source. On the other hand, a comprehensive scheme of taxation was inaugurated. The income tax, levied by virtue of the Sixteenth Amendment in 1913 and increased in 1916, was further increased by the act of October 3, 1917. Personal exemptions were reduced to $2,000 and $1,000 in the case of married and unmarried persons, and the rate graduated from 6 per cent on the first $4,000 above exemption to 67 per cent on incomes of

---

[1] The amounts subscribed and allotted are taken from the *Report of the Secretary of the Treasury, 1920*, tables on pp. 419 and 439.

over a million. In addition, this act inaugurated (1) a war excess profits tax on the incomes of corporations, partnerships, and individuals ranging from 15 to 60 per cent, depending on the amount of capital invested in the business; (2) additional taxes upon liquors, beverages, and tobacco; (3) taxes on luxuries and amusements; (4) war taxes on facilities furnished by public utilities; (5) war taxes on instruments and documents of various kinds; and (6) an increase on estate taxes.[1] The effect of these additions to taxation is shown by a comparison of the amounts raised by taxation immediately preceding and during the war, which were as follows: for the fiscal year ending June 30, 1914, $735,000,000; 1915, $692,000,000; 1916, $779,000,000; 1917, $1,118,000,000; 1918, $4,174,000,000; 1919, $4,648,000,000. Of the amount credited to internal revenue taxes, well over two-thirds was from the income and excess profits taxes. The legacy of these enormous war debts contracted by loans and taxation is destined to remain for many years a serious burden to the American people and a real factor in the cost of living. Of the large foreign war borrowers only England has as yet made a serious effort to liquidate her debts.

War-time Control of Industry and Commerce.—The policy of *laissez-faire* which had been gradually crumbling during the first two decades of the century completely collapsed under the stress of war conditions. Economic life in America was so radically affected by the war even before this country entered the conflict that it was felt necessary for the government to interfere actively in private business; with our entrance, centralized supervision and direction of production and distribution were absolutely essential to effective participation. Government control carried to an extent never before exercised here was effected through federal boards, commissions, or corporations, endowed in some instances with very large powers and sometimes aided by subordinate state or local bodies.

The demand for American products of all kinds, but especially foodstuffs and munitions, was enormously stimulated by the war.

---

[1] Certain of these taxes were reduced and others repealed by acts of February 24, 1917, November 23, 1921, and subsequent legislation.

Exportation, however, was hampered by inadequate shipping facilities. The great German steamship lines were driven from the sea at the very time that German submarines were destroying allied shipping faster than it could be replaced, and that considerable tonnage was being withdrawn from commerce for strictly war purposes. The tonnage of American vessels engaged in foreign trade in 1914 amounted to a little over a million and the lack of a merchant marine was most keenly felt. Government stimulation of shipping began as early as 1914, when the laws regarding registry and other matters were modified, and an act passed creating a Bureau of War Risk Insurance in the Treasury Department, the latter to insure American vessels and cargoes if insurance could not be provided otherwise on reasonable terms.[1] Two years later (September 7, 1916), Congress authorized the appointment of a United States Shipping Board to promote the development of the merchant marine and to regulate shipping. The functions of this board were increased after our entrance in the war and included the supervision of a vast ship-building program, the control of vessels under the jurisdiction of the government, and the training of men for service in the merchant marine. In April, 1917, the United States Emergency Fleet Corporation with a capital of $50,000,000 was organized under the supervision of this board to undertake the construction of merchant vessels. The efforts of the Emergency Fleet Corporation became a most important part of the war program, and through the expenditure of $1,000,000,000 it succeeded not only in building faster than the enemy could destroy, but in raising the tonnage of American vessels engaged in foreign trades from 2,191,000 in 1916 to 11,082,000 in 1921.[2]

The control and integration of economic resources for war purposes was carried out primarily through the Council of National Defense, authorized by an act of August 29, 1916, for the purpose of the coördination of industries and resources for the national security and welfare. The council consisted of the

---

[1] Extended by an act of October 6, 1917, to include compensation to sailors and their dependents in case of death or disability.

[2] See chap. xxv for a more detailed discussion of the merchant marine.

Secretaries of War, Navy, Interior, Agriculture, Commerce, and Labor, who were assisted by an advisory commission of seven experts, and by subordinate committees such as those on shipping, inland waterways, coal production, aircraft production, and munitions. With the expansion of war activities a most important subsidiary to the Council of National Defense was created on July 28, 1917, known as the War Industries Board. This board coöperated with the various committees of the council in regulating and promoting the production and purchase of war materials as well as in the distribution of credit, fuel, materials, and labor. As the War Industries Board was created to deal with problems of manufacture, so the War Trade Board was established a few weeks later to deal with commerce and foreign trade regulations. Supervision and control also seemed necessary in the realms of finance. By an act of April 5, 1918, a War Finance Corporation, with the Secretary of the Treasury as chairman, was created to distribute credit to essential war industries, and a Capital Issues Committee to pass upon proposed issues of stocks and bonds.

As the stock of foodstuffs and fuels declined in the warring nations, the latter came to depend so much upon the United States for these commodities that their production and conservation became a matter of extreme importance. By the Food Control Act passed on August 10, 1917, the government was given power to control "foods, fuels, fuel including fuel-oil and natural gas, and fertilizers and fertilizer ingredients, tools, utensils, implements, machinery, and equipment required for the actual production of foods, feeds, and fuel." It forbade hoarding, willful destruction, discrimination or unfair practices in sale and distribution, and gave the President power under certain conditions to purchase, store, and sell wheat and other commodities. Power was even granted to guarantee the price on prospective crops, and in accordance with this provision $2.26 a bushel for No. 1 northern spring wheat or its equivalent was set by the government. By licensing the manufacture, storage, and distribution of food products, effectual regulations limiting the use of sugar, wheat, meat, butter, and other foods were imposed. The people as a whole were stimulated to self-denial and to the use of substitutes, as well as to the production of greater amounts of foods, so that increasing quantities might be released for European countries.

The widest powers were also conferred in this act upon the President to fix the price and regulate the production and distribution of all kinds of food. The administration of fuel was inaugurated on August 23, 1917, and went to the extent of fixing prices and profits, of adjusting disputes between operators and workmen, of working out priority schedules according to the relative importance of the need, and of developing schemes for more economical production, distribution, and consumption.

One of the most important elements in the development of war-time control was the assumption on the part of the government of control over internal transportation facilities. Although actual government operation of the railroads was but a logical conclusion to a long period of increasing government supervision, it marked the most important step in our history as respects the participation of the federal authorities in private business. The government operation of railroads has already been discussed in Chapter XX. The return of the railroads to their owners under the Esch-Cummins law has by no means brought a final solution of the transportation problem.

Labor During War Time and Reconstruction.—The greatest immediate effect of the war upon labor was to create a shortage which was felt even before our participation. The normal flow of immigration declined from an annual average of about 662,100 during the three years 1912-14 to about 257,887 for the years 1915-18. At the same time thousands of those engaged in industry were called home to serve in the European armies or joined the military forces here. The supply of labor was consequently falling off just at the time when there was an abnormal expansion in industry. Although the gap in the labor ranks was partly filled by women, the shortage persisted, and as a natural consequence the hand of labor was strengthened. Wages were forced up and the increased buying power of labor in turn abnormally stimulated industries which were not essential to the war, but which further increased the labor shortage. While wages advanced rapidly until the early months of 1920, prices rose even more quickly so that it is doubtful if in the long run real wages were bettered by the war. But the rise in the cost of living further stimulated the demand for higher pay, already inevitable because of lack of labor.

Led by Gompers and the Executive Committee, who declared that "this is labor's war," the American Federation of Labor threw itself wholeheartedly into the war, removing restrictions and suspending regulations inimical to efficiency, but demanding that the standard of living be not lowered. Gompers was placed upon the Advisory Committee of National Defense and labor leaders were appointed to most of the war boards organized by the government. Early in 1918 the government created a War Labor Board and a War Labor Policies Board. The latter was to determine the general policies of the government toward hours, wages, and working conditions, and, as far as possible the relations between capital and labor. The War Labor Board was a judicial body to which disputes between employers and employees might be submitted. Some fifteen hundred disputes were adjudicated by this board, which in a number of cases prescribed methods through shop committees or otherwise by which further controversies might be adjusted. Matters of wage adjustment on the railroads were handled, after government control was inaugurated, by a Railroad Wage Commission appointed by the railroad administration. By means of this machinery, by wage increases, and by the loyalty of both leaders and rank and file, labor disturbances were greatly reduced during the war.

The contribution of labor to the winning of the war resulted in the recognition in the peace treaty of certain specific rights for which labor had long contended and the provision for a permanent organization to promote the international regulation of labor conditions. In accordance with the treaty an international labor conference was held in Washington in 1919 which drew up a program to be recommended to the League of Nations. Although the international position of labor was stronger after the war, the prosperity of American labor was decidedly checked and its unity broken after the conclusion of the conflict. The Soviet

Revolution in Russia, the socialist control in Germany, and the increased strength of the British Labor party, all contributed to give a new lease on life to the more radical element of American labor, a revival which was helped by a decided movement toward wage reductions following the armistice, and by the inevitable economic depression which commenced in 1920. Gompers and the leaders of the American Federation of Labor, on the one hand, bitterly condemned economic autocracy and demanded that the gains made during the war be retained, but on the other repudiated radicalism and insisted upon adhering in general to the old conservative policy. Nevertheless, the whole labor movement was restless and considerable discontent prevailed. The railroad workers now supported government ownership with the participation of employees in the management, and have precipitated various significant strikes since 1920. In 1919 there were nine labor disturbances in each of which over 90,000 men were involved. Over 4,000,000 men went out on strike in that year, notably in the building trades, in the iron and steel industry, in the mining of bituminous coal, and among textile workers, as well as numerous minor strikes in other lines. In most cases the post-war strikes were unsuccessful, but they illustrated not only the growing power of the more radical elements in the labor unions, but also the unwillingness of capital to allow the war gains to remain permanent.

Little of a constructive nature was accomplished by either state or federal authorities in settling the labor problem after the war. After prevailing upon the courts to issue in 1919 an injunction (under the terms of the Lever Fuel and Food Control act of 1917) ordering the union officials to recall the strike order in the bituminous coal fields, the government succeeded in patching up a truce in a conference at Washington of operators and coal miners, whereby work was resumed. In the fall of 1919 President Wilson called a National Industrial Conference, consisting of three groups—capital, labor, and the public—to discuss methods of "bettering the whole relationship between capital and labor." The opposition of the group representing capital to collective bargaining and to the right of labor to deal with employers through agents of their own choosing prevented the conference from agreeing on fundamentals, and it dissolved without accomplishing anything. A second labor conference called soon after, and representing industry as a whole rather than economic groups, recommended machinery for industrial conciliation, including national and regional boards, collective bargaining, and "shop committees," but no federal legislation has yet resulted. Kansas in 1920 initiated an interesting experiment by passing a law prohibiting strikes in the basic industries, creating a Court of Industrial Relations to investigate and settle disputes, and empowering the state to take over and operate any of the enumerated industries in which work is suspended in violation of the law. Although reformers expected much from this law, little was accomplished, for important parts were declared unconstitutional by the Supreme Court in 1923.

As prosperity gradually revived after the depression of 1919 and 1920, the labor situation was eased and the discontent lessened. At the same time, however, the natural reaction from the war-time conservatism, as well as the more radical influence of Europe, has caused large numbers of workmen to break away from the traditional political policy of the American Federation of Labor and swell the ranks of the American Labor party and the Farmer-Labor party. The war left American labor more self-conscious and the labor problem more acute.[1]

Expansion of Industry and Commerce.—The increased demand for American products which began with the opening of the European war was, of course, accentuated with the participation of the United States. As Europeans turned from peace-time pursuits the gap in production had to be filled elsewhere, and America served as a source of manufactured war supplies and, as in the days of the Napoleonic wars, of raw materials and foodstuffs. Producers of metals and other minerals were the first to feel the stimulus. The production of iron ore increased from 41,439,000 long tons in 1914 to 75,288,000 in 1917; the production of copper from 1,150,137,000 pounds to 1,886,120,-

---

[1] For a more detailed discussion of American labor, see chap. xxiv.

000 pounds, of zinc from 343,400 short tons to 584,600, of bituminous coal from 422,703,000 short tons to 551,790,000, and petroleum from 265,762,000 barrels to 335,315,000 Agricultural prosperity soon followed as the demand for cotton, wool, leather, and lumber increased. Cotton, which was a drug on the market in 1915 at 8½ cents a pound, rose to an average of 35.9 cents during 1920. The production of wheat, which had been 763,380,000 bushels in 1913, rose to 1,025,800,000 in 1915. Poor crops cut production in 1916 and 1917, but the output in both 1918 and 1919 was well over 900,000,000 bushels. Its selling price rose from 97 cents in 1913 to $2.73 in 1920. The corn crop, which amounted to 2,445,988,000 bushels in 1913, was pushed up to 3,065,233,000 in 1917, a record up to that time. Not only was the production of manufactured commodities and agricultural products stimulated, but many articles hitherto largely purchased abroad were now manufactured here in increased quantities. Of these should be mentioned dyes, potash, chemicals, scientific instruments, optical goods, and toys.

Since we were the largest producers of the chief raw materials and the most convenient source of foodstuffs, foreign purchases from us were extremely heavy during the war and immediately afterward. As a consequence, foreign commerce increased enormously, notwithstanding the activity of the German submarines. For a decade preceding the war American exports had surpassed imports by between $450,000,000 and $500,000,000, a favorable balance of trade which had been offset by the payment of interest and dividends on borrowed capital, by the payment of freight rates to European shippers, and by the expenditures of American travelers. The excess of exports over imports, which had amounted to $470,653,000 in the year ending June 30, 1914, jumped to $1,094,419,000 for 1915, to $2,135,599,000 in 1916, and to $3,630,693,000 in 1917. Exports, as will be seen from the accompanying table, considerably more than tripled between 1914 and 1920. As was natural, the great increase came in munitions and foodstuffs; thus the value of explosives exported rose from $6,272,197 in 1914, to $802,789,437 in 1917; of chemicals, dyes, drugs, etc., from $21,924,337 to $181,028,432; of iron and steel from $251,480,677 to $1,133,746,188; of meat products from $143,261,000 to $353,812,000; and wheat from $87,953,400 to $298,179,705.

FOREIGN TRADE OF THE UNITED STATES, 1914-1921 [1]
(IN MILLIONS OF DOLLARS)

Year *	Exports of domestic merchandise	Imports of merchandise	Excess of exports over imports	Percentage of agricultural exports	Percentage of manufactured exports
1914	2,329.7	1,893.9	435.8	48	47
1915	2,716.2	1,674.2	1,042.0	54	43
1916	4,272.2	2,197.9	2,074.3	36	62
1917	6,227.2	2,659.4	3,567.8	32	66
1918	5,838.7	2,945.7	2,893.0	39	58
1919	7,749.8	3,904.4	3,845.4	53	45
1920	8,080.5	5,278.5	2,802.0	43	52
1921	4,378.9	2,509.1	1,869.8	48	46

* Fiscal years ending June 30 to 1918; thereafter calendar years.

The Depression of 1920 and 1921.—The business boom of the war period with the accompanying inflation in prices and wages was followed in the later months of 1920 and during the entire year of 1921 by a corresponding period of industrial stagnation and deflation. Just as the wave of prosperity had been the result of conditions in Europe so was the deflation primarily the result of foreign influences. Exports and imports, which had reached unheard-of figures in 1918 and 1919, declined radically by 1921. The warring nations, impoverished by the struggle, staggering under crushing debts and with exchange rates decidedly against them, no longer had either the funds or the credit to make extensive purchases. This in itself would have held up sharply American production, but to it was added a "buyers' strike" on the part of the home consumer. The average American who had spent freely during the war now reacted strongly against the high cost of living and began to retrench in preparation for the uncertain future.

With these adverse factors at work the business cycle rapidly pursued its downward swing. Cancellation of orders or failure to book new ones reduced output and closed mills, causing reduced wages and unemployment. The depression, which had com-

[1] *Statistical Abstract*, 1921, table 482, pp. 840, 847, 849.

menced in the production of luxuries, especially silk and certain phases of the rubber and automobile industry, soon became quite general. In only a few industries, notably cotton manufacturing and the mining of bituminous coal, did business continue normal. Upon no group did the depression fall more heavily than upon the farmers. Encouraged by the war prices and the demand for foodstuffs, many had borrowed heavily to purchase land and equipment in order to increase production, only to be caught in a glutted market in which values were declining to a point below the cost of production. Wheat, which had sold for $2.15 a bushel in December, 1919, dropped to $1.443 in December, 1920; corn from $1.347 to $.677; oats from $.715 to $.472; and cotton from $.356 a pound to $.14 during the same period. Labor, which had become accustomed to a higher standard of living, was loath to drop back to previous wage scales, and, where strongly organized, succeeded in maintaining most of what was gained during the war period. Nevertheless, the number of those actively employed declined almost a third during the depression, and the average hourly wage as reported by the National Industrial Conference Board was as follows: July, 1914, 24.3 cents; peak, 1920, 62.1; December, 1921, 48.2. Capital as a whole probably suffered less than any other group, although, as in previous panics, the small business man was hit hard. Mercantile and industrial insolvencies, which had numbered 6,451 with liabilities of $113,291,237 in 1919, increased in 1921 to 19,652 with liabilities of $627,401,883. The depression of 1920 and 1921 differed from previous depressions in that the bank failures were kept at a minimum,[1] owing to the efficient working of the Federal Reserve system, and the deflation, although severe, did not reach the proportions of a panic.

The natural effect of this industrial collapse upon prices was to force them downward. The decline included most commodities, being especially noticeable in foodstuffs and clothing, but least apparent in fuel and rent. The partial cessation of normal building during the war, which caused a shortage of living ac-

---

[1] Nevertheless, banking failures numbered 119 in 1920 with liabilities of $50,708,300; and 383 in 1921 with liabilities of $167,849,555.

commodations, and the high cost of building materials, explain to some extent the failure of rents to decline. The explanation for the continued high cost of fuel is not so easy, but it was due among other factors to increased cost of labor and transportation and the continued high profits taken by capital. According to the United States Department of Labor the average index number of wholesale prices of typical commodities, using the figure for 1913 as 100, was 272 in May, 1920; 148 in June, 1921; 148 in May, 1922; 150 in June, 1922; 156 in December, 1922, and January, 1923; 153 in June, 1923; 151 in December, 1923.

By the end of 1922 deflation had run its course and industry was to a considerable extent adjusted to a peace basis. Increased industry on the part of labor and caution on the part of capital had placed the economic life on a sounder basis and increased the demand for all kinds of commodities. Most of the principal industries were again working at close to capacity, the railroads reported record business, and unemployment had virtually disappeared. Agriculture alone failed to respond to the renewed prosperity and the agricultural unrest continued an uncertain political and economic factor. That and the continued unsettled conditions abroad remained in 1923 as the chief adverse factors in what appeared to be the opening of a period of prosperity and normal conditions.

Results of the War—Problems of Reconstruction.—Although we are still too close to the great catastrophe of the World War to gauge with any degree of accuracy its effect upon the economic life of America, nevertheless it is quite evident that our social and economic problems have been enlarged and complicated rather than lessened by the conflict. In the first place, our international position economically, and consequently politically, was considerably altered. From a debtor nation struggling to keep out of world entanglements and but recently embarked on the road to financial imperialism, the United States emerged from the war the great creditor nation of the world, holding the whip hand financially and apparently determined irrevocably to a career of economic imperialism. The United States, whether it will or no, cannot avoid playing a leading rôle in solving the problems of the economic rehabilitation of the world. Not alone did the war force the nation to assume a larger international position with greater responsibilities, but it has introduced new problems at home and accentuated the old ones. The war left the usual legacy of debt, which still amounted in 1923 to almost $23,000,000,000, a stupendous amount even for a wealthy nation, and one which will necessitate increased taxation for years to come. Expenses of the federal government are now (1924) five times the pre-war figure, a situation brought about to no small extent by the increased appropriations for military purposes. It was hoped by many that the recent conflict was a "war to end war," and the renewed wave of militarism which has followed the conflict is depressing not only to humanitarians, but to the taxpayers of many impoverished nations. While the universal demand is for a reduction of taxation, the need of revenue continues with little abatement.

The inflation and business boom of the war period, as we have seen, was followed by deflation and depression, but industry as a whole has recovered, notwithstanding many adverse factors. The same has not been true with large agricultural groups, and the handling of the agricultural situation in all its ramifications of production, financing, and distribution is a problem, old, to be sure, in American history, but accentuated by the war, and is pressing insistently for solution. Closely connected with agriculture is the problem of transportation, complicated by the war and by new methods of motive power, as well as by an infinite number of old and new questions which must be solved to the satisfaction of capital, labor, and the public. The expenditure and waste of the war have made imperative a more courageous handling of the matter of the conservation of natural resources and the general integration of the resources of production and distribution.

Of all the domestic problems which the war has left for the future to solve, none is more important than that involving the relations between capital, labor, and the consumer as regards the distribution of the wealth produced. Both capital and organized labor were strengthened by the war, whereas the position of the consumer was weakened. The increased cost of living has pressed hard upon the last-named group and at the same time sharpened the demands of organized labor for a greater share in the created wealth. Since the war there have been strikes in many of the basic industries and in many public utilities. An obviously chaotic condition exists. While capital and labor (so called) fight bitterly over the profits of some particular industry, those not connected directly are usually forced to stand by, helpless to protect themselves. This incongruous situation must be remedied, but bettering the position of the public in the warfare between capital and labor is but a beginning of the solution. There yet remains the warfare itself. If the interests of labor and capital are the same, then some means must be discovered to reconcile the two groups. If their interests are fundamentally opposed, the warfare will go on until one or the other is the victor. The whole future of our economic, social, and political life depends chiefly upon the solution of this fundamental problem.

Conclusion.—For three centuries the drama of American history has been unfolding for us. We have seen the precarious settlements of the Atlantic Coast grow into a mighty and wealthy nation. We have seen one generation of frontiersmen after another push their way farther and farther west until they had conquered a continent and left for their children a heritage unsurpassed. We have seen a primitive agricultural people broaden their interests under the stimulus of limitless raw materials into a nation whose economic life has widened into almost every activity. And we have seen an economically dependent people achieve first political, then economic independence, until finally they have assumed a strong economic and political rôle and have become a nation which proved the decisive factor in the greatest of all wars. It has been the history of the opening and exploitation of a region enormously rich in raw materials and overflowing with possibilities. Our people have met the task with confidence, buoyancy, and optimism. A continent has been conquered, but the methods have been crude and wasteful. Much of value has been needlessly squandered and lost forever. Irreparable inroads have been made in our most valuable raw materials. As we have grown rapidly and almost chaotically into a mighty manufacturing nation, population has grown and concentrated in cities, economic groups have become more differentiated, and class feeling stronger. We are now experiencing the problems of the older industrial nations of Europe. Although there is room here for a much greater population and there is infinite wealth still waiting the hand of man, the time has long since arrived for a taking of stock and for a scientific and determined effort to solve the many economic problems pressing upon us.

NOTES FOR FURTHER REFERENCE

Short running accounts of the United States in the World War, with special reference to economics, are F. L. Paxson, *Recent American History* (1921), and I. Lippincott, *Economic Development of the United States* (1921). More detailed studies are to be found in the series, "Problems of War and Reconstruction," edited by F. G. Wickwire and including G. O. Smith, *The Strategy of Minerals* (1919) ; W. F. Willoughby, *Government Organization in War Time and After* (1919), and E. L. Bogart, *War Costs and Their Financing* (1921). See also G. B. Clarkson, *Industrial America in the World War* (1923), and George Creel, *How We Advertised America* (1920).

On the cost of the war, in addition to the work of Bogart cited above, see J. H. Hollander, *War Borrowing* (1919) ; E. L. Bogart, *Direct and Indirect Cost of the Great World War* (1919) ; E. R. A. Seligman, *The Cost of the War and How It Was Met*, Vol. IX, American Economic Review, December, 1919, and especially E. B. Rosa, *Expenditures and Revenues of the Federal Government*, in the Annals of the American Academy of Political and Social Science, Vol. XCV, No. 184, May, 1921.

On labor during the war read Mary Beard, *A Short History of the American Labor Movement* (1920), and Selig Perlman, *A History of Trade Unionism in the United States* (1922), Chaps. 10-15.

Post-war conditions in various phases are further treated in I. Lippincott, *Problems of Reconstruction* (1919) ; W. J. Cunningham, *American Railroads: Government Control and Reconstruction Problems* (1922) ; F. H. Dixon, *Railroads and Government: Their Relations in the United States, 1910-1921* (1922) ; *Inflation and High Prices* in Proceedings of the Academy of Political Science, Vol. IX, No. 1 (1920), and *The Money Problem, ibid.*, Vol. X, No. 2.

A valuable source book is that of J. M. Clark, W. H. Hamilton, and H. G. Moulton, *Readings in the Economics of War* (1918).

SELECTED READINGS

LIPPINCOTT, ISAAC, *Economic Development of the United States*, Chap. XXVII.
LIPPINCOTT, ISAAC, *Problems of Reconstruction*, Chaps. I-VII.
BOGART, E. L., *War Costs and Their Financing*, Chaps. III-V, VII, IX, XIV.
CUNNINGHAM, W. J., *American Railroads: Government Control and Reconstruction*, Chaps. I-XII.
WILLOUGHBY, W. F., *Government Organization of War Time and After*, Chap. XVI.

INDEX

A

Absentee landlordism:
  in early west, 132.
  since Civil War, 415*ff*.
  See also Tenancy.
Accidents, railroad, 472*f*.
Adams, John, 147 note, 160; quoted, 168*f*., 177, 178.
  Charles Francis, Jr., quoted, 460.
  John Quincy, 373.
Adams Express Company, 334.
Adamson Eight-Hour Act, 468, 602, 605.
Africa. *See* Slave trade.
Agricultural credit, 466*ff*.
  under Federal Reserve Act, 436.
  under Federal Farm Loan Act, 436*f*.
  under Agricultural Credits Act, 438.
Agricultural education:
  colleges and schools, 239, 430*ff*.
  experiment station, 440*ff*., 425, 431.
  fairs, 238, 382.
  government aid, 238*f*., 425, 430*ff*., 435*ff*.
  magazines, 238*f*., 431.
  societies, 170, 238, 419*ff*., 425, 431.
  Morrill Act, 239*f*., 430*f*.
Agricultural exports:
  colonial, 106*ff*., 141*ff*.; graph, 150, 152*ff*.
  from the Revolution to the Civil War, 223, 224, 225.
  during Civil War, 382*f*., 391*ff*.
  since Civil War, 689*ff*.
Agricultural labor:
  in the colonies, 67, 71*ff*.
  from the Revolution to the Civil War, 171, 175, 220.
  during the Civil War, 381*f*.
  since the Civil War, 429, 444, 446, 447.
  and machinery, 429.
  and tenancy, 444*f*.
  and the I. W. W., 429.
  See also Slavery.
Agricultural machinery:
  Indian, 60.
  Colonial tools and implements, 62.

from the Revolution to the Civil War, 233*ff*.
during the Civil War, 382.
inventions since the Civil War, 426*ff*.
and labor, 429.
Agricultural products, value of:
  compared with mineral, 16.
  compared with manufactured, 546.
  in 1920, 450.
Agricultural resources, 18*ff*.
Agricultural Revolution:
  and depression preceding American Revolution, 152.
  beginnings in America, 236*ff*., 423*ff*.
  beginnings in England, 236.
Agricultural societies:
  first founded, 170, 238.
  development of, 431.
  granger movement, 419*ff*., 425.
Agricultural unrest:
  in the Critical period, 179*ff*.
  and the Constitution, 181*f*.
  effect of westward expansion on, 216*ff*., 415*ff*.
  and the Greenbackers, 415*ff*.
  and the granger movement, 419*ff*., 425, 462*ff*.
  and the Populist party, 418, 420, 421.
  and the Democratic platform of 1896, 418.
  and the Nonpartisan League, 421, 425, 435.
  and the Farmer-Labor Party, 603, 605.
  and the farm bloc, 432.
  and railroads, 462*ff*.
Agriculture:
  Indian
    achievements of, 57.
    plants cultivated by, 58*f*.
    implements, 60.
    deficiency of, 60.
  Colonial
    in New England colonies, 60*ff*.
    in middle colonies, 64*ff*.
    in southern colonies, 66*ff*.
    in French colonies, 41*f*.
    in Spanish colonies, 38*f*.
    in New Netherlands, 43*f*.

"land butchery" in the colonies, 62, 65, 67, 71.
land tenure in the colonies, 52ff., 63, 68.
livestock in the colonies, 61, 62f., 66, 68, 89.
native plants, 58ff., 61.
plants imported into the colonies, 61.
tools and implements, 62.
scientific farming, unknown, 62, 71.
From Revolution to Civil War:
effect of Revolution on, 168f., 216.
effect on agriculture of westward movement, 216ff.
of invention of cotton gin, 220ff.
of labor-saving machinery, 233ff.
in New England, 232f.
in the south, 220ff.
in middle west, 228ff.
west of the Mississippi, 231f.
scientific farming, 236ff.
During Civil War, 381ff.
Since the Civil War, 401ff.
outstanding factors, 403.
agrarian discontent, 415, 418ff., 423ff., 435.
new machinery, 426ff.
education, 430ff.
government aid, 425, 430ff., 435ff.
irrigation, 438ff.
scientific farming, 440ff.
recent tendencies, 442ff.
in northeast, 444f.
in south, 445ff.
in north central states, 448f.
in far west, 449f.
See also
land tenure,
tenancy,
irrigation,
dry farming,
ranching,
"land butchery,"
farmer's frontier,
public lands,
land speculation.
Department of
government aid previous to founding, 239.
founded, 433, 653.
work of, 431ff.
Aguayo, Marquis of, 137.
Airplane:
progress of, 489.
mail service, 487.
Alabama, 2, 576.

Alamance, Battle of the, 134.
Alaska, 1, 2.
Albany, founded (Fort Orange), 44, 125.
Albany Congress, 128.
Albuquerque, 33.
Alien contract labor, 589.
Aldrich report of 1893, 385f., 606f., 672ff.
Aldrich-Vreeland Act of 1908, 509.
Aliens, in United States
See Immigration.
Alleghany Mountains, 3.
Allen, Ethan, 166.
Aluminum, production of, 16.
Amana Society, 348.
America, discovery of, 33ff.
American Express Company, 334.
American Federation of Labor, 592ff., 684.
"American Husbandry," quoted, 62, 66, 71, 112, 113.
American Multigraph Company, 663.
American Sugar Refining Company, 523, 535.
"American system," 213, 304.
American Telephone and Telegraph Company, 487f., 520.
American Tobacco case, 538.
Holding Company, 524.
Ames, Oakes, 459.
Amusements, 356f.
Anaconda, Montana, 15.
Ancram, N. Y., 62.
Animal life:
of North American continent, 7ff.
See also Livestock.
Animal products, 22f.
Anthracite coal:
early use of, 355.
deposits, 12.
monopoly of, 388, 532.
Anti-Slavery movement:
during Revolution, 164.
after Revolution, 370ff.
Anti-trust:
agitation, 531ff.
laws, 533ff.
Sherman Act, 534ff.
Clayton Act, 540f.
Federal Trade Commission, 540.
movement and Roosevelt, 536.
decisions, 537ff.
See also Combinations; monopoly.
Apaches, 57.
Appalachian Mountains, 2ff.
Arbitration and labor, 597.
Arctic Ocean, 2.
Atchison, Topeka, and Santa Fé Railroad, 457.
Arizona, 405, 563.

INDEX

Arkansas River, 2.
Arkwright, Richard, 220, 269.
Armories, 170f.
Arthur, C. A., 578.
Articles of Confederation, 177, 178, 192.
Asia:
  medieval trade route to, 29ff.
  products of, 27ff.
  relation to aborigines of America, 5.
Asiento Treaty, 39.
Assumption of state debts, 296ff.
Atlantic coast:
  cities of, 3.
  harbors of, 5, 102.
Atlantic coastal plain, 2.
Atlantic Ocean, 2, 5.
  shift of trade routes to, 33.
Atwater, W. O., 441.
Automobiles, 484ff., 549, 550.

B

Bacon, Nathaniel, 133.
Baily, Francis, quoted, 312.
Bakewell, Robert, 71, 236.
Balance of trade, 83, 103.
  and mercantilism, 140f., 155.
Balboa, Vasco de, 34.
Baldwin Locomotive Works, 281, 472.
Baltimore, growth, 341f.
Baltimore and Ohio Railroad, 328.
Bank:
  First National, 542.
  Bankers' Trust Company, 542.
  First United States, 298.
  of Massachusetts, 298.
  of New York, 298.
  of North America, 298.
  National City, 542.
  Second United States, 214, 300f.
Bank note circulation:
  See Currency.
Banking:
  colonial, 155.
  under the First and Second United States Bank, 298ff.
  Independent Treasury system, 301.
  National Banking system, 494f.
  Federal Reserve system, 509ff.
  international, 621ff.
  combinations, 541ff.
  See also Agricultural credit; Bank; Banks; Clearing house; Currency; Panics; "Money trust," Interlocking directorates; Depressions.
Banks:
  colonial, 155f.
  state, 300ff.
  national, 494f.
  "wild cat," 300ff.
  Federal Reserve, 509ff.
  Federal Farm Loan, 436ff.
  failures of, in 1861, 381.
  failures of, in 1919-21, 690.
Bankruptcies:
  in 1861, 381.
  in 1919-21, 690.
  See also Failures.
Baptist church, 354.
Barbary pirates, 178.
Barbed wire, 409.
Barley, production of, 19.
Barter, in colonies, 102f., 155.
Batchelder, Samuel, 274.
Beard, Charles A., quoted, 184, 655.
Beaver:
  furs, 80, 84.
  hats, 93, 144.
Becker, Carl, quoted, 140 note.
Beef Trust, 536.
Bell, Alexander G., 487.
Bellamy, Edward, 533.
Berkeley, Sir John, 51, 64.
Berkeley, Sir William, 69, 133.
Berlin-Bagdad Railway, 31 note.
Berlin Decree, 249ff.
Bernard, Governor, quoted, 98.
Berries, native to America, 8, 61.
Bessemer process, 275, 574.
Best Friend of Charleston, 326.
Beveridge, Albert J., quoted, 626.
Beverley, Robert, cited, 79.
Beverly, Mass., 185, 273, 275.
Biddle, Nicholas, 301.
de Bienville, Céleron, 135.
Big business. *See* Combinations.
Bigelow, Erastus B., 571.
Bimetallism, 500ff.
Bingham, Utah, 15.
Birkbeck, Morris, quoted, 198.
Birkinbine, John, quoted, 575f.
Birmingham, Alabama, 3, 14.
Bisbee, Arizona, 15.
Bishop's Program of Social Reconstruction, 664.
Bituminous coal:
  deposits, 12.
  monopoly of, 532.
Black Ball Line, 256.
Black Friday, 500.
Black Mountains, 3.
Blaine, J. J., quoted, 551, 623.
Blake, Lyman B., 383, 573.
Bland-Allison Act, 501.
Blast furnace, 280f., 284f., 573ff.
Blessing of the Bay, 81.
Block, Captain Adrian, 81.
Blockade:

of American coast in Continental Wars, 252.
of South in Civil War, 394*ff.*
Blockade running, during Revolution, 169.
Boas, Franz, 1 note.
Boats. *See* Shipping.
Bogart, E. L., 75.
Bonanza farms, 449.
Bonds:
flotation of, in Civil War, by the Federal Government, 493.
by the Confederate government, 399*f.*
flotation of, in World War, 680.
Boone, Daniel, 118, 164.
Boots and shoes, 573.
Boston Tea Party, 151.
Bounties:
colonial, 95*ff.*, 105*ff.*, 145*f.*, 169.
state, 171.
Bourgeoisie, influence of Commercial Revolution on, 35.
Bow Canal, 319.
Bowdoin, Governor, 179.
Boxer Rebellion, 628.
Boycott, colonial, 149, 151, 170.
Bradford, Governor William, quoted, 48*f.*
Braintree, Massachusetts, iron foundry in, 91.
Brewing, in colonies, 94.
Bridgeport, Conn., 102.
Bridgewater Canal, 318.
Brindley, James, 271, 318.
Brisbane, Albert, 349.
Britain. *See* England.
Brockton, Mass., 573.
Brook Farm, 350.
Brotherhood of Locomotive Engineers, 589.
Brown, John, 374.
Bryan, William J., 505, 625.
Buchanan, James, 375, 410.
Buck, S. J., quoted, 463.
Buck Stove and Range case, 597, 603.
Building:
colonial, 93*f.*
during Civil War, 389.
materials, 16.
Bullard, F. L., quoted, 599.
Burbank, Luther, 440.
Bureau of Corporations, 540.
Bureau of F o r e i g n and Domestic Commerce, 640.
Bureaus, government, 432*ff.*, 640.
Business depressions. *See* Depressions; Panics.
Business organization. *See* Combinations.

Bute, Lord, 148.
Butte, Montana, 15.
By-products:
of cotton, 565.
of meat packing, 567.
of copper mining, 406, 579.

C

Cabet, Etienne, 350.
Cabot, John, 34.
Cabral, Pedro Alvarez, 33.
Cahokia, 166.
Cairnes, J. F., table quoted, 363.
Calhoun, John C., attitude toward tariff, 304, 372.
California:
early settlement, 137.
acquisition, 209.
discovery of gold in, 210, 454.
products of, 4, 563.
agriculture in, 449.
Callender, Guy S., quoted, 154.
Calvert, Cecil and George, 51, 64, 124.
Canada:
exploration in, 34, 41, 120*f.*, 134.
colonial
under French, 41*ff.*, 85*f.*, 134*f.*
under British, 148, 151.
and refugee Tories, 162.
Canals:
English, 318.
early American, 318*ff.*
era of canal building, 318*ff.*
decline in importance of, 480*ff.*
Canning industry, 563, 566, 567, 568.
Cape Cod Canal, 313, 482.
Capital:
during Civil War, 387*ff.*
consolidation of, Chapter XXII.
and the government, 635*ff.*
foreign investment of, in the United States, 621.
American capital invested abroad, Chapter XXV.
and the New Imperialism, Chapter XXV.
Capitalism, and American Revolution, 161.
Capitalist class, 668*ff.*
Capitalization:
of railroads, 332, 458, 460.
See also Stock-watering.
Capitals, of states,
movement west of, during Revolution, 168.
Caravans, 29*ff.*
Carey Act, 439.
Carlisle Iron Works, 92.
Carlisle, John T., 579.

INDEX 699

Carolinas:
  settlement of, 51.
  agriculture of, 67ff.
  products of, 565.
Carpet and rug manufacture, 558, 571.
Carpini, 32.
Carriages, colonial, 93, 310f.
Carrier, Lyman, 8 note.
Carroll, Charles, of Carrollton, 326.
Carrying trade:
  struggle between England and Holland for, 141.
  British monopoly of, 142.
Cars, railway:
  early, 331.
  improvements in service, 472.
  refrigerator, 410.
  Pullman, 472.
  steel, 472.
  manufacture of, 281, 550.
Carter, James J., 353.
Carteret, Sir George, 51, 64.
Cartier, Jacques, 34, 37, 57; quoted, 58.
Cartwright, Edward, 270.
Catholic Church, 355.
Catskill Mountains, 3.
Cattle:
  colonial, 9, 61, 62f., 66, 68, 89.
  drives, 408f.
  ranges, 68.
  towns, 408.
  Bureau of Animal Industry, 409.
  combating diseases of, 440.
  improvement in breeds of, 237f.
  and the westward movement, 119, 206, 407ff.
Cattle raising:
  influence of climatic adaptability on, 10.
  centers of, 22.
  in the colonies, 9, 61, 62f., 66, 68, 89, 119.
  in the west, 407ff.
Cavaliers, 36.
Central Pacific Railroad, 454.
Champlain, Lake, 314, 320f., 482.
Champlain, Samuel, 57, 86.
Channing, Edward, cited, 111, 339.
Chapman, S. J., quoted, 392.
Charles I, of England, 50, 141.
Charles II, of England, 51, quoted, 133, 144.
Charleston, S. C., decline, 342.
Chartered companies, 46ff., 107.
Charters, Virginia, 46f.
Chartist Movement, 272.
Chatham, Earl of. *See* Pitt, William.
Cheap money and Gresham's Law. *See* Greenback Movement, Bimetallism.

Cheltenham, Icarian village at, 351.
Chemicals, 16, 582.
Cherokee Indians, 165.
Cherry Valley Massacre, 161.
Chesapeake, 249.
  and Ohio Canal, 323f., 481.
Cheney brothers, silk manufacturers, 279.
Chicago:
  development, 198.
  flour milling, 231.
  agricultural machinery. 235.
  meat packing before 1860, 230, 282.
  land speculation before 1840, 219.
  and canal transportation, 321, 329.
  railroads, 329.
  manufactures, 555, 556.
  Haymarket riots, 592.
  Pullman strike, 472.
Child labor, 290ff., 607ff.
China:
  early trade with, 185.
  imperialism in, 627ff.
  "Open Door" policy, 627ff.
  recent trade with, 645ff.
Chinese immigration, 589, 614.
Churchill, Winston, 536.
Church, economic outlook of, 664.
Cincinnati:
  flour manufacture in, 231.
  meat packing in, 199, 282.
  stove manufacture in, 281.
  engine manufacture in, 274.
  and canal transportation, 323.
  and railroads, 329.
  Society of the, 180.
Cities:
  before 1860, 340f.
  since 1860, 584f.
  effect of immigration on, 343.
  effect of transportation on, 308.
  street railways in, 323.
Civil War:
  economic causes of, Chap. XVI.
  economic conditions during, Chap. XVII.
  economic results of, in the north, 381ff.
  economic results of, in the south, 391ff.
  and the merchant marine, 260.
  cost of, 491f.
  financing of, 385, 492ff.
  effect on immigration, 381f.
  increase of public debt due to, 491ff.
  currency legislation during, 385, 491ff.
  and profiteering, 383, 389.
  depression of 1861, 381.

manufacturing in the north during, 383ff.
cost of living during, 385ff.
social background of, 389ff.
rôle of cotton in, 391ff.
and railroads,
building of railroads during, 387ff.
influence on railroads, 452.
railroads in south during, 398.
Clark, George Rogers, 166, 188, 190.
Clark, William. *See* Lewis and Clark Expedition.
Clark, Victor S., cited, 98; quoted, 275, 284f.
Clay, Henry, 191, 214.
and Second United States Bank, 30.
and his American system, 213, 304.
and Compromise Tariff, 306, 372.
imports Herefords, 237.
and Compromise of 1850, 373f.
Clayton Anti-Trust Act, 540, 603, 605, 642, 653.
Clayton-Bulwer Treaty, 629.
Clearing-house certificates, and panic of 1907, 508.
Cleaveland, Moses, 196.
Clermont, 271, 316.
Cleveland, founding of, 196.
Cleveland, Grover,
and trusts, 535.
and gold standard, 501, 504ff.
and tariff policy, 579.
and public lands, 411.
and imperialism, 622, 624.
and conservation, 657.
Climate, of United States, 5ff.
Clinton, General, 162.
Clinton, De Witt, 320, 323, 352.
Clipper ships, 258.
"Clipping" and "sweating" coins, 103.
Clothing industry:
in colonies, 93.
development of ready made clothing, 383, 572.
textiles, 227ff.
Coal:
deposits in United States, 12.
relation of, to steel industry, 3, 14, 280, 574.
production, 622.
See also Anthracite; Bituminous.
Coal industry, monopoly of, 532.
Coal strike of 1919, 686.
Coast line of United States, 5.
Coastal plain:
Atlantic, 2.
Pacific, 4.
Coastwise trade, 245.
Cod fisheries of New England, 87f.

Coinage:
in the colonies, 102ff., 155.
act of 1792, 298ff.
act of 1834, 301f.
act of 1873, 500.
act of 1878, 501f.
act of 1890, 502f.
present coins, 513.
Colbertism. *See* Mercantilism.
Cold storage, 410, 566.
Collective bargaining, 597.
Colleges, 353f.
during Civil War, 389.
Colonial agriculture. *See* Agriculture, colonial.
Colonial commerce. *See* Commerce, colonial.
Colonial industry. *See* Industry, colonial.
Colonial regulations, 38, 41, 43, 46, 95ff., 105ff., 140ff., 170.
See Navigation Acts.
Colonial system:
of England, 46ff., 105ff., 141ff.
of France, 41ff.
of Holland, 43ff.
of Spain, 38ff.
Colonization:
by private enterprise, 46.
by chartered companies, 42, 43, 46.
motives for, 35ff.
English, 46ff.
Dutch, 43ff.
French, 41ff.
Spanish, 38ff.
Colorado, industries of, 563.
Columbia River, 4.
Columbus, Christopher, 33f., 40; quoted, 33, 36.
Combination movement, Chap. XXII.
Combinations:
advantages of, 518ff.
anti-trust laws, 533ff.
banking, 541ff.
Clayton Act, 540f.
foreign markets and, 541.
industrial, 515ff.
interlocking directorates and, 524f., 541ff.
prices and profits, 531.
railroads, 468ff.
Sherman Act, 534ff.
Webb Act, 541.
See also "Community of Interest"; Holding companies; Mergers; Pools; Trusts.
Comet, 258.
Commerce, colonial, 102ff.
physiographic background, 102.
medium of exchange, 102ff.

commercial policy, 105*ff*.; to 1700, 106*ff*.
during 18th century, 109*ff*.
internal and coastwise, 113*f*., 143, 152, 482.
during Revolution, 172*f*.
during critical period, 185*f*., 243*f*.
to Civil War, 262*ff*.
since Civil War, 624*ff*.
assistance to foreign, 639*ff*.
Bureau of Foreign and Domestic, 640.
Constitutional provisions affecting control of, 184.
Department of, 640*ff*.
in American vessels, 243*ff*.
during Napoleonic Wars, 249*ff*.
on Great Lakes since Civil War, 482.
See also Exports; Domestic commerce; Foreign markets; Imports; Enumerated articles; Coastwise Trade.
Commerce Court, 468.
Commerce, domestic, in colonies, 113*f*.
to Civil War, 228*f*., 245, 312, 317, 321, 325.
since Civil War, 482.
Commerce, foreign, 242*ff*., 262*ff*., 634*ff*.
Commerce Reports, 641.
Commercial Crises. See Panics.
Commercial depressions. See Depressions; Panics.
Commercial Revolution, 31*ff*., 34*f*., 141.
Commercial rivalry between states, 178*f*.
Commercial treaties:
with France, 176*f*.
with Prussia, 177.
with Sweden, 177.
Commons, John R., quoted, 613.
Communism:
in the colonies, 48, 49.
before 1860, 347*ff*.
since 1860, 347*ff*.
Communication:
to Civil War, Chap. XIV.
since 1860, Chap. XX.
Community of interest, 470, 524*f*.
Competition:
industrial, 515*ff*.
among railroads, 461.
and the Clayton Act, 540*f*.
See also Combinations.
Compromise of 1850, 374.
Comstock Lode, 384.
Conestoga horses, 315.

Confederate States of America, 391*ff*.
Conference of Governors, 656.
Congress:
Continental, 151, 162, 173, 175.
under Articles of Confederation, 177*f*.
Confiscation, during Revolution, 163.
Connecticut River, 5.
Conservation:
economic, 655*ff*.
industrial, 655*ff*.
of man power, 656.
of natural resources, 411.
National Conservation Commission, 411.
of health, 658.
Constantinople, capture of, by Turks, 32.
Constitution, United States,
struggle for ratification, 180*ff*.
economic factors in, 184.
amendments to, 184, 515, 602.
Constitutions, States, 344*f*.
Continental army, 162.
Continental Congress, First, 151.
Continental Congress, Second, 162, 173, 175.
Continental system, 249*f*.
Cooke, Jay, 455, 500.
Cooper, Anthony Ashley, 51, 124.
Cooperage industry, importance in colonies, 88.
Coöperation:
in colonies, 64, 72.
farmers, 425*f*.
Copper:
resources, 14*f*.
production, 15, 405, 406.
mining of, in colonies, 93.
Cordillera Mountains, 1, 2, 4, 5.
Corn Laws, 106*ff*.
Corn:
importance of, to colonies, 10.
production of, 18.
cultivation of, by Indians, 58.
in south, 228.
in middle west, 229*ff*.
in New England, 233.
Kaffir, 433.
production during Civil War, 382.
since 1870, 621*f*.
Cornbury, Lord, quoted, 143.
Cornell, Ezra, 334.
Coronado, Francisco, 41, 57.
Corporations, 518*ff*.
Cortez, Hernando, 37.
Cost of living:
during Civil War, 385*ff*., 398.
and profits of monopoly, 531.
since Civil War, 606*ff*., 672*ff*.

Cotton:
  adaptability to south, 10.
  colonial production, 90f.
  effect of Industrial Revolution on, 220f.
  effect of invention of cotton gin on, 169f., 223ff.
  and westward movement, 224f.
  growth of production during Civil War, 391ff.
  culture under slavery, Chap. XVI.
  production since Civil War, 445ff., 622.
  by-products of, 565.
  value of crop, 20, 223f.
  Sea Island, 20, 221.
Cotton gin, 169f., 221ff.
Cotton manufacture, 275f., 565, 568.
Council of National Defense, 682ff.
Coxe, Tench, 99.
Credit Mobilier, 459, 500.
Creeks, 189, 191.
Crime of '73, 500.
Crimean War, 259.
Crises. *See* Panics.
Critical period:
  economic reorganization during, 175ff.
  weakness of central government during, 178ff.
  chaos in currency during, 179.
  struggle for the constitution, 180ff.
  revival of prosperity, 185.
  agriculture during, 216.
Crompton, Samuel, 220, 270.
Cromwell, Oliver, 141f.
Crusades, influence on medieval commerce, 29.
Cuba, 622, 624ff.
Cultivator, 427.
Cumberland Road, 314ff.
Cumberland Valley, 3.
Cunard steamship line, 257.
Currency:
  colonial, 102ff., 147f., 154f.
  in Revolution, 173f.
  under Articles of Confederation, 179f.
  in Constitution, 184.
  establishment of national system, 298ff.
  Coinage Act of 1834, 301f.
  Coinage Act of 1873, 500.
  Greenbacks, 497ff.
  National Bank Act of 1863, 494f., 508.
  Bland-Allison Act, 501f.
  Sherman Silver Purchase Act, 502f.
  Gold Standard Act of 1900, 508.
  Federal Reserve Act, 509ff.

Cutler, Manasseh, 194.
Cycle, business. *See* Depressions; Panics.

D

Dairy products:
  centres of, 22.
  in New England, 233.
  since Civil War, 445, 449.
Dakota, 405, 406, 415, 448.
Dale, Sir Thomas, 48, 66.
Dams, water-power, 23.
Danbury Hatters' case, 597.
Danish West Indies, purchased, 632.
Dartmouth College case, 462, 518 note.
Davis, John, 34, 37.
Dawes Act, 414.
Deane, Silas, quoted, 243.
De Bow, J. D. B., cited, 361f.
Debs, Eugene V., 598, 603, 604.
Debt, National:
  Revolutionary, 174f.
  funding of, 297f.
  since 1860, 492ff.
  increase due to World War, 678ff.
Debtors, 497ff.
Declaratory Act, 149.
Deere, John, 234.
Delaware and Chesapeake Canal, 322.
Delaware and Raritan Canal, 322.
Delaware Bay, settled by Dutch, 44.
Delaware River, 5.
Democracy:
  in Dutch colonies, 45.
  in England, 272.
  in Revolution, 163f.
  in formative period, 344ff.
Democratic party, platform of 1896, 418ff.
Department of Agriculture, 433.
Department of Commerce, 640.
Department of Labor, 640.
Depreciation, during Revolution, 174f.
Depressions, economic:
  preceding Revolution, 140, 148, 149, 152.
  agricultural, 416ff.
  of 1815-19, 304.
  of 1837, 302.
  of 1867, 303.
  of 1861, 381.
  of 1873, 499f., **652.**
  of 1893, 503f.
  of 1907, 507f.
  of 1914, 678.
  of 1920-21, 689ff.
Derby, Elias Hasket, **173, 244.**
Desert Land Act, 438.
De Tocqueville, Alexis, quoted, 587.

INDEX

Detroit, 329.
Dew, Thomas R., quoted, 361, 363.
Dewey, Davis R., quoted, 493.
De Witt Clinton, 327.
Diaz, Denis and Bartholomew, 33.
Diaz, Porfirio, 632.
Dingley Tariff, 6, 580.
Discontent:
 agricultural, 179, 181, 216, 462*ff*.
 industrial. *See* Labor Agitation.
Discovery. *See* Exploration.
Dismal Swamp Canal, 318*f*.
Distilling, 94, 230*f*.
Dodd, W. E., quoted, 70.
Domestic commerce. *See* Commerce, Domestic.
Dominican Republic, 631*f*.
Dongan, Governor, 86.
Douglas, Stephen A., 374, quoted, 624.
Drake, E. L., 384, 525.
Drake, Sir Francis, 36, 104.
Dred Scott decision, 374.
Drew, Daniel, 460.
Drugs, in medieval commerce, 28.
Dry farming, 441*f*.
Dunbar, Seymour, quoted, 326.
Dutch colonies, 43*ff*.
Dutch East-India Company, 43, 85.
Dutch West-India Company, 43*f*., 89.
Dutch settlers, 157.
Dyestuffs:
 importance in medieval commerce, 28.
 after World War, 581.

E

Eannes, Gil, 33.
Earle, E. M., 31 note.
East India Company, 245.
Eastern trade, in medieval commerce, 27*ff*.
Economic background of colonization, 27*ff*.
Economic depressions. *See* Depressions; Panics.
Edible plants, native in America, 8.
Education:
 agricultural, 238*f*.
 industrial, 430*f*.
 Morrill Act, 430*f*.
 colonial, 351.
 in formative period, 351*ff*.
 High Schools, 353.
 Colleges, 353.
Edward II., of England, 141.
Egberts, Egbert, 571.
Eggleston, J. C., quoted, 399.
Electric power, 23.
Electric railways, 484.

Electrical apparatus and machinery, 552, 577.
Electrical industries, 577.
Elizabeth, of England, 73, 141.
Ely, R. T., quoted, 286, 290, 592.
Emancipation, during Revolution, 164.
Embargo, 197, 223, 226, 249*ff*.
Emergency Tariff, 581.
Emigration. *See* Immigration.
Empress of China, 185, 244.
Encomienda, 40*ff*.
England:
 colonial policy of, 95*ff*., 105*ff*., 141*ff*.
 colonization by, 46*ff*.
 contest with France for America, 134*f*., 148.
 renewed imperialism after 1763, 148.
 and American Revolution, Chap. VIII.
 relations, during critical period, 176*ff*.
 industrial revolution, 269.
 relations with United States, during Napoleonic Wars, 246*ff*.
 War of 1812, 251*ff*.
 relations with United States, during Civil War, 260, 382, 391*ff*.
 merchant marine of, 257.
 as a market for American products, 102.
 "enumerated" articles, 97, 142*ff*., 148.
Entail, 53, 163.
Erie Canal, 318, 319*ff*., 480.
Erie Railroad, 328, 388, 460, 480, 504.
Esch-Cummins Act, 477*f*., 684.
Evans, Oliver, 274, 316, 325.
Excess profits tax, 681.
Expediting Act, 467, 536.
Exploration:
 inventions aiding, 27.
 motives for, 31*ff*.
Explorations:
 Dutch, 43*ff*.
 English, 34, 46*ff*.
 French, 34, 41*ff*.
 Portuguese, 32*ff*.
 Spanish, 33*ff*., 38*ff*., 136*f*.
Exports:
 colonial, 106*ff*., 141*ff*., graph, 150, 152*f*.
 See also Colonial Commerce; Colonial Regulations; Enumerated Articles; Mercantilism; Raw Materials.
 during critical period, 423*f*.
 until Civil War, 244*ff*.
 during Civil War, 382, 391*ff*., 394*f*.
 since 1860, 634*ff*.
 during World War, 688*f*.
 See also Commerce.
Expositions:

exhibition of American farm machinery, at Paris in 1855, 236.
Express, 333f.

F

Factory system:
  colonial beginnings, 89.
  introduction of, 272ff.
  development, 272ff.
Failures:
  bank, 381, 690.
  canal, 324f.
  railroads, 303, 499f., 503f.
  in 1861, 381.
  in 1919-1921, 690.
Fairfield, Conn., 161.
Fairs, 238, 382.
Fall line, cities of, 3.
Fall River, Mass., 102.
Faneuil, Peter, 75 note.
Far East, trade with, 185, 645ff.
Farmers' elevators, 426.
Farmer-Labor party, 603, 605.
Farmer's movements. See Agricultural Unrest.
Farming. See Agriculture.
Farming, scientific:
  unknown in colonies, 62, 71.
  introduction of, 236ff., 440ff.
Farm implements:
  Indian, 60.
  colonial, 62.
  to Civil War, 233ff.
Farm Loan Associations, 437.
Farm Loan Banks, 436ff.
Farm machinery, 233ff., 426ff.
Farm mortgages, 415ff., 424, 437ff., 445f.
Farnsworth, Ephraim, 334.
Federal Council of the Churches of Christ in America, 665.
Federal Farm Loan Banks, 436ff.
Federal Reserve Banking system, 436, 509f.
Federal Trade Commission, 540.
Federation of Labor: See American Federation of Labor; Labor Organizations.
Ferdinand and Isabella of Spain, 33.
Feudalism, 27.
  In Spanish America, 40ff.
  in French colonies, 42ff.
  in New Netherlands, 44ff.
  in English colonies, 51ff.
  and Baron Stiegel, 94.
  and American Revolution, 161, 163.
Field, Cyrus, 639.
Finance:
  and Constitution 184.

Hamilton's plan, 296ff.
Financing:
  of the Revolution, 173ff.
  of Civil War, 691ff.
  of World War, 678ff.
Financial crises. See Depressions; Panics.
First United States Bank, 298.
Fisheries:
  before settlement of United States, 9, 37.
  colonial, 10, 86ff.
  effect of the Revolution on, 176.
  development of, 262.
  present importance of, 22f.
Fitch, John, 186, 316.
Fite, E. D.:
  cited, 381, 382, 386.
  quoted, 392, 394.
Flax, 90.
Flint, Timothy, quoted, 202.
Florida, purchase of, 619.
Flour milling:
  westward extension, 231.
  roller process, 449.
Flying Cloud, 257.
Food:
  English policy as to import, 102.
  during Civil War, 382, 391ff.
  development of prepared food, 563, 566ff.
  Pure Food Law, 537, 653.
  control during World War, 683.
Food resources of America, 7ff.
Ford, Thomas, quoted, 219.
Fordney-McCumber Tariff, 581.
Foreign markets, and combinations, 541.
Foreign trade. See Commerce.
Foreigners. See Immigration.
Forest:
  resources, 16ff.
  exportation during colonial period, 80ff.
  conservations of, 411.
Fort Stanwix, 138.
Fortune, 80.
Fourier, Charles, 349f.
Fourteenth Amendment, 515, 601ff.
France:
  colonial system, 41ff.
  contest with England for America, 134ff.
  controversy during Napoleonic Wars, 247f.
  treaty, 177.
  See Colonial System.
Franklin, projected state of, 133.
Franklin, Benjamin, 154; quoted on slavery, 370.

INDEX

Free silver question. *See* Bimetallism; Currency.
Freight rates, 461, 473.
Freight, traffic, 470*ff*.
  See also Canals; Railroads; Rebates.
French and Indian War, 135, 148, 157, 190.
French Revolution, 162.
Frobisher, Sir Martin, 34, 37.
Frontier:
  the first, 121*ff*.
  life on, 205*ff*.
  influence on American character, 117*f*.
  towns, 123, 125.
  and American Revolution, 157, 164*ff*.
  farmers, 410*ff*.
  mining, 403*ff*.
  ranchers, 407*ff*.
  See Westward Movement.
Fruits:
  native to America, 8, 61.
  imported, 61*f*.
  raising of, 445, 449.
Frye, W. P., 642.
Fuel control, 683.
Fulton, Robert, 316*f*., 257, 271.
Funding Act of 1790, 175.
Fur trade:
  impetus to colonization, 37.
  importance to colonists, 83*ff*.
  English rivalry with France, 85*f*.
  influence on westward movement, 85*f*.
  French, 41*ff*., 85*f*.
  Dutch, 44, 85*f*.
  in Plymouth, 49.
  in 18th century, 108.
Furniture industry, colonial, 94*f*.

G

Gadsden Purchase, 209, 619.
Gallatin, Albert, report on internal improvements, 313*f*.
da Gama, Vasco, 33, quoted, 36.
Garrison, William L., 372.
Gas, illuminating, 355.
Gaspée, 151.
Gates, General, 66.
Genoa, Italy, 29*ff*.
Geographical discoveries. *See* Explorations.
Geographical influences on occupations, 10*f*.
George, Henry, 533.
George III., of England, 145.
Georgia:
  settlement, 52.
  western land claims, 191.
  forest products, 80, 83, 564*f*.
Germans:
  in colonies, 66, 72, 90, 124, 131, 157.
  and westward movement, 129*ff*., 166, 200.
Gibbons *v.* Ogden, 317.
Gilbert, Sir Humphrey, 46; quoted, 36.
Glass, colonial industry, 94.
Globe, Arizona, 15.
Gloucester, Mass., 102.
Gloves, localization of industry, 558.
Gold:
  in Spanish colonies, 39.
  search for, as a motive for colonization, 37.
  discovery of, in California, 210, 454.
  discovery of, in Colorado, 403*f*.
  production of, 15*f*.
  discovery of, in Alaska, 507.
  See also Currency.
Gold reserve, 501*ff*.
Gold standard, 506.
Golden Gate, 4.
Gompers, Samuel, 596, 603, 605, 685.
Gorges, Sir Fernando, 50, 51.
Gosnold, Bartholomew, 9.
Gould, Jay, 460.
Gould system of railroads, 470.
  strikes in, 592.
Government aid. *See* Subsidy.
Government ownership of railroads, 474*ff*.
Government regulation of business, 652*ff*.
  of industry, 652*ff*.
  of railroads, 465*ff*., 473*ff*.
Grain:
  elevators, 426, 464.
  exports, 239*ff*.
  transportation, 421*ff*.
  production, 448.
Granger movement, 419*ff*., 425, 462*ff*., 652.
Grand Turk, 185, 244.
Gray, Robert, 246.
Great Barrington, Mass., 92.
Great Britain. *See* England.
Great Lakes, 2, 4, 5, 482.
Great Northern railroad, 455*f*., 470.
Great Western, 257, 271.
Great War. *See* World War.
Greenback movement and party, 415*ff*., 497*ff*., 604.
Greenbacks:
  issue, 493*ff*.
  redeemed, 501*f*.
  See Currency.
Gregory Lode, 384.

H

Grenville, George, 148.
Guild system, of middle ages, 35, 38.
Gulf of Mexico, 2, 5.
Gulf states, 3, 5.
Gulf stream, 2.

H

Hamilton, Alexander, 147 note, 154; quoted, 169.
  assumption of states debts, 296ff.
  and national debt, 296ff.
  and United States bank, 298.
  report on currency, 297.
  and tariff, 304.
Hamilton, Lt.-Gov., 166.
Hammond, M. B., table quoted, 362, 396, cited, 391.
Hancock, John, 147 note, 160.
Haney, L. H., quoted, 518.
Hannay, David, quoted, 105.
Harbors, of United States, 5, 102.
Hard money. See Bimetallism.
Hard times. See Depressions; Panics.
Hargreaves, James, 220, 269, 270.
Harnden, W. F., 333.
Harrison, William Henry, 197.
Harriman system of railways, 470.
Harvesting machinery, 235ff.
Hat industry:
  colonial restrictions on, 144.
  labor and, 597.
Hatch Act, 425, 431, 441.
Haverhill, Mass., 125, 573.
Hawaiian Islands, 624.
Hawkins, Sir John, 36.
Hay, production of, 19.
Hay, John, "Open-Door" policy, 628f.
Hay-Herran treaty, 629.
Hay-Pauncefote Treaty, 629.
Hay-Varilla Treaty, 630.
Haymarket Riot, 592.
Hayti, 631f.
Haywood, William D., 598.
"Head-right," 52, 68.
Helper, H. R., quoted, 376.
Hemp, 227.
Henderson, Richard, 165.
Henry, Patrick, 161 note, 166; quoted, 181f., 366.
Henry VII., of England, 34, 141.
Hepburn, A. Barton, quoted, 495.
Hepburn Act, 467f., 653.
Hepburn v. Griswold, 497.
Herkimer, General, 166.
Hertz, Heinrich, 639.
Hides, 407ff.
High finance. See Speculation; Stock-watering.
High schools, 353.

Highways. See Roads, Internal Improvements.
Hill, James J., 455.
Hill system of railroads, 470.
Hogs, raising of, 9, 22, 61, 66, 68, 230, 382, 448.
Holding companies, 470, 523f.
Holland. See Netherlands, Dutch.
Hollidaysburg, Pa., 322.
Holmes, Justice, 601.
Home industries:
  in the colonies, 63, 89, 90, 93, 94.
  during Revolution, 170.
Homestead Act, 217, 382.
Hooker, Thomas, 36.
Hoosier element, 200.
Hoover, Herbert C., quoted, 666, 667.
Horses, 9, 22, 61, 62f., 66, 68, 237f.
Hosiery, 571.
Household manufactures, Chap. IV., 546, 566.
Houston Ship Canal, 482.
Howe, General, 162.
Howe, Elias, 383.
Hudson Bay, 2.
Hudson, Henry, 34, 43, 57.
Hudson River, 5.
  early steamboats on, 316f.
Hudson Valley, settled by Dutch, 44.
Hughes, Charles E., 629.
Huguenots, 125.
Hunter, Governor, 129.
Huntington, Collis P., 455.
Hussey, Obed, 235.
Hutchinson, Anne, banishment of, 36.
Hydro-electric development, 23, 557.

I

Icarian communities, 351.
Idaho, 405f., 563.
Illinois:
  settlement, 189ff.
  agriculture in, 228f.
  manufactures in, 561.
Illinois Central railroad, 388.
Illinois Valley, French in, 43.
Immigrants, effect of American climate on, 1, 6f.
Immigration:
  causes, 342f., 613.
  Colonial, 129ff.
  from 1820 to 1860, 342ff.
  during Civil War, 343.
  recent, 611ff.
  Chinese, 589, 614.
  Scotch-Irish, 344, 613.
  Irish, 343, 613.
  Japanese, 614.
  German, 343, 613.

INDEX

Scandinavian, 344.
Jewish, 613.
from southern and eastern Europe, 613.
and labor, 612*ff*.
restriction on, 614.
and land grant railroads, 413.
Imperialism, 618*ff*.
and Commercial Revolution, 35.
British, 148.
Old Imperialism, 618*f*.
New Imperialism, 619*ff*.
United States and New Imperialism, 621.
from Monroe Doctrine to Spanish American War, 622*ff*.
from inauguration of "Open Door" policy to acquisition of Virgin Islands, 627*ff*.
technique of, 633*f*.
and missionaries, 621.
Imports:
colonial, 106*ff*., 142*ff*.; graph 150, 152*f*.
during Revolution, 172*ff*.
from 1789 to 1860, 248, 250, 262*ff*.
since Civil War, 634*ff*.
Impressment of seamen, 247.
Income tax, 493, 580, 654.
Indentured servants, 71, 72 note, 73.
Independent treasury, 301.
Indiana, manufacturing, 562.
Indian Wars, 85, 127, 128, 137*f*., 189*ff*., 413*f*.
Indians:
agriculture of, 57*ff*.
industry of, 80 note.
fur trade of, 84*ff*.
and westward movement, 126*ff*., 137*f*., 189*ff*.
and settlement of far west, 413*f*.
Dawes Act, 414.
treatment by Spanish, 40*ff*.
and imperialism, 619.
Indigo:
in medieval commerce, 28.
introduction of, 69.
bounty on, 145.
decline of industry, 169.
Industrial accidents, 610*f*.
on railroads, 472*f*.
Industrial combinations. *See* Combinations; Consolidation.
Industrial Revolution:
and depressions preceding American Revolution, 151*ff*.
in England, 269*ff*.
in America, Chap. XII.
and labor problem, 287*ff*., 609.

and agriculture, 220*f*.
and imperialism, 619*ff*.
Industrial Workers of the World, 598*ff*.
and agricultural labor, 429.
Industrial Democracy, 661*ff*.
Industry:
colonial, 78*ff*.
beginnings of, 78.
Government aid, 95*ff*.
Government restriction, 97.
extent of, 98*ff*.
during Revolution, 170*ff*.
from 1789-1860, Chap. XII.
after Civil War, Chap. XXIII.
during World War, Chap. XXVII.
adjustments after World War, 691*ff*.
See also Building; Fisheries; Furniture Industry; Fur Trade; Glass; Iron and Steel Industry; Liquor; Lumbering; Milling; Shipping Industry; Textile Industries.
Inheritance laws, 163.
Inland Waterways. *See* Internal Improvements, Canals.
Inter-colonial trade. *See* Commerce, Inter-colonial.
Interlocking directorates, 524*f*.
and banks, 541*ff*.
and railroads, 470.
Internal improvements:
colonial, 309*ff*.
Gallatin's plan for, 313*f*.
before Civil War, 229*f*., 311*ff*.
since Civil War, 452*ff*.
See also Canals; Harbors; Rivers; Roads.
Interstate-Commerce Act, 466*ff*., 653.
Interstate Commerce Commission, 466*ff*.
Injunctions, 602*f*.
Inventions:
cotton gin, 169*f*., 221*ff*.
sewing machine and shoe machinery, 275, 573.
textile machinery, 269*ff*.
agricultural machinery, 233*ff*.
steam engine, 274, 326.
in transportation, 326, 329*ff*., 471*ff*.
in iron and steel manufactures, 275, 574*ff*.
telegraph, 334.
telephone, 487.
radio, 488.
wireless, 488.
Iowa, manufacturing, 562.
Ipswich, Mass., 194.
Irish, in colonies, 124.
Iron:

deposits, 13f.
resources, 14.
manufacturing, 280ff., 573ff.
Iron and steel industry, 14.
colonial, 91ff., 144f.
during Revolution, 170f.
before Civil War, 280ff.
since Civil War, 573ff., 622.
Iroquois Indians, 41, 57, 86, 127, 134.
Irrigation:
in colonial Pennsylvania, 65.
Carey Act, 439.
Reclamation Act, 439.
Isthmian Canal, 629ff.
Italy:
and medieval trade, 29ff.
emigration, 613.

J

Jackson, Andrew:
and United States Bank, 214, 301ff.
and Specie Circular, 303.
victories over Indians, 197.
and nullification, 372.
James I., of England, 67, 141.
James River, 5.
Jamestown, Virginia, settled, 47f., 66f.
Japan, 624, 627.
Japanese immigration, 615.
Jay, John, 178.
Jay Treaty (1795), 189, 247.
Jefferson, Thomas, quoted, 163 note;
164, 191, 192, 208, 213, 225, 237, 352; quoted on slavery, 370; on Missouri question, 372.
Jenks, Joseph, 92.
Jenks, J. W., quoted, 531; cited, 534.
Jerome, Arizona, 15.
Jewelry, 553, 556.
Jews. *See* Jewish Immigration, 613.
Johnson, Captain Edward, quoted, 62.
Johnson, Sir John, estate confiscated, 163.
Johnstown, N. Y., 558.
Jones, J. B., quoted, 398.
Jones Act of 1916, 627.
Joplin region, zinc in, 15.
Josselyn, John, 9.
Judd, Orange, 441.

K

Kaffir Corn, 433, 450.
Kalm, Peter, quoted, 65.
Kanawha, 188.
Kansas, 401, 408, 414.
Kansas Court of Industrial Relations, 687.
Kansas-Nebraska Act, 374.

Kaskaskia, 166.
Kay, John, 270.
Keating-Owen Bill, 602.
Kellogg, Royal S., quoted, 18 and note.
Kelly, Florence, quoted, note 601.
Kendrick, John, 246.
Kentucky:
settlement, 165.
enters Union, 190.
hemp growing in, 227.
tobacco growing in, 226.
horse breeding in, 237.
cattle breeding in, 228.
Keokuk, Iowa, dam, 23.
Kerosene, 13, 384.
Kidnaping, of labor, for colonies, 73.
King, Wilford I., quoted, 24 note;
cited, 671.
King John II., of Portugal, 33.
King Philip's War, 85, 127.
Knickerbocker Trust Company, 507.
Knight, V. E. C., case, 535.
Knights of Labor, 589ff.
Knit goods. *See* Textiles.
Knox *v.* Lee, 497.
Knox, William, 629.

L

Labor:
colonial, 71ff.
in northern colonies, 71, 72.
in middle colonies, 72.
in southern colonies, 67, 74.
during Revolution, 171f.
during critical period, 175f.
and industrial Revolution, 287ff., 586ff.
during Civil War, 385ff., 588f.
during World War, 684ff.
problems after World War, 692f.
migration of, to cities, 340f., 584ff.
and railroads, 468.
and public lands, 413.
and the courts, 600ff.
and politics, 587f., 603ff.
and immigration, 611ff.
Labor legislation:
Federal, 600ff.
Ten-Hour Act, 587.
importation of contract labor, 589.
in the Clayton Act, 603.
Adamson Act, 468.
Esch-Cummins Act, 477.
labor organizations and, 292, Chap. XXIV.
state, 600ff., 611.
Labor, child. *See* Child Labor.
Labor, of women, 288ff., 607ff.
Labor movement:

INDEX

causes of, 292f., 585f.
growth of, 292f., 586ff.
Labor organization:
  before 1860, 292f., 586ff.
  during Civil War., 588f.
  railroad brotherhoods, 589.
  National Labor Union, 589.
  Knights of St. Crispin, 590.
  "Mollie Maguires," 590 note.
  Knights of Labor, 590ff.
  American Federation of Labor, 593ff.
  I. W. W., 598f.
  policies of:
    on arbitration, 597.
    on boycotts, 597.
    closed shop, 597.
    strikes, 597.
    union label, 597.
    trade agreements, 597.
    criticisms of, 599f.
Labor leaders, 595f.
Labor Reform party, 603f.
Labor saving machinery, 233ff., 426ff.
La Follette, Robert, 471.
La Follette Seamen's Act, 246, 642f.
Laissez-faire:
  golden age of, 515.
  decline of, 652ff.
Lakes-to-the-Gulf Deep Waterway, 482.
Lambert, John, quoted, 150.
Lancaster, Pa., 90, 94, 311.
"Land butchery":
  in the colonies, 62, 65, 67, 71.
  to Civil War, 217.
Land grants, to railroads, 333f., 457f.
Land speculation:
  in colonies, 132.
  to Civil War, 218ff.
  since Civil War, 411ff.
Land tenure:
  in the colonies, 52ff., 63, 68.
  See Public Lands; Tenancy.
Lands, public. *See* Public Lands.
Lane, John, 234.
Large scale production:
  in agriculture, 449.
  in industry, Chap. XXIII.
Lawson, Thomas, 536.
Lead production, 15.
Leather manufacture:
  in the colonies, 93.
Lee, Ann, 347.
Legal tender, in colonies, 155.
Legal tender notes. *See* Currency.
de Leon, Ponce, 41.
Lenox, Mass., 92.
Leonard, Henry, 92.
Leopard, 249.

Levant trade, in medieval commerce, 30f.
Lewis and Clark Expedition, 208.
Lightner, O. C., quoted, 666.
Liliuokalani, Queen, 624.
Lincoln, Abraham, 374.
Linen, colonial, 90.
Liquor:
  manufacture of, in colonial New England, 94.
  and West Indian trade, 94.
  and African trade, 108ff.
  consumption of, 94.
  excise taxes on, 230f.
Lisbon, 33.
Literacy test, 615.
Little, Rev. Daniel, 171.
Livestock, 22.
  in colonies, 9, 61, 62f., 66, 68, 89, 119.
  colonial bounties, 61.
  improvements of, 237f.
  treatment of, 62f., 68.
  Bureau of Animal Industry, 409.
  value of, in southern states, 1860, 228.
Livingston, Philip, 92.
Livingston, Robert, 237.
Lloyd, Henry Demarest, 522, 523, quoted, 532.
Localization of industry, 282f., 555ff.
Lochner *v.* New York, 602.
Locke, John, 51, 124.
Loco-foco party, 587.
London Company, 46ff., 89.
Long Island, settled by Dutch, 44.
Looms:
  hand, 269ff.
  power, 269ff.
Lottery, government, during Revolution, 175.
Louis XIV. of France, 134.
Louisiana:
  French in, 43.
  purchase, 190f., 371f., 618.
  cane sugar in, 70, 226f.
  lumbering in, 564.
Louisville, Ky., 274, 282, 324.
Lowell, Mass., 283, 341.
Lowell, F. C., 274, 276.
Loyalists. *See* Tories.
Lucas, Eliza, 67f.
Lumber industry, 17f.
Lumbering:
  in the colonies, 80f.
  westward extension of, 564.
  in the south, 564ff.
Lybyer, A. H., quoted, 32.
Lynn, Mass.:

colonial iron works in, 91, 144f.
shoe manufacture in, 573.

M

Macadam, 271.
McAdoo, W. G., 475, 477.
McCormick, Cyrus, 235f.
McCulloch, Hugh, 498.
McKay, Gordon, 383.
MacKenzie River, 2.
McKinley Tariff, 502, 503, 579.
McKinley, William, 507.
  and trusts, 535.
  and imperialism, 624, 625.
McMaster, J. B., quoted, 198, 199, 250, 187; cited, 312.
McNeill, G. E., quoted, 590.
Machinery:
  agricultural, 233ff., 426ff.
  textile:
    prohibition of exports from England, 272f.
    effecting Industrial Revolution, 269ff.
  shoe, 573.
  steam, 270ff.
  electric, 554.
  water power, 28.
  interchangeable parts, 274.
  standardization, 553.
  characteristics of American, 234.
  See also Inventions.
Macon Bill, 251.
Madison, James, 251.
Magellan, Ferdinand, 34.
Maine, unsuccessful settlement, 47; settled, 50f.
Maine, battleship, 625.
Mail, 357, 486f.
Maize. *See* Corn.
Manhattan Island:
  trading port on, 43.
  settled, 44.
Mann, Horace, 352.
Mann-Elkins Act, 468, 653.
Manning, William, 235.
Manufacturing:
  colonial, 143ff.
    lumbering, 80f.
    shipbuilding, 10, 81ff., 142ff.
    naval stores, 83, 145.
    hats, 144.
    brewing, 94, 108.
    textile, 143f.
    iron, 144f.
  colonial aid and restrictions, 38, 41, 43, 46, 95ff., 105ff., 140ff., 145.
  extent of colonial, 96.
  during Revolution, 170ff.

  from Revolution to Civil War, 266ff.
    textiles, 275ff.
    metals, 280ff.
    distribution of, 282ff.
  during Civil War, 388ff., 397.
  after Civil War,
    growth of, 546ff.
    causes of, 549ff.
    characteristics of American, 532ff.
    localization of, 556ff.
    in northeast, 559ff.
    in middle west, 561ff.
    in south, 564ff.
    in far west, 562.
    of food, 566ff.
    of clothing, 569ff.
    of iron and steel, 573ff.
    and tariffs, 577ff.
Marblehead, Mass., 88.
Marconi, Guiglielmo, 639.
Marietta, Ohio, founded, 194.
Marion, Francis, 166.
Market:
  colonies as market for mother country, 37f.
  market for cotton, 220f.
Marquette, Father, 118.
Marshall, Chief Justice, 213, 301, 462, 518, 520.
Martineau, Harriet, quoted, 289f.
Marvin, W. L., quoted, 246, 253.
Maryland:
  settlement, 51.
  colonial agriculture, 64ff.
Mason, Captain John, 50, 51.
Mason, William, 274.
Massachusetts:
  settlement of, 48ff.
  colonial,
    agriculture, 60ff.
    industry, 78ff.
    commerce, 102ff.
  in the Revolution, 172f.
  agriculture, 232f., 444.
  beginnings of factory system in, 89.
  child labor in, 290ff., 608.
  compulsory education, 608.
Massachusetts Bay Company, 50.
Mauch-Chunk, Pa., 326.
Meat packing industry, 230, 410, 567.
Mechanics' Free Press, quoted, 291.
Medieval commerce, 27ff.
Mediums of exchange:
  colonial, 103, 155.
  See Currency; Coinage.
Mediterranean Sea, in medieval commerce, 29ff.
Menendez, Pedro, 41.

INDEX

Mennonites, 36, 124.
Mercantilism, 97f., 105ff., 140ff., 146ff.
 advantages and disadvantages of, 146f., 170.
Merchant marine:
 colonial, 81ff.
 during Revolution, 172ff.
 from Revolution to Civil War, 242ff.
 since Civil War, 260f., 642ff.
 the World War and after, 644f.
Merchant Marine Act of 1920, 645.
Mergers, 469, 524.
Merrimac River, 283, 319.
Metals. *See* Minerals.
Methodist Church, 354.
Mexican War, 373, 378, 619, 623.
Mexico, 632f.
Miami and Erie Canal, 323.
Michigan:
 agriculture in, 228ff.
 manufacturing, 562.
Middle Atlantic states:
 physiography, 11.
 agriculture in,
  colonial, 64f.
  since Civil War, 444.
Middle passage, 74f., 75 note.
Middlesex Canal, 319.
Milan Decree, 249ff.
Millennial Church, 347.
Milling, 231, 449.
Mineral products, value of, compared with agricultural, 16.
Mineral resources, 12ff.
Minerals. *See* Coal; Iron; Copper; Silver; Gold; Petroleum; Building Materials; Aluminum; Chemicals; Lead; Zinc.
Mining, and settlement of the west, 403ff.
Minimum wage acts, 610.
Minneapolis, Minn., 555, 556.
Minnesota:
 agriculture in, 238.
 regulation of railroads, 462.
Mint, colonial, 103.
Missionary enterprise, 621.
Mississippi:
 settlement, 201ff.
 cotton raising in, 234.
Mississippi River, 2, 4, 5, 118, 208.
 closing of, by Spain, 189f.
 traffic on, before Civil War, 224, 317.
 shrinkage of traffic since Civil War, 480f.
Mississippi Valley, 2, 4, 5.
 exploration of, 118, 208.

 conflicting international interests in, 189ff.
Missouri:
 settlement, 201ff.
 and slavery, 371.
 hemp growing in, 237.
 manufacturing, 562.
Missouri Compromise, 371ff.
Missouri River, 2, 4.
Mitchell, D. W., quoted, 218f.
Mitchel, John, 595ff.
Mitchel, W. C., quoted, 667, 668.
Mobile, Alabama, 226.
Mohawk Valley:
 settled by Dutch, 44, 125.
 agriculture in, 64, 65.
 control of, 86.
Molasses Act, 110, 111, 145f., 152.
Molasses, bounty on, 145.
Money. *See* Currency.
Money Trust, 541ff.
Monopolies:
 colonial,
  granted to trading companies, 39, 44, 107.
  French government, 43.
  Dutch, 44.
 of slave trade, 75.
 of tobacco, 145.
 railroad, 468ff.
 shoe machinery, 573.
 See also Combinations.
Monroe Doctrine, 622ff.
Montana, 405, 408, 415, 563.
Moody, John, quoted, 541.
Moore, John Bassett, quoted, 622.
Moravians, 36.
Morenci, Arizona, 15.
Morgan, J. P., 505, 541, 542.
Morgan system of railroads, 470.
Mormons, 351, 354, 403.
Morrill Act, 239, 390, 430f.
Morrill Tariff, 492, 577f.
Morris, Robert, 160, 196.
Morris Canal, 322.
Morse, S. F. B., 334.
Motor cars, 484ff.
Motor traffic, 484ff.
Muck-rakers, 536.
Mules, raising of, 22, 237.
Mulhall, M. G., cited, 547.
Munn *v.* Illinois, 464.

N

Nails, first cold cut, 171.
Nantucket, Mass., 88.
Napoleon:
 and Jefferson, 250.
 Louisiana Purchase, 190f.

Napoleonic Wars:
  effect of Continental System on American commerce, 249ff.
  English Orders in Council, 249.
  Berlin and Milan decrees, 249ff.
  Embargo and Non-intercourse, 249ff.
  American Neutrality, 247ff.
Narvaez, P. de, 41.
National Bank Act, 494f., 508.
National Bureau of Economic Research, quoted, 668ff.
National debt. *See* Debt, national.
National Labor Union, 589.
National Monetary Commission, 599.
Natural resources:
  extent of, 11ff.
  influence of, 7ff.
  conservation of, 655ff.
Natural gas, 355.
Naval stores:
  in colonial industry, 80, 83.
  bounties on, 145.
Navigation acts:
  English, 95ff., 105ff., 111, 141ff., 171;
  after Revolution, 176f.
  French, 43.
  Spanish, 38f., 106f.
  See also Mercantilism; Enumerated Articles; Commerce.
Nebraska, 408, 414, 415.
Negroes:
  free to Civil War, 104, 366.
  as tenant farmers in south, 445ff.
  See also Slavery.
Nelson, Admiral, 185.
Netherlands:
  influence of Commercial Revolution on, 35.
  explorations, 43f.
  commercial system of, 43f.
  colonial wars with England, 46.
  colonization of, 43f.
  treaty with, 177.
  See also Colonial System, Dutch.
Neutrality in Napoleonic Wars. *See* Rule of War of 1756.
Nevada, 404ff., 563.
New Amsterdam, 43, 46.
New Bedford, Mass., 88, 102, 268.
Newbold, Charles, 234.
Newburyport, Mass., 172f.
New England:
  physiography, 10f.
  colonization, 48ff.
  layout of villages, 64.
  colonial commerce, Chap. V.
  colonial agriculture, 60ff.
  colonial industry, 78ff.
  during Revolution, 158ff.
  attitude on War of 1812, 252.
  rise of manufacturing, 272ff.
  products of, 559f.
  agriculture to Civil War, 232f.
  after Civil War, 444f.
New England Confederation, 128.
New Englands First Fruits, quoted, 82.
Newfoundland, 1.
New Hampshire, 50.
New Harmony, 347.
New Haven, Conn., 102, 341.
New Jersey:
  settlement, 51.
  colonial agriculture, 64ff.
  colonial iron mining, 92.
  products of, 560f.
New London, Conn., 102.
New Mexico, 405, 415, 563.
New Netherlands, 44ff.
New Orleans, 197, 317.
New Orleans:
  founded, 43.
  importance of, to trade, 199, 224.
  trade of, 224.
Newport, Captain Christopher, 78.
Newport, R. I., 168.
Newspapers, 356.
New York Barge Canal, 482.
New York Central Railroad, 328.
New York City:
  during colonial period, 46.
  growth of, 341.
  shoe manufacture in, 573.
  water supply, 356.
  transportation, 356.
New York, New Haven & Hartford, 309, 471.
New York State:
  settlement, 43ff.
  colonial history, 43ff.
  during Revolution, 163, 164, 168.
  and Erie Canal, 318, 319ff., 480.
  and westward movement, 123, 125, 129, 196, 197, 200.
  products of, 560.
Newlands Act, 468.
Nicaragua, 631.
Nichols, J. B., 573.
Niles, Nathaniel, 170.
Non-importation agreements, 149, 151
Non-intercourse acts, 149, 197, 251.
Nonpartisan League, 421, 425, 435.
North Carolina:
  colonial agriculture, 68f.
  tobacco culture in, 226.
  rise of manufacturing, 565.
North Central States:
  agriculture in, 228ff., 448ff.
  manufacture in, 561f.

INDEX

713

North Dakota, 405, 408, 414, 415.
Northern Pacific Railroad, 455, 457, 470, 500, 504.
Northern Securities Company, 470, 537.
Northwest Ordinance, 191*ff.*, 370*f.*
Northwest Territory:
   annexed to Quebec by British, 151, 188.
   Proclamation of 1763, 138, 156*f.*, 164.
   settlement of, 164*f.*, 192, 194*ff.*
Norwalk, Conn., 161.
Noyes, John Humphrey, 348.
Nullification of South Carolina, 305*f.*

O

Oats, production of, 19, 228, 231.
Ocean freight, Chap. XI.
Ocean shipping, 81*ff.*, 172*ff.*, 242*ff.*, 260*f.*, 642*ff.*, 644*f.*
Oglethorpe, James, 136.
Ohio:
   settlement, 195*f.*
   manufactures, 562.
   agriculture, 228*ff.*, 448.
Ohio Canal, 323.
Ohio Company, 194*ff.*
"Ohio idea," 498.
Ohio valley, settlement of, 194*ff.*
Oil. *See* Petroleum.
Oil fields, 13.
Oklahoma, 415.
"Old Northwest," 164*ff.*, 194*ff.*, 379.
"Old Southwest," 201*ff.*
Oliver, James, 234.
Olmsted, Frederick Law, cited, 363.
Omaha, Neb., 410, 556.
de Oñate, Juan, 136.
Oneida Community, 348.
Onrest, 81.
Opechancanough's War, 127.
Open hearth process of steel making, 575.
Open door policy, 627*ff.*
Orders in Council, 249.
Ordinance of 1787. *See* Northwest Ordinance.
Oregon, acquisition of, 209; products of, 562*f.*
Oriental trade, Middle Ages, 27*ff.*
Osgood, H. L., quoted, 123.
Ottoman Turks, 32*f.*
Owen, Robert, 348*f.*

P

Pacific Ocean, 1, 2, 624*f.*
Pacific States, 4; agriculture in, 449*f.*

Packet lines, 256*ff.*
Paish, Sir George, cited, 620.
Panama Canal, 4, 630*ff.*
Panama, Isthmus of, 1, 2.
Pan-American Congress, 623.
Panics:
   1837, 302.
   1857, 303.
   1873, 499*f.*
   1893, 503*f.*
   1907, 507*f.*
   1920, 689*f.*
   See also Depressions.
Paper money:
   during colonial period, 103*f.*, 140, 154*f.*
   during Revolution, 173*ff.*
   during critical period, 179*f.*
   during Civil War, 385, 493*ff.*
   since Civil War, Chap. XXI.
   See also Currency.
Papin, Denis, 270.
Patents, 554.
Patent office, and agriculture, 238*f.*
Paternalism. *See* Government Aid; Regulation; Subsidy.
Paterson, N. J., 279, 556, 599.
Patroons, 45*f.*, 123.
Pawtucket, R. I., 185, 273.
Payne-Aldrich Tariff, 580.
Peace. *See* Treaties.
Peat, 12 note.
Peck, J. M., quoted, 206.
Peckham, Sir George, 37*f.*
Peik *v.* Chicago and Northwestern Railway Company, 464.
Penn, William, 51, 64, 124, 127.
Pennsylvania:
   resources of, 3.
   settlement of, 51.
   colonial agriculture, 64*ff.*
   colonial industries, 90, 92, 94, 98, 99.
   colonial commerce during Revolution, 163, 164, 168, 171, 173.
   manufacturing, 560.
Pennsylvania Canal, 321*f.*
Pennsylvania Railroad, 328, 388, 470.
Peonage in the south, 445*ff.*
Pepper, in medieval commerce, 27.
Pepperell estate, confiscated, 163.
Perry, Commodore, 624.
Petroleum:
   resources, 13.
   growth of industry, 622.
Peru, 37.
Philadelphia and Reading, 503, 504.
Philadelphia, Pa.:
   growth and importance, 341.
   water supply, 355.
   transportation, 356.

Philippine Act of 1902, 627.
Philippine Islands, 625ff.
 and Philippine Act of 1902, 627.
 and Jones Act, 627.
Philipse Manor, confiscated, 163.
Philips, Ulrich B., quoted, 360.
Piedmont Region:
 defined, 127.
 advance into, 128ff.
 men of, in Revolution, 166f.
Pilgrims, 48ff., 60.
Pinchot, Gifford, 656.
Pinckney, C. C., 190, 204.
Pine Tree shillings, 103.
Pioneers. See Westward movement.
Piracy, colonial, 104f.
Pitt, Hiram and John, 236.
Pitt, William, 148.
Pittsburgh, Pa., 3, 14, 341.
Pizarro, Francisco, 37.
Plantation system, 47ff., 67, 224ff.
Platforms, political, 415ff., 533, 539, 604.
Platt Amendment, 626.
Plows:
 colonial, 62.
 to Civil War, 234f.
Plymouth Company, 46ff.
Plymouth, Mass.:
 founded, 48.
 agriculture in, 60.
Polk, James K., 373.
Pollock v. Farmers' Loan and Trust Company, 580.
Polo, Marco, Maffeo, Nicolo, 32.
Pontiac, 137, 157.
Pools:
 railroad, 461, 469, 522.
 industrial, 522.
"Poor whites," 202, 564.
Population:
 during colonial period, 66, 108;
 slave, 110; 125ff., 157, 168.
 in "Old Northwest," 198ff., 228, 340.
 in "Old Southwest," 202ff., 340.
 in formative period, 338f.
 concentration in cities, 340f.
Population, center of:
 in 1830, 117.
 in formative period, 338f.
Populist party, 418, 420, 421.
Pork packing, 230, 410.
Portages, 120f.
Port Royal, 103.
Portland, Maine, 102.
Porto Rico, 625.
Portsmouth, N. H., 102.
Portuguese, explorations, 32ff.
Postal service, 356, 486.
Post Office, 356, 486f.

Potash, colonial manufacture of, 80.
Potatoes:
 cultivation, 20f.
 origin of, 59.
 in New York, 65.
Potter Law, 462.
Poultry:
 center of industry, 22.
 production, 9.
Powderly, T. V., 590f.
Precious stones, in medieval commerce, 28.
Presbyterian Church, 354.
Prester John, 36.
Prices:
 colonial, 82, 84.
 decline in, after Revolution, 175, 234.
 and monopolies, 5, 31.
 since Civil War, 605ff., 672ff.
Primogeniture, 53, 163.
Prince Henry, the Navigator, 32, 36.
Prince Rupert, 86.
Printing, in colonies, 95.
Privateering:
 colonial, 104f.
 during Revolution, 161, 170, 172f.
Proclamation of 1763, 138, 156, 164.
Profits:
 and monopoly, 5, 31.
 and business cycle, 668.
Progressive party, 421.
Propeller, screw, 257.
Proprietary colonies, 51.
Protection. See Tariffs.
Providence, R. I., 102, 123, 268.
Provincetown, Mass., 88.
Prussia, treaty with, 177.
Public lands:
 policy toward, to 1862, 191ff., 216f.;
 since 1862, 410ff.
 effect on agriculture, 216ff.
 effect on industry, 267.
 effect on labor, 288.
 and railroads, 333f., 413, 457f.
 effect on immigration, 413.
 and land grant railroads, 413.
 and land grant colleges, 430f.
 See also Morrill Act.
Public Lands Commission, 411.
Public utilities. See
 government ownership.
 government regulation,
 railroads,
 street railways,
Pueblo Indians, 57.
Puget Sound, 4.
Pujo Committee, 541ff.
Pullman cars, 472.
Pure Food Law, 537, 653.

INDEX

Q

Quakers, 36, 124.
Quartering Act, 149, 151.
Quebec:
  founded, 43.
  expeditions against, 103.
Quincy railroad, 326.
Quitrents, 52*f*., 68, 163.

R

Rabbeno, Ugo, quoted, 246.
Railroad brotherhoods, 589, 595, 596.
Railroad Labor Board, 478.
Railroads:
  significance of, 325.
  early, 326*ff*.
  problems of early building, 329*ff*.
  financing of, 332, 460.
  street railways, 333, 389.
  during Civil War, 388.
  development of, after Civil War, 452*ff*.
  transcontinental, 412*f*., 454*ff*.
  government aid to, 412*f*., 456*ff*.
  chaotic conditions and early abuses, 458*ff*.
  Granger movement and, 462*ff*.
  Granger cases and, 464.
  Interstate Commerce Act, 465*f*.
  Expediting Act, 467.
  Hepburn Act, 467.
  Mann-Elkins Act, 468.
  Newlands Act, 468.
  Adamson Act, 468.
  Transportation Act of 1920, 477*f*.
  consolidation of, 468*ff*.
  improvements of, 471*f*.
  rates and fares of, 473.
  and World War, 474*ff*.
  during reconstruction, 478*f*.
  capitalization, 332, 458, 460.
  stock-watering of, 460.
Rainfall, of United States, 6.
Raleigh, Sir Walter, 46.
Ranching:
  in Spanish America, 40*f*.
  in far west, 407*ff*.
Randolph, John, quoted, 305.
Rapp, George, 348.
Raw materials:
  America as a source for, 37.
  place of, in mercantile system, 141, 147.
Reapers, 42*f*., 235*f*.
Rebates, 461.
Reclamation Act, 439.
Reconstruction of the South, 424, 445*ff*.
  problems after World War, 690*ff*.

Red River, 2.
"Redemptioners," 73*f*.
Reformation, and motives for colonization, 36.
Refrigeration, 410, 566.
Regulation of industry. *See* Government regulation.
Regulators, 134.
Religion:
  a motive for colonization, 36.
  during revolution, 354.
  during formative period, 354*f*.
Religious qualifications, 164.
Renaissance, 27.
Repudiation of debts, 325.
Resources. *See* Natural resources.
Resumption of specie payments, 498, 501*f*.
Revolutionary War:
  economic causes of, 140*ff*.
  class divisions in, 160*ff*.
  financing of, 163, 173*ff*.
  economic conditions during, 162*ff*.
  frontier advance during, 164*ff*.
  effect on agriculture, 168.
  effect on industrial life, 170*ff*.
  commerce and privateering during, 172*ff*.
  economic reorganization after, 175*ff*.
Rhode Island:
  settlement, 51, 123.
  manufactures, 273, 275, 283.
Rhodes, J. F., quoted, 359, 396, 399.
Rice:
  production of, 19, 227.
  introduction of, 69.
  production during Revolution, 169.
Richard II., of England, 141.
Ripley, George Z., cited, 522.
River and harbor improvements, 317, 482.
Rivers:
  transportation, 317.
  navigable in United States, 5, 480.
Roads:
  colonial, 309*ff*.
  turnpikes and toll, 311*ff*.
  Gallatin's plan for, 313*f*.
  motor traffic and, 484.
  See Internal Improvements; Cumberland Road.
Robbstown, Pa., 198.
Robertson, James, 164*f*., 188, 190.
Rockefeller, John D., 525, 541.
Rocket, 271, 325.
Rocky Mountains, 4.
Rolfe, John, 66.
Roller process, in flour manufacturing, 449.
Roosevelt, Theodore:

and conservation movement, 411, 656f.
and the trusts, 536.
and panic of 1907, 508.
and railroads, 467.
and Panama Canal, 629f.
Rope. *See* Hemp.
Rosa, E. B., table quoted, 679.
Rotation of crops, 217.
Rowley, Mass., founded, 89.
Royal African Company, 75.
Rule of War of 1756, 247.
Rum:
  colonial manufacture, 94.
  and African slave trade, 108ff.
Rural life, 233, 428.
  and electric railways, 483.
  and automobiles, 485.
Russia, in China, 627f.
Russian Revolution of 1917, 162, 685.
Rutter, Thomas, 92.
Rye, protection of, 19, 228, 231.
Ryle, John, 279.

S

Sabotage, 599.
Sailing vessels, 81ff., Chap. XI.
St. Augustine, 41.
St. Lawrence River, 2, 5.
St. Louis, Mo.:
  founded, 43.
  milling in, 231, 237.
  packing in, 410.
  shoe manufacture in, 573.
  commercial importance, 341.
St. Mary's Falls Canal, 324, 481, 482.
St. Paul, Minn., 588.
Salem, Mass., 93, 103, 172, 268.
"Salutary neglect," 115, 147, 148.
Samoan Islands, 624.
Sandys, Sir Edwin, 48, 79.
San Francisco, founded, 137.
Santa Fé trail, 336.
Sault Ste. Marie, 324, 481, 482.
Savannah, Ga., decline, 342.
Savannah, 257.
Savary, James, 270.
Saw mills. *See* Lumbering.
Scherer, J. A. B., cited, 391; quoted, 392.
Schools. *See* Education.
Schwab, J. C., quoted, 400.
Scientific management, 658ff.
Scioto company, 196.
Scotch-Irish:
  in colonies, 90, 124, 157.
  and westward movement, 128ff., 166, 200.
Seaman, E. Z., table quoted, 369.

Seamen, 245f.
Searight, T. B., quoted, 315.
Second Hundred Years' War, 104, **134**, 618.
Second United States Bank, 300f.
Sectionalism, 377ff.
Seignories, 42ff.
Seligman, E. R. A., cited, 175.
Seneca Chief, 320.
Serfs, in Spanish America, 40ff.
Servants, indentured, 71ff.
Seven Years' War, 135, 148.
Sevier, John, 164, 188, 190.
Sewing machine, 281, 383.
Shaftsbury, Earl of, 51, **124**.
Shakers, 347.
Shaler, N. S., 4 note.
Shays' Rebellion, 168, **179**.
Sheep:
  importation of, 9.
  raising of, 22.
  bounties on, 61.
  in Pennsylvania, 66.
  growth of sheep raising, 89, 237.
  merino, 237.
Sheldon, George, quoted, 132.
Shenandoah valley, 3.
Sherman Anti-Trust Act, 534ff., 653.
Sherman Silver Purchase Act, 502f.
Ship building:
  colonial, 10, 81ff., 142.
  during Revolution, 171.
  from Revolution to Civil War, Chap. XI.
  during World War, 642ff., 682.
Shipping industry:
  colonial, 10, 81ff., 142ff.
  during Revolution, 242f.
  period of uncertainty, 1781-89, 243f.
  first period of growth and prosperity, 1789-1810, 244ff.
  War of 1812, 251ff.
  reaction from war, 253.
  second period of growth and prosperity, 1820-30, 254ff.
  period of overproduction and gradual decline, 1830-60, 256ff.
  effect of Civil War, 260f.
  decline after Civil War, 642ff.
  effect of World War, 682.
Shipping board, 682.
"Shoddy aristocracy," **383**.
Shoe industry, 573.
Shop committees, 662f.
Silk industry:
  experiments in, 69.
  founding of, 279.
  growth of, 572.
Silver:
  Comstock Lode, **16**.

INDEX 717

influence upon politics, 416*ff*., 500*ff*.
production of, 16.
decline in value, 500*ff*.
See Currency.
Silver Purchase Act, 502*f*.
Simons, A. M., quoted, 160*f*.
Sinclair, Upton, 536.
Sirius, 257.
Slater, Samuel, 185.
Slave Coast, 74*f*.
Slave population:
  colonial, 110*f*.
  from Revolution to Civil War, 361*ff*.
Slave trade:
  and the Asiento, 39.
  monopoly of, 75.
  colonial, 108*ff*., 110, 146.
  later, 362.
  internal, 363.
Slavery:
  introduction of, 74.
  during colonial period, 73*ff*.
  and British crown, 156.
  effect of cotton gin on, 223*f*.
  effect of western land on, 224.
  effect on southern land, 268*f*.
  effect on southern economic life, 368*f*., 375*f*.
  effect on southern morals, 369.
  cause of Civil War, Chap. XVI.
Smith, Capt. John, 107, 118;
  quoted, 36, 79, 84, 87.
Smuggling, 39.
  colonial, 105, 142*ff*., 147 note, 151.
  during Revolution, 172*f*.
Social life, Chap. XV.
Soils:
  of United States, 6.
  Bureau of, 434.
Soley, J. R., quoted, 254.
Soper, E. K., 12 note.
de Soto, Ferdinand, 41, 57.
South:
  and tariff, 304*ff*., 372.
  cotton raising in, 220*ff*.
  slavery in. *See* Slavery.
  manufactures before Civil War, 375*ff*., 383*f*.
  during Civil War, 397.
  agriculture before Civil War, 220*ff*.
  agriculture since Civil War, 445*ff*.
  manufacturing since Civil War, 564*ff*.
  railroads in, before Civil War, 328.
  railroads in, during Civil War, 398.
  See also Slavery; Cotton; Civil War, Lumbering.
South America, trade with, 363, 648*f*.
South Carolina:
  settlement, 51.

agriculture, 224, 227.
and nullification, 304*ff*., 372.
and manufacturing, 565.
South Dakota, 405*ff*.
Southern colonies:
  agriculture in, 66*ff*.
  labor in, 71*ff*.
Southern Pacific, 456*f*.
South Improvement Company, 526*f*.
Spain:
  colonial system, 38*ff*.
  possessions of after Revolution, 189.
  purchase of Florida from, 204.
  Treaty of San Lorenzo, 190, 204.
  Treaty of San Ildefonso, 190.
  war with, 1898, 621*f*.
Spanish-American War, 621*ff*.
Spanish Main, 39 note, 104.
Specie Circular of 1836, 303.
Specie payment, 498, 501*f*.
Speculation in railroads, 458*ff*.
Spices:
  importance of, in medieval commerce, 27*ff*.
  as a motive in exploration, 36.
Spinning inventions, 320, 570*f*.
Spotswood, Governor, 91.
Squanto, 60.
Stagecoach travel, 309*ff*.
Stamp Act, 149, 153.
Standard of living, 668*ff*.
Standardization of parts, 247, 553.
Standard Oil Company:
  and the railroads, 461.
  and consolidation, 522*ff*.
Stanford, Leland, 455.
State banks, 300*ff*., 494.
State:
  rivalries after Revolution, 178*ff*.
  debts, 297, 324.
  aid to internal improvements, 318*ff*., 324, 482.
  attempt to control railroads, 262*ff*.
  attempts to control monopolies, 533.
Steam engines, 270*ff*., 326.
Steam locomotives, 326.
Steamboats:
  invention of, 271, 316.
  on the ocean, 257*f*.
  on the Great Lakes, 317.
  on the western rivers, 317.
  national control of navigation, 317.
Steel:
  Bessemer process, 275, 574.
  increase and production, 280*f*., 573*ff*.
  open hearth process, 575.
  influence on railroads, 472.
Steffens, Lincoln, 536.
Stephenson, George, 271, 325.

Stevens, John, 274, 316, 325.
Stevens, Uriah S., 590.
Stiegel, Baron, 94.
Stock-watering of railroads, 460.
Stourbridge Lion, 326.
Stoves, 281.
Stowe, Harriet Beecher, 274.
Strasser, Adolph, 596.
Street railways, 333, 389, 483*ff*.
Strikes:
  coal strike of 1902, 596.
  Lawrence strike of 1912, 599.
  Paterson strike of 1913, 599.
  of 1919, 686.
  coal strike of 1919, 686.
Stuyvesant, Peter, 46.
Subsidy:
  English subsidy of colonial industry, 95*ff*., 145.
  French subsidy during Revolution, 174.
  of canals, 324.
  of railroads, 332*f*., 456*ff*.
  See Land Grants.
Sugar:
  colonial trade in, 107, 108, 110*f*.
  introduction in Louisiana, 70.
  production of sugar cane, 19, 226*f*.
  sugar beets, 20.
  Sugar Act of 1764, 111, 148*f*.
Sugar trust, 523, 535.
Sumter, General, 166.
Supreme Court:
  and water traffic, 317.
  and railroads, 462*ff*.
  and labor, 600*ff*.
Sweden, treaty with, 177.
Swedes, immigration of, 66, 123, 157.
Swine. *See* Hogs.
Sylvis, W. H., 589.
Symesbury, Conn., 93.
Symmes, Judge John Cleves, 196.

T

Taft, William H., 537.
Tanneries, 93.
Tarbell, Ida M., 536.
Tariff:
  state tariff acts, 177.
  of 1789, 296.
  of 1816, 304.
  of 1824, 305.
  of 1828, 305.
  of 1832, 305*f*.
  of 1833, 306.
  of 1842, 306.
  of 1846, 306.
  of 1857, 306.
  Civil War tariffs, 577*f*.
  of 1883, 579.
  of 1890, 579.
  of 1894, 580.
  of 1897, 580.
  of 1909, 580.
  of 1913, 581, 654.
  of 1920, 581.
  of 1922, 581*ff*.
Tariff Commission, 581.
Taunton, Mass., 91.
Taussig, F. W., quoted, 578.
Taxation:
  colonial, 140, 153*f*.
  during Revolution, 173.
  under Articles of Confederation, 178.
  in the Constitution, 184.
  during Civil War, 492*f*.
  during World War, 680*f*.
Taylor, Frederick W., 659*f*.
Telegraph, 334*f*.
Telephone, 487*ff*.
Telford, Thomas, 271.
Temperature, 16.
Tenancy:
  in early west, 131.
  since 1880, 444.
  in the south, 445*ff*.
Tennessee:
  settlement, 165*f*.
  enters Union, 190.
  products of, 576.
Tennessee valley, 3.
Territorial expansion:
  Florida, 204*f*.
  Louisiana, 190*f*.
  by Mexican War, 209.
  Oregon, 209.
  since 1898, 621*ff*.
  influenced by railroads, 209.
Texas:
  settlement, 136*f*.
  freedom from Mexico, 232, 619.
  enters United States, 232.
  products of, 232.
Textile industries:
  in colonies, 88*ff*.
  laws governing, 95*ff*.
  development before Civil War, 272*ff*.
  recent development, 568*ff*.
  in the south, 565.
Thresher, 236.
Timber. *See* Lumbering.
Tobacco:
  cultivation by Indians, 58*f*.
  colonial importance, 10.
  production, 1790, 169.
  decline after Revolution, 225*f*.
  exports of, 225.

INDEX

westward extension, 225f.
  value of crop, 20.
Toll bridges, 312.
Toll roads, 311ff.
Tom Thumb, 326.
Tories, 160ff., 164, 176, 180.
Townshend, Viscount Charles, 71, 148, 149, 236.
Townshend Acts, 148ff., 153.
Trade routes:
  medieval, 28ff.
  modern, importance of in World War, 31 note.
Trade unions. *See* Labor; Labor Legislation; Labor Movement; Labor Leaders.
Transcontinental railroads, 454.
Trans-Missouri Freight case, 469, 538.
Transportation:
  in the colonies, 309ff.
  by turnpikes, 311ff.
  by rivers, 316f.
  by canals, 318ff.
  by early railroads, 325ff.
  by street railways, 333.
  by automobile, 484f.
  See also Internal Improvements.
Transportation Act of 1920, 477f.
Transylvania Company, 165.
Travel. *See* Transportation; Internal Improvements.
Treaty:
  Asiento, 39.
  of Paris, 178, 188, 189.
  with Indians, 138, 189.
  early commercial, 176f.
  of 1783, 178, 188, 189.
  Jay (1795), 189, 247.
  of San Lorenzo (1795), 190, 204.
  of San Ildefonso (1800), 190.
Trent, W. P., quoted, 250.
Tripoli, 246.
Trolleys, 483ff.
Trumbull, Jonathan, 147 note.
Trusts, 470, 522f. *See* Combinations.
Tryon, Governor, 134, 161.
Tull, Jethro, 71, 236.
Turkey, wild, 9.
Turks, 32f.
Turner, F. J., quoted, 117, 119, 131, 168, 201.
Turnpikes, 311ff.
Turpentine, 83.
Twine binder, 427.

U

Underwood-Simmons tariff, 581.
Union Canal, 322.

Union Pacific Railroad, 388, 454, 455, 457, 504.
Unionism. *See* Labor; Labor Legislation; Labor Organization; Labor Leaders.
United Shoe Machinery Company, 573.
United States Bank:
  First, Second, 298, 300ff.
United Mine Workers, 595f.
United States:
  as a habitat for man, 5ff.
  geographic divisions, 1ff.
  production in, 635, 687ff.
United States Christian Commission, 390.
United States Sanitary Commission, 390.
United States Emergency Fleet Corporation, 682ff.
United States Steel Corporation, 520, 524.
Urban population, 340ff.
Utah, 406, 563.

V

Van Buren, Martin, 346, 586.
Vandalia, Illinois, 315.
Vanderbilt, Cornelius, 459.
Vegetable products, influence of, on early settlers, 7.
Vehicles, colonial, 93, 310f.
Venezuela Controversy, 623, 630.
Venice, 29ff.
Vermont:
  effort to found, 133.
  dispute over, 178.
  democracy in, 334.
Verrazano, G. da, 34.
Victoria, Magellan's ship, 34.
Village, layout of New England, 52f., 63f., 122.
Villard, Henry, 455.
Vincennes, 166.
Virginia:
  settlement, 46ff.
  agriculture, 66ff., 220, 224, 225, 226, 227.
  slavery, 362f.
Virgin Islands, 632.

W

Wabash and Erie Canal, 323.
Wages:
  in early mills, 287ff.
  during Civil War, 385ff.
  from 1865 to 1890, 672ff.
  from 1890 to 1914, 672ff.
  during World War, 610.

Walk-in-the-Water, 317.
Walker Tariff, 306.
Walpole, Robert, 147.
Waltham, Mass., 274, 289.
War Finance Corporation, 683.
War Industries Board, 683.
War Trade Board, 683.
War of 1812:
  causes, 247*ff*.
  history of, 251*ff*.
  effects on manufacturing, 276, 277, 278.
  attitude of New England, 252.
  results, 253*ff*.
Washington Conference, 629.
Washington, George, 160, 168, 257, 369.
  quoted, 169, 217.
Washington, products, 562*ff*.
Waste. *See* Conservation.
Watauga, 164*f*., 188.
Waterman, Capt. R. H., 258.
Water power, importance of, 23.
Water supply, 255.
Waterways:
  extent of, 5, 480.
  See Transportation.
Watkins, J. L., quoted, 392.
Watkins, G. S., quoted, note 602.
Watson, Elkanah:
  and county fairs, 238.
  and Erie Canal, quoted, 319.
Watt, James, 270, 271.
Wayne, General, quoted, 57, 189.
Wealth:
  of United States, 668*ff*.
  distribution of, 668*ff*.
Weather Bureau, 433.
Webb Act, 541, 641.
Webster, Daniel:
  opposes democracy, 346.
  and slavery, 373.
Weeden, John, 180.
Weare, Comptroller, quoted, 98.
Wells, David A., quoted, 147 note, 492*f*.
Wells-Fargo Company, 334.
Welsh, in Piedmont, 166.
Wentworth, Governor, 128, 163.
West Indies:
  Colonial trade with, 108*ff*.
  three-cornered trade, 74*f*., 108*ff*.
  British prohibitory acts, 145*f*.
  trade destroyed after Revolution, 176.
  recovery of trade, 185.
Western lands. *See* Public Lands.
Western Reserve, 195*ff*., 355.
Western Union, 335.
Westsylvania, projected state of, 165*f*.

Westward Movement:
  significance of, 117.
  stages of, 118, 206*ff*.
  routes of, 120, 199*ff*., 210.
  first frontier, 121.
  to fall line, 121*ff*.
  into the Piedmont, 128*ff*.
  and the Indians, 126*ff*., 189*ff*.
  opposition of easterners to, 131*ff*.
  absentee landlords and, 132.
  policy of England toward, 137, 189.
  during Revolution, 188*ff*.
  after close of Revolution, 188*ff*.
  and land policy of government, 191*ff*.
  into the Old Northwest, 164*ff*., 194*ff*.
  into the Old Southwest, 201*ff*.
  beyond the Mississippi, 208*ff*.
  effect upon East, 212*f*.
  influence of mining upon, 403*ff*.
  influence of ranching upon, 407*ff*.
  effect on agriculture, 216*ff*.
  of manufacturing, 554*f*.
Whaling:
  colonial, 86*ff*.
  later, 261.
Wheat:
  adaptability of western prairies to, 10.
  production of, 16*f*., 228, 231.
Wheeling, W. Va., 315.
Whigs, 160*ff*.
Whisky, 230*f*.
Whisky Rebellion, 231, 313.
White Mountains, 3.
Whitewater Canal, 323.
Whitman, Marcus, 118.
Whitney, Eli:
  quoted, 221*f*.
  invents the cotton gin, 220*ff*., 270, 371.
  pioneer in introducing interchangeable mechanism, 274.
Wild animals, 8*f*.
Wilkinson, Jeremiah, 171.
Willard, Daniel, 474.
William Demuth Company, 663.
William of Orange, 134.
Williams, Roger, 36, 51, 123.
Wilson-Gorman tariff, 580.
Wilson, Woodrow:
  cited, 375.
  quoted, 401, 635.
  and imperialism, 632*f*.
Winthrop, Governor John, 44, 81, 92.
Wire fence, 409.
Wireless, 488.
Wisconsin:
  income tax, 654.
  manufacturing, 562.

INDEX

Women:
  in early manufacturing, 289*ff*.
  in later industry, 609*ff*.
  labor legislation pertaining to, 601, 610.
Wood. *See* Lumber.
Wooden ships, 257*ff*. *See* Shipping.
Wood, Jethro, 234.
Wool:
  chief wool states, 22.
  colonial production, 88*ff*., 169.
Woolen manufacture:
  colonial, 89.
  Revolution to 1860, 277*f*.
  during Civil War, 383.
  since Civil War, 568*f*.
Workingmen's Compensation laws, 610*f*.
World War:
  effect on industry, 678, 681*ff*., 687*ff*.
  effect on agriculture, 678, 687*ff*.
  financing, 678*ff*.
  effect on labor, 684*ff*.
  government regulation during, 681*ff*.
  and government regulation of railroads, 473*ff*., 684.
Wright, Fanny, 349.
Wyoming, 405, 406, 563.
Wyoming Valley, Pa.:
  massacre, 161.
  Connecticut settlers attacked in, 178.
  Anthracite coal in, 3, 10, 12.

Y

Yeardly, Sir George, 48.

Z

Zinc, production of, 15.